Mathematical Methods in Nuclear Reactor Dynamics

NUCLEAR SCIENCE AND TECHNOLOGY

A Series of Monographs and Textbooks

CONSULTING EDITOR

V. L. PARSEGIAN

Chair of Rensselaer Professor
Rensselaer Polytechnic Institute
Troy, New York

---

1. John F. Flagg (Ed.)
   CHEMICAL PROCESSING OF REACTOR FUELS, 1961

2. M. L. Yeater (Ed.)
   NEUTRON PHYSICS, 1962

3. Melville Clark, Jr., and Kent F. Hansen
   NUMERICAL METHODS OF REACTOR ANALYSIS, 1964

4. James W. Haffner
   RADIATION AND SHIELDING IN SPACE, 1967

5. Weston M. Stacey, Jr.
   SPACE-TIME NUCLEAR REACTOR KINETICS, 1969

6. Ronald R. Mohler and C. N. Shen
   OPTIMAL CONTROL OF NUCLEAR REACTORS, 1970

7. Ziya Akcasu, Gerald S. Lellouche, and Louis M. Shotkin
   MATHEMATICAL METHODS IN NUCLEAR REACTOR DYNAMICS, 1971

In preparation:

John Graham
FAST REACTOR SAFETY

MATHEMATICAL METHODS IN NUCLEAR REACTOR DYNAMICS

Ziya Akcasu
UNIVERSITY OF MICHIGAN
ANN ARBOR, MICHIGAN

Gerald S. Lellouche
Louis M. Shotkin
BROOKHAVEN NATIONAL LABORATORY
UPTON, LONG ISLAND, NEW YORK

AP 1971

ACADEMIC PRESS New York and London

ACADEMIC PRESS, INC.
111 Fifth Avenue, New York, New York 10003

United Kingdom Edition published by
ACADEMIC PRESS, INC. (LONDON) LTD.
Berkeley Square House, London W1X 6BA

LIBRARY OF CONGRESS CATALOG CARD NUMBER: 71-137637

PRINTED IN THE UNITED STATES OF AMERICA

To the memory of Jack Chernick

Table of Contents

Preface

This book is derived from lectures given at the University of Michigan for the past several years by the first author. As such it has the qualities of a graduate text as well as a reference work. The book will prepare the reader to follow the current literature and approach practical and theoretical problems in reactor dynamics.

The reader is assumed to have adequate background in reactor physics and operational mathematics. Applied mathematicians and control engineers who are interested in the stability analysis and control of nuclear reactors will find sufficient background material on the physical aspects of reactor dynamics. Nuclear engineers interested in the more theoretical aspects of reactor dynamics are introduced to the basic concepts and tools of the mathematical theory of stability.

Because of the limitations on the size of the book, space-dependent kinetics, numerical methods for solving kinetic problems, and statistical methods in reactor analysis are not included. The topics which are treated in this book constitute a fairly complete study of point-reactor kinetics and linear and nonlinear reactor dynamics. The authors have tried not to duplicate the subject matter and approach of existing texts and have attempted to develop a consistent point of view in the treatment of each topic and in the derivation of the pertinent results.

The problems at the end of each chapter are an integral part of the text; in many cases the details of derivations are left as exercises. Furthermore, the problems have been chosen to make specific points where, in many cases, a discussion of a given point would have required a digression from the main argument. Practical problems in terms of the kinetic study of a particular reactor type should supplement these problems when the book is used as a text.

The authors thank Professor R. K. Osborn of the University of Michigan for allowing them to use his lecture notes in the discussion of feedback in Chapter 1. They are also grateful to the Reactor Theory Group at Brookhaven National Laboratory headed by Jack Chernick for providing the opportunity to complete this work. They gratefully acknowledge the typing of Miss Ellie Mitchell and Mrs. Elaine Taylor, as well as the other technical assistance received from Brookhaven National Laboratory.

CHAPTER 1

Kinetic Equations

Reactor dynamics is concerned with the time behavior of the neutron population in an arbitrary medium whose nuclear and geometric properties may vary in time. The first step in reactor dynamics is to introduce and define the macroscopic physical quantities and the dynamical variables that describe the medium and the neutron population in sufficient detail. The second step is to find time-dependent equations that interrelate the various dynamical variables in terms of the nuclear, thermal, and mechanical properties of the medium, and to determine the time evolution of the neutron population. These equations are the kinetic equations which we wish to discuss in this chapter. The last and most difficult step is to introduce analytical and numerical techniques in order to solve the kinetic equations either rigorously or approximately, and to extract all the information relevant to the performance and safety of the reactor as well as the power plant as a whole. The remaining chapters of this book are concerned with these aspects of reactor dynamics.

In obtaining the kinetic equations, we shall present the description of various physical phenomena influencing the temporal behavior of neutrons in detail to provide an adequate understanding of the physics of reactor dynamics, and to point out the interrelationships between various diverse phenomena in a sufficiently precise manner.

1.1. Transport Equation

The expected temporal behavior of the neutron population in an arbitrary medium is completely described by introducing a function of seven variables $n(\mathbf{r}, \mathbf{v}, t)$ which we shall refer to as the angular neutron density. It is defined as

$$n(\mathbf{r}, \mathbf{v}, t)\, d^3r\, d^3v$$

i.e., the expected number of neutrons in the volume element d^3r about $\mathbf{r}$ with velocities in d^3v about $\mathbf{v}$ at time t.

By defining the angular density as the expected, rather than the actual, number of neutrons in an element of volume in the phase space, we have excluded the possibility of describing the fluctuations in the neutron population. The actual number of neutrons is always an integer, and hence is discontinuous in time, whereas the expected number does not have to be an integer and is a continuous function of time. The description of fluctuations of the neutron population, commonly called reactor noise, requires probabilistic concepts. We shall not introduce these concepts in this book.

The next step after having introduced the angular density is to find an equation to determine its time evolution in terms of the nuclear and geometric properties of the medium. In vacuum, this equation is simple and readily obtained [1]:

$$[\partial n(\mathbf{r}, \mathbf{v}, t)/\partial t] + \mathbf{v} \cdot \nabla n(\mathbf{r}, \mathbf{v}, t) = S(\mathbf{r}, \mathbf{v}, t) \tag{1}$$

where $S(\mathbf{r}, \mathbf{v}, t)$ is the neutron source, defined as

$$S(\mathbf{r}, \mathbf{v}, t)\, d^3r\, d^3v\, dt$$

i.e., the number of neutrons inserted in d^3r at $\mathbf{r}$ and in d^3v at $\mathbf{v}$ in the time interval dt about t.

Equation (1) describes the streaming of neutrons on straight lines, and is the simplest form of the transport equation. It is a balance relation in which $-\mathbf{v} \cdot \nabla n\, d^3r\, d^3v$ is the rate of removal of neutrons in d^3r at $\mathbf{r}$ and d^3v at $\mathbf{v}$ due to streaming (leakage), and $S(\mathbf{r}, \mathbf{v}, t)\, d^3r\, d^3v$ is the production rate of neutrons in the same volume element. The solution of (1) is straightforward (see Problem 1 at the end of this chapter) [1]:

$$n(\mathbf{r}, \mathbf{v}, t) = n(\mathbf{r} - \mathbf{v}t, \mathbf{v}, 0) + \int_0^t dt'\, S[\mathbf{r} - \mathbf{v}(t - t'), \mathbf{v}, t'] \tag{2}$$

where $n(\mathbf{r}, \mathbf{v}, 0)$ is the initial distribution of neutrons.

It is assumed in (1) that there are no forces acting on the neutrons (no gravity), and no collisions between neutrons (Problem 2). We shall make these two assumptions consistently in this book unless stated otherwise.

As a result of the collisions of neutrons with atomic nuclei, the equation satisfied by the angular density in a material medium is extremely complicated. Before attempting to derive this equation, we must first introduce the appropriate functions to describe the effect of collisions on the evolution of neutrons quantitatively. These functions are related to the macroscopic nuclear properties of the medium.

We assume that the collisions of neutrons are instantaneous. Then, their effect on the motions of the neutrons can be described by specifying the mean free path between collisions as a function of the neutron velocity. The inverse of the mean free path is the macroscopic cross section $\Sigma(\mathbf{r}, \mathbf{v}, t)$, which is defined as

$$\Sigma(\mathbf{r}, \mathbf{v}, t)\, vn(\mathbf{r}, \mathbf{v}, t)\, d^3r\, d^3v$$

i.e., the mean number of collisions per second (collision rate) in d^3r at $\mathbf{r}$ for neutrons with velocities in d^3v at $\mathbf{v}$ at time t.

The origin of the time dependence of the macroscopic cross sections will be discussed later.

Depending on the type of the nuclear reaction taking place in a collision, a neutron may be captured, scattered, or may cause fission. Each of these events is characterized by a partial macroscopic cross section, which we denote by Σ_c, Σ_s, and Σ_f, respectively. Since the reaction rates associated with these events are additive, we have

$$\Sigma = \Sigma_c + \Sigma_s + \Sigma_f \tag{3}$$

The mathematical description of a scattering event requires the introduction of the differential scattering cross section $\Sigma_s(\mathbf{v} \to \mathbf{v}', \mathbf{r}, t)$, which is defined as

$$\Sigma_s(\mathbf{v} \to \mathbf{v}', \mathbf{r}, t)\, vn(\mathbf{r}, \mathbf{v}, t)\, d^3r\, d^3v\, d^3v'$$

i.e., the expected number of neutrons in d^3r at $\mathbf{r}$ scattered into d^3v' at $\mathbf{v}'$ per second at time t due to collisions of neutrons with velocities in d^3v at $\mathbf{v}$ in the same volume element at $\mathbf{r}$.

It follows from this definition that

$$\int \Sigma_s(\mathbf{v} \to \mathbf{v}', \mathbf{r}, t)\, d^3v' = \Sigma_s(\mathbf{r}, \mathbf{v}, t) \tag{4}$$

In a fission event, on the average, more than one neutron is produced. These neutrons are emitted at the point in space where the fission occurs. Some of them are emitted instantaneously (in time intervals of the order of 10^{-8} sec or shorter following the fission event), and are termed prompt neutrons [2]. The others are emitted with long time delays (of the order of seconds), and are called delayed neutrons. Figure 1.1.1 shows the origin of delayed neutrons. Their emission follows the deexcitation of certain fission fragments (^{87}Br in Figure 1.1.1)

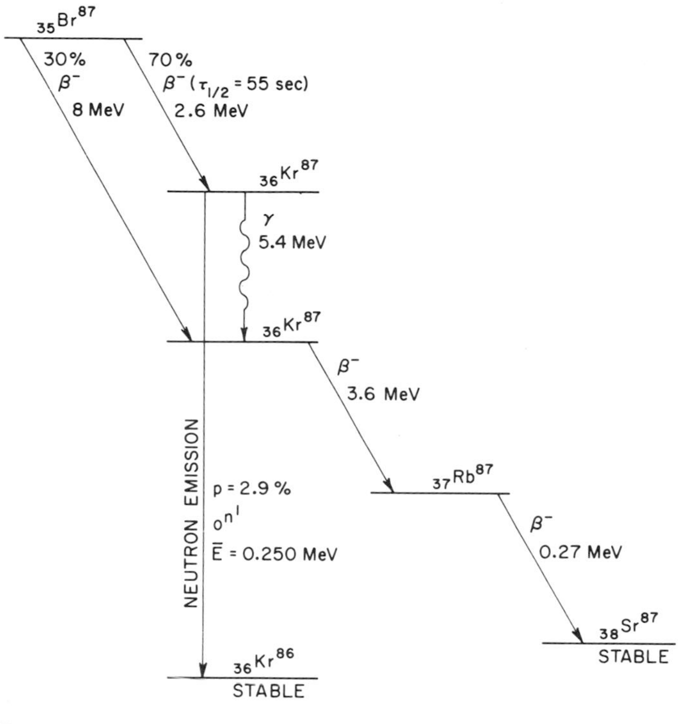

FIGURE 1.1.1. Decay scheme of a typical delayed neutron precursor.

by beta decay. These fission fragments are referred to as the delayed neutron precursors, or just precursor for brevity. A delayed neutron is emitted by the daughter nucleus (^{87}Kr in Figure 1.1.1) which is produced by the beta decay of the precursor. The long time delays distinguishing the delayed neutrons from the prompt neutrons are associated with the rather slow nuclear process of beta emission [2].

The migration of the precursors before the emission of a delayed neutron can be neglected in solid-fuel reactors, because they lose their kinetic energy very rapidly as a consequence of their large electric charge. They are stopped in a short distance from the point of their formation by fission (e.g., the range of light fission fragments in aluminum is approximately 1.4×10^{-3} cm [2]. It is therefore a good approximation to assume that, in a solid-fuel reactor, which we shall mainly be concerned with in this book, the delayed neutrons as well as the prompt neutrons are produced at the same point in space where the fission event takes place.

There are several distinct delayed neutron groups which are characterized by the decay constant of the responsible precursor in its beta decay, and by the fractional abundance of the delayed neutrons in fission due to this precursor.

The above discussion indicates that we need the following physical quantities to describe a fission event quantitatively from a reactor-dynamic point of view: λ_i, the decay constant of the ith kind of precursor; β_i, the fraction of fission neutrons due to the decay of the ith kind of precursor; β, the fraction of fission neutrons that are delayed, i.e.,

$$\beta = \sum_{i=1}^{6} \beta_i \tag{5}$$

a_i, the relative abundance of delayed neutrons in the ith group:

$$a_i = (\beta_i/\beta), \qquad \sum_{i=1}^{6} a_i = 1 \tag{6}$$

$f_i(v)$, the velocity distribution of delayed neutrons in the ith group; $f_0(v)$, the velocity distribution of the prompt fission neutrons (both $f_i(v)$ and $f_0(v)$ are normalized to unity as

$$\int d^3v\, f_j(v) = 1, \qquad j = 0, 1, 2, \ldots, 6), \tag{7}$$

and $\nu(v)$, the mean number of neutrons (prompt and delayed) per fission induced by a neutron of speed v.

The numerical values of the delayed neutron parameters λ_i, a_i, and β are listed in Tables 1–3 for various fissionable isotopes in fissions induced by thermal and fast neutrons with an average energy of about 1.8 MeV [4]. It is observed from these tables that the decay constants and the yields depend slightly on the incident neutron energy. For simplicity we shall ignore this dependence in the following derivations.

The energy spectra of the delayed neutrons in various groups, i.e., $f_i(v)$, are shown in Figure 1.1.2. Note that the curves are plotted for the mean number of delayed neutrons in a given group per unit energy,

TABLE 1

DELAYED NEUTRON PARAMETERS FOR THERMAL FISSION IN ^{235}U AND ^{239}Pu[a]

Parameter[b]	Fissionable isotope	
	^{235}U	^{239}Pu
λ_1	0.0124	0.0128
λ_2	0.0305	0.0301
λ_3	0.111	0.124
λ_4	0.301	0.325
λ_5	1.13	1.12
λ_6	3.00	2.69
a_1	0.033	0.035
a_2	0.219	0.298
a_3	0.196	0.211
a_4	0.395	0.326
a_5	0.115	0.086
a_6	0.042	0.044

[a] Data from Keepin *et al.* [3] and U. S. At. Energy Comm. Rep. [4].

[b] The λ's are decay constants, the a's are relative abundances.

TABLE 2

DELAYED NEUTRON PARAMETERS FOR FAST ($E_{\text{eff}} \sim 1.8$ MeV) FISSION IN ^{235}U, ^{238}U, ^{239}Pu, AND ^{240}Pu[a]

Parameter[b]	Fissionable isotope			
	^{235}U	^{238}U	^{239}Pu	^{240}Pu
λ_1	0.0127	0.0132	0.0129	0.0129
λ_2	0.0317	0.0321	0.0311	0.0313
λ_3	0.115	0.139	0.134	0.135
λ_4	0.311	0.358	0.331	0.333
λ_5	1.40	1.41	1.26	1.36
λ_6	3.87	4.02	3.21	4.04
a_1	0.038	0.013	0.038	0.028
a_2	0.213	0.137	0.280	0.273
a_3	0.188	0.162	0.216	0.192
a_4	0.407	0.388	0.328	0.350
a_5	0.128	0.225	0.103	0.128
a_6	0.026	0.075	0.035	0.029

[a] Data from Keepin *et al.* [3] and U. S. At. Energy Comm. Rep. [4].

[b] The λ's are decay constants, the a's are relative abundances.

TABLE 3

VALUES OF DELAYED NEUTRON FRACTION (β), MEAN NUMBER OF NEUTRONS PER FISSION (ν), AND MEAN NUMBER OF DELAYED NEUTRONS PER FISSION (n/F) FOR THERMAL AND FAST FISSIONS[a]

	Fissionable isotope			
Parameter	^{235}U	^{238}U	^{239}Pu	^{240}Pu
Thermal fission				
n/F	0.0158	—	0.0061	—
ν	2.43	—	2.87	—
β	0.0065	—	0.0021	—
Fast fission				
n/F	0.0165	0.0412	0.0063	0.0088
ν	2.57	2.79	3.09	3.3
β	0.0064	0.0148	0.0020	0.0026

[a] Data from Keepin [5].

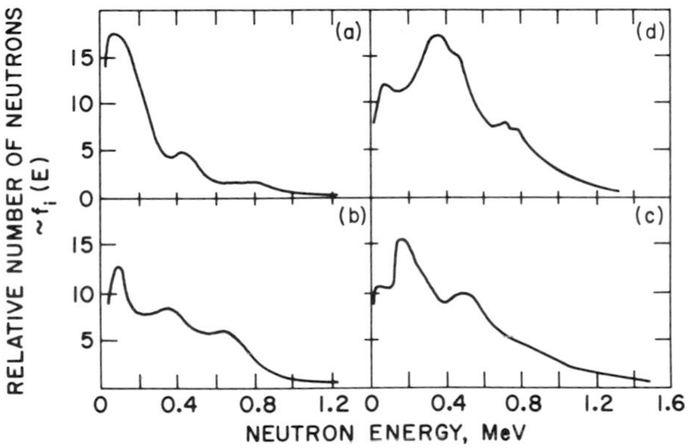

FIGURE 1.1.2. Delayed neutron spectra for various groups (the data represent a smoothed fit); replotted from Reference [5]. (a) $\tau_{1/2} \simeq 55$ sec, $\bar{E} = 0.250$ MeV; (b) $\tau_{1/2} \simeq 6$ sec, $\bar{E} = 0.405$ MeV; (c) $\tau_{1/2} \simeq 2$ sec, $\bar{E} = 0.450$ MeV; (d) $\tau_{1/2} = 22$ sec, $\bar{E} = 0.460$ MeV.

$f_i(E)$, rather than per unit volume in the velocity space. The curves are not normalized, but the $f_i(E)$ are normalized, by definition, to unity if they are interpreted as an energy distribution:

$$\int_0^\infty dE\, f_i(E) = 1 \tag{8}$$

The mean delayed neutron energies vary from 0.2 to 0.6 MeV. (The value is 0.250 MeV for the group due to ^{87}Br.) It is to be noted that the spectra of the delayed neutrons do not contain distinct peaks, although some pronounced structure can be observed. (For further discussion of the delayed neutron spectra, see Johnsen *et al.* [6] and Keepin [7].) The yield-weighted average of the delayed neutron spectra, i.e.,

$$\bar{f}_{\mathrm{d}} = \sum_{i=1}^{6} a_i f_i(E) \tag{9}$$

is plotted in Figure 1.1.3. The group-averaged mean energy is $\bar{E} = 0.43$ MeV.

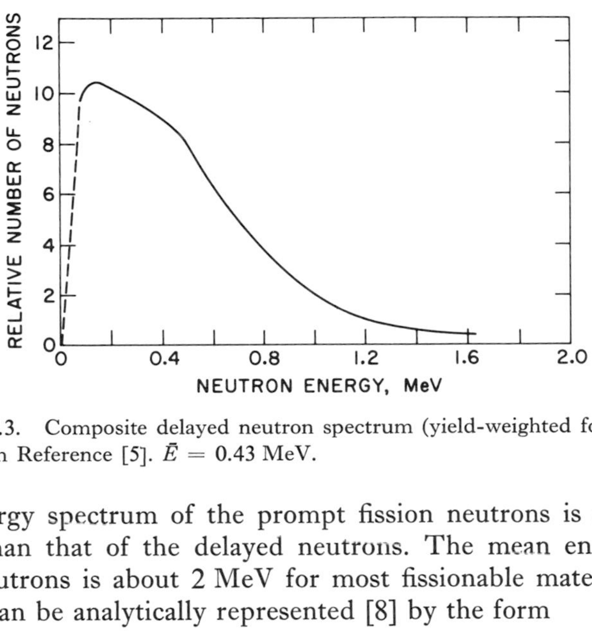

FIGURE 1.1.3. Composite delayed neutron spectrum (yield-weighted for all groups); replotted from Reference [5]. $\bar{E} = 0.43$ MeV.

The energy spectrum of the prompt fission neutrons is appreciably different than that of the delayed neutrons. The mean energy of the prompt neutrons is about 2 MeV for most fissionable materials. Their spectrum can be analytically represented [8] by the form

$$f_0(E) = ce^{-aE} \sinh(bE)^{1/2} \tag{10}$$

For ^{235}U, $c = 0.4527$, $a^{-1} = 0.965$, and $b = 2.29$ when E is measured in MeV [8]. The values of these constants change slightly for other nuclides [9].

We have implied in the above discussion that the fission neutrons (both delayed and prompt) are emitted isotropically in the laboratory system and that their spectra do not depend on the energy of the incident neutron causing fission. These implications are based on calculations using the evaporation model of fission [2] as well as on measurements [8, 9].

The mean number of neutrons per fission $\nu(v)$ depends appreciably on the energy of the incident neutron causing fission. The values of ν for various fission nuclides are listed in Table 3, as well as values for fissions induced by thermal and fast neutrons. The value of ν increases from 2.43 to 2.57 for ^{235}U when the energy of the incident neutron changes from thermal to an average energy of about 2 MeV. It is shown by Yiftah *et al.* [9] that the increase of ν with energy can be taken to be linear, i.e.,

$$\nu(E) = \nu_{\text{thermal}} + (d\nu/dE)\, E \tag{11}$$

The values of $d\nu/dE$ for various fission nuclides are obtained by the method of least squares in the U.S. At. Energy Comm. Rep. [4]. The results are listed in Table 4.

TABLE 4

DATA FOR THE EXPRESSION $\nu(E) = \nu_{\text{thermal}} + (d\nu/dE)E$ OBTAINED BY THE METHOD OF LEAST SQUARES FROM EXPERIMENTAL DATA [4]

	Fissionable isotope		
	^{235}U	^{238}U	^{239}Pu
ν_{thermal}	2.43	2.409	2.868
$d\nu/dE$ (MeV^{-1})	0.1346	0.1385	0.1106

In order to include the effect of the delayed neutrons quantitatively, we must introduce, in addition to the angular density, another funcion, the concentration of the dclayed neutron precursors

$$C_i(\mathbf{r}, t)\, d^3r$$

i.e., the expected number of fictitious precursors of ith kind in d^3r at $\mathbf{r}$ at time t which always decay by emitting a delayed neutron.

The above definition of $C_i(\mathbf{r}, t)$ requires some clarification: It is observed in Figure 1.1.1 that ^{87}Kr can decay by emitting either a beta particle or a neutron. The latter occurs with a probability of $p = 2.9\%$. Hence, the mean number of ^{87}Br per fission is not equal to the mean number of delayed neutrons per fission in its group. According to the above definition, $C_i(\mathbf{r}, t)$ corresponding to ^{87}Br is

$$C_{^{87}\text{Br}}(\mathbf{r}, t) = 0.0203\; {}^{87}\text{Br}(\mathbf{r}, t) \tag{12}$$

where $^{87}\mathrm{Br}(\mathbf{r}, t)$ is the actual concentration of ^{87}Br nuclei. Since the delayed neutron precursors are not observed experimentally, the characterization of their effect by a fictitious concentration is of no consequence from a practical point of view (this would not be the case, however, if we were interested in fluctuations of the neutron density).

We are now in a position to write down the transport equations in a multiplying medium:

$$\begin{aligned}\partial n(\mathbf{r}, \mathbf{v}, t)/\partial t = & -\mathbf{v} \cdot \nabla n(\mathbf{r}, \mathbf{v}, t) - v\Sigma(\mathbf{r}, \mathbf{v}, t)\, n(\mathbf{r}, \mathbf{v}, t) \\ & + \sum_j \left\{ f_0^j(v) \int d^3v' \, [\nu^j(\mathbf{r}, \mathbf{v}', t)(1 - \beta^j)\, v'\Sigma_f^j(\mathbf{r}, \mathbf{v}', t)\, n(\mathbf{r}, \mathbf{v}', t)] \right\} \\ & + \sum_{i=1}^{6} [\lambda_i f_i(v)\, C_i(\mathbf{r}, t)] \\ & + \int d^3v' \, [v'\Sigma_s(\mathbf{r}, \mathbf{v}' \to \mathbf{v}, t)\, n(\mathbf{r}, \mathbf{v}', t)] + S(\mathbf{r}, \mathbf{v}, t) \end{aligned} \tag{13}$$

$$\partial C_i(\mathbf{r}, t)/\partial t = \sum_j \int d^3v' \, [\beta_i^j \nu^j(\mathbf{r}, \mathbf{v}', t)\, v'\Sigma_f^j(\mathbf{r}, \mathbf{v}', t)\, n(\mathbf{r}, \mathbf{v}', t)] - \lambda_i C_i(\mathbf{r}, t) \tag{14}$$

where we have allowed the possibility of having more than one kind of fuel isotope, and distinguished them by the superscript j. The first equation, which involves the transport effects, is a mathematical statement of the neutron balance in an infinitesimal element of volume in the phase space. The remaining equations represent the balance relations for the precursors in an element of volume in the configuration space. Since the migration of the precursors has been neglected (Problem 3) we do not have the streaming term in Eq. (14).

The term $-\mathbf{v} \cdot \nabla n(\mathbf{r}, \mathbf{v}, t)\, d^3r\, d^3v$ in (13) denotes the removal of neutrons due to streaming, and is equal to the difference between the number of neutrons entering and emerging per second from the volume element $d^3r\, d^3v$ at $(\mathbf{r}, \mathbf{v})$.

The term $v\Sigma(\mathbf{r}, \mathbf{v}, t)\, n(\mathbf{r}, \mathbf{v}, t)\, d^3r\, d^3v$ is the number of neutrons in $d^3r\, d^3v$ that suffer a collision of any kind per second.

The third term on the right-hand side of Eq. (13) is the total number of prompt fission neutrons produced in $d^3r\, d^3v$ per second by fission events in d^3r at $\mathbf{r}$ in the configuration space where the fissions are induced by neutrons at all energies (integration on $\mathbf{v}'$).

The fourth term is the number of delayed neutrons emitted per second in d^3r at $\mathbf{r}$ by the delayed neutron precursor in all groups which are formed in fission events in d^3r in the past (migration of the precursors is neglected).

The fifth term is equal to the number of neutrons that are scattered into d^3v at $\mathbf{v}$ per second in scattering events in d^3r at $\mathbf{r}$ in configuration space at all energies.

Finally, the last term, $S(\mathbf{r}, \mathbf{v}, t)\, d^3r\, d^3v$ denotes the number of neutrons introduced into $d^3\mathbf{r}$ at $\mathbf{r}$ and d^3v at $\mathbf{v}$ by external neutron sources. The production rate of the external neutrons is independent of the chain reaction taking place in the reactor.

The second equation is the balance relation for the delayed neutron precursor of the ith kind. The first term on the right-hand side is the total number of delayed neutrons in the ith group at all energies produced per second in d^3r at $\mathbf{r}$ by all fissions. Since we assume that each delayed neutron precursor eventually emits one delayed neutron, then the number of delayed neutrons β_i and the number of precursors per fission will be the same. With this assumption, the term under consideration becomes the production rate of the delayed neutron precursors of the ith kind. Clearly, the second term in (14) is the decay rate of the ith precursor.

The rigorous solution of the transport equation is a formidable task even in the absence of delayed neutrons and feedback effects, and is possible only in highly simplified situations discussed elegantly by Case and Zweifel [1] in their book on linear transport theory (Problem 4). When the macroscopic cross sections are allowed to be explicit functions of time and to depend on the angular density itself, the task of obtaining exact solutions to the transport equation becomes hopelessly complicated. However, these equations provide a basis for simpler descriptions or models which may be treated analytically. At any rate, we must continue our discussion to formulate the dependence of the nuclear parameters of the medium on time in a quantitative and precise manner.

1.2. Feedback

It is convenient to express the macroscopic cross sections in terms of the effective microscopic cross sections when discussing their dependence on time:

$$\Sigma_j(\mathbf{r}, \mathbf{v}, t) = \sum_i N_i(\mathbf{r}, t)\, \sigma_{ji}(\mathbf{r}, \mathbf{v}, t), \qquad j = \mathrm{c, f, s} \tag{1}$$

$$\Sigma_\mathrm{s}(\mathbf{r}, \mathbf{v}' \to \mathbf{v}, t) = \sum_i N_i(\mathbf{r}, t)\, \sigma_{\mathrm{s}i}(\mathbf{r}, \mathbf{v}' \to \mathbf{v}, t) \tag{2}$$

where $N_i(\mathbf{r}, t)\, d^3r$ is the number of atomic nuclei of the ith kind in d^3r at $\mathbf{r}$ at time t.

In order to define the effective microscopic cross section $\sigma_{ji}(\mathbf{r}, \mathbf{v}, t)$, we must first introduce the effective area presented by a nucleus of type i to a neutron moving with a relative velocity $\mathbf{v} - \mathbf{V}$ for a jth type interaction: $\sigma_{ji}^{\infty}(|\mathbf{v} - \mathbf{V}|)$. Here, $\mathbf{V}$ is the velocity of the target nucleus and $\mathbf{v}$ is the velocity of the neutron in the laboratory frame (Figure 1.2.1). This

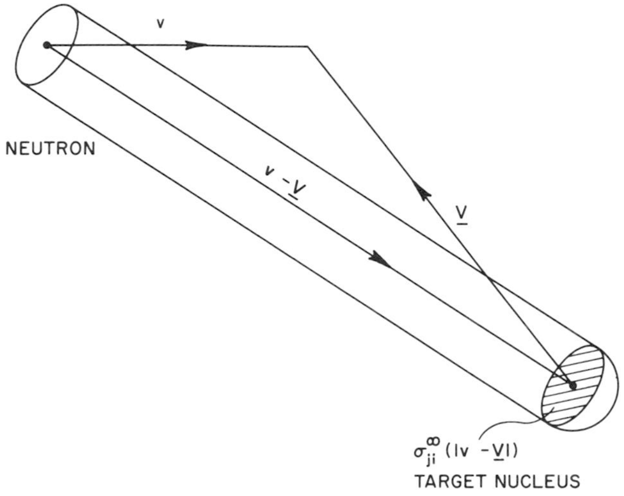

FIGURE 1.2.1. Collision of a neutron with a nucleus.

quantity is the microscopic cross section when the target nucleus is at rest. It depends only on the relative speed of the neutron. Figure 1.2.1 shows that a neutron moving with a relative velocity $\mathbf{v} - \mathbf{V}$ suffers collisions of type j with the nuclei of kind i at a rate of

$$|\mathbf{v} - \mathbf{V}|\, \sigma_{ji}^{\infty}(|\mathbf{v} - \mathbf{V}|)\, N_i(\mathbf{r}, t)\, \mathscr{M}_i(\mathbf{V}, \mathbf{r}, t)\, d^3V \tag{3}$$

where we have introduced the velocity distributions of the target nuclei as $\mathscr{M}_i(\mathbf{V}, \mathbf{r}, t)\, d^3V$, the fraction of nuclei of type i having velocities in d^3V about $\mathbf{V}$ at position $\mathbf{r}$ and time t.

The total reaction rate per neutron is obtained by integrating (3) over all target velocities $\mathbf{V}$ and summing over all species i. The result must be equal to $v\Sigma_j(\mathbf{r}, \mathbf{v}, t) = \sum_i N_i(\mathbf{r}, \mathbf{v}, t)\, v\sigma_{ji}(\mathbf{r}, \mathbf{v}, t)$. Hence, we obtain the relationship between $\sigma_{ji}(\mathbf{r}, \mathbf{v}, t)$ and $\sigma_{ji}^{\infty}(|\mathbf{v} - \mathbf{V}|)$:

$$\sigma_{ji}(\mathbf{r}, \mathbf{v}, t) = \int d^3V\, [(|\mathbf{v} - \mathbf{V}|/v)\, \sigma_{ji}^{\infty}(|\mathbf{v} - \mathbf{V}|)\, \mathscr{M}_i(\mathbf{V}, \mathbf{r}, t)] \tag{4}$$

Similar arguments lead to the following definition of the effective differential scattering cross section:

$$\sigma_{si}(\mathbf{r}, \mathbf{v}' \to \mathbf{v}, t) = \int d^3V\, [(|\, \mathbf{v}' - \mathbf{V}\,|/v')\, \sigma_{si}^{\infty}(\mathbf{v}' \to \mathbf{v}, \mathbf{V})\, \mathscr{M}_i(\mathbf{V}, \mathbf{r}, t)] \tag{5}$$

(The definition of $\sigma_{si}^{\infty}(\mathbf{v}' \to \mathbf{v}, \mathbf{V})$ will not be needed in subsequent discussions.)

The origin of the $\mathbf{r}$ and t dependences of $\nu^i(\mathbf{r}, \mathbf{v}, t)$ in the transport equation (1.1, Eq. 13) can now be explained. We can easily verify that $\nu^i(\mathbf{r}, \mathbf{v}, t)$ must be defined by

$$\nu^i(\mathbf{r}, \mathbf{v}, t) = \frac{\int d^3V\, [|\, \mathbf{v} - \mathbf{V}\,|\, \nu^i(|\, \mathbf{v} - \mathbf{V}\,|)\, \sigma_{fi}^{\infty}(|\, \mathbf{v} - \mathbf{V}\,|)\, \mathscr{M}_i(\mathbf{V}, \mathbf{r}, t)]}{\int d^3V\, [|\, \mathbf{v} - \mathbf{V}\,|\, \sigma_{fi}^{\infty}(|\, \mathbf{v} - \mathbf{V}\,|)\, \mathscr{M}_i(\mathbf{V}, \mathbf{r}, t)]} \tag{6}$$

We recall that $\nu^i(|\, \mathbf{v} - \mathbf{V}\,|)$ is the mean number of neutrons per fission induced by a neutron with a relative speed $|\, \mathbf{v} - \mathbf{V}\,|$. The experimental evidence showing the energy dependence of ν^i was discussed in the previous section. If this energy dependence is ignored, then (6) yields

$$\nu^i(\mathbf{r}, \mathbf{v}, t) = \nu^i \tag{7a}$$

Similarly, if the neutron speed v is much higher than V for the values of $\mathbf{V}$ where $\mathscr{M}_i(\mathbf{V}, \mathbf{r}, t)$ is appreciable, then (6) yields

$$\nu^i(\mathbf{r}, \mathbf{v}, t) = \nu^i(|\, \mathbf{v}\,|) \tag{7b}$$

In the rest of this book, we shall ignore the distinction between $\nu^i(\mathbf{r}, \mathbf{v}, t)$ and $\nu^i(|\, \mathbf{v}\,|)$ whenever the former appears explicitly in the equations.

Equation (1) indicates that the time dependence of the macroscopic cross sections is implicit in $\sigma_{ji}(\mathbf{r}, \mathbf{v}, t)$ and $N_i(\mathbf{r}, t)$, which may be separately time dependent. The time dependence of these factors is determined by different physical phenomena, and these will be discussed separately.

A. Variations in the Microscopic Cross Sections

Consider first the microscopic cross section $\sigma_{ji}(\mathbf{r}, \mathbf{v}, t)$ defined by (4). Clearly, the time dependence of this quantity is due to the variations in $\mathscr{M}_i(\mathbf{V}, \mathbf{r}, t)$; hence, the problem here is to deduce an appropriate characterization of the temporal dependence of $\mathscr{M}_i$. We shall assume that $\mathscr{M}_i$ is the atomic velocity distribution that is characteristic of a system of atoms in thermal equilibrium at a temperature T. The latter is allowed to be a function of time (local thermal equilibrium). With this assumption, we obtain an explicit form for $\mathscr{M}_i(\mathbf{V}, \mathbf{r}, t)$ as

$$\mathscr{M}_i(\mathbf{V}, \mathbf{r}, t) = [M_i/2\pi kT]^{3/2} \exp(-\tfrac{1}{2}M_i V^2/k_\mathrm{B}T) \tag{8}$$

where M_i is the mass of the atoms of ith kind and $T(\mathbf{r}, t)$ is the temperature of the medium at $\mathbf{r}$ at time t. We note that the temperature is the same for all types of atoms in the same element of volume at $\mathbf{r}$ at t as a result of the assumed thermal equilibrium. Equation (8) is, of course, the Maxwell–Boltzmann distribution, in which k_B is the Boltzmann constant. Thus, the macroscopic cross sections are time- and position-dependent through their dependence on temperature. Using (4) and (5), we get

$$\sigma_{ji}(v, T) = \int d^3V\, [(|\,\mathbf{v} - \mathbf{V}\,|/v)\, \sigma_{ji}^{\infty}(|\,\mathbf{v} - \mathbf{V}\,|)\, \mathscr{M}_i(V, T)] \tag{9a}$$

$$\sigma_{\mathrm{s}i}(\mathbf{v}' \to \mathbf{v}, T) = \int d^3V\, [(|\,\mathbf{v}' - \mathbf{V}\,|/v')\, \sigma_{\mathrm{s}i}^{\infty}(\mathbf{v}' \to \mathbf{v}, \mathbf{V})\, \mathscr{M}_i(V, T)] \tag{9b}$$

In these equations, $\sigma_{ji}^{\infty}(|\,\mathbf{v} - \mathbf{V}\,|)$ represents the theoretical reaction cross sections, whose precise forms are generally not known. The microscopic cross sections measured in the laboratory and presented in BNL-325 [10] are $\sigma_{ji}(v, T)$ at a specified temperature. The utility of (9a) and (9b) lies in the fact that they display the temperature dependence of $\sigma_{ji}(v, T)$ in an analytical form. Using these equations, it is possible to estimate the cross section at a desired temperature from the measured data at a different temperature [11]. For this purpose, we rewrite (9a) dropping the subscripts (j, i), changing $\mathbf{V}$ to $\mathbf{s} = \mathbf{v} - \mathbf{V}$, and introducing $R(v, T) \equiv v\sigma(v, T)$, $R^{\infty}(v) = v\sigma^{\infty}(v)$:

$$R(v, T) = \int d^3s\, [R^{\infty}(s)\, \mathscr{M}(|\,\mathbf{v} - \mathbf{s}\,|, T)] \tag{10}$$

Taking the three-dimensional Fourier transform of both sides with respect to $\mathbf{v}$, we obtain

$$\bar{R}(k, \theta) = \bar{R}^{\infty}(k) \exp(-k^2\theta) \tag{11}$$

where the bars denote the Fourier transforms, and $\theta \equiv k_\mathrm{B}T/2M$. (Note that $\bar{R}(k, \theta)$ and $\bar{R}^{\infty}(k)$ depend on $|\,\mathbf{k}\,|$ because $R(v, T)$ and $R^{\infty}(v)$ are functions of $|\,\mathbf{v}\,|$.)

In obtaining (11), we have used

$$\int d^3V\, [(\exp i\mathbf{k} \cdot \mathbf{V})\, \mathscr{M}(V, \theta)] = \exp(-k^2\theta) \tag{12}$$

(Problem 5). Evaluating (11) at two different temperatures and eliminating $\bar{R}^{\infty}(k)$, we find

$$\bar{R}(k, \theta') = \bar{R}(k, \theta) \exp[-k^2(\theta' - \theta)] \tag{13}$$

The inverse Fourier transform yields [11] (Problem 6)

$$v\sigma(v, \theta') = \int d^3s \left\{ \frac{s\sigma(s, \theta)}{[4\pi(\theta' - \theta)]^{3/2}} \exp - \frac{|\mathbf{v} - \mathbf{s}|^2}{4(\theta' - \theta)} \right\} \tag{14}$$

which is the desired relation, where $\sigma(v, \theta)$ is obtained from the cross-section data at θ. For small temperature changes $\Delta\theta$ about θ, we can expand (13) as

$$\bar{R}(k, \theta + \Delta\theta) \cong \bar{R}(k, \theta)[1 - k^2 \Delta\theta] \tag{15}$$

whose inverse transform in the limit of $\Delta\theta \to 0$ yields

$$\begin{aligned} \partial\sigma(v, \theta)/\partial\theta &= (1/v) \nabla_{\mathbf{v}}^2 R(v, \theta) \\ &= (1/v^2)(\partial^2/\partial v^2)\, v^2\sigma(v, \theta) \end{aligned} \tag{16}$$

Equation (16) is useful in the calculation of the temperature-reactivity coefficient, which we shall discuss in Chapter 5. In order to illustrate how (9) can be used to extract the temperature dependence explicitly once the form of $\sigma_{ji}^{\infty}(|\mathbf{v} - \mathbf{V}|)$ is known, we consider three special cases (dropping the subscripts for the moment): (a) $(1/v)$-cross section, i.e., $\sigma^{\infty}(r) = (\lambda_0/v)$; (b) constant cross section, i.e., $\sigma^{\infty}(r) = \sigma_0$; (c) one-level resonance cross section (Breit–Wigner one-level formula [2, 11, 12])

$$\sigma_{\mathrm{c}}^{\infty}(E) = \sigma_0 \frac{\Gamma_\gamma}{\Gamma} \left[1 + \left(\frac{E - E_0}{\Gamma/2}\right)^2\right]^{-1} \tag{17}$$

$$\sigma_{\mathrm{sr}}^{\infty}(E) = \sigma_0 \frac{\Gamma_n}{\Gamma} \left[1 + \left(\frac{E - E_0}{\Gamma/2}\right)^2\right]^{-1} \tag{18}$$

where $\sigma_c^{\infty}(E)$ and $\sigma_{\mathrm{sr}}^{\infty}(E)$ are the capture and resonance scattering cross sections, and where σ_0 is the peak value of the resonance (capture plus scattering), E_0 is the center-of-mass energy determining the position of the resonance peak, E is the kinetic energy corresponding to the relative velocity between neutron and nucleus, namely

$$E = \tfrac{1}{2}[Mm/(M + m)]\, |\mathbf{v} - \mathbf{V}|^2$$

(m neutron mass), Γ_γ and Γ_{n} are the radiative capture and neutron widths, and $\Gamma = \Gamma_{\mathrm{n}} + \Gamma_\gamma$ is the total width.

Case (a) describes the velocity dependence of most absorption processes (radiative capture and fission) for low-energy neutrons (thermal neutrons). Substituting $\sigma^{\infty}(|\mathbf{v} - \mathbf{V}|) = \lambda_0/|\mathbf{v} - \mathbf{V}|$ in (9a), we obtain

$$\sigma_{\mathrm{a}}(v, T) = \lambda_0/v \tag{19}$$

indicating that the effective microscopic cross section, which takes into account the thermal motion of the target nuclei, also obeys a $(1/v)$ law, and is independent of temperature.

Case (b) characterizes the low-energy potential elastic scattering cross section. The associated effective microscopie cross section is clearly temperature-dependent:

$$\sigma_{\mathrm{sp}}(v, T) = \sigma_0(M/2\pi kT)^{3/2} \int d^3V\, [(|\mathbf{v} - \mathbf{V}|/v) \exp(-\tfrac{1}{2}MV^2/kT)] \tag{20}$$

The indicated integral is readily performed (Problem 7):

$$\sigma_{\mathrm{sp}}(v, T) = \sigma_0\{[(\exp -x^2)/x\sqrt{\pi}] + [1 + (1/2x^2)] \operatorname{erf}(x)\} \tag{21a}$$

$$x \equiv v/V_{\mathrm{p}} \tag{21b}$$

where V_{p} is the most probable speed in a Boltzmann distribution,

$$V_{\mathrm{p}} = (2kT/M)^{1/2} \tag{22}$$

and $\operatorname{erf}(x)$ is the error function, defined by

$$\operatorname{erf}(x) = (2/\sqrt{\pi}) \int_0^x dy \exp -y^2 \tag{23}$$

It is interesting to note that, in the cases of large and small neutron speeds compared to V_{p}, (21a) reduces to

$$\sigma_{\mathrm{sp}}(v, T) \to \sigma_0 \qquad (v \gg V_{\mathrm{p}}) \tag{24a}$$

$$\sigma_{\mathrm{sp}}(v, T) \to 2\sigma_0 V_{\mathrm{p}}/v\sqrt{\pi} = (\sigma_0/v)(8kT/\pi M)^{1/2} \qquad (v \ll V_{\mathrm{p}}) \tag{24b}$$

Denoting the mass number of the moderator in which neutrons are diffusing by A, we can express V_{p} as

$$V_{\mathrm{p}} = 2200/\sqrt{A} \quad \text{meters/second} \tag{25}$$

where $2200 = (2kT/m)^{1/2}$ for $T = 293°\mathrm{K}$.

According to (24), the potential scattering cross section begins to be appreciably dependent on the reactor temperature at smaller neutron speeds in heavy moderators then in light moderators.

Case (c) takes into account that the compound nucleus resulting from neutron capture may rid itself of some or all of its excitation energy by either γ-ray or neutron emission. The probabilities of these modes of decay are Γ_γ/Γ and Γ_n/Γ, respectively. The effective microscopic cross sections $\sigma_c(v, T)$ and $\sigma_{sr}(v, T)$ can be obtained by substituting (17) and (18) into (9a). These cross sections enter importantly in the discussion of the temperature dependence of the fast fission factor and the resonance escape probability. The effect of the nuclear motion is the broadening of the resonance lines (Doppler broadening). We shall not digress to discuss these points any further because they are not directly related to the derivation of the kinetic equations, which is the main objective of this chapter. The reader is referred to the literature for further discussion [2, 11, 12].

In conclusion, we find that the effective microscopic cross sections depend on the neutron speed and the local temperature. The temperature dependence is expressed in a functional form (9) involving the theoretical reaction cross section, and the local temperature parametrically. The value of the cross section at a given time t is determined only by the value of the temperature at the same instant of time. It is possible to compute the value of the cross section at a temperature T' from its measured values as a function of neutron energy at a different temperature.

B. Variations in the Atomic Concentrations

We now focus our attention on the time dependence of the atomic concentrations, $N_i(\mathbf{r}, t)$. They may change with time for two reasons (we are not considering here either the changes by external means nor those due to phase transition, i.e., liquid to vapor). First, these quantities are temperature dependent. Their dependence on temperature is more appreciable in the liquid and gaseous state of matter than in the solid state. However, even in the solid state, the decrease in the atomic concentrations with increase in temperature due to thermal expansion can have a significant effect upon reactor operation.

Second, atomic concentrations change because of the continual occurrence of radiative capture, fission, and β-decay in the reactor. For example, the production as fission fragments of high-absorption-cross section, radiative nuclei such as ^{135}Xe leads to important temporal variations in the reactor. Also, fuel nuclei are "burning up," and low-absorption-cross section, relatively stable fission products are accumulating so long as the reactor is in operation.

It follows from these remarks that the formulation of reactor kinetics with feedback requires a description of the temporal and spatial variations of the temperature in the reactor and a knowledge of the rates of gain and loss of nuclear species.

Production of Heat Energy by Nuclear Processes

The formulation of the first of these problems requires an investigation of the heat transfer equation in the reactor:

$$\mu\{[\partial T(\mathbf{r}, t)/\partial t] + \mathbf{V}(\mathbf{r}, t) \cdot \nabla T(\mathbf{r}, t)\} - \nabla \cdot K \nabla T(\mathbf{r}, t) = H(\mathbf{r}, t) \qquad (26)$$

where μ is the heat content per unit volume of the medium, $\mathbf{V}$ is the hydrodynamic velocity, K is the heat conductivity, and $H(\mathbf{r}, t)$ is the distributed heat source. The quantities μ and K are, in general, functions of position and temperature. In the solid regions of the reactor, $\mathbf{V} = 0$, and the heat transfer is purely conductive. In the fluid regions, $\mathbf{V} \neq 0$ and is governed by the hydrodynamic equations expressing the conservation of momentum. If phase transitions can occur in the reactor (such as boiling of the coolant), additional equations are needed to describe the energy and mass conservation in each phase. It is clear from these comments that a general discussion of the temporal and spatial behavior of the temperature without reference to specific systems is not informative. The distributed heat source in (26) can still be discussed in general terms because it represents the production of heat energy in the reactor by fundamental nuclear processes which are the same in all reactor types.

The fundamental nuclear sources of heat energy are fission and radiative capture events. The former is generally by far the most important because the relative probabilities of these events are almost the same (Σ_c/Σ_f is of the order of unity), and the average release of nuclear energy in the form of γ-rays is about 2 MeV per radiative capture, whereas it is approximately 200 MeV per fission. Hence, the energy production due to the radiative capture is only of the order of 1 or 2% of the energy release by fission.

The nuclear energy released per fission is accounted for approximately as follows:

Kinetic energy of fission fragments	165 MeV
Kinetic energy of neutrons	5 MeV
Prompt γ's	5 MeV
Delayed γ's	6 MeV
β's	5 MeV
Neutrino energy	11 MeV
TOTAL	197 MeV

THE KINETIC ENERGY OF FISSION FRAGMENTS. The fission fragments are relatively massive ($A \sim 100$) and generally quite highly charged, so that their ranges (distances of travel before coming to rest) are very short. It is thus reasonable to assume that these particles give up their energy to the ambient atoms roughly at the position at which they were created by the fission process. This assumption is excellent for a solid-fuel reactor but may be less so in a gaseous-fuel reactor. With this assumption, we may express the distributed heat source due to the fission fragments by

$$H_{\mathrm{f}}(\mathbf{r}, t) = W_{\mathrm{f}} \int d^3v \, [v\Sigma_{\mathrm{f}}(\mathbf{r}, v, t)\, n(\mathbf{r}, \mathbf{v}, t)] \tag{27a}$$

where

$$\Sigma_{\mathrm{f}}(\mathbf{r}, v, t) = \sum_j \Sigma_{\mathrm{f}}^{\,j}(\mathbf{r}, v, t) \tag{27b}$$

and W_{f} is the heat energy in an appropriate unit (e.g., joules) per fission which corresponds to the 165-MeV kinetic energy of the fission fragments.

Although the energy carried by the fission fragments constitutes the largest portion of the fission energy, there is still a considerable amount of energy per fission ($\sim 15\%$) to be accounted for, some significant fraction of which may well appear as heat energy somewhere within the system. We may account for the details of the conversion of some of this "fission energy" to "heat energy" within the concepts already developed. However, the uncertainties in the concept of temperature itself as well as in the heat transfer and the accompanying hydrodynamic equations do not warrant such a detailed quantitative description of the heat sources in an actual power reactor. Since the heat sources due to the neutrons, β-rays, and γ-rays may play an appreciable role in the temporal behavior of the reactor in some more idealized situations, we shall present only a brief qualitative discussion of their contribution to the heat source.

THE KINETIC ENERGY OF NEUTRONS. The kinetic energy of the neutrons is dissipated by scattering collisions, and it is reasonable to expect that the recoil energy of the struck nuclei is largely transferred to the surrounding atoms roughly at the points at which the collision with the neutron took place. Consequently, the distributed heat source associated with the neutron moderation is given by

$$H_n(\mathbf{r}, t) = W_n \int d^3v \int d^3v' \, [\Sigma_{\mathrm{s}}(\mathbf{r}, \mathbf{v}' \to \mathbf{v}, t)\, v' n(\mathbf{r}, \mathbf{v}', t)\, \tfrac{1}{2}m(v'^2 - v^2)] \tag{28}$$

Of course, only elastic scatterings are envisaged here, since for inelastic scattering, some of the energy lost by the neutron remains with the struck nucleus as internal excitation which ultimately appears in the system in the form of γ-rays. The heat generation due to neutron moderation can be combined with that due to the slowing down of fission fragments because they both depend on the instantaneous value of $n(\mathbf{r}, \mathbf{v}', t)$ at t. The situation will be different in the case of heat produced by β- and γ-rays.

The expected 11 MeV per fission that appears in the form of neutrinos may be disregarded, for it is surely unavailable in the reactor, as the interaction mean free paths of these particles with anything are enormously greater than conceivable reactor dimensions.

We now consider the remaining 16 MeV per fission which appears in the system in the form of prompt and delayed γ-rays and β-rays. The task of dealing with the spatial and temporal dependence of the heat sources generated by the "thermalization" of these "particles" is somewhat complicated. The amounts of energy carried by these particles are too great to ignore altogether; but at the same time, it is beyond the scope of this presentation to provide more than an extremely formal and essentially suggestive treatment of the problems involved.

THE KINETIC ENERGY OF THE β-PARTICLES. We consider first the conversion of energy of the β-particles to thermal energy in the ambient medium. The mean energy of the β-rays emitted by the fission fragments is sufficiently low; hence, it is reasonable to assume that they give up most of their energy by collisions rather than by electromagnetic radiation. We may also assume that in most reactor situations their ranges are probably sufficiently short to allow the assumption of "thermalization" at the point of birth. Hence, so far as the spatial dependence of this contribution to the heat energy of the system is concerned, it seems reasonable to identify it with the spatial distribution of fissions. But the temporal dependence does not lend itself well to such a simplification. A given fission may produce any one or more of a considerable variety of β-emitters, each characterized by its own mean lifetime. We can define a function $f_\beta(\tau, E)$ such that $f_\beta(\tau, E)\, dE\, d\tau$ is the expected number of β-rays produced with energies in dE about E in $d\tau$ about τ following a fission at $\tau = 0$. Then, the heat source in the system due to the β-rays may be expressed as

$$H_\beta(\mathbf{r}, t) = W_\beta \int_0^\infty d\tau \int d^3v \int_0^\infty dE[v\Sigma_f(\mathbf{r}, v, t-\tau)\, n(\mathbf{r}, \mathbf{v}, t-\tau)\, Ef_\beta(\tau, E)] \tag{29}$$

THE ENERGY CARRIED BY THE γ-RAYS. The γ-rays cannot be dealt with so easily because they cannot be assumed to "thermalize" at the point at which they are produced because of their large mean free path in matter. They transfer their energy to the medium as a consequence of scattering collisions (Compton scattering) and absorptions. A straightforward approach to this problem is to write down an appropriate transport equation for the photon number density similar to (1.1, Eq. 13) with a γ-ray source consisting of prompt fission γ's, delayed fission γ's, the γ-rays resulting from the radiative capture of neutrons, and finally, γ-rays due to the inelastic scattering of neutrons.

The angular neutron density enters the photon transport equation through the source term. The solution of this transport equation is at least as difficult as solving the neutron transport equation in its full generality. Once the angular photon density is determined at each point, the contribution of the γ-rays to the heat source, which we denote by $H_\gamma(\mathbf{r}, t)$, can be obtained by summing the heat energies resulting from pair production, photoelectric absorption, and the moderation of the photons by Compton scattering.

The complete distributed heat source $H(\mathbf{r}, t)$ appearing in the heat transfer equation (26) due to the release of nuclear energy in the fission process is obtained as

$$H(\mathbf{r}, t) = H_\mathrm{f} + H_\mathrm{n} + H_\beta + H_\gamma \tag{30}$$

where, of course, the dominant term is $H_\mathrm{f}(\mathbf{r}, t)$ due to the fission fragments. We shall consider only this term in the description of the temperature feedback.

The Production and Burnup of Nuclear Species

We now turn our attention to the time changes in the atomic concentrations due to radiative capture, fission, and β-decay.

We first consider the production, decay, and burnup of some fission fragments that have large capture cross sections as well as relatively high yields such that they influence the temporal behavior of a reactor appreciably. The most important of these are ^{135}I, ^{135}Xe, and ^{149}Sm, which we shall discuss separately.

Iodine-135 is produced as a decay product of ^{135}Te, which is a direct fission product,

$$^{135}_{52}\mathrm{Te} \xrightarrow[\beta^-]{(2\mathrm{min})} {}^{135}_{53}\mathrm{I} \xrightarrow[\beta^-]{(6.7\mathrm{hr})} {}^{135}_{54}\mathrm{Xe} \xrightarrow[\beta^-]{(9.2\mathrm{hr})} {}^{135}_{55}\mathrm{Cs} \tag{31}$$

where the numbers in the parenthesis are the half-lives of the indicated β-decays. Denoting the atomic concentrations of ^{135}Te, ^{135}I, and ^{135}Xe by Te(**r**, t), I(**r**, t), and Xe(**r**, t), respectively, we obtain the following balance relations:

$$\partial(\mathrm{Te})/\partial t = y_{\mathrm{Te}} \int d^3v\, [v\Sigma_f(\mathbf{r}, v, t)\, n(\mathbf{r}, \mathbf{v}, t)] - \lambda_{\mathrm{Te}}\mathrm{Te}(\mathbf{r}, t)$$
$$- \mathrm{Te}(\mathbf{r}, t) \int d^3v\, [\sigma_{a\mathrm{Te}}(v, T)\, vn(\mathbf{r}, \mathbf{v}, t)] \tag{32}$$

$$\partial \mathrm{I}/\partial t = \lambda_{\mathrm{Te}}\mathrm{Te}(\mathbf{r}, t) - \mathrm{I}(\mathbf{r}, t) \left[\lambda_{\mathrm{I}} + \int d^3v\, [\sigma_{a\mathrm{I}}(v, T)\, vn(\mathbf{r}, \mathbf{v}, t)]\right] \tag{33}$$

$$\partial(\mathrm{Xe})/\partial t = \lambda_{\mathrm{I}}\mathrm{I}(\mathbf{r}, t) - \mathrm{Xe}(\mathbf{r}, t) \left\{\lambda_{\mathrm{Xe}} + \int d^3v\, [\sigma_{a\mathrm{Xe}}(v, T)\, vn(\mathbf{r}, \mathbf{v}, t)]\right\}$$
$$+ y_{\mathrm{Xe}} \int d^3v\, [\Sigma_f(\mathbf{r}, v, t)\, vn(\mathbf{r}, \mathbf{v}, t)] \tag{34}$$

where y_{Te} is the number of ^{135}Te per fission, $= 6.1 \times 10^{-2}$; y_{Xe} is the number of atoms of ^{135}Xe per fission, $= 0.19 \times 10^{-2}$; λ_{Te} is the decay constant of ^{135}Te, $= 5.8 \times 10^{-3}\ \mathrm{sec}^{-1}$; λ_{I} is the decay constant of ^{135}I, $= 2.88 \times 10^{-5}\ \mathrm{sec}^{-1}$; λ_{Xe} is the decay constant of ^{135}Xe, $= 2.10 \times 10^{-5}\ \mathrm{sec}^{-1}$; $\sigma_{a\mathrm{Te}}$ is the absorption cross section of ^{135}Te; $\sigma_{a\mathrm{I}}$ is the absorption cross section of ^{135}I; and $\sigma_{a\mathrm{Xe}}$ is the absorption cross section of ^{135}Xe.

We observe from (31) that the half-life of ^{135}Te ($\sim$2 min) is much shorter than the half-lives of ^{135}I and ^{135}Xe. It is a good approximation to treat ^{135}I as a direct fission fragment with a yield $y_{\mathrm{I}} = y_{\mathrm{Te}} = 0.061$. We shall use this approximation consistently in this book.

Samarium-149 is the β-decay product of $^{149}_{61}$Pm, which is a direct fission product. It is also produced by the β-decay of $^{149}_{61}$Nd:

$$^{149}_{60}\mathrm{Nd} \xrightarrow[\beta-]{(1.7\mathrm{hr})} {}^{149}_{61}\mathrm{Pm} \xrightarrow[\beta-]{(47\mathrm{hr})} {}^{149}_{62}\mathrm{Sm} \tag{35}$$

Again the half-life of ^{149}Nd is relatively short, so that we can treat ^{149}Pm as a direct fission product. Denoting the concentration of ^{149}Pm and ^{149}Sm by Pm(**r**, t) and Sm(**r**, t), we obtain

$$\partial(\mathrm{Pm})/\partial t = y_{\mathrm{Pm}} \int d^3v\, [v\Sigma_f(\mathbf{r}, v, t)\, n(\mathbf{r}, \mathbf{v}, t)]$$
$$- \mathrm{Pm}(\mathbf{r}, t) \left\{\lambda_{\mathrm{Pm}} + \int d^3v\, [\sigma_{a\mathrm{Pm}}(v, T)\, vn(\mathbf{r}, \mathbf{v}, t)]\right\} \tag{36}$$

$$\partial(\mathrm{Sm})/\partial t = \lambda_{\mathrm{Pm}}\mathrm{Pm}(\mathbf{r}, t) + y_{\mathrm{Sm}} \int d^3v\, [\Sigma_f(\mathbf{r}, v, t)\, vn(\mathbf{r}, \mathbf{v}, t)]$$
$$- \mathrm{Sm}(\mathbf{r}, t) \int d^3v\, [\sigma_{a\mathrm{Sm}}(v, T)\, n(\mathbf{r}, \mathbf{v}, t)] \tag{37}$$

where the meaning of the symbols should be clear, and $y_{\mathrm{Pm}} = 0.014$ atoms/fission, y_{Sm} is small, and $\lambda_{\mathrm{Pm}} = 3.56 \times 10^{-6}\ \mathrm{sec}^{-1}$. (Note that ^{149}Sm is stable, hence (37) has no decay term.)

Next we consider, as an example, the burnup of $^{235}_{92}$U and the conversion of $^{238}_{92}$U into $^{239}_{94}$Pu to illustrate time changes in the concentration of fissionable nuclei. The conversion of ^{238}U into ^{239}Pu takes place as

$$ {}_0n^1 + {}^{238}_{92}\mathrm{U} \rightarrow {}^{239}_{92}\mathrm{U} \xrightarrow[\beta^-]{(23.5\mathrm{min})} {}^{239}_{93}\mathrm{Np} \xrightarrow[\beta^-]{(2.3\mathrm{days})} {}^{239}_{94}\mathrm{Pu} \tag{38}$$

With the obvious notation, we obtain

$$\partial \mathrm{U}_{25}/\partial t = -\mathrm{U}_{25}(\mathbf{r}, t) \int d^3v\, [v\sigma_{a25}(v, T)\, n(\mathbf{r}, \mathbf{v}, t)] \tag{39}$$

$$\partial \mathrm{U}_{28}/\partial t = -\mathrm{U}_{28}(\mathbf{r}, t) \int d^3v\, [\sigma_{a28}(v, T)\, vn(\mathbf{r}, \mathbf{v}, t)] \tag{40}$$

$$\partial(\mathrm{Np})/\partial t = -(\partial \mathrm{U}_{28}/\partial t) - \mathrm{Np}(\mathbf{r}, t) \left\{\lambda_{\mathrm{Np}} + \int d^3v\, [\sigma_{a\mathrm{Np}}(v, T)\, vn(\mathbf{r}, \mathbf{v}, t)]\right\} \tag{41}$$

$$\partial(\mathrm{P_u})/\partial t = \lambda_{\mathrm{Np}}\mathrm{Np}(\mathbf{r}, t) - \mathrm{Pu}(\mathbf{r}, t) \int d^3v\, [\sigma_{a\mathrm{Pu}}(v, T)\, vn(\mathbf{r}, \mathbf{v}, t)] \tag{42}$$

where $\lambda_{\mathrm{Np}} = 3.48 \times 10^{-6}$/sec. Since the half-life of ^{239}U is small, we have neglected its time derivative in Eq. (41).

The description presented above is meant to be only representative. The inclusion of other important nuclidic chains in a given reactor with known fuel composition and design details can easily be achieved by the concepts and examples introduced in this section.

1.3. Kinetic Equations

In the previous section, we developed an analytical description of the temporal behavior of the neutron population in the presence of feedback effects, and presented the equations interrelating the angular neutron density, delayed neutron precursor density, temperature, and atomic concentrations of the influential nuclear species. We shall refer to these equations as the "kinetic equations," which we summarize here in a compact form and with some slight change in notation. It is more convenient for the later discussions to express the angular density $n(\mathbf{r}, \mathbf{v}, t)$ in terms of the lethargy u and the unit vector $\mathbf{\Omega}$, namely $n(\mathbf{r}, u, \mathbf{\Omega}, t)$, where u is the kinetic energy of the neutron ($E = mv^2/2$)

in the lethargy scale [11] [$u \equiv \log(E_0/E)$, where E_0 is a reference energy such that there are no neutrons with $E > E_0$], and $\mathbf{\Omega}$ is the unit vector denoting the direction of motion of the neutron ($\mathbf{\Omega} = \mathbf{v}/v$). The definition of $n(\mathbf{r}, u, \mathbf{\Omega}, t)$ follows from

$$n(\mathbf{r}, u, \mathbf{\Omega}, t)\, du\, d\Omega = n(\mathbf{r}, \mathbf{v}, t)\, d^3v \tag{1}$$

which leads to

$$n(\mathbf{r}, u, \mathbf{\Omega}, t) = \tfrac{1}{2}v^3 n(\mathbf{r}, \mathbf{v}, t) \tag{2}$$

In this new notation, Eqs. (13) and (14) of Section 1.1 become

$$\begin{aligned}\partial n(\mathbf{r}, u, \mathbf{\Omega}, t)/\partial t = &-\mathbf{\Omega} \cdot \nabla v(u)\, n(\mathbf{r}, u, \mathbf{\Omega}, t) - \Sigma(\mathbf{r}, u, t)\, v(u)\, n(\mathbf{r}, u, \mathbf{\Omega}, t)\\ &+ \int du' \int d\Omega' \Big\{\sum_j [f_0{}^j(u)/4\pi]\, \nu^j(u')(1 - \beta^j)\, \Sigma_{\mathrm{f}}{}^j(\mathbf{r}, u', t)\\ &+ \Sigma_{\mathrm{s}}(\mathbf{r}, u' \to u, \mathbf{\Omega} \cdot \mathbf{\Omega}', t)\Big\}\, v(u')\, n(\mathbf{r}, u', \mathbf{\Omega}', t)\\ &+ S(\mathbf{r}, u, \mathbf{\Omega}, t) + \sum_{i=1}^{6} \lambda_i [f_i(u)/4\pi]\, C_i(\mathbf{r}, t)\end{aligned} \tag{3a}$$

where $f_0{}^j(u)$ and $f_i(u)$ are normalized to unity as

$$\int_0^\infty du\, f_{\mathrm{s}}{}^j(u) = 1, \qquad s = 0, 1, \ldots, 6 \tag{3b}$$

$$\partial C_i(\mathbf{r}, t)/\partial t = -\lambda_i C_i(\mathbf{r}, t) + \int du \Big[\sum_j \beta_i{}^j \nu^j(u)\, v(u)\, \Sigma_{\mathrm{f}}{}^j(\mathbf{r}, u, t)\, n(\mathbf{r}, u, t)\Big] \tag{4a}$$

where we have defined

$$n(\mathbf{r}, u, t) \equiv \int d\Omega\, n(\mathbf{r}, u, \mathbf{\Omega}, t) \tag{4b}$$

which we shall refer to as the "scalar" neutron density (we prefer not to introduce a new symbol for it. The arguments of $n(\mathbf{r}, u, \mathbf{\Omega}, t)$ and $n(\mathbf{r}, u, t)$ will distinguish between the angular and scalar neutron densities). In (3) and (4), we have used the isotropy of the medium [$\mathscr{M}(\mathbf{V}, T)$ in (1.2, Eq. 9) is a function of $|\mathbf{V}|$ only] explicitly by implying that the cross sections depend only on u but not on $\mathbf{\Omega}$ (note that $\Sigma_{\mathrm{s}}(\mathbf{r}, \mathbf{v}' \to \mathbf{v}, t)$ depends only on $|\mathbf{v}|$, $|\mathbf{v}'|$, and the angle between them in an isotropic medium).

The temperature of the medium is determined by (1.2, Eq. 26), which now reads

$$\mu\{[\partial T(\mathbf{r}, t)/\partial t] + \mathbf{V}(\mathbf{r}, t) \cdot \nabla T(\mathbf{r}, t)\} - \nabla \cdot K\, \nabla T(\mathbf{r}, t)$$
$$= W_{\mathrm{f}} \int du\, [v(u)\, \Sigma_{\mathrm{f}}(\mathbf{r}, u, t)\, n(\mathbf{r}, u, t)] \tag{5a}$$

where

$$\Sigma_{\mathrm{f}}(\mathbf{r}, u, t) = \sum_j \Sigma_{\mathrm{f}}{}^j(\mathbf{r}, u, t) \tag{5b}$$

The concentrations of ^{135}I and ^{135}Xe satisfy

$$\partial \mathrm{I}(\mathbf{r}, t)/\partial t = y_{\mathrm{I}} \int du\, [v(u)\, \Sigma_{\mathrm{f}}(\mathbf{r}, u, t)\, n(\mathbf{r}, u, t)]$$
$$- \mathrm{I}(\mathbf{r}, t) \left\{\lambda_{\mathrm{I}} + \int du\, [v(u)\, \sigma_{\mathrm{aI}}(u, T)\, n(\mathbf{r}, u, t)]\right\} \tag{6}$$

$$\partial[\mathrm{Xe}(\mathbf{r}, t)]/\partial t = \lambda_{\mathrm{I}} \mathrm{I}(\mathbf{r}, t) - \mathrm{Xe}(\mathbf{r}, t) \left\{\lambda_{\mathrm{Xe}} + \int du\, [v(u)\, \sigma_{\mathrm{aXe}}(u, T)\, n(\mathbf{r}, u, t)]\right\}$$
$$+ y_{\mathrm{Xe}} \int du\, [v(u)\, \Sigma_{\mathrm{f}}(\mathbf{r}, u, t)\, n(\mathbf{r}, u, t)] \tag{7}$$

where we have treated ^{135}I as a direct fission product (cf the comment after 1.2, Eq. 34). The equations for ^{149}Pm and ^{149}Sm become

$$\partial[\mathrm{Pm}(\mathbf{r}, t)]/\partial t = y_{\mathrm{Pm}} \int du\, [v(u)\, \Sigma_{\mathrm{f}}(\mathbf{r}, u, t)\, n(\mathbf{r}, u, t)]$$
$$- \mathrm{Pm}(\mathbf{r}, t) \left\{\lambda_{\mathrm{Pm}} + \int du\, [v(u)\, \sigma_{\mathrm{aPm}}(u, T)\, n(\mathbf{r}, u, t)]\right\} \tag{8}$$

$$\partial[\mathrm{Sm}(\mathbf{r}, t)]/\partial t = \lambda_{\mathrm{Pm}} \mathrm{Pm}(\mathbf{r}, t) + y_{\mathrm{Sm}} \int du\, [v(u)\, \Sigma_{\mathrm{f}}(\mathbf{r}, u, t)\, n(\mathbf{r}, u, t)]$$
$$- \mathrm{Sm}(\mathbf{r}, t) \int du\, [v(u)\, \sigma_{\mathrm{aSm}}(u, T)\, n(\mathbf{r}, u, t)] \tag{9}$$

Similar equations can be written for the production, decay, and burnup of other fission fragments whenever they are considered to be important in a reactor unde consideration. For the burnup and conversion of fissionable nuclei, we write a representative equation as

$$\partial N_j(\mathbf{r}, t)/\partial t = -\lambda_j N_j(\mathbf{r}, t) + \int du\, \{v(u)[\sigma_{\mathrm{a}i}(u, T)\, N_i(\mathbf{r}, t)$$
$$- N_j(\mathbf{r}, t)\, \sigma_{\mathrm{a}j}(u, T)]\, n(\mathbf{r}, u, t)\} \tag{10}$$

where the ith element is burned up and converted into the jth element, whose decay constant is λ_j. One can write as many equations of this type as called for in a specified type of reactor.

Kinetic equations (3)–(10) represent a set of coupled, nonlinear, integrodifferential equations, and provide a basis for detailed numerical calculations as well as for simplified models which can be handled analytically. This set contains more information about the average temporal behavior of a reactor than one is interested in, and is much too complex even for machine calculations. The main task of reactor dynamics is then to simplify this set and to develop various analytical and numerical techniques in order to extract from it the relevant information concerning the performance and safety of a reactor. This will be the main objective of the remaining chapters of the book.

Boundary and Initial Conditions

The boundary conditions for various dependent variables such as the angular neutron density, temperature, and atomic concentrations can be easily determined from the physical nature of the quantity under consideration. It is clear that the angular density $n(\mathbf{r}, u, \mathbf{\Omega}, t)$ must be continuous, positive, and finite everywhere in the medium in the absence of point external sources. If point sources are present, the angular density is not finite at the point in space where the source is located. To specify the boundary conditions, we assume that the medium is surrounded by a convex surface in a vacuum. We also assume that there are no neutrons incident on the outher surface from ourside. Then, $n(\mathbf{r}, u, \mathbf{\Omega}, t)$ satisfies

$$n(\mathbf{r}_s, u, \mathbf{\Omega}, t) = 0 \qquad \text{for} \quad \mathbf{\Omega} \cdot \mathbf{n} < 0 \tag{11}$$

where $\mathbf{r}_s$ is a point on the outer boundary, and $\mathbf{n}$ is the outward normal to the surface at $\mathbf{r}_s$. We shall refer to (11) as the "regular" boundary condition to distinguish it from the adjoint boundary condition introduced in Chapter 2. At the interface of two media with different nuclear properties, $n(\mathbf{r} + R\mathbf{\Omega}, u, \mathbf{\Omega}, t + R/v)$ is a continuous function of R about $R = 0$, where $\mathbf{r}$ is a point on the interface [13]. The boundary conditions appropriate to various approximate descriptions of neutron populations such as the diffusion and P_N-approximations will not be stated here (cf [2, 11, 13]). They will be mentioned and used when they are needed.

The boundary condition for temperature $T(\mathbf{r}, t)$ cannot be stated in general terms without reference to particular reactor types. It requires a detailed description of the thermal aspects of the reactor. We shall mention these in the examples discussed in other chapters. They are always obtained by requiring that $T(\mathbf{r}, t)$ and $K\,\nabla T(\mathbf{r}, t)$ (the thermal current) are continuous across the interface of two media with different thermal properties.

In addition to the boundary condition, one has to specify the initial conditions of all the dependent variables in order to specify a unique solution of the kinetic equations for $t > 0$.

1.4. Reduced Forms of the Kinetic Equations

A. Diffusion Approximation

An inspection of the precursor and feedback equations (4)–(10) of the previous section shows that they contain only the scalar neutron density $n(\mathbf{r}, u, t)$. The angular density $n(\mathbf{r}, u, \mathbf{\Omega}, t)$ enters only the transport equation (1.3, Eq. 3). This observation suggests that one may integrate the transport equation over angle to express the kinetic equations only in terms of the scalar flux, namely

$$\phi(\mathbf{r}, u, t) = v(u)\, n(\mathbf{r}, u, t) \tag{1a}$$

This is possible if we adopt Fick's law to relate the neutron current, defined by

$$\mathbf{J}(\mathbf{r}, u, t) = \int d\Omega\, [\mathbf{v}(u)\, n(\mathbf{r}, u, \mathbf{\Omega}, t)] \tag{1b}$$

to the scalar flux

$$\mathbf{J}(\mathbf{r}, u, t) = -D(\mathbf{r}, u, t)\, \nabla\phi(\mathbf{r}, u, t) \tag{2}$$

where $D(\mathbf{r}, u, t)$ is the diffusion coefficient [2, 11, 13, 14]. This approximation enables one to replace $\nabla \cdot \mathbf{J}$, which arises upon the angle integration of the term $\mathbf{\Omega} \cdot \nabla v(u)\, n(\mathbf{r}, u, \mathbf{\Omega}, t)$ in the transport equation, by $-\nabla \cdot D\, \nabla\phi$, and is known as the diffusion approximation (the limitations of this approximation are discussed in the cited references).

With these remarks, we obtain from (1.3, Eq. 3)

$$\begin{aligned} \partial n(\mathbf{r}, u, t)/\partial t = {} & \nabla \cdot D(\mathbf{r}, u, t)\, \nabla\phi(\mathbf{r}, u, t) - \Sigma(\mathbf{r}, u, t)\, \phi(\mathbf{r}, u, t) \\ & + \int du' \Big[\sum_j f_0{}^j(u)\, \nu^j(u')(1 - \beta^j)\, \Sigma_f{}^j(\mathbf{r}, u', t) \\ & + \Sigma_s(\mathbf{r}, u' \to u, t)\Big]\, \phi(\mathbf{r}, u', t) \\ & + S(\mathbf{r}, u, t) + \sum_{i=1}^{6} \lambda_i f_i(u)\, C_i(\mathbf{r}, t) \end{aligned} \tag{3}$$

where

$$S(\mathbf{r}, u, t) \equiv \int d\Omega\, S(\mathbf{r}, u, \mathbf{\Omega}, t) \tag{4}$$

$$\Sigma_s(\mathbf{r}, u' \to u, t) \equiv \int d\Omega\, \Sigma_s(\mathbf{r}, u' \to u, \mathbf{\Omega} \cdot \mathbf{\Omega}', t) \tag{5}$$

$$D(\mathbf{r}, u, t) \equiv \{3[\Sigma(\mathbf{r}, u, t) - \bar{\mu}(\mathbf{r}, u, t)\, \Sigma_s(\mathbf{r}, u, t)]\}^{-1} \tag{6}$$

The quantity $\bar{\mu}(\mathbf{r}, u, t)$ in (6) is the average of the cosine of the angle of rotation in the laboratory system and is defined by

$$\bar{\mu}(\mathbf{r}, u, t) = 2\pi \int_{-1}^{+1} d\mu \left[\mu \int_0^\infty \Sigma_s(\mathbf{r}, u \to u', \mu, t)\, du'\right] \Big/ \Sigma_s(\mathbf{r}, u, t) \tag{7}$$

B. Multigroup Description of the Energy Dependence

The energy dependence of the kinetic equations can be treated very conveniently in the multigroup representation [15]. To this end, we integrate (3) over u between u_j and u_{j+1} and introduce

$$\phi^j(\mathbf{r}, t) \equiv \int_{u_j}^{u_{j+1}} du\, \phi(\mathbf{r}, u, t) \tag{8a}$$

$$\Sigma^j(\mathbf{r}, t) \equiv (1/\phi^j) \int_{u_j}^{u_{j+1}} \Sigma(\mathbf{r}, u, t)\, \phi(\mathbf{r}, u, t) \tag{8b}$$

$$\Sigma_s^{ij}(\mathbf{r}, t) \equiv (1/\phi^j) \int_{u_j}^{u_{j+1}} du \left\{ \int_{u_i}^{u_{i+1}} du' \left[\Sigma_s(\mathbf{r}, u' \to u, t)\, \phi(\mathbf{r}, u', t)\right] \right\} \tag{8c}$$

$$S^j(\mathbf{r}, t) \equiv \int_{u_j}^{u_{j+1}} S(\mathbf{r}, u, t) \tag{8d}$$

$$\nu^{(m)j} \Sigma_f^{(m)j}(\mathbf{r}, t) \equiv (1/\phi^j) \int_{u_j}^{u_{j+1}} du \left[\nu^{(m)}(u)\, \Sigma_f^{(m)}(\mathbf{r}, u, t)\, \phi(\mathbf{r}, u, t)\right] \tag{8e}$$

where the superscript (m) designates the kind of the fissionable nucleus,

$$f_s^j \equiv \int_{u_j}^{u_{j+1}} du\, f_s(u), \qquad s = 0, 1, \ldots, 6, \tag{8f}$$

$$D^j(\mathbf{r}, t) \equiv \left\{ \int_{u_j}^{u_{j+1}} du \left[D(\mathbf{r}, u, t)\, \nabla\phi(\mathbf{r}, u, t)\right] \right\} [\nabla\phi^j(\mathbf{r}, t)]^{-1} \tag{8g}$$

it is assumed here that $\phi(\mathbf{r}, u, t)$ is separable in $\mathbf{r}$ and u in the interval $[(u_j, u_{j+1}),]$

$$1/v^j \equiv \left[\int_{u_j}^{u_{j+1}} du\, n(\mathbf{r}, u, t)\right][\phi_j]^{-1} \tag{8h}$$

With this notation, (3) takes the following form:

$$(1/v^j)\,\partial\phi^j/\partial t = \nabla \cdot D^j\, \nabla\phi^j - \Sigma^j\phi^j + \sum_{i=0}^{N-1}\left[\sum_m f_0^{(m)j}\nu^{(m)i}\Sigma_{\mathrm{f}}^{(m)i}(1-\beta^{(m)}) + \Sigma_{\mathrm{s}}^{ij}\right]\phi^i$$
$$+ S^j + \sum_{i=1}^{6}\lambda_i f_i{}^j C_i \tag{9}$$

where we have omitted the argument $(\mathbf{r}, t)$ of ϕ^j, D^j, Σ^j, Σ_{s}^{ij}, $\Sigma_{\mathrm{f}}^{(m)i}$, S^j, and C_i. Here, N is the number of groups. We choose $u_0 = 0$, which corresponds to an energy such that there are no neutrons whose energies are greater than this energy (e.g., 20 MeV). We may choose u_N as ∞, which corresponds to zero energy. The matrix Σ^{ij} is the transfer matrix, which represents up- and down-scattering of neutrons (namely Σ_{s}^{ij} is the up-scattering cross section from u_i to u_j if $i > j$ and the down-scattering cross section if $i < j$, provided we adopt the convention $u_i \gtrless u_j$ for $i \gtrless j$). We also note that, in (9),

$$\Sigma_{\mathrm{s}}{}^j = \sum_{i=0}^{N-1}\Sigma_{\mathrm{s}}^{ji} \tag{10}$$

and that the diagonal terms cancel between $-\Sigma_{\mathrm{s}}{}^j\phi^j$ and $\sum_{i=0}^{N-1}\Sigma_{\mathrm{s}}^{ij}\phi^i$.

In the multigroup form, the precursor equations (4) of Section 1.3 become

$$\partial C_i/\partial t = -\lambda_i C_i + \sum_{j=0}^{N-1}\sum_m \beta_i^{(m)}\nu^{(m)j}\Sigma_{\mathrm{f}}^{(m)j}\phi^j \tag{11}$$

Similar changes can be made in the equations describing feedback, i.e., in (5)–(10) of Section 1.3. As examples, we rewrite the ^{135}Xe and ^{135}I equations explicitly in multigroup form:

$$\partial \mathrm{I}/\partial t = y_{\mathrm{I}}\sum_{j=0}^{N-1}\Sigma_{\mathrm{f}}{}^j\phi^j - \mathrm{I}\left[\lambda_{\mathrm{I}} + \sum_{j=0}^{N-1}\sigma_{\mathrm{aI}}^j\phi^j\right] \tag{12}$$

$$\partial(\mathrm{Xe})/\partial t = \lambda_{\mathrm{I}}\mathrm{I} - \mathrm{Xe}\left[\lambda_{\mathrm{Xe}} + \sum_{j=0}^{N-1}\sigma_{\mathrm{aXe}}^j\phi^j\right] + y_{\mathrm{Xe}}\sum_{j=0}^{N-1}\Sigma_{\mathrm{f}}{}^j\phi^j \tag{13a}$$

where

$$\Sigma_{\mathrm{f}}^{j} = \sum_{m} \Sigma_{\mathrm{f}}^{(m),j} \tag{13b}$$

The definition of σ_{aI}^{j} and $\sigma_{\mathrm{aXe}}^{j}$ follow from (8b).

The combination of the diffusion and multienergy group descriptions leads to a coupled set of nonlinear differential equations for the dependent variables $\phi^{j}(\mathbf{r}, t)$, $C_i(\mathbf{r}, t)$, $T(\mathbf{r}, t)$, and $N_j(\mathbf{r}, t)$. This form of the kinetic equations is particularly suitable to machine calculations by which the space- and energy-dependent dynamical behavior of a reactor can be realistically assessed. However, the numerical solution of the multigroup space-dependent kinetic equations is at present prohibitive in terms of computer time for all but the simplest problems. We can extract a considerable amount of qualitative information from the kinetic equations by analytical techniques before returning to numerical computations. (A description of multigroup techniques in reactor physics and discussion of the choice of group constants can be found in the work of Clark and Hansen [16].)

C. Slowing-Down Diffusion Equation and Fermi Age Model

The purpose of this section is to rewrite Eq. (3) in terms of the slowing-down density $q(\mathbf{r}, u, t)$ and to develop the continuous slowing-down diffusion equations (often called the Fermi age model); this will be very useful for introducing some elementary notions of reactor dynamics in homogeneous bare systems, even though it has very limited utility in realistic reactor-dynamic problems [2, 11].

The slowing-down density is defined as follows:

$$q(\mathbf{r}, u, t)\, d^3r\, dt$$

i.e., the net average number of neutrons slowing down past lethargy u in d^3r at r and in dt about t. It follows from this definition that $q(\mathbf{r}, u, t)$ can be related to the scalar flux as

$$\begin{aligned} q(\mathbf{r}, u, t) = & \int_0^u du' \int_u^\infty du'' \, [\Sigma_{\mathrm{s}}(\mathbf{r}, u' \to u'', t)\, \phi(\mathbf{r}, u', t)] \\ & - \int_u^\infty du' \int_0^u du'' \, [\Sigma_{\mathrm{s}}(\mathbf{r}, u' \to u'', t)\, \phi(\mathbf{r}, u', t)] \end{aligned} \tag{14a}$$

and satisfies

$$q(\mathbf{r}, 0, t) = q(\mathbf{r}, \infty, t) = 0 \tag{14b}$$

The two terms in (14a) account for the down- and up-scattering of neutrons past u. Differentiating (14a) with respect to u, and substituting $\int du'[\Sigma_s(\mathbf{r}, u' \to u, t)\, \phi(\mathbf{r}, u', t)]$ from the resulting equations (Problem 8a) into (3), we find

$$\partial n(\mathbf{r}, u, t)/\partial t = \nabla \cdot D(\mathbf{r}, u, t)\, \nabla\phi(\mathbf{r}, u, t) - \Sigma_a(\mathbf{r}, u, t)\, \phi(\mathbf{r}, u, t)$$
$$+ \sum_j f_0{}^j(u)(1 - \beta^j) \int du'\, [\nu^j(u')\, \Sigma_f{}^j(\mathbf{r}, u', t)\, \phi(\mathbf{r}, u', t)]$$
$$- [\partial q(\mathbf{r}, u, t)/\partial u] + S(\mathbf{r}, u, t) + \sum_{i=1}^{6} \lambda_i f_i(u)\, C_i(\mathbf{r}, t) \qquad (15)$$

This equation is equivalent to (3) in the sense that it does not contain any more approximations than (3) does. It differs from (3) in the scattering integral terms, which are replaced here by $\partial q(\mathbf{r}, u, t)/\partial u$. The purpose of this procedure is that, under very restrictive assumptions, $q(\mathbf{r}, u, t)$ can be taken to be approximately proportional to $\phi(\mathbf{r}, u, t)$, and (15) can be written in terms of $q(\mathbf{r}, u, t)$ (slowing-down equation) as a partial differential equation first order in t and u and second order in $\mathbf{r}$, rather than as an integrodifferential equation.

Before displaying these steps, we point out that, if we integrate (15) over u in the interval (u_j, u_{j+1}), we get the group equation (9) in which the terms containing the transfer matrix are replaced by $q(\mathbf{r}, u_j, t) - q(\mathbf{r}, u_{j+1}, t)$.

The "thermal" group is obtained by integrating (3) or (15) in the interval (u_{th}, ∞). Here, u_{th} corresponds to the thermal cutoff energy E_{th}, which is chosen such that (a) all the fission neutrons (prompt and delayed) are born with energies greater than E_{th}, (b) the atoms of the medium can be assumed to be at rest for neutrons whose energies lie in $E > E_{\text{th}}$ (consequently, these neutrons always slow down whenever they suffer a scattering collision), and (c) neutrons with energies in the thermal group cannot gain energies above E_{th} when they suffer an up-scattering collision with the atomic nuclei (note that the motion of the atoms is not ignored for neutrons in the thermal group).

Condition (a) implies that $f_0{}^j(u) = f_i(u) = 0$ for $u > u_{\text{th}}$. Using this condition, and integrating (15) in the interval (u_{th}, ∞) we obtain

$$(1/v^{\text{th}})\, \partial\phi_{\text{th}}(\mathbf{r}, t)/\partial t = \nabla \cdot D_{\text{th}}(\mathbf{r}, t)\, \nabla\phi_{\text{th}}(\mathbf{r}, t) - \Sigma_a^{\text{th}}(\mathbf{r}, t)\, \phi_{\text{th}}(\mathbf{r}, t)$$
$$+ q_{\text{th}}(\mathbf{r}, t) + S_{\text{th}}(\mathbf{r}, t) \qquad (16)$$

where $q_{\text{th}}(\mathbf{r}, t) \equiv q(\mathbf{r}, u_{\text{th}}, t)$ and where we have used $q(\mathbf{r}, \infty, t) \equiv 0$. The thermal-group constants appearing in (16) are defined in (8) with

$u_{N-1} = u_{\text{th}}$ and $u_N = \infty$. (Clearly, (16) could be obtained directly from the group equations (9) with the same assumption.) The thermal-group equation plays an important role in the discussion of thermal reactors. In this equation, $S_{\text{th}}(\mathbf{r}, t)$ is the external thermal source, which is almost always zero because known neutron sources emit neutrons with energies greater than E_{th}. Then, $q_{\text{th}}(\mathbf{r}, t)$ is the only thermal neutron source in (16), which is the number of neutrons slowing down past u_{th} per unit volume per second.

By breaking up the integration in the fission term in (15) into thermal and fast energy groups, we obtain

$$\begin{aligned} \partial n(\mathbf{r}, u, t)/\partial t &= \nabla \cdot D(\mathbf{r}, u, t)\, \nabla\phi(\mathbf{r}, u, t) - \Sigma_{\text{a}}(\mathbf{r}, u, t)\, \phi(\mathbf{r}, u, t) \\ &\quad + \sum_j f_0^j(u)(1 - \beta^j)\, \epsilon^j(\mathbf{r}, t)\, \nu_{\text{th}}^j \Sigma_{\text{f}}^{\text{th},j}(\mathbf{r}, t)\, \phi_{\text{th}}(\mathbf{r}, t) \\ &\quad - [\partial q(\mathbf{r}, u, t)/\partial t] + S(\mathbf{r}, u, t) + \sum_{i=1}^{6} \lambda_i f_i(u)\, C_i(\mathbf{r}, t) \end{aligned}$$

$$u < u_{\text{th}} \qquad (17)$$

where we have introduced $\epsilon^j(\mathbf{r}, t)$ as

$$\epsilon^j(\mathbf{r}, t) = \left\{ 1 + \frac{\int_0^{u_{th}} du' \, [\Sigma_{\text{f}}^j(\mathbf{r}, u', t)\, \nu^j(u')\, \phi(\mathbf{r}, u', t)]}{\nu_{\text{th}}^j\, \Sigma_{\text{f}}^{\text{th},j}(\mathbf{r}, t)\, \phi_{\text{th}}(\mathbf{r}, t)} \right\} \qquad (18)$$

which is the "fast fission" factor [11]. If $\epsilon^j(\mathbf{r}, t)$ can be estimated *a priori*, then (17) becomes a partial differential equation in $\phi(\mathbf{r}, u, t)$, provided $q(\mathbf{r}, u, t)$ can be approximated to be proportional to $\phi(\mathbf{r}, u, t)$. Such a relation can be obtained (Problem 8b) as

$$q(\mathbf{r}, u, t) \simeq \xi(\mathbf{r}, u, t)\, \Sigma(\mathbf{r}, u, t)\, \phi(\mathbf{r}, u, t) \qquad (19a)$$

where

$$\xi(\mathbf{r}, u, t) \equiv \sum_i [\Sigma_{\text{s}}^i(\mathbf{r}, u, t)/\Sigma_{\text{s}}(\mathbf{r}, u, t)]\, \xi^i \qquad (19b)$$

and where ξ^i is the mean logarithmic energy loss per collision for the ith kind of scatterer.

The assumptions inherent in (19a) are numerous. To mention a few: (a) the slowing down of fast neutrons ($u < u_{\text{th}}$) is due to elastic scattering, (b) the scattering is isotropic in the center-of-mass system, (c) the collision density is a slowly varying function of lethargy over the maximum lethargy increment per collision, (d) the neutron source at u is weak, (e) the reactor is large, etc. (For a detailed discussion of the validity as well as utility of (19), the reader is referred, for example, to Weinberg and Wigner [2] and Meghreblian and Holmes [11].) Equations (16), (17), and (19) are referred to as the Fermi age model. The validity of this model is restricted entirely almost to bare homogeneous systems. Its main usefulness lies in the fact that it provides closed-form analytical solutions in such systems, by which some concepts of reactor dynamics (e.g., fast lifetime) can be introduced in a restricted sense. We shall illustrate these points in the subsequent section, where we discuss simple kinetic models.

1.5. Simple Kinetic Models

The purpose of this section is to present solutions of some simple kinetic problems in order to clarify some of the points made in the previous sections (e.g., boundary conditions, Fermi age model) and to provide concrete examples at this stage for some more abstract tools to be introduced in Chapter 2. This presentation is meant to be a brief review of the methods for solving simple boundary-value problems.

A. One-Group Diffusion Model

We first consider an infinite, bare slab reactor in the absence of delayed neutrons and feedback effects, and with constant cross sections in position and time. We assume only one kind of fissionable isotope, and treat the energy dependence in a one-group approximation. Furthermore, we adopt the diffusion approximation. Then, the temporal behavior of the reactor is described by

$$(1/v)\,\partial\phi(x,t)/\partial t = [D(\partial^2/\partial x^2) - \Sigma_\mathrm{a} + \nu\Sigma_\mathrm{f}]\,\phi(x,t) + S(x,t) \tag{1}$$

which is obtained from (1.4, Eq. 9) with the above simplifying assumptions and changes in notation. This equation can be regarded as the thermal group (cf 1.4, Eq. 16) in which the fission neutrons are born thermal. If the thickness of the slab, including the extrapolated distance, is $\tilde{a}$ [11, 14], then the appropriate boundary conditions are $\phi(0, t) = \phi(\tilde{a}, t) = 0$, and $\phi(x, t)$ is finite and nonnegative (note that these boundary conditions for the scalar flux are approximate, as opposed to the exact boundary conditions mentioned in Section 1.3 for the angular density). We wish to solve this problem for the initial condition $\phi(x, 0) \equiv \phi_0(x)$, where $\phi_0(x)$ is a known function of x, and the source $S(x, t)$ has arbitrary position and time dependences.

We first consider the eigenvalue problem

$$[d^2\psi_n(x)/dx^2] + B_n{}^2\psi_n(x) = 0 \tag{2a}$$

$$\psi_n(0) = \psi_n(\tilde{a}) = 0 \tag{2b}$$

This is a homogeneous equation, and has nontrivial solutions only for some particular values of $B_n{}^2$ (called eigenvalues). We present the eigenvalues and the corresponding eigenfunctions:

$$\psi_n(x) = (2/\tilde{a})^{1/2} \sin B_n x, \qquad n = 0, 1, 2,..., \tag{3a}$$

$$B_n = (n + 1)\, \pi/\tilde{a}, \tag{3b}$$

The "scalar" product of the eigenfunctions is defined as

$$\langle \psi_n \mid \psi_m \rangle \equiv \int_0^{\tilde{a}} dx\, [\psi_n(x)\, \psi_m(x)] = \delta_{nm} \tag{4}$$

where we consider only real functions at present and use the ket notation, whose convenience becomes apparent when we deal with eigenvectors. The second equality in (4) implies that the eigenfunctions $\psi_n(x)$ (or eigenkets $\mid \psi_n\rangle$) form an orthonormal set. This set is complete, i.e., any functions (for restrictions, see Churchill [17]) can be expanded into this set. In particular, the Dirac delta function has an expansion as

$$\delta(x - x_0) = \sum_{n=0}^{\infty} \psi_n(x)\, \psi_n(x_0) \tag{5a}$$

$$= (2/\tilde{a}) \sum_{n=0}^{\infty} \sin B_n x \sin B_n x_0 \tag{5b}$$

[The relations (5) are referred to as the closure property.] With this machinery, we write the desired solution as

$$\phi(x, t) = \sum_{n=0}^{\infty} b_n(t)\, \psi_n(x) \tag{6}$$

and

$$S(x, t) = \sum_{n=0}^{\infty} S_n(t)\, \psi_n(x), \qquad S_n(t) = \langle \psi_n \mid S(x, t)\rangle \tag{7}$$

Here, the $S_n(t)$ are known functions of time. Substituting (6) into (1) yields for $b_n(t)$

$$\dot{b}_n(t) + \lambda_n b_n(t) = vS_n(t) \tag{8a}$$

where

$$\lambda_n \equiv Dv[B_n{}^2 - B_m{}^2], \qquad B_m{}^2 \equiv (\nu\Sigma_{\mathrm{f}} - \Sigma_{\mathrm{a}})/D \tag{8b}$$

The solution to (8a) is (using the Laplace transform in time):

$$b_n(t) = (\exp -\lambda_n t)\, b_n(0) + \int_0^t dt' \,[(\exp -\lambda_n t')\, vS_n(t-t')] \tag{9}$$

where the $b_n(0)$ have to be determined from the initial distribution $\phi_0(x)$; if we let $t = 0$ in (6),

$$\phi_0(x) = \sum_{n=0}^{\infty} b_n(0)\, \psi_n(x), \tag{10a}$$

which yields

$$b_n(0) = \langle \psi_n \mid \phi_0 \rangle \tag{10b}$$

The complete solution is

$$\phi(x, t) = \sum_{n=0}^{\infty} \left(\langle \psi_n \mid \phi_0 \rangle (\exp -\lambda_n t) + \int_0^t dt' \,\{[\exp -\lambda_n (t-t')]\, vS_n(t')\} \right) \psi_n(x) \tag{11}$$

Discussion

We first consider the source-free case, i.e., $S(x, t) \equiv 0$. Then the problem becomes an initial-value problem. We draw attention to the following points:

1. The λ_n in (8b) are ordered as $\lambda_0 < \lambda_1 < \lambda_2 \cdots$.
2. The solution can be written as

$$\phi(x, t) = e^{-\lambda_0 t} \left[\langle \psi_0 \mid \phi_0 \rangle\, \psi_0(x) + \sum_{n=1}^{\infty} \langle \phi_0 \mid \psi_n \rangle\, e^{-(\lambda_n - \lambda_0)t} \psi_n(x) \right] \tag{12}$$

Since $\lambda_n - \lambda_0 > 0$ for all n, $\phi(x, t)$ approaches asymptotically

$$\phi(x, t) \to \langle \psi_0 \mid \phi_0 \rangle\, e^{-\lambda_0 t} \psi_0(x) \tag{13}$$

Here, $\psi_0(x)$ is called the "fundamental" mode, and in this problem it is given explicitly by

$$\psi_0(x) = (2/\tilde{a})^{1/2} \sin(\pi x/\tilde{a}) \tag{14}$$

This eigenfunction, which corresponds to the algebraically lowest eigenvalue, is everywhere nonnegative, and it determines the asymptotic flux shape. Hence, after long times, $\phi(x, t)$ rises or decays exponentially without changing its spatial distribution. For this reason, $\psi_0(x)$ is also called the "persisting" mode.

3. λ_0 is called the "inverse" reactor period. Its sign determines the asymptotic time behavior of the reactor. The reactor is said to be

$$\text{supercritical if} \quad \lambda_0 < 0 \quad \text{or} \quad \pi/\tilde{a} < B_m \tag{15a}$$

$$\text{critical if} \quad \lambda_0 = 0 \quad \text{or} \quad \pi/\tilde{a} = B_m \tag{15b}$$

$$\text{subcritical if} \quad \lambda_0 > 0 \quad \text{or} \quad \pi/\tilde{a} > B_m \tag{15c}$$

Note that the criticality condition is independent of the initial conditions. When $\lambda_0 = 0$, $\phi(x, t)$ approaches a "steady state,"

$$\phi(x, t \to \infty) = \langle \psi_0 \mid \phi_0 \rangle \, \psi_0(x) \tag{16}$$

4. The terms for $n \geqslant 1$ in (12) are called "higher spatial" modes. They decay faster than the fundamental.

The relative rate of decay in (12) is determined by the magnitude of $\lambda_n - \lambda_0$. For the first higher mode, we have

$$\lambda_1 - \lambda_0 = 3Dv(\pi^2/\tilde{a}^2) \tag{17}$$

which indicates that the higher modes decay relatively faster in small reactors ($\tilde{a}$ is small) than in large reactors. We may expect, in the light of this result, that the spatial effects will be more important in large reactors.

Next, we consider the response of the reactor to an external neutron source. We assume that the reactor is critical ($\lambda_0 = 0$) and operated at steady state for $t \leqslant 0$. Then, the initial flux $\phi_0(x) \sim \psi_0(x)$. The solution follows from (11) as

$$\phi(x, t) = \left[\langle \psi_0 \mid \phi_0 \rangle + v \int_0^t dt' \; S_0(t')\right] \psi_0(x)$$

$$+ v \sum_{n=1}^{\infty} \psi_n(x) \int_0^t dt' \; [(\exp -\lambda_n t') \, S_n(t - t')] \tag{18}$$

In this case, the following remarks are pertinent:

1. If the source distribution does not contain higher spatial modes, i.e., $S_n = 0$ for $n \geqslant 1$, $S_0 \neq 0$, then the higher modes in the flux are not excited.

2. If the source is constant in time, then the magnitude of the fundamental increases linearly in time (the response would remain finite if the reactor were subcritical).

Other applications of the one-speed diffusion model are given in Problems 9–12.

B. Diffusion of Neutrons in an External Field

In this application, we again consider an infinite, homogeneous, bare slab reactor, as indicated in Figure 1.5.1. We assume that the

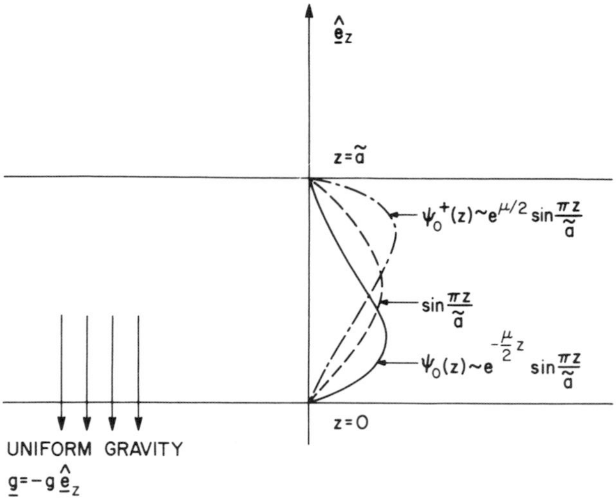

FIGURE 1.5.1. Neutron distribution under the influence of a uniform force field.

neutrons move under the influence of a uniform gravitational field parallel to the z axis in the negative z direction. We want to investigate the position and time dependences of the flux for $t > 0$ for a given initial distribution $\phi_0(z)$.

Our main objective here is to introduce the concepts of adjoint flux and neutron importance using perhaps the simplest non-self-adjoint reactor problem.

The one-speed diffusion equation in the presence of gravity can be written as

$$\frac{1}{v}\frac{\partial\phi(z,t)}{\partial t} = D\frac{\partial^2}{\partial z^2}\phi(z,t) + (\nu\Sigma_f - \Sigma_a)\phi(z,t) + \mu D\frac{\partial\phi(z,t)}{\partial z} \qquad (19)$$

where the last term accounts for the gravitational correction, and $\mu = 3g/v^2$ (Problems 2 and 13).

To solve this problem, we consider the following eigenvalue problem:

$$[(d^2/dz^2) + \mu(d/dz) + B_n^2]\, \psi_n(z) = 0 \tag{20}$$

with the usual boundary conditions $\psi_n(0) = \psi_n(\tilde{a}) = 0$. (We do not consider neutrons escaping the reactor in the positive z direction and returning back under the influence of gravity. The above boundary conditions can be visualized as those for a reactor wrapped with an infinite absorber.) The solution of (20) can be shown to be (Problem 14)

$$\psi_n(z) = (2/\tilde{a})^{1/2}\, e^{-\mu z/2} \sin[(n+1)\pi z/\tilde{a}] \tag{21a}$$

$$B_n^2 = (n+1)^2(\pi/\tilde{a})^2 + (\mu^2/4), \qquad n = 0, 1, 2, \ldots, \tag{21b}$$

where we have chosen the normalization $(2/\tilde{a})^{1/2}$ anticipating the final result (Figure 1.5.1). Because of the exponential factor, these eigenfunctions do not form an orthonormal set, i.e., $\langle \psi_n | \psi_m \rangle \neq \delta_{nm}$. However, by comparing (21a) to (3a), we can guess that the set $\psi_n^+(z)$, defined as

$$\psi_n^+(z) = (2/\tilde{a})^{1/2}\, e^{+\mu z/2} \sin[(n+1)\pi z/\tilde{a}] \tag{22}$$

is orthonormal to $\psi_n(z)$ in the sense that

$$\langle \psi_n^+ \mid \psi_m \rangle = \delta_{nm} \tag{23}$$

We shall refer to $\psi_n^+(z)$ as the "adjoint" set. They satisfy the following equation:

$$[(d^2/dz^2) - \mu(d/dz) + B_n^2]\, \psi_n^+(z) = 0 \tag{24}$$

which differs from (20) in the sign of the first derivative. This equation is called the "adjoint" equation. Although the adjoint eigenfunctions satisfy the same boundary conditions in this problem as $\psi_n(z)$, in general they satisfy different boundary conditions, which are referred to as the "adjoint" boundary conditions (cf Chapter 2).

The solution of the initial-value problem is similar to that presented in Section 1.5A once a complete biorthonormal set is found. We just present the final result:

$$\phi(z, t) = \sum_{n=0}^{\infty} \langle \psi_n^+ \mid \phi_0 \rangle\, e^{-\lambda_n t} \psi_n(z) \tag{25}$$

where

$$\lambda_n = Dv[(\pi/\tilde{a})^2(n+1)^2 + (\mu^2/4) - B_m^2] \tag{26}$$

DISCUSSION

1. The criticality condition is

$$\lambda_0 = (\pi/\tilde{a})^2 + (\mu^2/4) - B_m{}^2 = 0 \tag{27}$$

If the reactor is just critical in the absence of gravity ($\mu = 0$), then the reactor becomes subcritical, i.e., $\lambda_0 > 0$, when gravity is present.

2. Let us suppose that the reactor is critical, and find $\phi(z, t)$ due to a neutron introduced at z' at $t = 0$ (actually, neutrons are introduced in a thin slab at z'). We assume that there are no neutrons for $t \leqslant 0$. The initial distribution in this case is $\phi_0(z) = \delta(z - z')$. Using (25) and $\langle \psi_0{}^+ \mid \phi_0 \rangle = \psi_0{}^+(z')$, we find

$$\phi(z, t) \to \psi_0{}^+(z')\,\psi_0(z) \qquad \text{as} \quad t \to \infty \tag{28}$$

This result indicates that the magnitude of the asymptotic distribution due to a neutron introduced at z' is proportional to the value of the adjoint flux at z'. The fission rate in the entire reactor a long time following the introduction of the neutron at $t = 0$ is obtained from (28) as

$$I(z', t \to \infty) = \psi_0{}^+(z') \int_0^{\tilde{a}} dz\, [\Sigma_f \psi_0(z)] \tag{29}$$

This quantity is known as the "neutron importance" (cf Chapter 2). Hence, neutron importance is proportional to the adjoint fundamental mode. (These concepts will be introduced in a more abstract sense in Chapter 2.)

C. Fermi Age Model

In this section, we will use the Fermi age model introduced in Section 1.4C to include the energy dependence in the mathematical model. More precisely, we shall investigate the time-dependent behavior of the flux in a bare, homogeneous reactor following a uniform step change in the thermal cross sections at $t = 0$. The reactor will be assumed to be critical for $t \leqslant 0$ with a steady-state flux $\phi_0(\mathbf{r})$. Delayed neutrons and feedback effects will not be considered.

The Fermi age equations (1.4, Eqs. 16 and 17) reduce for $t > 0$ to

$$\partial n(\mathbf{r}, u, t)/\partial t = D(u)\, \nabla^2 \phi(\mathbf{r}, u, t) - \Sigma_a(u)\, \phi(\mathbf{r}, u, t) - [\partial q(\mathbf{r}, u, t)/\partial u]$$
$$+ f(u)\, \nu_{th} \Sigma_f^{th} \phi_{th}(\mathbf{r}, t) \tag{30a}$$

$$(1/v_{\text{th}})\, \partial\phi_{\text{th}}(\mathbf{r}, t)/\partial t = D_{\text{th}}\, \nabla^2\phi_{\text{th}}(\mathbf{r}, t) - \Sigma_a^{\text{th}}\phi_{\text{th}}(\mathbf{r}, t) + q_{\text{th}}(\mathbf{r}, t) \tag{30b}$$

$$q(\mathbf{r}, u, t) \simeq \xi\Sigma(u)\, \phi(\mathbf{r}, u, t) \tag{30c}$$

In (30a), we have assumed only one kind of fissionable nucleus, and neglected fast fission. In addition, we shall assume that all fission neutrons are born monoenergetically at $u = 0$ [i.e., $f(u) = \delta(u)$].

The initial conditions are

$$\phi(\mathbf{r}, u, 0) = \phi_0(\mathbf{r}, u) \tag{31a}$$

$$\phi_{\text{th}}(\mathbf{r}, 0) = \phi_0(\mathbf{r}) \tag{31b}$$

where $\phi_0(\mathbf{r}, u)$ and $\phi_0(\mathbf{r})$ denote the steady-state distribution of the fast and thermal fluxes for $t \leqslant 0$. The boundary conditions are: (a) $\phi(\mathbf{r}, u, t)$ and $\phi_{\text{th}}(\mathbf{r}, t)$ nonsingular and nonnegative for all $u > 0$ and $\mathbf{r}$, and (b) $\phi(\tilde{\mathbf{R}}, u, t) = \phi_{\text{th}}(\tilde{\mathbf{R}}, t) = 0$, where $\tilde{\mathbf{R}}$ denotes a point on the extrapolated outer surface of the reactor.

The cross sections appearing in (30) are those appropriate to the perturbed reactor. Since the perturbation is introduced as a uniform step change, the perturbed cross sections are independent of position and time.

Using (30c) and (30a), we obtain

$$[1/v(u)\, \xi\Sigma(u)](\partial/\partial t)\, q(\mathbf{r}, u, t) = [D(u)/\xi\Sigma(u)]\, \nabla^2 q(\mathbf{r}, u, t) - [\Sigma_a(u)/\xi\Sigma(u)]\, q(\mathbf{r}, u, t) - [\partial q(\mathbf{r}, u, t)/\partial u] \tag{32}$$

The initial condition for $q(\mathbf{r}, u, t)$ in u is

$$q(\mathbf{r}, 0, t) = \nu_{\text{th}}\Sigma_f^{\text{th}}\phi_{\text{th}}(\mathbf{r}, t) \tag{33}$$

which implies that the fission neutrons begin slowing down immediately after their birth.

Following Meghreblian and Holmes [11], we look for a separable solution of the form

$$q(\mathbf{r}, u, t) = F(\mathbf{r})\, Q(u)\, e^{\mu t} \tag{34a}$$

$$\phi_{\text{th}}(\mathbf{r}, t) = F(\mathbf{r})\, \phi_{\text{th}}\, e^{\mu t} \tag{34b}$$

where $F(\mathbf{r})$ satisfies the Helmholtz equation:

$$\nabla^2 F(\mathbf{r}) + B^2 F(\mathbf{r}) = 0 \tag{35}$$

with the boundary condition $F(\tilde{\mathbf{R}}) = 0$. The solution of this equation for a specified geometry of the reactor determines both $F(\mathbf{r})$, within a normalization factor, and B^2, which is termed the geometric buckling. For a sphere, for example, $B^2 = (\pi/\tilde{R})^2$. The function $Q(u)$ and the quantities μ and ϕ_{th} are to be determined by requiring (34) to satisfy (30b) and (32). We shall leave the steps as an exercise (Problem 15), and present the results:

$$Q(u) = \nu_{\text{th}}\Sigma_{\text{f}}^{\text{th}}\phi_{\text{th}} \exp\left\{-\int_0^u du' \left[\frac{D'}{\xi\Sigma'} B^2 + \frac{\Sigma_{\text{a}}'}{\xi\Sigma'} - \frac{\mu}{v'\xi\Sigma'}\right]\right\} \tag{36}$$

where primes denotes the values of the indicated quantities at u'. Substituting

$$q_{\text{th}}(\mathbf{r}, t) = F(\mathbf{r})\, Q(u_{\text{th}})\, e^{\mu t}$$

into (30b) using the value of $Q(u_{\text{th}})$ from (36), we obtain the following equation for μ:

$$\frac{\mu}{v_{\text{th}}(D_{\text{th}}B^2 + \Sigma_{\text{a}}^{\text{th}})} = -1 + \frac{\nu^{\text{th}}\Sigma_{\text{f}}^{\text{th}}}{D_{\text{th}}B^2 + \Sigma_{\text{a}}^{\text{th}}}$$

$$\times \exp\left\{-\int_0^{u_{\text{th}}} du' \left[\frac{D'B^2}{\xi\Sigma'} + \frac{\Sigma_{\text{a}}'}{\xi\Sigma'} - \frac{\mu}{v'\xi\Sigma'}\right]\right\} \tag{37}$$

This result is simplified by introducing the following standard notation:

$$l_{\text{F}} \equiv \int_0^{u_{\text{th}}} du\, [1/\xi v(u)\, \Sigma(u)] \tag{38a}$$

$$l_{\text{th}} \equiv 1/v^{\text{th}}(\Sigma_{\text{a}}^{\text{th}} + D_{\text{th}}B^2) \tag{38b}$$

$$p_{\text{th}} \equiv \exp\left\{-\int_0^{u_{\text{th}}} [\Sigma_{\text{a}}(u)/\xi\Sigma(u)]\, du\right\} \tag{38c}$$

$$\tau_{\text{th}} \equiv \int_0^{u_{\text{th}}} du\, [D(u)/\xi\Sigma(u)] \tag{38d}$$

$$k_{\text{eff}} \equiv \nu_{\text{th}}\Sigma_{\text{f}}^{\text{th}} p_{\text{th}}(\exp -B^2\tau_{\text{th}})/(\Sigma_{\text{a}}^{\text{th}} + D_{\text{th}}B^2) \tag{38e}$$

The result is

$$1 + \mu l_{\text{th}} = k_{\text{eff}} e^{-\mu l_{\text{F}}} \tag{39}$$

In these equations, l_{F} is the lifetime of fast neutrons, l_{th} is the lifetime of thermal neutrons, p_{th} is the resonance escape probability during slowing-down to thermal, τ_{th} is the neutron age to thermal, and k_{eff} is the multiplication constant. The exponential function $\exp(-\tau_{\text{th}}B^2)$ is the fast nonleakage probability.

Equation (39) yields μ. In the case of $l_F\mu \ll 1$, it can be approximated ($e^{-\mu l_F} \approx 1 - \mu l_F$) by

$$\mu = \delta k/(l_F + l_{th}) \tag{40a}$$

where

$$\delta k = k_{eff} - 1 \tag{40b}$$

and is called the "excess reactivity."

The expression for the thermal flux is obtained by combining the foregoing results:

$$\phi_{th}(\mathbf{r}, t) = \phi_0(\mathbf{r}) \exp\{[\delta k/(l_F + l_{th})]\, t\} \tag{41}$$

We draw attention to the following points:

1. The thermal flux is separable in time and space:

$$\phi_{th}(\mathbf{r}, t) = P(t)\, \phi_0(\mathbf{r}) \tag{42a}$$

where the time function $P(t)$ is given by

$$P(t) = \exp\{-\, [\delta k/(l_F + l_{th})]\, t\} \tag{42b}$$

2. $P(t)$ satisfies

$$dP(t)/dt = (\delta k/l)\, P(t) \tag{43a}$$

where l is the neutron "lifetime," i.e.,

$$l \equiv l_F + l_{th} \tag{43b}$$

which includes both the slowing-down time l_F and the time the neutrons spend as thermal neutrons before they escape or are absorbed in the reactor.

Equation (43a) is the "point-kinetic" equation in the absence of delayed neutrons.

Concepts (such as space–time separability) and quantities (such as the lifetime and the multiplication factor) which have been introduced above using the Fermi age model will be reexamined in the next chapter under more realistic conditions.

PROBLEMS

1. Find the solution of Eq. (1) of Section 1.1.
   [Hint: Let $\mathbf{r}' = \mathbf{r} - \mathbf{v}t$ and $t' = t$ and show that

$$\partial n(\mathbf{r}', \mathbf{v}, t')/\partial t' = S(\mathbf{r}' + \mathbf{v}t', \mathbf{v}, t')]$$

2. Show that, in the presence of external forces, Eq. (1) of Section 1.1 is replaced by

$$[\partial n(\mathbf{r}, \mathbf{v}, t)/\partial t] + \mathbf{v} \cdot \nabla n(\mathbf{r}, \mathbf{v}, t) + \mathbf{a} \cdot \nabla_{\mathbf{v}} n(\mathbf{r}, \mathbf{v}, t) = S(\mathbf{r}, \mathbf{v}, t)$$

where $\mathbf{a}(\mathbf{r}, \mathbf{v}, t)$ is the acceleration due to the external forces acting on a neutron at $\mathbf{r}$ with a velocity $\mathbf{v}$ at time t.

3. Write down a transport equation for the delayed neutron precursors assuming that precursors are stopped whenever they suffer a collision, and they are emitted isotropically and monoenergetically with speed V.

4. Show that the solution of the transport equation in a purely absorbing medium (i.e., $\Sigma_s = \Sigma_f = 0$), namely

$$[\partial n(\mathbf{r}, \mathbf{v}, t)/\partial t] + \mathbf{v} \cdot \nabla n(\mathbf{r}, \mathbf{v}, t) + v\Sigma(\mathbf{r}, \mathbf{v})\, n(\mathbf{r}, \mathbf{v}, t) = S(\mathbf{r}, \mathbf{v}, t)$$

is [1]

$$\begin{aligned} n(\mathbf{r}, \mathbf{v}, t) = {} & \exp\left[-\int_0^t v\Sigma(\mathbf{r} - \mathbf{v}(t-\tau), \mathbf{v})\, d\tau\right] n(\mathbf{r} - \mathbf{v}t, \mathbf{v}, 0) \\ & + \int_0^t d\tau' \left\{\exp\left[\int_t^{\tau'} v\Sigma(\mathbf{r} - \mathbf{v}(t-\tau), \mathbf{v})\, d\tau\right]\right. \\ & \left. \times\, S(\mathbf{r} - \mathbf{v}(t-\tau'), \mathbf{v}, \tau')\right\} \end{aligned}$$

(Hint: Use the transformation in Problem 1.)

5. Show that

$$\int d^3V\, [(\exp i\mathbf{k} \cdot \mathbf{V})\{[\exp(-V^2/4\theta)]/(4\pi\theta)^{3/2}\}] = \exp(-k^2\theta)$$

(Hint: Reduce the left-hand side to

$$-(4\pi/k)[1/(4\pi\theta)^{3/2}](d/dk)\, \mathrm{Re}\, \tfrac{1}{2} \int_{-\infty}^{+\infty} dV \exp[ikV - (V^2/4\theta)]$$

and complete the square in the exponential.)

6. Obtain (1.2, Eq. 14) from (1.2, Eq. 13).
(Hint: Use the convolution theorem for Fourier transforms, and let $\theta \to \theta' - \theta$ in the formula of Problem 5.)

7. Use

$$\int d^3s\,[s\exp(-|\mathbf{v}-s|^2\alpha)]$$

$$= (\pi/\alpha v)(\exp -\alpha v^2)\int_{-\infty}^{+\infty} ds\,\{s^2\exp[-\alpha(s^2+2vs)]\}$$

and complete the square on the exponent (cf 1.2, Eq. 21).

8. (a) Using (1.4, Eq. 14a), verify the following relation:

$$\partial q(\mathbf{r}, u, t)/\partial u = \Sigma_s(\mathbf{r}, u, t)\,\phi(\mathbf{r}, u, t) - \int_0^\infty du'\,[\Sigma_s(\mathbf{r}, u' \to u, t)\,\phi(\mathbf{r}, u', t)]$$

(b) Assuming isotropic elastic scattering in the center-of-mass system for fast neutrons as the slowing-down mechanism, assuming the atoms of the medium to be at rest, and expanding ($\mathbf{r}$ and t dependences are omitted)

$$\Sigma_s(u')\,\phi(u') \cong \Sigma_s(u)\,\phi(u) + (u'-u)(\partial/\partial u)[\Sigma_s(u)\,\phi(u)]$$

one obtains for $q(\mathbf{r}, u, t)$ directly from its definition

$$q(\mathbf{r}, u, t) \cong \xi(\mathbf{r}, u, t)\,\Sigma_s(\mathbf{r}, u, t)\,\phi(\mathbf{r}, u, t) + (\partial/\partial u)[a(\mathbf{r}, u, t)\,\Sigma_s(\mathbf{r}, u, t)\,\phi(\mathbf{r}, u, t)]$$

where

$$a(\mathbf{r}, u, t) \equiv \sum_i \frac{\Sigma_s{}^i(\mathbf{r}, u, t)}{\Sigma_s(\mathbf{r}, u, t)}\left[\frac{\alpha_i \ln^2 \alpha_i}{2(1-\alpha_i)} - \xi_i\right]$$

and $\alpha_i \equiv (A_i - 1)^2/(A_i + 1)^2$. Substitute $q \approx \xi\Sigma_s\phi$ as a first approximation in the exact equation

$$(1/v)\,\partial\phi/\partial t = \nabla\cdot D\,\nabla\phi - \Sigma_a\phi + S(\mathbf{r}, t) - (\partial q/\partial u)$$

and obtain $\partial(\xi\Sigma_s\phi)/\partial u$. Then, approximate a by $-\xi^2$ and treat $\xi(\mathbf{r}, u, t)$ as a slowly varying function of u to eliminate $\partial(a\Sigma_s\phi)/\partial u$ in the expression of q. Verify the result:

$$q(\mathbf{r}, u, t) \approx \xi(\mathbf{r}, u, t)\{\Sigma(\mathbf{r}, u, t)\,\phi(\mathbf{r}, u, t) - \nabla\cdot D(\mathbf{r}, u, t)\,\nabla\phi(\mathbf{r}, u, t) - S(\mathbf{r}, u, t) + [1/v(u)][\partial\phi(\mathbf{r}, u, t)/\partial t]\}$$

Discuss the approximation needed to reduce this form to (1.4, Eq. 19).

9. Using the one-speed diffusion model, find the flux $\phi(x, t)$ in a bare, homogeneous, subcritical slab reactor of thickness $\tilde{a}$ (including extrapolation distance) due to a plane source located at the center of the slab and emitting neutrons isotropically as

$$S(t) = S_0(1 + m \sin \omega t) \text{ neutrons cm}^{-2} \text{ sec}^{-1}.$$

10. Solving the one-speed diffusion equation in an infinite, nonmultiplying, homogeneous medium, find the flux $\phi(x, t)$ for $x \neq 0$ and $t > 0$ due to an isotropic plane source $S(x, t)$ neutrons cm^{-2} sec^{-1} assuming

(a) $S(x, t) = \delta(x)\,\delta(t)$

(b) $S(x, t) = \delta(x)(1 + m \sin \omega t)$

In (a), find the time after the source burst at a point $x > 0$ at which $\phi(x, t)$ attains its maximum. In (b), $\phi(x, t)$ represents a damped wave. Find the speed of propagation of the wave, the wavelength, and the damping coefficient (e-folding distance of the amplitude of the wave).

11. Solving the one-group diffusion equation in a nonmultiplying, homogeneous sphere of radius $\tilde{R}$ (including the extrapolation distance), find the flux $\phi(r, t)$ due to an isotropic point source located at the center and emitting neutrons as a function of time as (neutrons cm^{-3} sec^{-1})

(a) $S(\mathbf{r}, t) = \delta(\mathbf{r})\,\delta(t)$

(b) $S(\mathbf{r}, t) = \delta(\mathbf{r})(1 + m \sin \omega t)$

12. A bare, homogeneous slab reactor is critical and operating in steady state with flux $\phi_0(x)$ for $t \leqslant 0$. At t, the absorption cross section in one-half of the slab is changed by an amount $\delta\Sigma_a$ (step change in time). Find the flux $\phi(x, t)$ as a function of x and time using the one-speed diffusion model.

13. Show that the one-group diffusion equation in the presence of external forces becomes

$$(1/v)\,\partial\phi(\mathbf{r}, t)/\partial t = D\,\nabla^2\phi(\mathbf{r}, t) - (3D/v^2)\,\mathbf{a}(\mathbf{r}, t) \cdot \nabla\phi(\mathbf{r}, t) + (\nu\Sigma_f - \Sigma_a)\,\phi(\mathbf{r}, t)$$

where $\nabla \cdot \mathbf{a} = 0$ is assumed.

(Hint: Integrate the transport equation with gravity over $\mathbf{v}$ assuming that $\mathbf{a}$ depends on $\mathbf{r}$ and t only, to obtain

$$(1/v)\,\partial\phi/\partial t = -\nabla\cdot\mathbf{J} - (\Sigma_{\mathrm{a}} - \nu\Sigma_{\mathrm{f}})\,\phi$$

Next, multiply it by $\mathbf{v}$ and then integrate over $\mathbf{v}$ to show that, in the diffusion approximation,

$$\mathbf{J} \cong -(1/3\Sigma_{\mathrm{tr}})\,\nabla\phi + \mathbf{a}(\phi/v^2\Sigma_{\mathrm{tr}})$$

where $\mathbf{v}$ is the average group velocity and Σ_{tr} is the transport cross section [11].)

14. Find the solution of

$$[(d^2/dz^2) + \mu(d/dz) + B_n^{\,2}]\,\psi_n = 0$$

(Hint: Let $\psi_n = \exp(-\mu z/2)\,u_n(z)$ and obtain

$$u_n'' + (B_n^{\,2} - \tfrac{1}{4}\mu^2)\,u_n = 0$$

with $u_n(0) = u_n(\tilde{a}) = 0$.)

15. Verify (1.5, Eq. 36).
(Hint: The solution of

$$y' + f(x)\,y = g(x)$$

is

$$y(x) = y(0)\exp\left[-\int_0^x f(x')\,dx'\right] + \int_0^x dx' \left\{\exp\left[-\int_{x'}^x dx''\,f(x'')\,dx''\right]\right\} g(x').)$$

16. (a) Estimate the ratio of the concentration of delayed neutron precursors to that of neutrons in a critical reactor for which $l = 10^{-4}$ sec, $\beta = 0.007$, and $\lambda = 0.08$ (one group of delayed neutrons).

(b) Estimate the actual concentration of bromine nuclei per neutron in a critical reactor.

(c) Estimate the number of neutrons per cubic centimeter in a reactor when the flux is $10^8\ \mathrm{cm}^{-2}\,\mathrm{sec}^{-1}$.

(d) Estimate the probability in (c) that all the neutrons present at a time instant t will disappear without causing a fission (this is the probability that a reactor shuts itself down when it is still critical), ignoring delayed neutrons.

REFERENCES

1. K. M. Case and P. F. Zweifel, "Linear Transport Theory." Addison-Wesley, Reading, Massachusetts, 1967.
2. A. M. Weinberg and E. P. Wigner, "The Physical Theory of Neutron Chain Reactors." Univ. of Chicago Press, Chicago, Illinois, 1958.
3. G. Keepin, T. Wimelt, and R. Zeigler, Delayed neutrons from fissionable isotopes of uranium, plutonium, and thorium. *Phys. Rev.* **107**, 1044 (1957).
4. Reactor physics constants. ANL-5800. U.S. At. Energy Comm., Div. of Tech. Inform., Argonne Nat. Lab., Argonne, Illinois, July 1963.
5. G. R. Keepin, Delayed fission neutron data in reactor physics and design. *Proc. Panel Delayed Fission Neutrons, Vienna, Aril* 1967. IAEA, Vienna, 1968.
6. T. Jahnsen, A. C. Papas, and T. Tuneal, Delayed neutron emission, theory, and precursor systematics. *Proc. Panel Delayed Fission Neutrons, Vienna, April* 1967. IAEA, Vienna, 1968.
7. G. R. KEEPIN, "Physics of Nuclear Kinetics." Addison-Wesley, Reading, Massachusetts, 1965.
8. L. Cranberg, G. Frye, N. Nereson, and L. Rosen, Fission neutron spectrum of ^{235}U. *Phys. Rev.* **103**, 662 (1956).
9. S. Yiftah, D. Okrent, and P. A. Moldauer, "Fast Reactor Cross Sections." Pergamon, 1960.
10. Neutron cross sections. BNL-325, 2nd ed. Brookhaven Nat. Lab., Upton, New York, 1965.
11. R. V. Meghreblian and D. K. Holmes, "Reactor Analysis." McGraw-Hill, New York, 1960.
12. J. M. Blatt and V. F. Weisskopf, "Theoretical Nuclear Physics." Wiley, New York, 1952.
13. B. Davison, "Neutron Transport Theory." Oxford Univ. Press, London and New York, 1958.
14. J. R. Lamarsh, "Introduction to Nuclear Reactor Theory." Addison-Wesley, Reading, Massachusetts, 1966.
15. A. F. Henry, Computation of parameters appearing in the reactor kinetic equations. WAPD-142. 1955.
16. M. Clark, Jr., and K. F. Hansen, "Numerical Methods of Reactor Analysis." Academic Press, New York, 1964.
17. R. V. Churchill, "Fourier Series and Boundary Value Problems." McGraw-Hill, New York, 1941.

CHAPTER 2

Point Kinetic Equations

The kinetic equations discussed in the previous chapter describe the time behavior of the neutron population in too much detail for most of the practical uses of reactor dynamics. In many applications, we are interested only in the dominant features of the time behavior of the neutron population, such as the variation of the total number of neutrons or the total power generation in the medium as a function of time. Details such as the angular dependence of the neutrons are almost never needed in reactor dynamics. In some applications, even the spatial distribution of the neutron population is not of great interest. It is, therefore, desirable to cast the basic kinetic equations into a simpler form which contains only the dominant aspects of the neutron population of practical interest, but leaves out all the undesired details.

The purpose of this chapter is to obtain a set of equations which will describe the time behavior of the total power generated in the medium. These equations are called the "point reactor kinetic" (or simply point kinetic) equations. They were first obtained in a systematic manner by Hurwitz [1] in 1949, and later by Ussachoff [2] in 1955 and by Henry [3, 4] in 1955 using adjoint fluxes. More recently, Gyftopoulos [5] derived these equations in 1964 with some modifications in Henry's approach, and Becker [6] reformulated them in 1968 using a variational principle and extending Lewin's derivations [7] of 1960.

Our derivation in this chapter will follow Henry's derivation. The Gyftopoulos modification and Becker's formulation will be discussed at the end of the chapter.

2.1. Mathematical Preliminaries

The purpose of this section is to introduce the mathematical concepts, terminology, and symbols which will be used in the derivation of the point kinetic equations in the following sections.

A. Scalar Product

The "scalar" product of two complex functions $\psi(\mathbf{r}, \mathbf{v})$ and $\phi(\mathbf{r}, \mathbf{v})$ is a complex number $\langle\psi \mid \phi\rangle$ defined by

$$\langle\psi \mid \phi\rangle \equiv \int_R d^3r \int d^3v \, [\psi^*(\mathbf{r}, \mathbf{v}) \, \phi(\mathbf{r}, \mathbf{v})] \tag{1}$$

where ψ^* is the complex conjugate of ψ. We note that the integration in the configuration space is extended only over the volume of the reactor, whereas the velocity integration is extended over the entire velocity space. In some applications, the symbol $\mid \phi\rangle$ will denote a vector whose components are functions of $\mathbf{r}, \mathbf{v}$; i.e., $\mid \phi\rangle = \text{col}[\phi_1(\mathbf{r}, \mathbf{v}),..., \phi_N(\mathbf{r}, \mathbf{v})]$. The definition of the scalar product of two vectors $\mid \phi\rangle$ and $\mid \psi\rangle$ is

$$\langle\psi \mid \phi\rangle \equiv \int_R d^3r \int d^3v \left[\sum_i \psi_i^*(\mathbf{r}, \mathbf{v}) \, \phi_i(\mathbf{r}, \mathbf{v})\right] \tag{2}$$

The symbol $\langle\psi \mid$ can be interpreted as a row vector whose components are the complex conjugate of the column $\mid \psi\rangle$. Then, (2) implies both a scalar product of two vectors in the usual sense, and an integration over $\mathbf{r}$ and $\mathbf{v}$.

If α_1 and α_2 are two complex numbers, the following relations follow from (1) and (2) (these relations are used in the subsequent manipulation without explanation):

$$\langle\alpha_1\psi \mid \phi\rangle = \alpha_1^*\langle\psi \mid \phi\rangle \tag{3a}$$

$$\langle\psi \mid \alpha_1\phi_1 + \alpha_2\phi_2\rangle = \alpha_1\langle\psi \mid \phi_1\rangle + \alpha_2\langle\psi \mid \phi_2\rangle \tag{3b}$$

$$\langle\psi \mid \phi\rangle^* = \langle\phi \mid \psi\rangle \tag{3c}$$

B. Adjoint Operators

In order to prepare for our discussion of orthonormal modes, we need to introduce the concept of adjoint operator. Let O be an operator operating on a function $\phi(\mathbf{r}, \mathbf{v})$ such that $O\phi$ is another function of $\mathbf{r}$

and $\mathbf{v}$. It may be an integral or differential operator acting on both variables $\mathbf{r}$ and $\mathbf{v}$.

An operator O^+ is defined to be the "adjoint" of O if

$$\langle O^+\psi \mid \phi\rangle = \langle \psi \mid O\phi\rangle \tag{4}$$

holds for all $\phi(\mathbf{r}, \mathbf{v})$ and $\psi(\mathbf{r}, \mathbf{v})$.

When the integration over $\mathbf{r}$ in the definition of scalar product is extended over a finite region in the configuration space (as is the case in reactor applications where the finite region is the volume of the reactor), one has to include the boundary conditions in the definition of the adjoint operator. We consider the functions $\phi(\mathbf{r}, \mathbf{v})$ that satisfy the regular boundary condition (cf Section 1.3), namely $\phi(\mathbf{r}, \mathbf{v}) = 0$ for $\hat{\mathbf{n}} \cdot \mathbf{v} < 0$, where $\mathbf{r}$ is on the outer surface of the reactor. We shall show presently that (4) can be satisfied for differential operators only if the functions $\psi(\mathbf{r}, \mathbf{v})$ are allowed to satisfy a different boundary condition, which is referred to as the "adjoint" boundary condition. We shall denote these functions by $\phi^+(\mathbf{r}, \mathbf{v})$. The following examples will illustrate the method of finding the adjoint of an operator, and the associated adjoint boundary condition.

Example 1. Differential operator. The adjoint of $\mathbf{v} \cdot \nabla$ is $-\mathbf{v} \cdot \nabla$ with the adjoint boundary condition

$$\phi^+(\mathbf{r}, \mathbf{v}) = 0 \qquad \text{for} \quad \hat{\mathbf{n}} \cdot \mathbf{v} > 0, \quad \mathbf{r} \in S \tag{5}$$

The proof is as follows:

$$\begin{aligned}\langle \phi^+ \mid O\phi\rangle &\equiv \int d^3v \int_R d^3r\, [\phi^{+*}(\mathbf{r}, \mathbf{v})\, \mathbf{v} \cdot \nabla\phi(\mathbf{r}, \mathbf{v})] \\ &= \int_0^\infty dv\, v^2 \int d\Omega \int d^3r\, [\phi^{+*}(\mathbf{r}, \mathbf{v})\, \nabla \cdot \mathbf{v}\phi(\mathbf{r}, \mathbf{v})]\end{aligned}$$

where we have used $\mathbf{v} \cdot \nabla\phi = \nabla \cdot \mathbf{v}\phi$, which follows from the fact that the gradient operator operates only $\mathbf{r}$. Using Green's theorem,† we obtain

$$\begin{aligned}\langle \phi^+ \mid O\phi\rangle = \int_0^\infty dv\, v^2 \Big\{ &\int d\Omega \int_S ds\, [\hat{\mathbf{n}} \cdot \mathbf{v}\phi^{+*}\phi] \\ &- \int d\Omega \int_R d^3r\, [(\mathbf{v} \cdot \nabla\phi^+)^*\, \phi]\Big\}\end{aligned}$$

† By Green's theorem, $\int_R d^3r\, (\nabla \cdot \mathbf{A}) = \int_s ds\, (\hat{\mathbf{n}} \cdot \mathbf{A})$. Let $A = \mathbf{v}\phi_1\phi_2$ and use $\nabla \cdot \mathbf{v}\phi_1\phi_2 = \phi_1\mathbf{v} \cdot \nabla\phi_2 + \phi_2\mathbf{v} \cdot \nabla\phi_1$.

where s is the outer surface of the reactor. Since ϕ^+ satisfies (5), the surface integral vanishes because either ϕ^+ or ϕ will always be zero on the surface in integrating over $\mathbf{\Omega}$. The remaining term can be recognized as $\langle -\mathbf{v} \cdot \nabla\phi^+ \mid \phi \rangle$, which proves the assertion.

Example 2. Integral operator. The adjoint of

$$\int d^3v' \, [v'\Sigma_s(\mathbf{v}' \to \mathbf{v}, \mathbf{r})] \tag{6a}$$

is

$$\int d^3v' \, [v\Sigma_s(\mathbf{v} \to \mathbf{v}', \mathbf{r})] \tag{6b}$$

which is obtained by interchanging the $\mathbf{v}$ and $\mathbf{v}'$ in the integrand. Note that there is no restriction on the adjoint functions ϕ^+ on the boundaries. The proof starts with the definition of the scalar product.

$$\langle \phi^+ \mid O\phi \rangle \equiv \int d^3r \int d^3v \, \phi^{+*}(\mathbf{r}, \mathbf{v}) \int d^3v' \, [v'\Sigma_s(\mathbf{v}' \to \mathbf{v}, \mathbf{r}) \, \phi(\mathbf{r}, \mathbf{v}')]$$

If the order of integration over $\mathbf{v}$ and $\mathbf{v}'$ is interchanged, the right-hand side becomes

$$\int d^3r \int d^3v \, \phi(\mathbf{r}, \mathbf{v}) \int d^3v' \, [v\Sigma_s(\mathbf{v} \to \mathbf{v}', \mathbf{r}) \, \phi^{+*}(\mathbf{r}, \mathbf{v}')] \equiv \langle O^+\phi^+ \mid \phi \rangle$$

It follows from this example as a special case that the adjoint of

$$f_0(v) \int d^3v' \, [\Sigma_f(\mathbf{r}, v') \, v'] \tag{7a}$$

is

$$v\Sigma_f(\mathbf{r}, v) \int d^3v' \, f_0(v') \tag{7b}$$

If the adjoint of an operator is identical to itself, namely $O^+ \equiv O$, then it is called "self-adjoint." It is easy to show, for example, that $\nabla \cdot D\nabla$ is self-adjoint for functions $\phi(\mathbf{r})$ that vanish on the outer surface (Problem 1).

The above definitions can be extended to matrix operators and to their adjoints. A matrix operator $O \equiv [O_{ij}]$ is a square matrix whose elements O_{ij} are operators as defined above. If $\mid \phi \rangle$ is a column vector, then $O \mid \phi \rangle$ is also a column vector, defined by

$$O \mid \phi \rangle = \mid O\phi \rangle \equiv \mathrm{col}[O_{1j}\phi_j, \ldots, O_{Nj}\phi_j] \tag{8}$$

with the summation convention on j in each element of $| O\phi\rangle$. The adjoint of O is a square matrix defined by

$$O^+ \equiv [O_{ji}^+] \tag{9}$$

which is obtained by interchanging the columns and rows of O and taking the adjoints of each element. The proof is left as an exercise (Problem 2). The following relations are self-evident:

$$\langle O^+\phi^+ | \phi\rangle = \langle \phi^+ | O | \phi\rangle = \langle \phi^+ | O\phi\rangle \tag{10}$$

These equalities define the symbol $\langle \phi^+ | O | \phi\rangle$, which will be used extensively in the following.

C. Eigenvalue Problem

The problem of finding the nontrivial solutions of a homogeneous equation of the form

$$O | \phi_n\rangle = \lambda_n | \phi_n\rangle \tag{11}$$

with certain boundary and regularity conditions on the solutions is called an eigenvalue problem. The set of complex numbers λ_n for which (11) has nontrivial solutions are called the eigenvalues and the corresponding solutions $| \phi_n\rangle$ are called the eigenfunctions of the operator O.

In our applications, the eigenfunctions will satisfy the regular boundary condition.

For a given operator O, we can also define an adjoint eigenvalue problem as

$$O^+ | \phi_n{}^+\rangle = \mu_n | \phi_n{}^+\rangle \tag{12}$$

with the adjoint boundary conditions. The eigenfunctions and the eigenvalues of O and O^+ are related to each other. The following properties are proven in standard texts [8, 9] on functional analysis in certain function spaces and for certain classes of operators. In our analysis, these properties will always be satisfied.

1. The sets of eigenvalues $\{\lambda_n\}$ and $\{\mu_n\}$ are complex conjugates of each other, i.e., for a given λ_n, there is a μ_n such that $\mu_n = \lambda_n{}^*$.
2. The eigenfunctions $\{| \phi_n\rangle\}$ and $\{| \phi_n{}^+\rangle\}$ form a biorthogonal set, i.e.,

$$\langle \phi_m{}^+ | \phi_n\rangle = \delta_{mn} \tag{13}$$

where δ_{mn} is the Kronecker delta.

3. $\{|\phi_n\rangle\}$ and $\{|\phi_n^+\rangle\}$ are complete, i.e., any vector $|\phi\rangle$ can be expanded into $\{|\phi_n\rangle\}$ as

$$|\phi\rangle = \sum_{n=0}^{\infty} a_n |\phi_n\rangle \tag{14a}$$

where

$$a_n = \langle\phi_n^+ | \phi\rangle \tag{14b}$$

4. The eigenvalues of a self-adjoint operator are real and positive, and its eigenfunctions form a complete orthonormal set.

5. The components of $|\phi_n\rangle$ satisfy (closure property)

$$\delta(\mathbf{r} - \mathbf{r}')\,\delta(\mathbf{v} - \mathbf{v}')\,\delta_{ij} = \sum_{n=0}^{\infty} \phi_{ni}(\mathbf{r}', \mathbf{v}')\,\phi_{nj}(\mathbf{r}, \mathbf{v}) \tag{15}$$

(Problem 3).

2.2. Stationary Reactor and the Multiplication Factor

The purpose of this section is to introduce the concept of criticality and neutron cycle, and to define the multiplication factor in a stationary noncritical reactor.

We start with the basic kinetic equations (1.3, Eq. 3) and (1.3, Eq. 4), which can be written in a compact way by defining the operators:

$$\begin{aligned} L \equiv &-\mathbf{\Omega} \cdot \nabla v(u) - \Sigma(\mathbf{r}, u, t)\, v(u) \\ &+ \int du' \int d\Omega'\, [v(u')\, \Sigma_s(\mathbf{r}, u' \to u, \mathbf{\Omega} \cdot \mathbf{\Omega}', t)] \end{aligned} \tag{1}$$

$$M_0 \equiv \sum_j [f_0{}^j(u)/4\pi] \int du' \int d\Omega'\, [v(u')\, \nu^j(u')(1 - \beta^j)\, \Sigma_f{}^j(\mathbf{r}, u', t)] \tag{2}$$

$$M_i \equiv \sum_j [f_i(u)/4\pi] \int du' \int d\Omega'\, [\beta_i{}^j \nu^j(u')\, v(u')\, \Sigma_f{}^j(\mathbf{r}, u', t)] \tag{3}$$

$$H \equiv L + M_0 \tag{4}$$

as follows:

$$\partial n/\partial t = Hn + \sum_{i=1}^{6} \lambda_i f_i C_i + S \tag{5a}$$

$$\partial(C_i f_i)/\partial t = M_i n - \lambda_i f_i C_i \tag{5b}$$

where we have suppressed the arguments, and multiplied (5b) by $f_i(u)$

for convenience. Here $C_i(r, t)$ has been redefined to absorb $1/4\pi$. The physical meaning of these operators, which motivates their introduction, can be deduced from their definitions: L describes the transport, absorption, and scattering of neutrons in a nonmultiplying medium; M_0 determines the rate of production of prompt neutrons when it operates on the angular density; hence, it can be called the prompt-production operator. Similarly, M_i can be called the precursor-production operator, which determines the rate of production of the delayed neutron precursors of the ith type when it operates on $n(\mathbf{r}, u, \mathbf{\Omega}, t)$. Finally, H describes neutrons in a multiplying medium in the absence of delayed neutrons; it is called the Boltzmann operator. We note that the operators H and M_i depend on the composition of the medium, and describe the medium completely. They are, in general, time dependent, insofar as the cross sections are functions of time. We shall assume for the present that the properties of the medium do not depend on the neutron population, i.e., feedback is not present. In this case, the kinetic equations (5) are linear. The effect of feedback will be considered at the end of this chapter. In the absence of feedback, the time dependence of H and M_i is explicit, and is due to the changes introduced externally in the composition and configuration of the medium.

In this section, we shall focus our attention on a reactor with time-independent cross sections (stationary reactor), in the absence of external neutron sources, i.e., $S(\mathbf{r}, \mathbf{v}, t) \equiv 0$. The angular neutron density in a stationary reactor is still a function of time, and either increases or decreases. A source-free, multiplying medium with time-independent cross sections is "critical" if it can support a stationary neutron population $N_0(\mathbf{r}, u, \mathbf{\Omega})$ which is not zero everywhere. It is supercritical or subcritical if $n(\mathbf{r}, u, \mathbf{\Omega}, t)$ is increasing or decreasing respectively in time.

The time behavior involved in this definition is the asymptotic behavior at long times following initial disturbances. The neutron population is a function of time even in a critical reactor during the transients because criticality is a property of the medium, but not of the neutron population.

The steady angular neutron density $N_0(\mathbf{r}, u, \mathbf{\Omega})$ in a critical reactor satisfies the following equation:

$$\mathscr{H}_0 N_0(\mathbf{r}, u, \mathbf{\Omega}) = 0 \tag{6a}$$

which is obtained from (5) by setting the time derivatives to zero and eliminating $C_{i0}(r)$. Here, $\mathscr{H}_0$ is defined by

$$\mathscr{H}_0 \equiv H + \sum_{i=1}^{6} M_i \equiv L + M \tag{6b}$$

where M is the modified multiplication operator:

$$M \equiv \sum_j [f^j(u)/4\pi] \int du' \int d\Omega' \, [\nu^j(u') \, \Sigma_f^j(\mathbf{r}, u') \, v(u')] \tag{7}$$

in which $f^j(u)$ is defined by

$$f^j(u) \equiv (1 - \beta^j) \, f_0^j(u) + \sum_{i=1}^{6} \beta_i^j f_i(u) \tag{8}$$

We note that M differs from M_0 in the energy distribution of the fission neutrons, i.e., $f_0(u)$ in M_0 is replaced by $f(u)$ in M. The latter is the weighted average of the prompt and delayed neutron energy distributions. The energy dependence of the second term in (8), the yield-averaged delayed neutron spectrum, is shown in (Figure 1.1.3).

The steady-state equation is a homogeneous equation which is to be solved with the regular boundary conditions requiring $N_0(\mathbf{r}, u, \boldsymbol{\Omega})$ to be continuous and positive everywhere in the reactor volume (cf Section 1.3). The condition which must be satisfied by $\mathscr{H}_0$ (i.e., by the material and geometric properties of the medium) such that $\mathscr{H}_0 N_0 = 0$ will have a nontrivial solution is referred to as the criticality condition.

We now introduce the concept of effective multiplication factor by considering a noncritical stationary reactor. Since the reactor is not critical, $(L + M) \, N_0(\mathbf{r}, u, \boldsymbol{\Omega}) = 0$ has no other solution but the trivial one $N_0(\mathbf{r}, u, \boldsymbol{\Omega}) \equiv 0$. Let us suppose that we modify the mean number of neutrons per fission ν^j for each isotope by multiplying them by $(1/k_{\text{eff}})$, keeping all the other nuclear and geometric properties the same. This modification is equivalent to multiplying the production operator M by $(1/k_{\text{eff}})$. Adjusting the value of the positive number k_{eff}, we can always associate a critical fictitious system with a given noncritical reactor. The proper value for k_{eff} is determined by requiring that

$$[L + (1/k_{\text{eff}}) \, M] \, N_0(\mathbf{r}, u, \boldsymbol{\Omega}) = 0 \tag{9}$$

have a nontrivial solution. It is clear that k_{eff} must be unity if the actual reactor is already critical.

In order to explain the physical meaning of k_{eff}, which has been introduced as a formal mathematical device, we mentally break, following Ussachoff [2], the continuous chain process into neutron cycles. We begin a cycle by introducing neutrons into the reactor with the space and velocity distributions appropriate to the steady state. These neutrons disappear eventually through absorption and leakage, marking the end of a cycle. Some of the absorbed neutrons cause fission, and produce neutrons which originate the next cycle.

We shall show that k_{eff} is the ratio of the number of fission neutrons emitted in a given cycle to the number of fission neutrons emitted in the preceding cycle. Therefore, k_{eff} can be identified as the *effective multiplication factor*.

Let us denote the number of fission neutrons emitted in the critical reactor per second per unit volume at $\mathbf{r}$ and $\mathbf{v}$ by $Q(\mathbf{r}, u, \mathbf{\Omega})$,

$$Q(\mathbf{r}, \mathbf{v}) \equiv MN_0(\mathbf{r}, u, \mathbf{\Omega}) \tag{10}$$

Let us inject instantaneously $Q(\mathbf{r}, \mathbf{v})/k_{\text{eff}}$ neutrons into the actual noncritical reactor at $t = 0$. Assume that there are no neutrons in this reactor prior to $t = 0$. The angular density $q(\mathbf{r}, u, \mathbf{\Omega}, t)$ of the neutrons that originate directly from $Q(\mathbf{r}, \mathbf{v})/k_{\text{eff}}$ will satisfy the following time-dependent equation:

$$\partial q/\partial t = Lq \tag{11}$$

with the initial condition that $q(\mathbf{r}, u, \mathbf{\Omega}, 0) = Q(\mathbf{r}, u, \mathbf{\Omega})/k_{\text{eff}}$. We note that (11) describes only the removal of neutrons by leakage and absorption, as implied by the definition of a neutron cycle. By integrating (11) from 0 to ∞ and taking into account the initial condition and the fact that $q(\mathbf{r}, u, \mathbf{\Omega}, t) = 0$ as $t \to \infty$, we obtain

$$[Q(\mathbf{r}, u, \mathbf{\Omega})/k_{\text{eff}}] + \int_0^\infty Lq(\mathbf{r}, u, \mathbf{\Omega}, t)\, dt = 0 \tag{12a}$$

or

$$[Q(\mathbf{r}, u, \mathbf{\Omega})/k_{\text{eff}}] + L \int_0^\infty q(\mathbf{r}, u, \mathbf{\Omega}, t)\, dt = 0 \tag{12b}$$

Substituting $Q(\mathbf{r}, u, \mathbf{\Omega})$ from (10) in (12b), and comparing the resulting equation to (9), we find that the time-integrated angular density $\int_0^\infty q(\mathbf{r}, u, \mathbf{\Omega}, t)\, dt$ is equal to $N_0(\mathbf{r}, u, \mathbf{\Omega})$. [It may seem at first sight that there is a mismatch of dimensions because of the time integration on $q(\mathbf{r}, u, \mathbf{\Omega}, t)$. However, this is not the case, because $q(\mathbf{r}, u, \mathbf{\Omega}, t)$ has the same dimension as $Q(\mathbf{r}, u, \mathbf{\Omega})$, which has the dimension of a source.] Hence, we can also write (10) as

$$Q(\mathbf{r}, u, \mathbf{\Omega}) = \int_0^\infty Mq(\mathbf{r}, u, \mathbf{\Omega}, t)\, dt \tag{13}$$

which indicates that $Q(\mathbf{r}, u, \mathbf{\Omega})$ is the total number of fission neutrons in the actual reactor produced directly by the $Q(\mathbf{r}, u, \mathbf{\Omega})/k_{\text{eff}}$ neutrons in the interval $(0, \infty)$, which were introduced instantaneously at $t = 0$.

Thus, the ratio of the number of neutrons in the subsequent cycle, i.e., $Q(\mathbf{r}, u, \mathbf{\Omega})$ to that of neutrons in the preceding cycle, i.e., $Q(\mathbf{r}, u, \mathbf{\Omega})/k_{\text{eff}}$, is indeed equal to k_{eff}.

We would like to emphasize that the source neutrons, $Q(\mathbf{r}, u, \mathbf{\Omega})/k_{\text{eff}}$, are introduced into the actual reactor with the same distributions in velocity and position as they have in the stationary reactor. Hence, the effective multiplication factor computed by (9) is called the static multiplication factor [10].

2.3. Adjoint Angular Density and Neutron Importance

In this section, we shall introduce the concept of adjoint angular density [2, 11, 12] in a source-free critical reactor, which plays an important role in reactor theory. We have already defined the adjoint of an operator in Section 2.1. Using this definition and the examples in that section, we find that the adjoint of the operator $\mathscr{H}_0$ is

$$\mathscr{H}_0^+ \equiv \mathbf{\Omega} \cdot \nabla v(u) - \Sigma(\mathbf{r}, u)\, v(u) + \int du' \int d\Omega' \, \{\Sigma_s(\mathbf{r}, u \to u', \mathbf{\Omega} \cdot \mathbf{\Omega}') + \sum_j [f^j(u')/4\pi]\, \nu^j(u)\, \Sigma_f^j(\mathbf{r}, u)\}\, v(u) \tag{1}$$

In (1), the sign of the streaming operator has been changed, and $(u, \mathbf{\Omega})$ and $(u', \mathbf{\Omega}')$ in the integrands have been interchanged.

We define the "adjoint" angular density [11, 12] $N_0^+(\mathbf{r}, u, \mathbf{\Omega})$ as the nontrivial solution of

$$\mathscr{H}_0^+ N_0^+(\mathbf{r}, u, \mathbf{\Omega}) = 0 \tag{2}$$

with the adjoint boundary condition

$$N_0^+(\mathbf{r}, u, \mathbf{\Omega}) = 0 \qquad \text{for} \quad \hat{\mathbf{n}} \cdot \mathbf{\Omega} > 0, \quad \mathbf{r} \in S. \tag{3}$$

In order to attach a physical meaning to the adjoint angular density, we consider the following problem [2]. Suppose a neutron is injected into a critical reactor at $t = 0$ at the space point $\mathbf{r}'$ with a velocity $\mathbf{v}'$, and assume that there are no neutrons in the reactor prior to $t = 0$. We want to determine the time-dependent angular density $n(\mathbf{r}, u, \mathbf{\Omega}, t)$ as a function of $\mathbf{r}$ and $\mathbf{v}$ for all subsequent times, and in particular as $t \to \infty$. For the time being, we shall ignore the delayed neutrons for the sake of simplicity. Then, $n(\mathbf{r}, u, \mathbf{\Omega}, t)$ satisfies

$$\partial n/\partial t = Hn \tag{4}$$

with the initial condition

$$n(\mathbf{r}, u, \mathbf{\Omega}, 0) = \delta(\mathbf{r} - \mathbf{r}')\,\delta(u - u')\,\delta(\mathbf{\Omega} - \mathbf{\Omega}') \tag{5}$$

Equation (4) is obtained from the kinetic equations (2.2, Eqs. 5a and 5b) by omitting the delayed neutrons and the source term. In order to find the solution of (4), suppose that it is possible to find the eigenfunctions of the operator H (Boltzmann operator) by solving the following equation:

$$H\phi_n = \omega_n \phi_n \tag{6}$$

with the regular boundary conditions. We assume that the eigenvalues of ω_n can be arranged in increasing order of the magnitude of their real parts in case they are complex. The eigenvalues and the eigenfunctions of the Boltzmann operator are not known in general, except for some very special cases, e.g., the one-speed model with isotropic scattering and plane symmetry [13]. In general, the eigenvalue spectrum contains a discrete set and a continuum, as is the case in space- and/or energy-dependent kinetic problems, and, therefore, the index n takes on discrete and continuous values. Since the purpose of the present discussion is only to understand the physical meaning of the adjoint angular density, we shall not get involved in questions of existence, completeness, etc., and instead proceed formally. However, the assumptions introduced in this formal presentation can be justified in many practical particular cases, e.g., the one-speed diffusion model to be discussed subsequently.

Since the Boltzmann operator is not self-adjoint, we have to consider the adjoint eigenvalue problem also, i.e.,

$$H^{+}\phi_n^{+} = \omega_n^{*}\phi_n^{+} \tag{7}$$

so that $\{\phi_n\}$ and $\{\phi_n^{+}\}$ will form a complete biorthonormal set. Then, we can expand the time-dependent angular density $n(\mathbf{r}, u, \mathbf{\Omega}, t)$ in the functions $\phi_n(\mathbf{r}, u, \mathbf{\Omega})$ as

$$n(\mathbf{r}, u, \mathbf{\Omega}, t) = \sum_{n=0}^{\infty} a_n(\mathbf{r}', u', \mathbf{\Omega}', t)\,\phi_n(\mathbf{r}, u, \mathbf{\Omega}) \tag{8}$$

where the expansion coefficients are of course given by $a_n = \langle \phi_n^{+} \mid n \rangle$. Substituting (8) into (4) and using (6), we obtain a_n as

$$a_n(\mathbf{r}', u', \mathbf{\Omega}', t) = a_n(\mathbf{r}', u', \mathbf{\Omega}', 0)\, e^{\omega_n t} \tag{9}$$

The initial values $a_n(\mathbf{r}', u', \mathbf{\Omega}', 0)$ must be determined by the initial condition on $n(\mathbf{r}, u, \mathbf{\Omega}, t)$:

$$\delta(\mathbf{r} - \mathbf{r}')\,\delta(\mathbf{\Omega} - \mathbf{\Omega}')\,\delta(u - u') \equiv \sum_{n=0}^{\infty} a_n(\mathbf{r}', u', \mathbf{\Omega}', 0)\,\phi_n(\mathbf{r}, u, \mathbf{\Omega})$$

Multiplying both sides by $\phi_n^{+}(\mathbf{r}, u, \mathbf{\Omega})$ and forming scalar products, we get $a_n(\mathbf{r}', u', \mathbf{\Omega}', 0) = \phi_n^{+}(\mathbf{r}', u', \mathbf{\Omega}')$. Thus,

$$n(\mathbf{r}, u, \mathbf{\Omega}, t) = \sum_{n=0}^{\infty} \phi_n^{+}(\mathbf{r}', u', \mathbf{\Omega}')\,\phi_n(\mathbf{r}, u, \mathbf{\Omega})\,e^{\omega_n t} \tag{10}$$

From the physical consideration that in a critical reactor the density cannot increase indefinitely, we assert that the ω_n have negative real parts for $n \neq 0$, and $\omega_0 = 0$. The last equality follows from the fact that the reactor is critical, and hence $H\phi_0 = 0$ has a unique nontrivial solution. It is also clear that the eigenfunction ϕ_0 corresponding to $\omega_0 = 0$ is the steady-state angular density $N_0(\mathbf{r}, u, \mathbf{\Omega})$. Thus, the coefficients of all the higher modes in (10) decay exponentially in time, and the asymptotic angular density is obtained as

$$n_\infty(\mathbf{r}', u', \mathbf{\Omega}'; \mathbf{r}, u, \mathbf{\Omega}) = N_0^{+}(\mathbf{r}', u', \mathbf{\Omega}')\,N_0(\mathbf{r}, u, \mathbf{\Omega}) \tag{11}$$

where we have shown the dependence of n_∞ on $\mathbf{r}'$, u', $\mathbf{\Omega}'$ explicitly. We now introduce the concept of "importance."

The "importance" of a neutron injected into a critical reactor at $\mathbf{r}'$ with a lethargy u' in the direction of $\mathbf{\Omega}'$ is the total number of fissions per second in the entire reactor at a long time following the injection of the neutron at $t = 0$.

The importance function is readily obtained from (11) by multiplying both sides by $\Sigma_f(\mathbf{r}, u)\,v(u)$ and integrating over $\mathbf{r}$ and $\mathbf{v}$:

$$I(\mathbf{r}', u', \mathbf{\Omega}') = N_0^{+}(\mathbf{r}', u', \mathbf{\Omega}')\langle v\Sigma_f \mid N_0\rangle \tag{12}$$

We conclude from this result that the adjoint angular density $N_0^{+}(\mathbf{r}', u', \mathbf{\Omega}')$ is proportional to the importance of neutrons at $\mathbf{r}'$ moving with a lethargy u' in the direction of $\mathbf{\Omega}'$ in sustaining the chain reaction in the reactor. The proportionality constant in (12) is obtained with the normalization of $N_0^{+}(\mathbf{r}, u, \mathbf{\Omega})$ and $N_0(\mathbf{r}, u, \mathbf{\Omega})$ as $\langle N_0^{+} \mid N_0\rangle = 1$. In general, Eq. (12) can be written as

$$I(\mathbf{r}', u', \mathbf{\Omega}') = N_0^{+}(\mathbf{r}', u', \mathbf{\Omega}')\langle v\Sigma_f \mid N_0\rangle / \langle N_0^{+} \mid N_0\rangle \tag{13}$$

which is clearly independent of the choice of normalization for N_0^{+} and N_0.

Since the eigenvalue problem $H^+N_0^+ = 0$ yields N_0^+ with an indeterminate factor, we can make only the following statement unambiguously: The adjoint angular density is a measure of the "relative importance" of neutrons at two different positions with two different velocities in the reactor.

The foregoing analysis can easily be extended to include the delayed neutrons. In this case, the time-dependent angular density satisfies the full set of kinetic equations

$$\partial n/\partial t = Hn + \sum_{i=1}^{6} \lambda_i(f_i C_i) \tag{14a}$$

$$\partial(f_i C_i)/\partial t = M_i n - \lambda_i(C_i f_i) \tag{14b}$$

which are to be solved with the initial conditions $n(\mathbf{r}, u, \mathbf{\Omega}, 0) = \delta(\mathbf{r} - \mathbf{r}')\,\delta(\mathbf{\Omega} - \mathbf{\Omega}')\,\delta(u - u')$ and $C_i(\mathbf{r}, 0) = 0$ for all $i = 1, 2, \ldots, 6$. Equations (14) can be expressed in a compact way by defining a matrix operator $\mathscr{K}$ by

$$\mathscr{K} = \begin{bmatrix} H & \lambda_1 & \lambda_2 & \cdots & \lambda_6 \\ M_1 & -\lambda_1 & 0 & & 0 \\ M_2 & 0 & & & \\ \vdots & & & & \\ M_6 & 0 & 0 & \cdots & -\lambda_6 \end{bmatrix} \tag{15}$$

and a column vector

$$|\,\psi(t)\rangle \equiv \begin{bmatrix} n(\mathbf{r}, u, \mathbf{\Omega}, t) \\ C_1(\mathbf{r}, t)\, f_1(u) \\ \vdots \\ C_6(\mathbf{r}, t)\, f_6(u) \end{bmatrix} \tag{16}$$

as

$$\mathscr{K}\,|\,\psi(t)\rangle = \partial\,|\,\psi(t)\rangle/\partial t \tag{17}$$

We assume again that it is possible to find the eigenvectors and the eigenvalues of (cf Section 2.1)

$$\mathscr{K}\,|\,\phi_n\rangle = \omega_n\,|\,\phi_n\rangle \tag{18a}$$

$$\mathscr{K}^+\,|\,\phi_n^+\rangle = \omega_n^*\,|\,\phi_n^+\rangle \tag{18b}$$

where the adjoint matrix operator $\mathscr{K}^+$ is given explicitly by

$$\mathscr{K}^+ = \begin{bmatrix} H^+ & M_1^+ & \cdots & M_6^+ \\ \lambda_1 & -\lambda_1 & \cdots & 0 \\ \vdots & \vdots & & \vdots \\ \lambda_6 & 0 & \cdots & -\lambda_6 \end{bmatrix} \tag{19}$$

(See Problem 4 for an illustrative example.) The eigenvectors $|\,\phi_n\rangle$ are referred to as period modes [14] or, more frequently, as ω-modes [15–18].

We expand $|\,\psi(t)\rangle$ in terms of $\{|\,\phi_n\rangle\}$ as follows:

$$|\,\psi(t)\rangle = \sum_{n=0}^{\infty} a_n(\mathbf{r}', u', \mathbf{\Omega}', 0)\,|\,\phi_n\rangle\, e^{\omega_n t} \tag{20}$$

where

$$a_n(\mathbf{r}', u', \mathbf{\Omega}', 0) = \langle \phi_n{}^+\,|\,\psi(0)\rangle \tag{21}$$

Since

$$|\,\psi(0)\rangle = \mathrm{col}[\delta(\mathbf{r} - \mathbf{r}')\,\delta(u - u')\,\delta(\mathbf{\Omega} - \mathbf{\Omega}'), 0,..., 0] \tag{22}$$

we find

$$a_n(\mathbf{r}', u', \mathbf{\Omega}', 0) = N_n{}^+(\mathbf{r}', u', \mathbf{\Omega}') \tag{23}$$

where $N_n{}^+(\mathbf{r}, u, \mathbf{\Omega})$ is the first component of the eigenvector $|\,\phi_n{}^+\rangle$. Hence,

$$|\,\psi(t)\rangle = \sum_{n=0}^{\infty} N_n{}^+(\mathbf{r}', u', \mathbf{\Omega}')\,|\,\phi_n\rangle\, e^{\omega_n t} \tag{24}$$

or, taking the first component of (24),

$$n(\mathbf{r}, u, \mathbf{\Omega}, t) = \sum_{n=0}^{\infty} N_n{}^+(\mathbf{r}', u', \mathbf{\Omega}')\, N_n(\mathbf{r}, u, \mathbf{\Omega})\, e^{\omega_n t} \tag{25}$$

where $N_n(\mathbf{r}, u, \mathbf{\Omega})$ is the first component of the vector. It is assumed that the ω_n all have negative real parts with increasing order in magnitude. The criticality of the reactor implies that $\omega_0 = 0$, with the corresponding eigenvector satisfying $\mathscr{H}\,|\,\phi_0\rangle = 0$. The first component of $|\,\phi_0\rangle$, i.e., $N_0(\mathbf{r}, u, \mathbf{\Omega})$ satisfies, as can be readily seen from (14), $\mathscr{H}_0 N_0(\mathbf{r}, u, \mathbf{\Omega}) = 0$. Similarly, $\mathscr{H}_0{}^+ N_0{}^+(\mathbf{r}, u, \mathbf{\Omega}) = 0$. Hence, the asymptotic distribution is

$$n_\infty(\mathbf{r}', u', \mathbf{\Omega}'; \mathbf{r}, u, \mathbf{\Omega}) = N_0{}^+(\mathbf{r}', u', \mathbf{\Omega}')\, N_0(\mathbf{r}, u, \mathbf{\Omega})$$

which is identical to (11). The definition of the importance is therefore the same as in the case with no delayed neutrons, provided $N_0(\mathbf{r}, u, \mathbf{\Omega})$ and $N_0{}^+(\mathbf{r}, u, \mathbf{\Omega})$ are interpreted as the steady-state angular density and its adjoint with the delayed neutrons.

2.4. Reduction of the Kinetic Equations

We are now in a position to transform the basic kinetic equations into a more tractable form, from which the point reactor kinetic equations can be deduced by making certain approximations. Such a precise derivation of the point reactor kinetic equation is needed for various reasons; for example, we must take into account the delayed neutrons produced by fast fission, and consider the fact that the delayed neutrons do not have the same energy distribution as the prompt neutrons. Furthermore, the definition of other quantities, such as the generation time, external source, etc., must be sufficiently precise to allow them to be computed for a given reactor composition and geometry. It is therefore essential to establish the point reactor kinetic equations so that the underlying assumptions restricting the validity of these equations and the precise physical meaning of the quantities appearing in them will be clear.

For this purpose, we partition the angular neutron density [2, 3] $n(\mathbf{r}, u, \mathbf{\Omega}, t)$ into a shape function $\phi(\mathbf{r}, u, \mathbf{\Omega}, t)$ and a time function $P(t)$ such that

$$n(\mathbf{r}, u, \mathbf{\Omega}, t) = P(t)\,\phi(\mathbf{r}, u, \mathbf{\Omega}, t) \tag{1}$$

This separation of $n(\mathbf{r}, u, \mathbf{\Omega}, t)$ into the product of two new unknown functions is not unique. However, later, we shall impose further normalization restrictions to make the choice of $P(t)$ and $\phi(\mathbf{r}, u, \mathbf{\Omega}, t)$ unique.

We substitute (1) in the basic kinetic equations (2.2, Eq. 5):

$$P(\partial\phi/\partial t) + \phi(dP/dt) = PH(t)\,\phi + \sum_{i=1}^{6} \lambda_i f_i C_i + S \tag{2a}$$

$$\partial(f_i C_i)/\partial t = PM_i(t)\phi - \lambda_i f_i C_i \tag{2b}$$

We have written the time dependence of the operators $H(t)$ and $M_i(t)$ explicitly, to remind ourselves that the reactor parameters are explicit functions of time. We neglect feedback effects for present purposes, and assume that the geometry of the reactor is unchanged.

We imagine a critical, source-free "reference" reactor of the same geometry, and with similar nuclear properties. In fact, if the actual reactor is critical in the initial or final state, then the reference reactor can be chosen as the actual reactor in that particular state. Being critical, the reference reactor is completely described by the stationary Boltzmann operator $\mathscr{H}_0$. We assume that the steady-state angular density $N_0(\mathbf{r}, u, \mathbf{\Omega})$ and its adjoint $N_0^+(\mathbf{r}, u, \mathbf{\Omega})$ are known. Multiplying (2a) and (2b) by

$N_0^+(\mathbf{r}, u, \mathbf{\Omega})$ and, integrating the resulting equations over $\mathbf{r}$, u, and $\mathbf{\Omega}$, we obtain, in the scalar product notation,

$$\langle N_0^+ \mid \phi\rangle(dP/dt) + P(d/dt)\langle N_0^+ \mid \phi\rangle = P\langle N_0^+ \mid H(t) \mid \phi\rangle + \sum_{i=1}^{6} \lambda_i\langle N_0^+ \mid f_i C_i\rangle + \langle N_0^+ \mid S\rangle \quad (3a)$$

$$d\langle N_0^+ \mid f_i C_i\rangle/dt = P\langle N_0^+ \mid M_i(t)\phi\rangle - \lambda_i\langle N_0^+ \mid f_i C_i\rangle \quad (3b)$$

We must now impose a "normalization" condition on the shape function to ensure uniqueness, which we choose as (other choices will be discussed later in Section 2.5).

$$(d/dt)\langle N_0^+ \mid \phi\rangle \equiv (d/dt)\int_R d^3r \int du \int d\mathbf{\Omega}\, [N_0^+(\mathbf{r}, u, \mathbf{\Omega})\, \phi(\mathbf{r}, u, \mathbf{\Omega}, t)] = 0 \quad (4)$$

The physical interpretation of this condition is as follows: $N_0^+(\mathbf{r}, u, \mathbf{\Omega})$ is proportional to the importance of neutrons (cf 2.3, Eq. 12). Hence, $\langle N_0^+ \mid \phi\rangle$ is the total importance of neutrons in the reference reactor with a distribution function $\phi(\mathbf{r}, u, \mathbf{\Omega}, t)$. According to (4), the shape function must be so chosen that the total importance in the reference reactor will remain constant in time even though $\phi(\mathbf{r}, u, \mathbf{\Omega}, t)$ itself may vary slowly in time locally. An alternative interpretation of the normalization condition (4) can be obtained by expanding $\phi(\mathbf{r}, u, \mathbf{\Omega}, t)$ in the eigenfunctions of $\mathscr{H}_0$,

$$\phi(\mathbf{r}, u, \mathbf{\Omega}, t) = \langle N_0^+ \mid \phi\rangle\, N_0 + \sum_{n=1}^{\infty} \langle N_n^+ \mid \phi\rangle\, N_n \quad (5)$$

Note that all the expansion coefficients are functions of time. The condition (4) requires the coefficient of the fundamental mode [i.e., the projection of ϕ on $N_0(\mathbf{r}, u, \mathbf{\Omega})$ in the function space spanned by $N_n(\mathbf{r}, u, \mathbf{\Omega})$] to be independent of time. It is also clear from this interpretation that the normalization condition (4) does not imply that the shape function be independent of time locally.

This discussion enables us to clarify the physical meaning of the time function. Multiplying both sides of (1) by N_0^+, we find that

$$P(t) = \langle N_0^+ \mid n\rangle/\langle N_0^+ \mid \phi\rangle \quad (6)$$

which states that $P(t)$ is the ratio of the total importance of neutrons with a distribution $n(\mathbf{r}, u, \mathbf{\Omega}, t)$ to the importance of those neutrons that have a distribution $\phi(\mathbf{r}, u, \mathbf{\Omega}, t)$. The denominator of (6) is constant in time, and can be scaled as unity. Then, $P(t)$ becomes the instantaneous value of the total importance of the neutron population in the actual reactor which is necessary to sustain a chain reaction in the reference reactor. Since $\langle N_0^+ \mid n \rangle$ is the coefficient of the fundamental mode in the expansion of $n(\mathbf{r}, u, \boldsymbol{\Omega}, t)$, $P(t)$ is simply the time dependence of the fundamental mode. It is emphasized that $P(t)$ is not the total number of neutrons in the reactor volume at the time t. Also notice that $P(t)$ is independent of the normalization of the adjoint angular density $N_0^+(\mathbf{r}, u, \boldsymbol{\Omega})$ as indicated by (6).

We now return to Eqs. (3a) and (3b), and observe that the normalization condition (4) removes the second term on the left of (3a). To introduce the concept of perturbation, i.e., the deviations of the reactor parameters of the actual reactor from those of the reference reactor, we define a perturbation operator $\delta\mathscr{H}(t)$ as

$$\delta\mathscr{H}(t) \equiv H(t) + \sum_{i=1}^{6} M_i(t) - \mathscr{H}_0 \equiv L(t) + M(t) - \mathscr{H}_0 \tag{7}$$

More explicitly,

$$\delta\mathscr{H}(t) \equiv -v(u)\,\delta\Sigma(\mathbf{r}, u, t) + \int du' \int d\boldsymbol{\Omega}'\, \delta\{\Sigma_{\mathrm{s}}(\mathbf{r}, u' \to u, \boldsymbol{\Omega}\cdot\boldsymbol{\Omega}', t)$$
$$+ \sum_j [f^j(u)/4\pi]\, \nu^j(u')\, \Sigma_{\mathrm{f}}^j(\mathbf{r}, u', t)\}\, v(u') \tag{8}$$

where $\delta\Sigma_j(\mathbf{r}, u, t)$ measures the variations of the cross sections about their reference values, i.e.,

$$\delta\Sigma_j(\mathbf{r}, u, t) \equiv \Sigma_j(\mathbf{r}, u, t) - \Sigma_{j0}(\mathbf{r}, u)$$

where the subscript j denotes a, f, or s. Substituting $H(t)$ from (7) into (3a), and observing that $\langle N_0^+ \mid \mathscr{H}_0 \mid \phi \rangle = \langle \mathscr{H}_0^+ N_0^+ \mid \phi \rangle = 0$ for any function ϕ with the regular boundary conditions (this is a crucial point in the derivation), we obtain the desired form of the kinetic equations:

$$dP/dt = [(\rho(t) - \bar{\beta})/l]\, P(t) + \sum_{i=1}^{6} \lambda_i \bar{C}_i(t) + \bar{S}(t) \tag{9a}$$

$$d\bar{C}_i/dt = (\bar{\beta}_i/l)\, P(t) - \lambda_i \bar{C}_i(t) \tag{9b}$$

with the following definitions:
Reactivity:

$$\rho(t) \equiv (1/F)\langle N_0^+ \mid \delta\mathscr{H}(t) \mid \phi \rangle \tag{10}$$

Effective delayed neutron fraction:

$$\bar{\beta}_i \equiv (1/F)\langle N_0^+ \mid M_i \mid \phi\rangle \tag{11}$$

$$\bar{\beta} = \sum_{i=1}^{6} \bar{\beta}_i \tag{12}$$

Effective concentration of delayed neutron precursors:

$$\bar{C}_i(t) \equiv (1/Fl)\langle N_0^+ \mid f_i C_i\rangle \tag{13}$$

Effective source:

$$\bar{S}(t) \equiv (1/Fl)\langle N_0^+ \mid S\rangle \tag{14}$$

Mean prompt generation time:

$$l \equiv (1/F)\langle N_0^+ \mid \phi\rangle \tag{15}$$

Normalization factor:

$$F \equiv \langle N_0^+ \mid M(t) \mid \phi\rangle \tag{16}$$

and

$$M(t) \equiv \sum_j \int du' \int d\mathbf{\Omega}' \{[f^j(u)/4\pi]\, \nu^j(u')\, \Sigma_f^j(\mathbf{r}, u', t)\, v(u')\} \tag{17}$$

The physical meaning of the symbols and the reason for naming them as indicated will be discussed presently. First, however, the following general remarks about the equations (9) are in order. (a) These equations are *exact* and completely equivalent to the basic kinetic equations in a different form. The quantities appearing in this equation still contain the unknown shape function $\phi(\mathbf{r}, u, \mathbf{\Omega}, t)$, which can only be solved through the original space- and energy-dependent kinetic equation. The advantage of this new form lies in the fact that it lends itself easily to various physical approximations which are used to investigate the time behavior of $P(t)$, this quantity being one of the most important aspects of the neutron population in reactor dynamics. These approximations will be discussed in the following section. (b) The choice of the normalization factor F is arbitrary in the sense that the kinetic equations (9) are entirely independent of it. However, the magnitudes of ρ, $\bar{\beta}_i$, and l depend on F. The particular definition used in (17) is chosen so that the reactivity can approximately be interpreted as $(k_{\text{eff}} - 1)/k_{\text{eff}}$ in terms of the effective multiplication constant defined in (2.2, Eq. 9), as will be shown later. (c) The arbitrariness in the choice of F implies that the quantities ρ, l, $\bar{\beta}_i$ cannot be defined as physical quantities in an absolute sense, and that only the ratios (ρ/l) and $(\bar{\beta}_i/l)$ can be defined unambiguously. This conclusion has the important consequence that only the ratios of these quantities can be measured experimentally for a given reactor.

Effective Delayed Neutron Fraction

We can express the effective delayed neutron fraction more explicitly by substituting M_i in (11) from (2.2, Eq. 3) and using the normalization in (17):

$$\bar{\beta}_i = \sum_j \Big(\beta_i^j \int_R d^3r \Big\{ \int du \int d\Omega \, [N_0^+(\mathbf{r}, u, \mathbf{\Omega}) f_i(u)]$$

$$\times \int du' \int d\Omega' \, [\nu^j(u') \, v(u') \, \Sigma_f^j(\mathbf{r}, u', t) \, \phi(\mathbf{r}, u', \mathbf{\Omega}', t)] \Big\} \Big)$$

$$\div \sum_j \Big(\int_R d^3r \Big\{ \int du \int d\Omega \, [N_0^+(\mathbf{r}, u, \mathbf{\Omega}) f^j(u)$$

$$\times \int du' \int d\Omega' \, [\nu^j(u') \, v(u') \, \Sigma_f^j(\mathbf{r}, u', t) \, \phi(\mathbf{r}, u', \mathbf{\Omega}', t)] \Big\} \Big) \quad (18)$$

The distinction between $\bar{\beta}_i$ and β_i^j, the actual delayed neutron fraction from the jth isotope, is apparent in (18). It is the ratio of the importance of all the delayed neutrons of the ith group emitted per second in the entire reactor to the importance of all fission neutrons, delayed or prompt, emitted per second in the entire reactor. Since the energy spectrum of the delayed neutrons $f_i(u)$, is much lower (a few hundred keV) than that of prompt neutrons (a few MeV), their importance in a thermal reactor is greater (smaller leakage probability) than the importance of prompt neutrons. Therefore, $\bar{\beta}_i/\beta_i^j$ is greater than unity, and the difference may be 20 or 30% [3]. Explicit formulas for calculating $\bar{\beta}_i$ in the age-diffusion approximation will be given later.

Effective Source

The definition of the effective source follows from (14) as

$$\bar{S}(t) = \langle N_0^+ \mid S \rangle / \langle N_0^+ \mid \phi \rangle \quad (19)$$

which indicates that $\bar{S}(t)$ is proportional to the importance of all the external neutrons introduced per second in the entire reactor. If the distribution of the source neutrons in position and velocity is the same as $\phi(\mathbf{r}, u, \boldsymbol{\Omega}, t)$, the shape function, at all times, i.e., $S(\mathbf{r}, \mathbf{v}, t) = S_0(t)\,\phi(\mathbf{r}, u, \boldsymbol{\Omega}, t)$, then $S(t) = S_0(t)$. We can also interpret $\bar{S}(t)$ as the coefficient of the fundamental mode in the expansion of $S(\mathbf{r}, u, \boldsymbol{\Omega}, t)$ in terms of the eigenfunctions of the Boltzmann operator $\mathscr{H}_0$.

Mean Prompt Generation Time

The definition of l is given by (15) as

$$l = \langle N_0^+ \mid \phi\rangle / \langle N_0^+ \mid M \mid \phi\rangle = \langle N_0^+ \mid n\rangle / \langle N_0^+ \mid M \mid n\rangle \tag{20}$$

Hence, it is the ratio of the total importance of all neutrons in the reactor at time t to the importance of all fission neutrons produced per second in the entire reactor. In other words, the total amount of importance created by the fission neutrons in the mean prompt generation time l is equal to the instantaneous value of the importance of neutrons present at time t.

Reactivity

The concept and the definition of reactivity in a reactor whose nuclear properties are changing continuously in time require closer attention. Various definitions [19, 20] of reactivity are obtained by appropriate choice of the shape function ϕ in

$$\rho(t) = \langle N_0^+ \mid \delta\mathscr{H} \mid \phi\rangle / \langle N_0^+ \mid M \mid \phi\rangle$$

Method 1. The crudest, and the simplest, approximation is to assume the shape function to be proportional to the steady-state distribution $N_0(\mathbf{r}, u, \boldsymbol{\Omega})$ in the critical reference reactor. If we denote the proportionality constant by $(1/P_0)$, this approximation implies

$$n(\mathbf{r}, u, \boldsymbol{\Omega}, t) \approx [P(t)/P_0]\, N_0(\mathbf{r}, u, \boldsymbol{\Omega}) \tag{21a}$$

The normalization condition (4) is automatically satisfied. Within the limitation of the perturbation approximation, we may interpret $P(t)/P_0$ as

$$P(t)/P_0 \cong \langle v\Sigma_{f0} \mid n\rangle / \langle v\Sigma_{f0} \mid N_0\rangle \tag{21b}$$

where $\Sigma_{f0}(\mathbf{r}, u)$ is the fission cross section in the reference reactor. Since $P(t)$ is approximately proportional to the instantaneous power $\langle v\Sigma_{f0} \mid n\rangle$, it is often referred to as the reactor power (this point will be discussed further in 2.5). The kinetic parameters ρ, $\bar{\beta}_i$, and l are independent of the proportionality constant $1/P_0$. However, the effective source (cf Eq. 19) depends on P_0 as

$$\bar{S}(t) \cong [\langle N_0^+ \mid S\rangle / \langle N_0^+ \mid N_0\rangle] P_0 \tag{21c}$$

so that it has the right dimension in the kinetic equations. The mean prompt generation time l and the effective delayed neutron fractions $\bar{\beta}_i$ become independent of time in the case of the constant-shape-function approximation if the perturbation does not affect the multiplication operators M and M_i. However, it is consistent with the first-order perturbation approximation to ignore the changes in $\Sigma_f^j(\mathbf{r}, u, t)$ in calculating $\bar{\beta}_i$, l, and F, even though they do change.

Although it is the crudest, the first-order perturbation approximation is the only approximation technique which allows further analytical treatment of the point kinetic equations.

Method 2. The second method of approximation consists in choosing the shape function as proportional to that solution of

$$[L(t) + (1/k_{\text{eff}})\, M(t)] N_{k_{\text{eff}}}(\mathbf{r}, u, \mathbf{\Omega}, t) = 0 \tag{22}$$

that is everywhere positive within the volume of the reactor ("adiabatic" approximation). This equation describes a critical reactor that possesses identical nuclear properties to those of the actual reactor at the instant t, and is made critical by a fictitious value of k_{eff} if the instantaneous configuration is not already stationary (cf 2.2, Eq. 9). Clearly, both k_{eff} and $N_{k_{\text{eff}}}(\mathbf{r}, u, \mathbf{\Omega}, t)$ will depend on time parametrically because the stationary equation (22) will be different at different times. [The function $N_{k_{\text{eff}}}(\mathbf{r}, u, \mathbf{\Omega}, t)$ should not be confused with the time-dependent angular density $n(\mathbf{r}, u, \mathbf{\Omega}, t)$.] Since the parameters of the perturbed reactor are known as a function of time, we can determine the time dependence of k_{eff} and $N_{k_{\text{eff}}}(\mathbf{r}, u, \mathbf{\Omega}, t)$ by solving (22) for each t.

We can now compute the reactivity $\rho(t)$ by replacing the shape function $\phi(\mathbf{r}, u, \mathbf{\Omega}, t)$ in (21) by $N_{k_{\text{eff}}}(\mathbf{r}, u, \mathbf{\Omega}, t)$, recalling that

$$\delta\mathscr{H} = L(t) + M(t) - \mathscr{H}_0 \tag{23}$$

(cf Eqs. 7 and 8), and using $M(t)\, N_{k_{\text{eff}}} = -k_{\text{eff}} L(t)\, N_{k_{\text{eff}}}$. The result is

$$\rho(t) = (k_{\text{eff}} - 1)/k_{\text{eff}} \tag{24}$$

indicating that the original identification of the symbol $\rho(t)$ as reactivity is indeed consistent with the conventional definition of reactivity in terms of the effective multiplication factor. The reactivity defined with respect to the stationary distribution $N_{k_{\text{eff}}}(\mathbf{r}, u, \boldsymbol{\Omega}, t)$ is referred to as the "static" reactivity [14, 19, 20].

A word of caution is needed at this point. As soon as we choose the shape function as $N_{k_{\text{eff}}}(\mathbf{r}, u, \boldsymbol{\Omega}, t)$, which depends on time parametrically, we have no guarantee in general that the normalization condition (4) is satisfied, because there is no reason for $\langle N_0^+ \mid N_{k_{\text{eff}}} \rangle$ to be independent of time, unless the nuclear properties of the perturbed reactor are time-independent. This question does not arise in the perturbation analysis because $N_0(\mathbf{r}, u, \boldsymbol{\Omega})$ is by its definition independent of time. However, if the variations in the shape function $N_{k_{\text{eff}}}(\mathbf{r}, u, \boldsymbol{\Omega}, t)$ are slow, we can still use the point kinetic equations without additional terms. The possibility of allowing a time-dependent normalization will be discussed in Section 2.5.

λ-Modes

The solutions of the eigenvalue problem defined by

$$L\phi_\lambda = -(1/\lambda)\, M\phi_\lambda \tag{25}$$

are referred to as reactivity modes or simply λ-modes. The adjoint equation is

$$L^+\phi_\lambda{}^+ = -(1/\lambda^*)\, M^+\phi_\lambda{}^+ \tag{26}$$

The orthonormality relation for the functions ϕ_λ and $\phi_\lambda{}^+$ is

$$\langle \phi_\lambda{}^+ \mid M \mid \phi_\lambda \rangle = \delta_{\lambda\lambda'} \tag{27}$$

There is one eigenfunction, designated by ϕ_{λ_0}, that is positive everywhere throughout the reactor volume. The corresponding eigenvalue λ_0 is real and positive. The stationary distribution $N_{k_{\text{eff}}}(\mathbf{r}, u, \boldsymbol{\Omega}, t)$ and the effective multiplication factor k_{eff} in (22) correspond to $\phi_{\lambda_0}(\mathbf{r}, u, \boldsymbol{\Omega})$ and λ_0. It therefore follows that the second method of approximation is equivalent to replacing the shape function in $\rho(t)$ by the lowest-reactivity mode at each instant of time. The time-dependent angular density is approximately given by $n(\mathbf{r}, u, \boldsymbol{\Omega}, t) = P(t)\, N_{k_{\text{eff}}}(\mathbf{r}, u, \boldsymbol{\Omega}, t)$, where $P(t)$ is obtained from the solution of the point kinetic equations (9) with the time-dependent $\rho(t)$, $\bar{\beta}_i(t)$ and $l(t)$. The latter two quantities are also determined by using $N_{k_{\text{eff}}}(\mathbf{r}, u, \boldsymbol{\Omega}, t)$ as the shape function in their definitions (18) and (20). Both the solution of (22) for $N_{k_{\text{eff}}}(\mathbf{r}, u, \boldsymbol{\Omega}, t)$ and (9) for $P(t)$ in general require machine calculations.

Method 3. A third method of approximation is achieved by choosing the shape function as the fundamental ω-mode defined by (2.3, Eqs. 18).

By eliminating the delayed neutron precursor densities in these equations, we obtain the following alternative definition of the ω-modes:

$$\left[H + \sum_{i=1}^{6} M_i\right] N_n(\mathbf{r}, u, \boldsymbol{\Omega}) = \omega_n \left[1 + \sum_{i=1}^{6} \frac{M_i}{\lambda_i + \omega_n}\right] N_n(\mathbf{r}, u, \boldsymbol{\Omega}) \qquad (28)$$

The orthonormality relation for the functions $N_n(\mathbf{r}, u, \boldsymbol{\Omega})$ and their adjoint follows from (2.3, Eq. 19) as

$$\langle N_n^+ \mid N_m \rangle + \sum_{i=1}^{6} \frac{\lambda_i \langle N_m^+ \mid M_i \mid N_n \rangle}{(\lambda_i + \omega_n)(\lambda_i + \omega_m)} = \delta_{nm} \qquad (29)$$

which can also be obtained directly from (28) by multiplying both sides by $N_n^+(\mathbf{r}, u, \boldsymbol{\Omega})$, integrating over $\mathbf{r}$, u, and $\boldsymbol{\Omega}$, and making use of the adjoint equation.

The fundamental ω-mode is the one that corresponds to the algebraically largest eigenvalue ω_0, and will be denoted by $N_{\omega_0}(\mathbf{r}, u, \boldsymbol{\Omega})$. It is the asymptotic distribution in a stationary reactor following an initial perturbation (cf 2.3, Eq. 25). The asymptotic time dependence is described by $\exp[\omega_0 t]$. The inverse of ω is called the "asymptotic" (or stable) reactor period.

The fundamental ω-mode corresponding to the instantaneous configuration at time t of the actual reactor under consideration will be denoted by $N_{\omega_0}(\mathbf{r}, u, \boldsymbol{\Omega}, t)$. In contrast to the definition of the fundamental reactivity mode $N_{k_{\text{eff}}}(\mathbf{r}, u, \boldsymbol{\Omega}, t)$, we do not, in the present case, adjust the number of neutrons per fission to make the instantaneous configuration critical. Therefore, $N_{\omega_0}(\mathbf{r}, u, \boldsymbol{\Omega}, t) \exp[\omega_0(t + \tau)]$ would be the asymptotic behavior ($\tau \to \infty$) of the neutron population, had the core properties of the actual reactor remained unchanged for $\tau \geqslant 0$. Clearly, this asymptotic behavior would be different at different t.

The third method of approximation then consists in replacing the shape function $\phi(\mathbf{r}, u, \boldsymbol{\Omega}, t)$ in the definition of $\bar{\beta}_i$, l, and $\rho(t)$ by $N_{\omega_0}(\mathbf{r}, u, \boldsymbol{\Omega}, t)$. The reactivity is obtained from [21] as

$$\rho(t) = \omega_0 \left[l + \sum_{i=1}^{6} \frac{\bar{\beta}_i}{\lambda_i + \omega_0}\right] \qquad (30)$$

where

$$\bar{\beta}_i(t) = \langle N_0^+ \mid M_i \mid N_{\omega_0} \rangle / \langle N_0^+ \mid M \mid N_{\omega_0} \rangle \qquad (31)$$

$$l(t) = \langle N_0^+ \mid N_{\omega_0} \rangle / \langle N_0^+ \mid M \mid N_{\omega_0} \rangle \qquad (32)$$

Equation (30) is referred to conventionally as the inhour equation, which relates the reactivity to the inverse reactor period ω_0.

The time behavior of the neutron population is approximated by $P(t)\, N_{\omega_0}(\mathbf{r}, u, \mathbf{\Omega}, t)$ in the third method, where $P(t)$ is to be determined from the point kinetic equations with the time-dependent parameters l and $\bar{\beta}_i$ defined above.

The reactivity defined by (30), using the persisting neutron distribution $N_{\omega_0}(\mathbf{r}, u, \mathbf{\Omega}, t)$ is referred to as the dynamic reactivity [21].

Discussion

By operating on (28) by $\langle N_0^+ \mid$ and integrating over $\mathbf{r}$, u, and $\mathbf{\Omega}$, we obtain a generalized version of the inhour equation (30) as

$$\left(\frac{\rho}{l}\right)_n = \omega_n \left[1 + \sum_{i=1}^{6} \left(\frac{\bar{\beta}_i}{l}\right)_n \frac{1}{\lambda_i + \omega_n}\right] \tag{33}$$

where

$$(\rho/l)_n \equiv \langle N_0^+ \mid H + \sum_{i=1}^{6} M_i \mid N_n\rangle / \langle N_0^+ \mid N_n\rangle \tag{34a}$$

$$(\bar{\beta}_i/l)_n \equiv \langle N_0^+ \mid M_i \mid N_n\rangle / \langle N_0^+ \mid N_n\rangle \tag{34b}$$

Its solution yields the time constants ω_n associated with all the higher modes N_n as well as with the fundamental N_{ω_0}. It was pointed out by Henry [17] that the six other roots $\omega_{n_1}, \ldots, \omega_{n_6}$ of the inhour equation for particular values of $(\rho/l)_n$ and $(\bar{\beta}_i/l)_n$ are not in general eigenfunctions of (28). The reason is that the eigenfunctions N_n are different functions of $\mathbf{r}$, u, and $\mathbf{\Omega}$ for different eigenvalues ω_n. Therefore, $\omega_{n_1}, \ldots, \omega_{n_6}$, which correspond to the same eigenfunction, cannot be eigenvalues. However, some of the eigenfunctions fall into clusters of seven such that the eigenfunctions N_{n_J}, $J = 0, 1, \ldots, 6$, in the nth cluster are approximately the same function of $\mathbf{r}$, u, and $\mathbf{\Omega}$ even though the corresponding eigenfunctions ω_{n_J} are different. Since the parameters $(\rho/l)_{n_J}$ and $(\bar{\beta}_i/l)_{n_J}$ will be approximately the same for a given cluster, the seven roots of the inhour equation computed with any one of these N_{n_J} will yield approximately the seven eigenvalues ω_{n_J} belonging to the cluster. Six of these seven eigenfunctions, labeled ω_{n_i}, $i = 1, \ldots, 6$, are connected with the six precursor decay constants λ_i (delayed eigenvalues [19]). They are all negative and slightly larger in absolute magnitude than the corresponding λ_i for large n. The seventh eigenvalue ω_{n_0} can have values between $-\lambda_1$ and infinity (prompt eigenvalues). Problem 4 illustrates these properties of the ω-modes in the one-speed diffusion approximation with one group of delayed neutrons.

It was also demonstrated by Henry [17], considering the case of the two-group P-1 approximation with isotropic scattering applied to a bare slab, that the eigenfunctions representing higher angular and energy modes cannot be identified with a cluster (Problem 4). The neutron-density component of these eigenvectors are strongly dependent on the eigenvalues. It is concluded that [17] the inhour equation is useful only if N_{n_J} for $J = 0, 1, \ldots, 6$ are all approximately the same functions of $\mathbf{r}$, u, and $\mathbf{\Omega}$. For the fundamental, this is usually the case [14], and hence the remaining six roots of (30) can be identified as the delayed eigenvalues. (For further discussion of the ω-modes see Henry [17] and Gozani [19].)

We conclude this section with the following remark. The reactivity [strictly speaking, $\rho(t)/l$] becomes a linear functional of the perturbations when it is computed in the constant-shape approximation. Only in this case are the various reactivity changes resulting from perturbations in different nuclear properties and locations additive. In all other approximations, the shape function itself varies with the perturbation and thus the reactivity fails to be additive.

2.5. Alternative Derivations of the Point Kinetic Equations

The reduction of the kinetic equation into point kinetics as described in the previous section contains two crucial steps: (a) multiplication of the kinetic equation (2.2, Eqs. 5) by the adjoint angular density $N_0^+(\mathbf{r}, u, \mathbf{\Omega})$ appropriate to a critical source-free reference reactor, and integration over $\mathbf{r}$, u, and $\mathbf{\Omega}$; and (b) separation of the angular density as a product of a time and a shape function, $n(\mathbf{r}, u, \mathbf{\Omega}, t) = P(t)\ \phi(\mathbf{r}, u, \mathbf{\Omega}, t)$ such that ϕ satisfies the following normalization condition:

$$(d/dt)\langle N_0^+ \mid \phi\rangle = 0 \tag{1}$$

The first step enables one to express the reactivity (cf 2.4, Eq. 10) in terms of $\delta\mathscr{H}$, which characterizes the deviations of the nuclear properties of the reactor from those of the reference reactor. The normalization condition (1) leads to the interpretation of $P(t)$ as the time variation of the coefficient of the fundamental mode in the expansion of $n(\mathbf{r}, u, \mathbf{\Omega}, t)$ into the set $\{N_n(\mathbf{r}, u, \mathbf{\Omega})\}$. Although $P(t)$ satisfies the point kinetic equations, it is not directly interpretable as the total instantaneous power in the reactor or the output of a detector characterized by a cross section $\Sigma_D(\mathbf{r}, u)$, which are more significant from the safety and experimental points of view. The total power is given by

$$P(t) = \langle v\Sigma_f \mid n \rangle w_f \tag{2}$$

where $\Sigma_f(\mathbf{r}, u, t)$ is the instantaneous fission cross section. The detector output is obtained as

$$I_D(t) = \langle v\Sigma_D \mid n \rangle \tag{3}$$

We can investigate these cases in general by introducing a "weight" function $w(\mathbf{r}, u, t)$ and choosing the time function $P(t)$ proportional to $\langle w \mid n \rangle$. Since $n = P\phi$, we must impose the following normalization condition on ϕ

$$(d/dt)\langle w \mid \phi \rangle = 0 \tag{4}$$

so that $P(t) \sim \langle w \mid n \rangle$, i.e.,

$$P(t) = \langle w \mid n \rangle / \langle w \mid \phi \rangle \tag{5}$$

Following the steps described in the previous section, we can show that $P(t)$ satisfies the point kinetic equations (cf 2.4., Eqs. 9) with the following identification of the parameters when w is constant:

$$(\rho/l) \equiv \langle w \mid H + \sum_i M_i \mid \phi \rangle / \langle w \mid \phi \rangle \tag{6}$$

$$(\bar{\beta}_i/l) \equiv \langle w \mid M_i \mid \phi \rangle / \langle w \mid \phi \rangle \tag{7}$$

The approximation introduced in the previous section to replace $\phi(\mathbf{r}, u, \mathbf{\Omega}, t)$ by a known function can also be used here. We have already pointed out before that the normalization condition (4) is not satisfied in general once the shape function ϕ is replaced by the fundamental λ- or ω-modes (in fact, it is not satisfied even in the case of the perturbation approximation if the weighting function w is allowed to be a function of time). This difficulty led Gyftopoulos [5] to derive point kinetic equations that do not require (4). If we relax (4), let

$$w_1(t) = (d/dt) \log\langle w \mid \phi \rangle \tag{8}$$

and assume that w is constant in time, we obtain the modified kinetic equations due to Gyftopoulos:

$$dP/dt = [(\rho - \bar{\beta})/l]\, P + \sum_{i=1}^{6} \lambda_i \bar{C}_i + \bar{S} - Pw_1(t) \tag{9a}$$

$$d\bar{C}_i/dt = (\bar{\beta}_i/l)\, P(t) - \lambda_i \bar{C}_i - w_1(t)\, \bar{C}_i \tag{9b}$$

This set contains the additional term $w_1(t)$, which was assumed to be zero previously. It is interesting to note that (9) can be reduced to the standard form [5, 6] if we let

$$\rho_1 = \rho - w_1(t)l \tag{10a}$$

$$\lambda_{i_1} = \lambda_i + w_1(t) \tag{10b}$$

Thus, the conventional form of the point kinetic equations is obtained, provided different precursor decay constants are used in the power and precursor equations.

The additional term $w_1(t)$ in (9) vanishes exactly if the shape function is independent of time. The effect of the shape changes has been investigated by Gyftopoulos [5]. He concluded that it can be ignored only if the shape function changes by a small amount over a long period of time. If the transients under consideration involve fast shape changes, one has either to change the definition of reactivity as indicated in (10), or attack the problem as a space–energy–time-dependent problem in the first place.

Further refinements in the derivation of the point kinetic equations have been presented by Becker [6] using variational principles. In his derivations, he allows not only the flux shape, but also the weighting function to vary with time. The additional terms disappear if appropriate normalization conditions are imposed on the shape function. If these conditions are found to be inconvenient to work with, then more general forms of the kinetic equations such as those due to Gyftopoulos discussed above can be used.

2.6. Point Reactor Kinetic Equations with Feedback

In this section, we shall derive the point reactor kinetic equations with feedback starting from the general description presented above. The motivation and the method of approach are the same as those presented in Section 2.4 in obtaining the point kinetic equations in the absence of feedback.

We begin our analysis by observing that the cross sections are now functionals of the angular density $n(\mathbf{r}, u, \mathbf{\Omega}, t)$, i.e., with $j = \text{a, f, or s}$,

$$\Sigma_j(\mathbf{r}, u, t) = \Sigma_j(\mathbf{r}, u, t; [n])$$

where the bracket denotes the functional dependence on n. Consequently, the operators H and M_i appearing in (2.2, Eq. 5) are also functionals of n. Hence, our starting equations in operator form are

$$\partial n/\partial t = H[n]\, n + \sum_{i=1}^{6} \lambda_i f_i C_i + S, \tag{1a}$$

$$\partial(f_i C_i)/\partial t = M_i[n]\, n - \lambda_i f_i C_i\,, \qquad i = 1, 2,..., 6 \tag{1b}$$

We now consider a stationary reference reactor supporting a neutron distribution characterized by $N_0(\mathbf{r}, u, \boldsymbol{\Omega})$. Since a reactor is never truly stationary when the burnup and buildup of the various nuclear species are included (cf 1.3, Eqs. 6–10), we must either assume that the reference reactor is operated at zero power level, and hence free from all the feedback effects, or ignore the long-term changes in the nuclear species due to burnup and buildup by irradiation. In the first case, the reference reactor is critical in the absence of feedback effects, and represents a cold, clean reactor free from fission products. In the second case, the reference reactor is critical in the presence of all the feedback effects except for those arising from the depletion of the fuel and continuous buildup of stable isotopes; the effects of the burnable poisons, such as ^{135}Xe, are still included. It is more realistic to visualize the reference reactor as in the second case, because then the reference distribution $N_0(\mathbf{r}, u, \boldsymbol{\Omega})$ can be chosen as the steady-state distribution in the actual reactor at the operating power level before the perturbations are introduced. Since this distribution includes the initial feedback effects, a perturbation analysis based on it is better justified than choosing $N_0(\mathbf{r}, u, \boldsymbol{\Omega})$ as the steady-state distribution in a reactor critical in the absence of feedback.

The steady-state distribution $N_0(\mathbf{r}, u, \boldsymbol{\Omega})$ can be obtained in principle by solving the set of coupled nonlinear integrodifferential equations derived in the previous section. In operator form, we can formulate this problem as

$$\mathscr{H}_0[N_0]\, N_0 = 0 \tag{2}$$

where we recall (cf 2.2, Eq. 6b)

$$\mathscr{H}_0[N_0] \equiv H[N_0] + \sum_{i=1}^{6} M_i[N_0] \tag{3a}$$

or, more explicitly,

$$\mathscr{H}_0[N_0] \equiv -\mathbf{v}\cdot\boldsymbol{\nabla} - v(u)\,\Sigma(\mathbf{r}, u, [N_0]) + \int du' \int d\Omega'\, v(u')$$
$$\times \left\{\Sigma_{\rm s}(\mathbf{r}, u' \to u, \boldsymbol{\Omega}'\cdot\boldsymbol{\Omega};\, [N_0]) + \sum_j [f^j(u)/4\pi]\, \nu^j(u')\, \Sigma_{\rm f}^j(\mathbf{r}, u';\, [N_0])\right\} \tag{3b}$$

We note that the operator $\mathscr{H}_0[N_0]$ has the same structure as the steady-state Boltzmann operator defined in Section 2.3. The presence of feedback modifies only the energy and space dependences of the cross sections in the expression of $\mathscr{H}_0[N_0]$ but does not affect its form. Hence, its adjoint, $\mathscr{H}_0^+[N_0]$ can be obtained by the same procedure as described in Section 2.3. Using $\mathscr{H}_0^+[N_0]$, we define the adjoint angular density as the solution of

$$\mathscr{H}_0^+[N_0]\, N_0^+ = 0 \tag{4}$$

with adjoint boundary conditions. In the following analysis, $N_0(\mathbf{r}, u, \mathbf{\Omega})$ and $N_0^+(\mathbf{r}, u, \mathbf{\Omega})$ will be assumed to be known functions of $\mathbf{r}$, u, and $\mathbf{\Omega}$.

In order to obtain the appropriate point kinetic equations with feedback, we introduce the shape and time functions as $n(\mathbf{r}, u, \mathbf{\Omega}, t) = P(t)\,\phi(\mathbf{r}, u, \mathbf{\Omega}, t)$, multiply (1) by N_0^+, integrate over $\mathbf{r}$, u, and $\mathbf{\Omega}$, and use the normalization condition $d\langle N_0^+ \mid \phi\rangle/dt = 0$. We obtain the following kinetic parameters:

$$\rho/l \equiv \langle N_0^+ \mid \delta\mathscr{H}[n] \mid \phi\rangle/\langle N_0^+ \mid \phi\rangle \tag{5a}$$

$$\bar{\beta}_i/l \equiv \langle N_0^+ \mid M_i[n] \mid \phi\rangle/\langle N_0^+ \mid \phi\rangle \tag{5b}$$

$$\bar{C}_i \equiv \langle N_0^+ \mid f_i C_i\rangle/\langle N_0^+ \mid \phi\rangle \tag{5c}$$

$$\bar{S} \equiv \langle N_0^+ \mid S\rangle/\langle N_0^+ \mid \phi\rangle \tag{5d}$$

where

$$\delta\mathscr{H}[n] \equiv \mathscr{H}[n] - \mathscr{H}_0[N_0] \tag{6}$$

We note the difference between (5) and the corresponding equations (10), (11), (13), and (14) of Section 2.4 in the absence of feedback. In (5), the perturbation operator $\delta\mathscr{H}[n]$ depends on $P(t)$ as well as on the shape function ϕ, whereas in the absence of feedback, $\delta\mathscr{H}$ was only an explicit function of time.

Equation (6) represents the difference between the nuclear properties of the reference reactor at steady state and those of the actual reactor at time t, with the feedback effects being included in both cases. This difference may be due to the changes in the cross sections resulting from feedback effects, or due to the changes introduced externally in the atomic composition of the reactor, e.g., by moving the control rods. In order to separate the various contribution to $\delta\mathscr{H}[n]$, we express (6) as the sum of the partial derivatives of $\mathscr{H}_0[N_0]$ evaluated at equilibrium:

$$\delta\mathscr{H}[n] \simeq \sum_i \left[\delta N_i^{\text{ext}}(t)\,\frac{\partial\mathscr{H}_0}{\partial N_{i0}} + \delta N_i^{\text{c}}[n]\,\frac{\partial\mathscr{H}_0}{\partial N_{i0}}\right] + \delta T[n]\,\frac{\partial\mathscr{H}_0}{\partial T_0} \tag{7}$$

where

$$\delta N_i^{\text{ext}}(t) \equiv N_i(\mathbf{r}, T_0, t; [N_0]) - N_{i0}(\mathbf{r}) \tag{8}$$

$$\delta N_i^{\text{c}}[n] \equiv N_i(\mathbf{r}, T_0; [n]) - N_{i0}(\mathbf{r}) \tag{9}$$

$$\delta T[n] \equiv T(\mathbf{r}; [n]) - T_0(\mathbf{r}) \tag{10}$$

Here, $N_{i0}(\mathbf{r})$ and $T_0(\mathbf{r})$ are the equilibrium concentration of the ith nucleus and the local temperature at $\mathbf{r}$, respectively, and $\mathscr{H}_0$ is defined in (3b). (Note that the partial derivatives remove the streaming term in this definition.) Substituting (7) into (5a), we break up the reactivity into three parts:

$$\rho/l = (\delta\rho_{\text{ext}}/l) + (\delta\rho_{\text{c}}/l) + (\delta\rho_T/l) \tag{11}$$

where

$$\delta\rho_{\text{ex}}/l \equiv (1/\langle N_0^+ \mid \phi\rangle)\langle N_0^+ \mid \sum_i \delta N_i^{\text{ext}}(\partial\mathscr{H}_0/\partial N_{i0}) \mid \phi\rangle \tag{12}$$

$$\delta\rho_{\text{c}}/l \equiv (1/\langle N_0^+ \mid \phi\rangle)\langle N_0^+ \mid \sum_i \delta N_i^{\text{c}}(\partial\mathscr{H}_0/\partial N_{i0}) \mid \phi\rangle \tag{13}$$

$$\delta\rho_T/l \equiv (1/\langle N_0^+ \mid \phi\rangle)\langle N_0^+ \mid \delta T(\partial\mathscr{H}_0/\partial T_0) \mid \phi\rangle \tag{14}$$

The terms in (11) represent, respectively, the external reactivity changes, reactivity feedback due to changes in atomic concentrations, and reactivity feedback due to temperature variations.

The meaning of the scalar product in the presence of temperature feedback requires clarification, because temperature changes affect the size of the reactor. Although one can use perturbation theory involving boundary variations as discussed by Morse and Feshbach [22] to take into account the size changes, we prefer to ignore them in this book because their effect on reactivity in large power reactors is negligible. In small, bare cores such as in the SNAP reactor, where the size changes constitute an appreciable contribution to feedback reactivity, one can calculate the reactivity more directly (Problem 5) than by using the perturbation approximation, as a result of the simplicity of the system.

In order to complete the derivation of the point kinetic equations with feedback, we must also consider $(\bar{\beta}_i/l)$ defined in (5b). We observe that it is a functional of $n(\mathbf{r}, u, \mathbf{\Omega}, t)$ through

$$M_i[n] = \sum_j \Big(\beta_i^j N_j(\mathbf{r}, T, t; [n]) \times \int du' \int d\Omega' \, \{\nu^j(u')\, v(u')[f_i(u)/4\pi]\, \sigma_f^j(u', T)\}\Big) \tag{15}$$

It is consistent with the perturbation approximation to replace (15) by its equilibrium value

$$M_i[N_0] = \sum_j \Big(\beta_i{}^j N_j(\mathbf{r}, T_0 ; [N_0]) \times \int du' \int d\Omega' \{[f_i(u)/4\pi]\, \nu^j(u')\, v(u')\, \sigma_f{}^j(u'\ T_0)\}\Big) \tag{16}$$

The shape function $\phi(\mathbf{r}, u, \mathbf{\Omega}, t)$ appearing in (11)–(14) may be chosen in one of the ways discussed in Section 2.4. Here, we use the first-order perturbation approximation, which implies (cf 2.4, Eq. 21)

$$n(\mathbf{r}, u, \mathbf{\Omega}, t) \approx [P(t)/P_0]\, N_0(\mathbf{r}, u, \mathbf{\Omega}) \tag{17}$$

where $N_0(\mathbf{r}, u, \mathbf{\Omega})$ is the angular neutron density at equilibrium.

To conclude this section, we summarize the above results in a form which will be used later. The point kinetic equations in the presence of feedback can be written within the framework of first-order perturbation theory as

$$\dot{P}(t) = [\{\delta\rho_{\text{ext}}(t) + \delta\rho_{\text{f}}[P] - \bar{\beta}\}/l]\, P(t) + \sum_{i=1}^{6} \lambda_i \bar{C}_i(t) + \bar{S}(t) \tag{18a}$$

$$\dot{\bar{C}}_i(t) = (\bar{\beta}_i/l)\, P(t) - \lambda_i \bar{C}_i(t), \qquad i = 1, 2, \ldots, 6 \tag{18b}$$

with the following identification of the parameters:

$$\bar{\beta}_i/l = (1/\langle N_0{}^+ \mid N_0\rangle)\langle N_0{}^+ \mid M_i[N_0] \mid N_0\rangle \tag{19a}$$

$$\bar{S}(t) = P_0\langle N_0{}^+ \mid S\rangle/\langle N_0{}^+ \mid N_0\rangle \tag{19b}$$

$$\bar{C}_i(t) = (P_0/\langle N_0{}^+ \mid N_0\rangle)\langle N_0{}^+ \mid f_i C_i\rangle \tag{19c}$$

$$l = \langle N_0{}^+ \mid N_0\rangle/\langle N_0{}^+ \mid M[N_0] \mid N_0\rangle \tag{19d}$$

$$\delta\rho_{\text{ext}}(t)/l = (1/\langle N_0{}^+ \mid N_0\rangle)\langle N_0{}^+ \mid \sum_j \delta N_j^{\text{ext}}(\partial\mathscr{H}_0/\partial N_{j0}) \mid N_0\rangle \tag{20}$$

$$\delta\rho_{\text{f}}[P]/l = (\delta\rho_{\text{c}}[P]/l) + (\delta\rho_T[P]/l) \tag{21}$$

$$\delta\rho_{\text{c}}[P]/l = (1/\langle N_0{}^+ \mid N_0\rangle)\langle N_0{}^+ \mid \sum_j \delta N_j{}^{\text{c}}(\partial\mathscr{H}_0/\partial N_{j0}) \mid N_0\rangle \tag{22}$$

$$\delta\rho_T[P]/l = (1/\langle N_0{}^+ \mid N_0\rangle)\langle N_0{}^+ \mid \delta T(\partial\mathscr{H}_0/\partial T_0) \mid N_0\rangle \tag{23}$$

A comparison of (18) with (2.4, Eqs. 9) representing the point kinetics in the absence of feedback reveals that all the feedback effects in (18) are accounted for by the feedback functional $\delta\rho_{\text{f}}[P]$. The computation of this functional for various reactor types will be discussed in Chapter 5.

It is to be noted that the reactivity ρ can be expressed as the superposition of the external and feedback reactivities only in the first-order perturbation theory. In general, an external change in the atomic concentration will affect not only the external reactivity, but also the feedback reactivity as a result of the changes in the shape function.

2.7. Calculation of Kinetic Parameters in the Diffusion Approximation

The definitions of the kinetic parameters given by Eqs. (19)–(23) of the preceding section involve the angular density $N_0(\mathbf{r}, u, \mathbf{\Omega})$ and its adjoint $N_0^+(\mathbf{r}, u, \mathbf{\Omega})$. In the diffusion approximation, one calculates only the scalar flux $\phi_0(\mathbf{r}, u)$ and its adjoint $\phi_0^+(\mathbf{r}, u)$ [or the corresponding scalar neutron densities $N_0(\mathbf{r}, u)$ and $N_0^+(\mathbf{r}, u)$] as discussed in Section 1.4. It is therefore desirable to express the definition of the kinetic parameters in terms of the scalar fluxes only within the framework of the diffusion approximation. For this purpose, we expand $N_0(\mathbf{r}, u, \mathbf{\Omega})$ and $N_0^+(\mathbf{r}, u, \mathbf{\Omega})$ in spherical harmonics, and treat the anisotropic terms as being small. The first two terms in these expansions are (Problem 7)

$$vN_0(\mathbf{r}, u, \mathbf{\Omega}) = (1/4\pi)[\phi_0(\mathbf{r}, u) + 3\mathbf{J}_0(\mathbf{r}, u) \cdot \mathbf{\Omega}] + \cdots \tag{1a}$$

$$vN_0^+(\mathbf{r}, u, \mathbf{\Omega}) = (1/4\pi)[\phi_0^+(\mathbf{r}, u) + 3\mathbf{J}_0^+(\mathbf{r}, u) \cdot \mathbf{\Omega}] + \cdots \tag{1b}$$

where $\mathbf{J}_0(\mathbf{r}, u)$ and $\mathbf{J}_0^+(\mathbf{r}, u)$ are the current vector and its adjoint. Substituting these into the definition of $(\bar{\beta}_i/l)$, $\bar{S}(t)$ and $\bar{C}_i(t)$ and ignoring the terms containing the products of currents, we obtain

$$\bar{\beta}_i/l = (1/\langle N_0^+ \mid N_0\rangle)\langle N_0^+ \mid \overline{M}_i[N_0] \mid N_0\rangle \tag{2a}$$

$$\bar{S}(t) = (P_0/\langle N_0^+ \mid N_0\rangle)\left\langle N_0^+ \,\middle|\, \int d\Omega\, S(\mathbf{r}, u, \mathbf{\Omega}, t)\right\rangle \tag{2b}$$

$$\bar{C}_i(t) = (1/\langle N_0^+ \mid N_0\rangle)\langle N_0^+ \mid f_i C_i\rangle \tag{2c}$$

$$l = \langle N_0^+ \mid N_0\rangle/\langle N_0^+ \mid \overline{M}[N_0] \mid N_0\rangle \tag{2d}$$

where

$$\overline{M}_i[N_0] \equiv \int d\Omega\, M_i[N_0] = \sum_j f_i(u) \int_0^\infty du'\, [\beta_i^j \nu^j(u')\, v(u')\, \Sigma_{f0}^j(\mathbf{r}, u')] \tag{3a}$$

$$\overline{M}[N_0] \equiv \int d\Omega\, M[N_0] = \sum_j f^j(u) \int_0^\infty du'\, [\nu^j(u')\, v(u')\, \Sigma_{f0}^j(\mathbf{r}, u')] \tag{3b}$$

$$\langle N_0^+ \mid N_0\rangle \equiv \int_0^\infty du \int_R d^3r\, [N_0^+(\mathbf{r}, u)\, N_0(\mathbf{r}, u)] \tag{4}$$

The last equation defines the scalar product for functions of u and $\mathbf{r}$. This definition is implied in (2).

The calculation of the reactivity in the diffusion approximation using (2.6, Eqs. 20–23) is not as straightforward as above. If we write $\mathscr{H}_0$ explicitly in these equations using (2.6, Eq. 3b), we encounter the following integral:

$$\int_0^\infty du' \int d\Omega \int d\Omega' \, [\Sigma_s(u' \to u, \mathbf{\Omega}' \cdot \mathbf{\Omega}) \, v(u') \, N_0^+(u, \mathbf{\Omega}) \, N_0(u', \mathbf{\Omega}')]$$
$$- \int d\Omega \, [v(u) \, \Sigma(u) \, N_0^+(u, \mathbf{\Omega}) \, N_0(u, \mathbf{\Omega})] \tag{5}$$

where we omitted the variable $\mathbf{r}$ in the arguments for the present discussion. Substituting (1) into (5) and keeping the terms containing the products of currents, we obtain (Problem 8)

$$(1/4\pi)\Big[\int_0^\infty du' \, \{\Sigma_s(u' \to u) \, N_0^+(u) \, \phi_0(u')\} - \Sigma(u) \, N_0^+(u) \, \phi_0(u)$$
$$- [3/v(u)] \, \Sigma_{tr}(u) \, \mathbf{J}^+(u) \cdot \mathbf{J}(u)\Big] \tag{6}$$

where Σ_{tr} is the transport cross section, defined by $\Sigma_{tr} = \Sigma - \bar{\mu}\Sigma_s$, $\bar{\mu}$ being the average cosine of the angle of deflection in the laboratory system in a scattering event. To compress the notation, we define

$$\Gamma_0 \equiv -\Sigma_0(\mathbf{r}, u) + \int_0^\infty du' \left[\Sigma_{s_0}(\mathbf{r}, u' \to u) + \sum_j f^j(u) \, \nu^j(u') \, \Sigma_{f_0}^j(\mathbf{r}, u')\right] \tag{7}$$

and obtain the following expression for the reactivity $\rho = \delta\rho_{ext} + \delta\rho_f$ in the diffusion approximation:

$$\rho = (1/\langle N_0^+ \mid \overline{M}[N_0] \mid N_0\rangle)[\langle N_0^+ \mid \delta\Gamma_0 \mid \phi_0\rangle - \langle \mathbf{J}_0^+ \mid (3\delta\Sigma_{tr_0}/v) \mid \mathbf{J}_0\rangle] \tag{8}$$

The scalar products in this equation are defined by (2d). This expression can be cast into a more conventional form by using

$$\mathbf{J}_0^+ = D_0 \, \nabla\phi_0^+ \tag{9a}$$

$$\mathbf{J}_0 = -D_0 \, \nabla\phi_0 \tag{9b}$$

$$D_0 = 1/3\Sigma_{tr_0} \tag{9c}$$

and noting that $3\delta\Sigma_{tr} = -\delta D_0/D_0^2$. The result is

$$\rho = (1/\langle N_0^+ \mid \overline{M}[N_0] \mid N_0\rangle)[\langle N_0^+ \mid \delta\Gamma_0 \mid \phi_0\rangle - \langle \nabla N_0^+ \mid \delta D_0 \mid \nabla\phi_0\rangle] \tag{10}$$

The various contributions to reactivity discussed in the previous section can be obtained from (3) or (10) by considering the partial derivatives of Γ_0 and Σ_{tr_0} (or D_0).

The energy integrals appearing in the definition of the kinetic parameters, Eqs. (2) and (8), can be reduced to summations by adopting the multigroup formalism introduced in Section 1.4. We shall illustrate the application of these formulas by considering simple reactor models.

Example 1. One-group diffusion model. In the one-group diffusion model, $\phi(\mathbf{r})$ satisfies

$$\nabla \cdot D_0 \nabla\phi_0 + (\nu\Sigma_{\mathrm{f}_0} - \Sigma_{\mathrm{a}_0})\phi_0(\mathbf{r}) = 0 \tag{11}$$

in the absence of external sources. The cross sections are allowed to be position-dependent. The operators $\overline{M}_i[N_0]$, $\overline{M}[N_0]$, and Γ_0 defined in Eqs. (3) and (7) reduce to

$$\overline{M}_i[N_0] = \beta_i v\Sigma_{\mathrm{f}_0} \tag{12a}$$

$$\overline{M}[N_0] = \nu v\Sigma_{\mathrm{f}_0} \tag{12b}$$

$$\Gamma_0 = -\Sigma_{\mathrm{a}_0} + \nu\Sigma_{\mathrm{f}_0} \tag{12c}$$

From (2), we find

$$l = \left[\int_R d^3r\, \phi_0^2(r)\right] \Big/ v \int_R d^3r\, [\nu\Sigma_{\mathrm{f}_0}\phi_0^2(r)] \tag{13}$$

which reduces to $l = (1/v\nu\Sigma_{\mathrm{f}_0})$ when the fission cross section is independent of position. In the one-speed model, $\bar{\beta}_i = \beta_i$. The reactivity follows from (8) as

$$\rho = \left\{\int_R d^3r\, [3\delta\Sigma_{\mathrm{tr}_0} \mid \mathbf{J}_0 \mid^2 - \delta(\Sigma_{\mathrm{a}_0} - \nu\Sigma_{\mathrm{f}_0})\,\phi_0^2]\right\} \Big/ \nu \int_R d^3r\, (\Sigma_{\mathrm{f}_0}\phi_0^2) \tag{14}$$

or from (10) as

$$\rho = -\left\{\int_R d^3r\, [\delta D_0 \mid \nabla\phi_0 \mid^2 + \delta(\Sigma_{\mathrm{a}_0} - \nu\Sigma_{\mathrm{f}_0})\,\phi_0^2]\right\} \Big/ \nu \int_R d^3r\, (\Sigma_{\mathrm{f}_0}\phi_0^2) \tag{15}$$

[See Problems 6, 9, and 10 for some simple applications of (14) and (15).]

Example 2. Modified one-group diffusion model. A more realistic model, which takes into account the slowing down process, is the age-diffusion model (cf Section 1.5C). The thermal flux satisfies the following equations [23] in this model if we assume instantaneous slowing down of the fission neutrons to thermal energies:

$$(1/v)\,\partial\phi(\mathbf{r},t)/\partial t = \nabla\cdot D(\mathbf{r},t)\,\nabla\phi(\mathbf{r},t) - \Sigma_a(\mathbf{r},t)\,\phi(\mathbf{r},t) + p(1-\beta)\,\nu[\exp(-B_g^2\tau_p)]\,\Sigma_f(\mathbf{r},t)\,\phi(\mathbf{r},t) + \sum_{i=1}^{6} p\lambda_i[\exp(-B_g^2\tau_i)]\,C_i(\mathbf{r},t) \tag{16}$$

$$\partial C_i(\mathbf{r},t)/\partial t = \nu\beta_i\Sigma_f(\mathbf{r},t)\,\phi(\mathbf{r},t) - \lambda_i C_i(\mathbf{r},t) \tag{17}$$

where B_g is the geometric buckling, τ_p and τ_i are the ages of the prompt and delayed neutrons, and p is the resonance escape probability, which is the same for both the delayed and prompt neutrons because the resonances occur at lower energies than the delayed neutron energy spectrum. It is assumed that there is only one species of fissionable material. The first and second terms in the right-hand side of (16) are the leakage and absorption rates of the thermal neutrons. The last two terms account for the production rate of the thermal neutrons due to the fission process and to the decay of delayed neutron precursors. The $\phi\Sigma_f(1-\beta)\,\nu$ in the third term gives the number of prompt neutrons per second per unit volume due to fission events caused by the thermal neutrons. The factor $p\exp[-B_g^2\tau_p]$ is the probability that a prompt fission neutron will escape the resonances and will not leak out during the slowing down to thermal energy. The interpretation of Eq. (17) and the last term in (16) is self-evident. These equations can be written in the standard form given (2.3, Eqs. 5) by defining

$$M_0 \equiv \nu(1-\beta)\,\Sigma_f p\exp(-B_g^2\tau_p)v \tag{18a}$$

$$M_i \equiv \beta_i\nu\Sigma_f p\exp(-B_g^2\tau_i)v \tag{18b}$$

$$L = -(\Sigma_a - \nabla\cdot D\nabla)v \tag{18c}$$

$$M = \nu\Sigma_f p[(1-\beta)[\exp(-B_g^2\tau_p)] + \sum_{i=1}^{6}\beta_i\exp(-B_g^2\tau_i)]v \tag{18d}$$

if we absorb the factor $pv\exp(-B_g^2\tau_i)$ in $C_i(\mathbf{r},t)$.

We assume that the unperturbed reactor is a bare, homogeneous reactor, i.e., D_0, Σ_{a_0} and Σ_{f_0} are independent of position, and that the perturbation does not affect the slowing-down properties of the medium, i.e., τ_p, τ_i, and p are unchanged. We also assume that the perturbation is introduced uniformly in the reactor, i.e., δD, $\delta\Sigma_a$, and $\delta\Sigma_{\mathrm{f}}$ are independent of position. These approximations are not essential for the calculation of $\bar{\beta}_i$, l, and ρ, but they simplify the analysis, and allow us to express these quantities in a more familiar form.

The effective delayed neutron fraction $\bar{\beta}_i$ follows from (2a) and (2d) as

$$\bar{\beta}_i/\beta_i = \{\exp[-B_{\mathrm{g}}^2(\tau_i - \tau_{\mathrm{p}})]\}\Big/\Big\{(1-\beta) + \sum_{i=1}^{6} \exp[-B_{\mathrm{g}}^2(\tau_i - \tau_{\mathrm{p}})]\Big\} \quad (19)$$

which reduces to

$$\bar{\beta}_i/\beta_i = [(1-\beta)\exp\{-B_{\mathrm{g}}^2(\tau_{\mathrm{p}} - \tau_{\mathrm{d}}) + \beta]^{-1} \quad (20)$$

if we assume that the ages of the delayed neutron groups are the same, namely, $\tau_i \approx \tau_{\mathrm{d}}$ for all i.

The mean generation time l follows from (2d) by substituting M from (18d):

$$(1/l) = v\nu p\Sigma_{\mathrm{f}0}\Big\{(1-\beta)[\exp(-B_{\mathrm{g}}^2\tau_{\mathrm{p}})] + \sum_{i=1}^{6} \beta_i \exp(-B_{\mathrm{g}}^2\tau_i)\Big\} \quad (21)$$

where Σ_{f_0} is the unperturbed fission cross section. We can cast (21) into a more conventional form by using the fact that the unperturbed reactor is critical, namely

$$D_0\nabla^2\phi_0 - \Sigma_{\mathrm{a}_0}\phi_0 = -\nu p\Sigma_{\mathrm{f}_0}\Big[(1-\beta)[\exp(-\tau_{\mathrm{p}}B_{\mathrm{g}}^2)] + \sum_{i=1}^{6} \beta_i \exp(-\tau_i B_{\mathrm{g}}^2)\Big]\phi_0 \quad (22)$$

Thus, (21) becomes

$$l = l_0/(1 + L^2B_{\mathrm{g}}^2) \quad (23)$$

where $l_0 = 1/\Sigma_{\mathrm{a}_0}v$ and $L = (D_0/\Sigma_{\mathrm{a}_0})^{1/2}$. The reactivity follows from (10) as in (15). The presence of delayed neutrons and the inclusion of the fast nonleakage probabilities are accounted for in this expression through the fission cross section in the denominator and the criticality condition. To emphasize this point, we express the effective multiplication factor $k_{\mathrm{eff}} = 1/(1-\rho)$. We can verify that

$$\begin{aligned} k_{\mathrm{eff}} &= \langle\phi_0 \mid M \mid \phi_0\rangle/\langle\phi_0 \mid \Sigma_{\mathrm{a}} - D\nabla^2 \mid \phi_0\rangle v \\ &= k_\infty\Big\{(1-\beta)[\exp(-B_{\mathrm{g}}^2\tau_{\mathrm{p}})] + \sum_i \beta_i \exp(-B_{\mathrm{g}}^2\tau_i)\Big\}\Big/(1 + L^2B_{\mathrm{g}}^2) \end{aligned} \quad (24)$$

where $k_\infty = \nu p\Sigma_{\mathrm{f}}/\Sigma_{\mathrm{a}}$, i.e., the infinite medium multiplication factor.

PROBLEMS

1. (a) Show that $\nabla \cdot D(\mathbf{r})\, \nabla$ is self-adjoint for functions $\phi(\mathbf{r})$ that vanish on the outer surface of the reactor.
(b) Show that the adjoint of $(D\nabla^2 + \mathbf{a} \cdot \nabla + B_n{}^2)$ is $(D\nabla^2 - \mathbf{a} \cdot \nabla + B_n{}^2)$ for the same class of functions. (cf Chapter 1, Problem 14.)

2. Show that, if $O = [O_{ij}]$ is a matrix operator, its adjoint is $O^+ = [O_{ji}^+]$. Hint: Consider $O \mid \phi\rangle = \operatorname{col}[O_{in}\phi_n]$ and $O^+ \mid \phi^+\rangle = \operatorname{col}[O_{ni}^+\phi_n{}^+]$ with summation convention on n. Verify (summation on i and n)

$$\begin{aligned}\langle \phi^+ \mid O\phi\rangle &= \langle \phi_i{}^+ \mid O_{in}\phi_n\rangle \\ &= \langle O_{in}^+\phi_i{}^+ \mid \phi_n\rangle \\ &= \langle O_{ni}^+\phi_n{}^+ \mid \phi_i\rangle = \langle O^+\phi^+ \mid \phi\rangle\end{aligned}$$

3. Find the expansion of the column vector $\mid I\rangle$ whose elements are all zero except for $I_i(\mathbf{r}, \mathbf{v}) = \delta(\mathbf{r} - \mathbf{r}')\, \delta(\mathbf{v} - \mathbf{v}')$ and prove

$$\delta(\mathbf{r} - \mathbf{r}')\, \delta(\mathbf{v} - \mathbf{v}')\, \delta_{ij} = \sum_{n=0}^{\infty} \phi_{ni}(\mathbf{r}', \mathbf{v}')\, \phi_{nj}(\mathbf{r}, \mathbf{v})$$

where ϕ_{nj} are the components of the eigenvector $\mid \phi_n\rangle$.

4. Using the one-group diffusion equation with one group of delayed neutrons, show that the ω-modes for an infinite, bare, slab reactor of thickness $\tilde{a}$ are

$$\mid \phi_{n_J}\rangle = \begin{bmatrix} 1 \\ (\nu\Sigma_f + \beta)/(\lambda + \omega_{n_J}) \end{bmatrix} A_{n_J} \sin[(n + 1)(\pi/\tilde{a})\, x]$$

where ω_{n_J} are the two roots of

$$\frac{(k_{\text{eff}})_n - 1}{(k_{\text{eff}})_n} = \omega_{n_J} \left[\frac{1}{\nu\Sigma_f v} + \frac{\beta}{\lambda + \omega_{n_J}}\right]$$

$$(k_{\text{eff}})_n = \frac{\nu\Sigma_f/\Sigma_a}{1 + L^2[(\pi/a)(n + 1)]^2}, \qquad n = 0, 1, 2,\ldots$$

$$A_{n_J} = \left(\frac{2}{\tilde{a}}\right)^{1/2} \frac{1}{1 + [\nu\Sigma_f\beta/(\lambda + \omega_{n_J})]^2}$$

Hint: Use

$$\mathscr{K} = \begin{bmatrix} D(d^2/dx^2) + (1 - \beta)\, \nu\Sigma_f - \Sigma_a & \lambda \\ \nu\Sigma_f\beta & -\lambda \end{bmatrix}, \qquad \mid \phi_{n_J}\rangle = \begin{bmatrix} N_{n_J}(x) \\ C_{n_J}(x) \end{bmatrix}$$

and solve

$$\mathscr{K} \mid \phi_{n_J}\rangle = \omega_{n_J} \mid \phi_{n_J}\rangle$$

with the boundary condition $N_{n_J}(0) = N_{n_J}(\tilde{a}) = 0$. Observe that $N_{n_J}(x) = A_{n_J} \sin[(n+1)(\pi/\tilde{a})\, x]$ has the same spatial distribution for $J = 1, 2$ and a fixed n. Verify the orthogonality

$$\langle \phi_{n_J} \mid \phi_{n'J'}\rangle = \delta_{nn'}\delta_{JJ'}$$

5. (a) Find the reactivity change in an initially critical, homogeneous, spherical, bare reactor using the one-group diffusion approximation when the reactor temperature is raised uniformly by ΔT. Assume that the temperature change affects only the density of the medium as $N = N_0(1 - \alpha T)$.
   (b) Calculate the reactivity change in the same reactor using (2.6, Eq. 23) and compare the results.
6. A thin, absorbing rod is inserted into a bare, uniform, cubic reactor of edge H, along the z axis (see Figure P6). Using first-order

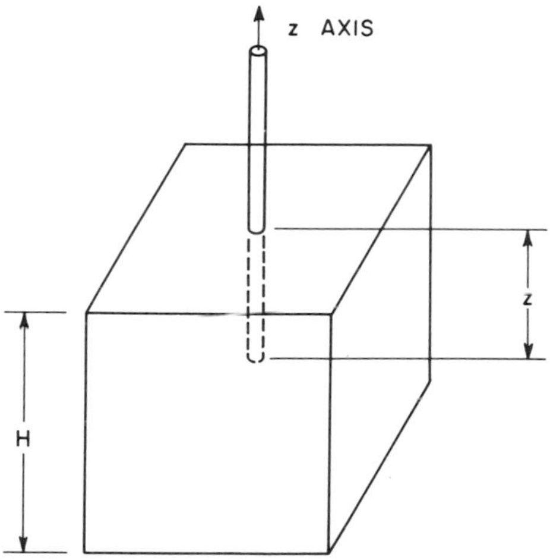

FIGURE P6.

perturbation theory in the absence of delayed neutrons, show that the reactivity change due to the rod is

$$\rho(z) = \rho(H)[(z/H) - (1/2\pi)\sin(2\pi z/H)]$$

7. The spherical harmonic expansion of the angular flux [18] is

$$\phi(\mathbf{\Omega}) = \phi_0 Y_{00}(\mathbf{\Omega}) + \sum_{m=0,\pm1} Y_{1m}(\mathbf{\Omega})\,\phi_{1m} + \cdots$$

where the expansion coefficients ϕ_{lm} are $\int Y_{lm}^* \phi(\mathbf{\Omega})\, d\Omega$. Let Ω_x, Ω_y, Ω_z be the Cartesian components of the unit vector $\mathbf{\Omega}$. (a) Show that the "spherical" components of $\mathbf{\Omega}$, defined as

$$\Omega_0 = \Omega_z, \qquad \Omega_\pm = \mp(1/\sqrt{2})(\Omega_x \pm i\Omega_y)$$

are given by

$$\Omega_m = (3/4\pi)^{1/2}\, Y_{1m}, \qquad m = 0, \pm1$$

(b) Using the spherical components of two vectors **A** and **B**, show that their scalar product is

$$\mathbf{A} \cdot \mathbf{B} = \sum_m A_m B_m^*, \qquad m = 0, \pm1$$

(c) Show that

$$\sum_m Y_{1m}\phi_{1m} = (3/4\pi)\,\mathbf{J} \cdot \mathbf{\Omega}$$

noting that the spherical components of the current **J** are $(4\pi/3)^{1/2}\,\phi_{1m}$.

8. Hint: First, verify

$$\int d\Omega\, \Sigma_s(u' \to u, \mathbf{\Omega}' \cdot \mathbf{\Omega})\, \mathbf{\Omega} = \mathbf{\Omega}' \Sigma_{s1}(u' \to u)$$

where

$$\Sigma_{s1}(u' \to u) = 2\pi \int_{-1}^{+1} d\mu\, [\mu \Sigma_s(u' \to u, \mu)]$$

and then use

$$\int_0^\infty du'\, [\mathbf{J}(u')\, \Sigma_{s1}(u' \to u)] \cong \mathbf{J}(u)\, \Sigma_s(u)\, \bar{\mu}$$

$$\Sigma_s(u)\, \bar{\mu} \equiv \int_0^\infty du' \int d\Omega\, \Sigma_s(u \to u', \mu)\, \mu$$

9. (a) Calculate the reactivity in an initially critical, bare, homogeneous, spherical reactor of radius R using the one-group diffusion model due to a concentric spherical void of radius r.

(b) Calculate the reactivity using the first-order perturbation theory assuming $r \ll R$, and compare it to the result of (a).
Hint: Use (2.7, Eq. 14) in (b).

10. Find the reactivity change in a perfectly reflected homogeneous, slab reactor of thickness a when the absorption cross section is changed uniformly in the slab by an amount $\delta\Sigma_a$.

REFERENCES

1. J. Hurwitz, Jr., *Nucleonics* **5**, 62 (1949).
2. L. N. Ussachoff, *Proc. Int. Conf. Peaceful Uses At. Energy, Geneva,* 1955, P/656. Columbia Univ. Press, New York, 1955.
3. A. F. Henry, Computation of parameters appearing in the reactor kinetics equation. WAPD-142. December 1955, Westinghouse Atomic Power Division, Pittsburgh, Pennsylvania.
4. A. F. Henry, Application of reactor kinetics to the analysis of experiments. *Nucl. Sci. Eng.* **3**, 52 (1958).
5. E. P. Gyftopoulos, *in* "The Technology of Nuclear Reactor Safety" (T. J. Thompson and J. G. Beckerly, eds.), Vol. 1, pp. 175–204. M.I.T. Press, Cambridge, Massachusetts, 1964.
6. M. Becker, A generalized formulation of point nuclear reactor kinetic equations. *Nucl. Sci. Eng.* **31**, 458 (1968).
7. J. Lewins, *J. Nucl. Energy Part A* **12**, 108 (1960).
8. R. Courant and D. Hilbert, "Methods of Mathematical Physics," Vol. I. Wiley (Interscience), New York, 1953.
9. B. Friedman, "Principles and Techniques of Applied Mathematics." Wiley, New York, 1956.
10. E. E. Gross and J. H. Marable, *Nucl. Sci. Eng.* **7**, 281 (1960).
11. A. Weinberg, *Amer. J. Phys.* **20**, 401 (1952).
12. R. Ehrlich AND H. Hurwitz, *Nucleonics* **12**, 23 (1954).
13. K. M. Case and P. F. Zweifel, "Linear Transport Theory." Addison-Wesley, Reading, Massachusetts, 1967.
14. E. R. Cohen, *Proc. U. N. Int. Conf. Peaceful Uses At. Energy, 2nd, Geneva,* 1958, A/conf., P/629. United Nations, New York, 1958.
15. A. F. Henry, *Trans. Amer. Nucl. Soc.* **6**, 212 (1963).
16. A. F. Henry, *Trans. Amer. Nucl. Soc.* **9**, 235 (1965).
17. A. F. Henry, *Nucl. Sci. Eng.* **20**, 338 (1964).
18. R. V. Meghreblian and D. K. Holmes, "Reactor Analysis." McGraw-Hill, New York, 1960.
19. T. Gozani, The concept of reactivity and its application to kinetic measurements. *Nukleonik* **5**, 55 (1963).
20. N. Corngold, *Trans. Amer. Nucl. Soc.* **7**, 211 (1964).
21. E. E. Gross and J. H. Marable, *Nucl. Sci. Eng.* **7**, 281 (1960).
22. P. M. Morse and H. Feshbach, "Methods of Theoretical Physics," p. 1038. McGraw-Hill, New York, 1953.
23. S. Glasstone and M. C. Edlund, "Nuclear Reactor Theory." Van Nostrand, Princeton, New Jersey, 1952.

CHAPTER 3

Exact Solutions of the Point Kinetic Equations without Feedback

In order to gain insight into the dynamic behavior of a reactor under accidental or programmed perturbations, one must examine the solutions of the point kinetic equations for various reactivity insertions. Knowledge of the time response of a reactor to various perturbations is of importance for safety considerations, as well for the interpretation of kinetic experiments designed to measure macroscopic reactor parameters. For example, it is important to know the total energy release during reactor shutdown in a loss of coolant accident in order to be able to predict possible damage to the core. As another example; we can determine the ratio $\bar{\beta}/l$ from a pile-oscillator experiment if we can predict, analytically, the response of the reactor to a sinusoidal reactivity insertion.

This chapter will be devoted to a discussion of the exact methods for solving the feedback-free point kinetic equations for a number of reactivity insertions. The approximate methods will be considered in Chapter 4. Although the known exact solutions of the point kinetic equations are, except for the solution for a step reactivity change, much too complicated for practical calculations, they are of interest from a mathematical point of view, and they may be used to test the validity of the various approximate methods of solution presented in Chapter 4.

3.1. Standard and Integrodifferential Forms of the Point Kinetic Equations

The point kinetic equations (2.4, Eqs. 9) can be written as

$$(l/\beta)\, \dot{P}(t) = [k(t) - 1]\, P(t) + \sum_{i=1}^{6} \lambda_i C_i(t) + (l/\beta)\, S(t) \tag{1}$$

$$\dot{C}_i(t) = a_i P(t) - \lambda_i C_i(t), \qquad i = 1, 2,..., 6 \tag{2}$$

where we have introduced the following symbols:

$$a_i = \beta_i/\beta, \qquad \sum_{i=1}^{6} a_i = 1 \tag{3a}$$

$$k(t) = \rho(t)/\beta \tag{3b}$$

$$C_i(t) = \bar{C}_i(t)\, l/\beta \tag{4}$$

Furthermore, we have omitted the bars on $\bar{\beta}_i$, $\bar{\beta}$, and $\bar{S}(t)$ for simplicity in the notation, with the convention that β_i, β, and $S(t)$ in (1) and (2) will henceforth denote the effective delayed neutron fractions and the effective source unless otherwise specified.

The relative delayed neutron fractions a_i are rather insensitive to the type of reactor under consideration, as can be seen from their definition in the first-order perturbation approximation (cf. 2.4, Eq. 11)

$$a_i = \langle N_0^+ \mid M_i \mid N_0 \rangle \Big/ \langle N_0^+ \mid \sum_{i=1}^{6} M_i \mid N_0 \rangle \tag{5}$$

Since the energy spectra of the delayed neutrons in various groups are not much different (cf. Figure 1.1.2), we can approximate (5) by letting $f_i \approx \bar{f}_d$, and thereby show that a_i is approximately equal to the ratio of the actual delayed neutron fractions.

The quantity $k(t)$ defined by (4) is a measure of reactivity in terms of the effective delayed neutron fraction, and is called the reactivity in dollars, i.e., $\rho = \beta$ implies $k = 1\$$.

We shall refer to the form of the point kinetic equations as in (1) and (2) as the standard form.

Integrodifferential Form of the Kinetic Equations

In most of the applications, we are not interested in the time behavior of the delayed neutron precursor densities because they are not observable quantities. It is often expedient to eliminate them from (1) and (2), and to obtain a single equation containing $P(t)$ only. Solving (2) for $C_i(t)$ in terms of $P(t)$ by using integrating factors, we get

$$C_i(t) = \{\exp[-\lambda_i(t - t_0)]\}\left\{C_i(t_0) + \int_{t_0}^{t} a_i P(t') \exp[\lambda_i(t' - t_0)]\, dt'\right\} \tag{6}$$

Let us assume that

$$\lim_{t_0 \to -\infty} C_i(t_0)\, e^{\lambda_i t_0} = 0 \tag{7}$$

which is certainly true for any physical system. Then, (6) reduces to

$$C_i(t) = \int_{-\infty}^{t} a_i \{\exp[-\lambda_i(t - t')]\}\, P(t')\, dt' \tag{8a}$$

or

$$C_i(t) = \int_0^{\infty} a_i e^{-\lambda_i u} P(t - u)\, du \tag{8b}$$

In this way, the initial conditions are included in (8a), since

$$C_i(t_0) = \int_{-\infty}^{t_0} a_i \{\exp[-\lambda_i(t_0 - t')]\}\, P(t')\, dt'$$

Substitution of (8b) into (1) yields the integrodifferential form:

$$(l/\beta)\dot{P} = (k - 1)P + \int_0^{\infty} D(u)\, P(t - u)\, du + (l/\beta)S \tag{9}$$

where

$$D(u) \equiv \sum_{i=1}^{6} \lambda_i a_i e^{-\lambda_i u} \tag{10}$$

The function $D(u)$ is called the delayed neutron kernel, and is normalized to unity

$$\int_0^{\infty} D(u)\, du = 1 \tag{11}$$

It follows from (10) and (11) that $D(u)\, du$ is the probability that a delayed neutron will be emitted in du about u following a fission event at $u = 0$.

3.2. Inverse Method for Solving Kinetic Problems

Exact solutions of the point kinetic equations in closed form are known only for a few types of reactivity insertions. However, Eq. (9) indicates that we can solve the kinetic equations exactly for $k(t)$ where the variations of the reactor power $P(t)$ are known arbitrary functions of time. This solution is

$$k(t) = 1 + \frac{l}{\beta}\frac{d[\ln P(t)]}{dt} - \int_0^\infty du \left[D(u)\frac{P(t-u)}{P(t)}\right] - \frac{l}{\beta}\frac{S(t)}{P(t)} \qquad (1)$$

which is the basis of the so-called "inverse" method. This method is important for several reasons: (a) In reactor operation, the time dependence of the applied reactivity required to yield a specified power variation must be known in order to program the control-rod motion [1]. (b) The interpretation of measured power responses in transient analysis in terms of the reactivity changes provides information about the feedback mechanism in the reactor [2–4]. This may be accomplished by first inserting a known reactivity and measuring the power variation. Then, by using the inverse method for this power variation, one gets a reactivity function which is the sum of the known input plus unknown feedback effects. (c) The knowledge of the applied reactivity required to obtain a known power response can be used to check the validity of the approximate solutions of the kinetic equations for a known reactivity insertion.

The purpose of this section is to illustrate the application of the inverse method, and to indicate the type of information provided by this technique.

A. Periodic Power Variation

Let us assume that $P(t)$ is a periodic function of time with a period T, i.e., $P(t) = P(t + nT)$, where $n = 0, \pm 1, \ldots$, and investigate the nature of the reactivity insertion necessary to produce this response [5] in the absence of external sources, i.e., $S(t) \equiv 0$ in (1).

By considering $k(t)$ and $k(t + T)$, it immediately follows from (1) that $k(t)$ is also a periodic function of time with the same period. The right-hand side of (1) remains unchanged when t is replaced by $t + T$, by virtue of the periodicity of $P(t)$. This conclusion holds also in the presence of a constant external source.

We shall now show that the average of $k(t)$ over a period, i.e.,

$$k_{\text{av}} \equiv (1/T) \int_0^T k(t)\, dt \tag{2}$$

is always negative. Integrating the right-hand side of (1) in the interval $(0, T)$, we find

$$k_{\text{av}} = -(1/T) \int_0^\infty du\, D(u) \left(\int_0^T \{[P(t-u)/P(t)] - 1\}\, dt \right) \tag{3}$$

where we have used the fact that $\log P(t) = \log P(t+T)$, and $D(u)$ is normalized to unity (cf 3.1, Eq. 11). We shall now demonstrate that

$$(1/T) \int_0^T [P(t-u)/P(t)]\, dt \geqslant 1 \tag{4}$$

Following Smets [5] we consider this integral as a sum of the products of positive numbers $a_j = P(t_j - u)$ and $b_j = 1/P(t_j)$, and make use of a theorem [6] regarding sums of the form $\sum a_j b_j$. This theorem, which was first used in reactor dynamics by Ergen [7], states that the sum is greatest if the series $(a_1, a_2, \ldots, a_n)$ and $(b_1, b_2, \ldots, b_n)$ are similarly ordered, i.e., if the largest a is multiplied by the largest b the next largest a is multiplied by the next largest b, and so on. The sum is minimum if these series are oppositely ordered, i.e., the largest a is multiplied by the smallest b, the next largest a is multiplied by the next smallest b, and so on. An intuitive interpretation of this theorem is given by Ergen [7] by regarding the a's as distances along a rod to hooks, and the b's as weights suspended from the hooks. The maximum static moment with respect to one end of the rod is obtained by hanging the heaviest weights on the hooks farthest from that end. Any unordered sum of pairs of a's and b's lies between the foregoing extremum values.

Since $P(t)$ is assumed to be periodic, $P(t_j - u)$ runs through the same values as $P(t_j)$ in a different order. Therefore, the summation corresponding to the integration in (4) can be regarded as an ordered sum of a_j and b_j. The oppositely ordered sum is obtained when $u = 0$ because then $b_j = 1/P(t_j)$ is smallest when $a_j = P(t_j)$ is the largest. Thus, by virtue of the cited theorem, we have

$$(1/T) \int_0^T [P(t-u)/P(t)]\, dt \geqslant (1/T) \int_0^T [P(t)/P(t)]\, dt = 1 \tag{5}$$

which demonstrates the assertion.

We now return to (3), and observe that the integrand is positive, and thus $k_{av} < 0$, as a result of (5).

As a more specific example, let us calculate the reactivity input for a pure sinusoidal output [8], i.e., $P(t) = P_0 + p_1 \sin \omega t$. Substituting the latter into (1), we obtain

$$k(t) = \frac{p_1}{P_0} \frac{1}{|Z(i\omega)|} \frac{\sin(\omega t - \phi)}{1 + (p_1/P_0) \sin \omega t} \tag{6a}$$

$$\phi = \arg Z(i\omega) \tag{6b}$$

where

$$Z(i\omega) \equiv \left[i\omega \left(\frac{l}{\beta} + \sum_{j=1}^{6} \frac{a_j}{\lambda_j + i\omega} \right) \right]^{-1} \tag{7}$$

which is called the zero-power transfer function of the reactor, and will be discussed in more detail in Chapter 4. We observe from (6a) that the reactivity insertion that gives rise to a pure sinusoidal power variation is periodic, but not sinusoidal, for large power variations. The average of $k(t)$ is obtained from (6a) as (Problem 2)

$$k_{av} = -\operatorname{Re}[1/Z(i\omega)]\{[1 - (p_1^2/P_0^2)]^{-1/2} - 1\} \tag{8}$$

which can be approximated, when $p_1 \ll P_0$, by

$$k_{av} = -(p_1/P_0)^2 \operatorname{Re}[1/Z(i\omega)] \tag{9}$$

where

$$\operatorname{Re}[1/Z(i\omega)] \equiv \sum_{j=1}^{6} [\omega^2 a_j/(\lambda_j^2 + \omega^2)] \geqslant 0 \tag{10}$$

We note that the negative reactivity bias is proportional to the square of the relative amplitude of the sinusiodal power oscillations, when the oscillations are small. This result will be obtained again in Chapter 4 when we investigate approximate solutions of the point kinetic equations.

Reactivity after a Positive Power Excursion

As a second application, we consider a positive power excursion as indicated in Figure 3.2.1.

We want to show that the reactivity is negative at the time t_0 when

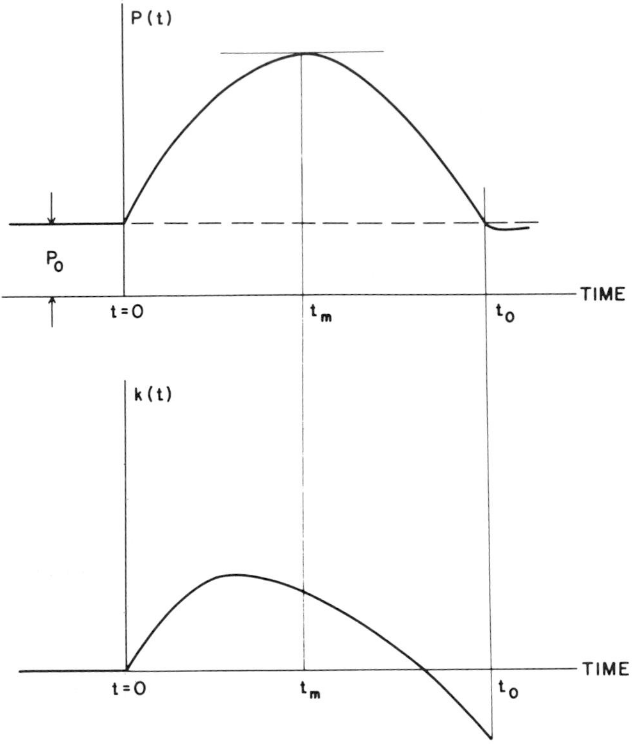

FIGURE 3.2.1. Reactivity variation during a power excursion.

the power returns to its initial value P_0. It is assumed that the reactor power was constant and equal to P_0 prior to $t = 0$. Hence, $k(t) = 0$ for $t < 0$. With these remarks, we obtain $k(t_0)$ from (1) as

$$k(t_0) = \frac{l}{\beta} \frac{1}{P_0} \frac{dP(t)}{dt} \bigg|_{t=t_0} - \int_0^{t_0} du\, D(u) \left[\frac{P(t_0 - u)}{P_0} - 1 \right] \tag{11}$$

The integral term is positive because $P(t) \geqslant P_0$ in the interval $(0, t_0)$ (positive excursion). The first term is nonpositive because the slope at $t = t_0$ is either zero or negative. Hence, $k(t_0) < 0$, i.e., a positive excursion that returns to the original power level tends to produce a negative reactivity. We can in fact compute the time dependence of this reactivity by assuming that the duration of the excursion is very short such that $D(u)$ does not vary appreciably in the interval $(u, u + t_0)$.

We also assume that $P(t) \equiv P_0$ for $t > t_0$. Then, for $t \geqslant t_0$, (11) reduces to

$$k(t) = -\int_0^t du \, \{D(u)[P(t-u) - P_0]/P_0\}$$

$$= -\int_0^{t_0} du \, \{D(t-u)[P(u) - P_0]/P_0\}$$

$$\approx -(I/P_0) \sum_{i=1}^{6} a_i \lambda_i e^{-\lambda_i(t-t_0)}, \qquad t \geqslant t_0 \tag{12a}$$

where

$$I = \int_0^{t_0} du \, [P(u) - P_0] \tag{12b}$$

Here, I is the excess energy release in the excursion. It follows from (12), which was obtained by Corben [4], that the negative reactivity produced by a short, positive power excursion, or fluctuation, tends to zero as t becomes large. The magnitude of the negative reactivity at $t = t_0$ can be approximately estimated from (12a) as

$$k(t_0) = -(I/P_0)\, \lambda^* \tag{13}$$

where λ^* is the mean decay constant, defined by

$$\lambda^* \equiv \sum_{i=1}^{6} a_i \lambda_i \tag{14}$$

For ^{235}U, Keepin and Wimmett's (see Keepin *et al.* [9]) data (see Section 1.1, Table I) yield $\lambda^* = 0.46$. Equation (13) indicates that the negative reactivity at the end of a short power pulse is equal to the ratio of the excess energy of the pulse to the energy (P_0/λ^*) produced at the steady state in an interval of $(1/\lambda^*) \simeq 2.2$ sec.

Another interesting observation is that $k(t_m) > 0$, where t_m is the time at which the power excursion attains its first maximum. Since $dP(t)/dt \,|_{t=t_m} = 0$, we obtain from (1)

$$k(t_m) = -\int_0^{\infty} du \, D(u)\{[P(t_m - u)/P(t_m)] - 1\} \tag{15}$$

Since $P(t_m) > P(t_m - u)$ for all u in $(0, \infty)$ by virtue of the fact that $P(t_m)$ is the first maximum, the integrand is always negative, and hence $k(t_m) > 0$. It is interesting to note that $k(t_m)$ would be zero had the delayed neutrons not been present, i.e., $D(u) \equiv 0$.

B. Reactivity for $P(t) = P_0 \exp(\alpha t^2)$

As a last example, we shall determine the reactivity insertion required for a power change given as

$$\begin{aligned} P(t) &\equiv P_0 & t &< 0 \\ &\equiv P_0 \exp \alpha t^2, & t &\geqslant 0 \quad (\alpha > 0) \end{aligned}$$

using the inverse method. Substituting $P(t)$ into Eq. (1), we obtain

$$\begin{aligned} k(t) = 1 + (l/\beta)\, 2\alpha t - (\exp -\alpha t^2) \sum_{i=1}^{6} a_i(\exp -\lambda_i t) \\ - \sum_{i=1}^{6} a_i \lambda_i \int_0^t du \exp[+\alpha u^2 - 2\alpha t u - \lambda_i u] \end{aligned} \tag{16}$$

The last integral can be simplified as follows:

$$\begin{aligned} I_i(t) &\equiv \int_0^t du \exp[\alpha u^2 - 2\alpha t u - \lambda_i u] \\ &= (\exp -\alpha z_i^2) \int_0^t \exp[\alpha(u - z_i)^2]\, du \\ &= (\exp -\alpha z_i^2) \int_{\lambda_i/2\alpha}^{z_i} du \exp(\alpha u^2) \end{aligned} \tag{17}$$

where $z_i \equiv t + (\lambda_i/2\alpha)$. Integration by parts yields the following expansion:

$$I_i(t) = \frac{1}{2\alpha z_i} - \frac{\exp \alpha(\lambda_i/2\alpha)^2}{\lambda_i} \exp(-\alpha z_i^2) + \cdots \tag{18}$$

which indicates that $I_i(t) \to 0$ as $(1/t)$. Keeping only the leading term in (18), and substituting it into (16), we obtain, for large t,

$$k(t) \simeq 1 + (l/\beta)\, 2\alpha t - \sum_{i=1}^{6} [a_i \lambda_i/(\lambda_i + 2\alpha t)] \tag{19}$$

where the neglected terms decay as $\exp(-\alpha t^2)$.

It is observed that $k(t)$ approaches a linear function $1 + (2\alpha l t/\beta)$ after the transient effects disappear, as shown in Figure 3.2.2.

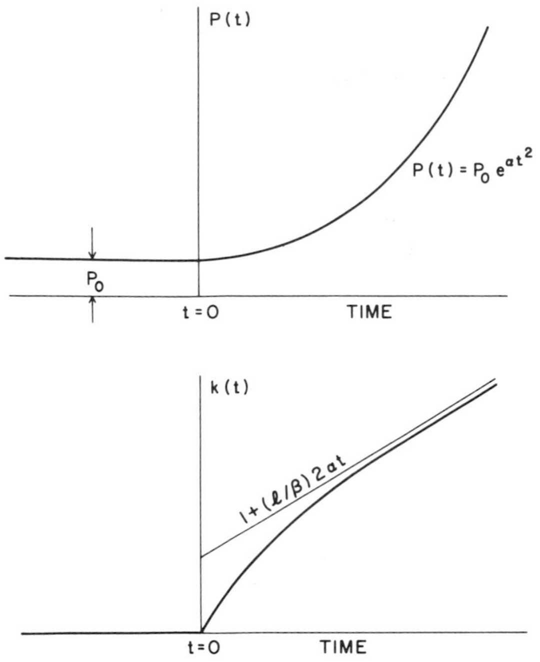

FIGURE 3.2.2. Reactivity insertion required for $P(t) = P_0 \exp(\alpha t^2)$.

This result indicates that the response of the reactor power to a ramp reactivity insertion $k(t) = \gamma t$ should behave as $\exp(\beta\gamma t^2/2l)$ for large times. We shall return to this point in the subsequent section (see Problems 3–5 for further applications of the inverse method).

3.3. Exact Solution of Reactor Kinetic Equations for a Known Reactivity Insertion

In this and subsequent sections, we shall be concerned with determining the time variation of the power $P(t)$ when the reactivity insertion $k(t)$ is a known function of time. The present section considers those reactivity insertions for which exact solutions are available [10]. These are defined, for $t > 0$, by

$$k(t) = k_0 \qquad \text{(step)}$$

$$k(t) = k_0 + \gamma t \qquad \text{(ramp)}$$

$$k(t) = k_0 - k_2 \exp(-\gamma t) \qquad \text{(exponential)}$$

$$k(t) = k_0 - (1/\gamma t) \qquad \text{(reciprocal)}$$

where γ is positive. In all cases, $k(t) = 0$ for $t < 0$. Because of their importance in reactor operation, we shall consider only the step and ramp reactivity insertions in detail.

A. Step Reactivity Insertion

The response of reactor power to a step reactivity insertion at $t = 0$ can be obtained by taking the Laplace transform of (3.1, Eqs. 1 and 2)

$$\bar{P}(s) = \{(l/\beta)\, P(0) + \sum_{i=1}^{6} [\lambda_i C_i(0)/(\lambda_i + s)] + (l/\beta)\, \bar{S}(s)\}/[Y(s) - k_0] \tag{1}$$

where $\bar{P}(s)$ and $\bar{S}(s)$ are the Laplace transforms of the reactor power $P(t)$ and the external source $S(t)$, respectively. The $P(0)$ and $C_i(0)$ are the initial values of $P(t)$ and $S(t)$, respectively. The function $Y(s)$ is defined by

$$Y(s) \equiv s\left\{(l/\beta) + \sum_{i=1}^{6} [a_i/(\lambda_i + s)]\right\} \tag{2}$$

and is the inverse of the zero-power transfer functions $Z(s)$ introduced by (3.2, Eq. 7). In most of the kinetic experiments involving a step reactivity insertion, the reactor is operated, prior to the insertion of the reactivity, at a constant power level P_0, either in a critical state, or in a subcritical state sustained by an external source. Hence,

$$P(t) \equiv P_0, \qquad t < 0 \tag{3a}$$

$$C_i(t) \equiv a_i P_0/\lambda_i, \qquad t < 0 \tag{3b}$$

Furthermore, the initial values $P(0)$ and $C_i(0)$ are equal to their steady-state values given by (3). Substituting $P(0) = P_0$ and $C_i(0)\, \lambda_i = a_i P_0$, we simplify (1):

$$\bar{P}(s) = \{1/[Y(s) - k_0]\}\{[Y(s)/s]\, P_0 + (l/\beta)\, \bar{S}(s)\} \tag{4}$$

The inverse transform of $\bar{P}(s)$ can be evaluated easily by finding the zeros of the denominator, namely

$$k_0 = s\left\{(l/\beta) + \sum_{i=1}^{6} [a_i/(\lambda_i + s)]\right\} \tag{5}$$

which is known as the inhour equation, as mentioned previously (see, 2.4, Eq. 30). Let the roots of the inhour equation be denoted by ω_j.

Then, $P(t)$ can be written as

$$P(t) = P_0 \sum_{j=0}^{6} \frac{k_0 e^{\omega_j t}}{\omega_j Y'(\omega_j)} + \frac{l}{\beta} \sum_{j=0}^{6} \frac{\bar{S}(\omega_j)\, e^{\omega_j t}}{Y'(\omega_j)}$$

$$+ \frac{l}{\beta} \sum_{i=0}^{6} \frac{e^{\mu_i t}}{Y(\mu_i) - k_0} \operatorname{Res}[\bar{S}(s)]_{s=\mu_i} \tag{6}$$

where $Y'(s) = dY(s)/ds$, and the μ_i are poles of $\bar{S}(s)$, which are assumed to be simple. In obtaining (6), we have used the fact that the only singularities of $\bar{P}(s)$ are the roots of the inhour equation, and the poles of $\bar{S}(s)$. We point out that there is no singularity at the origin because $sZ(s)$ is finite at $s = 0$.

In evaluating the residues of $[Y(s) - k_0]^{-1}$ at $s = \omega_j$, we have used the fact that $Y(s)$ is a ratio of two polynomials, and therefore the residue can be calculated by evaluating the derivative of the denominator with respect to s at the simple poles $s = \omega_j$, i.e.,

$$\operatorname{Res}[Y(s) - k_0]^{-1}_{s=\omega_j} = [1/Y'(\omega_j)]^{-1} \tag{7}$$

We shall first consider a step reactivity insertion into a critical reactor in the absence of external sources.

Step Reactivity Insertion into a Critical Reactor

The time response $P(t)$ in this case is obtained from (6) by setting $\bar{S}(s) = 0$:

$$P(t) = k_0 P_0 \sum_{j=0}^{6} \frac{1}{\omega_j} \left[\frac{l}{\beta} + \sum_{i=1}^{6} \frac{a_i \lambda_i}{(\lambda_i + \omega_j)^2} \right]^{-1} e^{\omega_j t} \tag{8}$$

where we have written $Y'(\omega_j)$ explicitly. It is clear that the behavior of the reactor for $t > 0$ depends on the roots ω_j of the inhour equation. They satisfy certain interesting sum rules, obtained by Henry [11], which we obtain here in an easier way from (8). The first sum rule follows immediately by evaluating (8) at $t = 0$:

$$\sum_{j=0}^{6} \frac{1}{\omega_j} \left[\frac{l}{\beta} + \sum_{i=1}^{6} \frac{\lambda_i a_i}{(\omega_j + \lambda_i)^2} \right]^{-1} = \frac{1}{Y(\omega_j)} \tag{9}$$

To obtain the second sum rule, we consider the first of the kinetic equations

$$(l/\beta)\,\dot{P}(t) = (k_0 - 1)\,P(t) + \sum_{i=1}^{6} \lambda_i C_i(t), \qquad t > 0$$

and evaluate the time derivative of $P(t)$ at $t = 0^+$:

$$\dot{P}(t)\,|_{t=0^+} = (\beta/l)\,k_0 P_0 \tag{10}$$

Comparing this to the derivative of (8), we establish the following relation:

$$\beta/l = \sum_{j=0}^{6} [1/Y'(\omega_j)]$$

or

$$\sum_{j=0}^{6} \left[1 + \sum_{i=1}^{6} \frac{\beta_i}{l} \frac{\lambda_i}{(\lambda_i + \omega_j)^2}\right]^{-1} = 1 \tag{11}$$

A third relation is obtained by considering $C_i(t)$, which can easily be shown to be

$$C_i(t) = P_0 \sum_{j=0}^{6} \frac{a_i}{\lambda_i + \omega_j} \frac{k_0}{\omega_j Y'(\omega_j)}\, e^{\omega_j t} \tag{12}$$

and, using the fact that $\dot{C}_i(0) = 0$,

$$\sum_{j=0}^{6} \frac{a_i}{\lambda_i + \omega_j} \frac{k_0}{Y'(\omega_j)} = 0$$

or, more explicitly,

$$\sum_{j=0}^{6} \frac{1}{\lambda_i + \omega_j} \left[1 + \sum_{r=1}^{6} \frac{\beta_r}{l} \frac{\lambda_r}{(\lambda_r + \omega_j)^2}\right]^{-1} = 0, \qquad i = 1, 2, \ldots, 6 \tag{13}$$

We can obtain a final sum rule from (13) by writing it for two different values of i and subtracting the resulting equations for $l \neq l'$:

$$\sum_{j=0}^{6} \frac{1}{(\lambda_l + \omega_j)(\lambda_{l'} + \omega_j)} \left[1 + \sum_{r=1}^{6} \frac{\beta_r}{l} \frac{\lambda_r}{(\lambda_r + \omega_j)^2}\right]^{-1} = 0 \tag{14}$$

We shall now briefly discuss the roots of the inhour equation with the help of Figure 3.3.1, which displays $Y(s)$ as a function of real arguments.

The following conclusions about the properties of the roots of the inhour equation ω_j can be drawn from Figure 3.3.1.

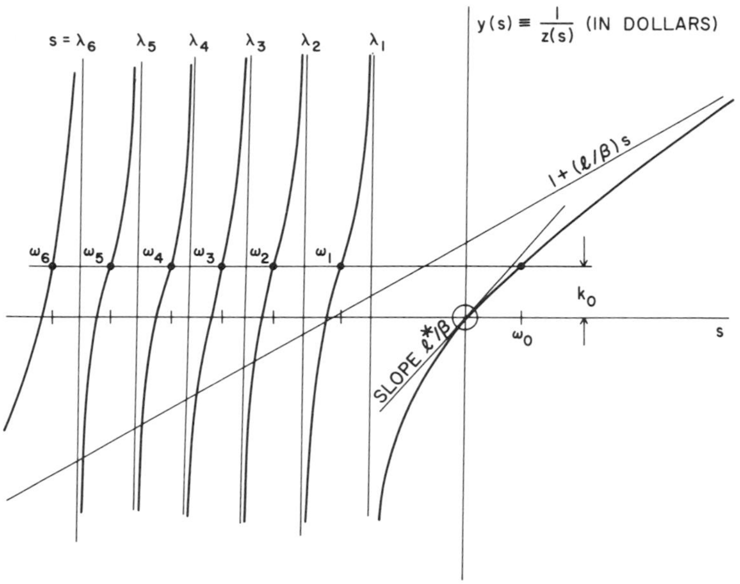

FIGURE 3.3.1. Location of the roots of the inhour equation.

1. There are $m + 1$ roots, where m is the number of delayed neutron groups, which is usually taken to be six. All the roots are real, and m of them are always negative and less than $-\lambda_1$ regardless of the sign and magnitude of k_0. The remaining root, denoted by ω_0, is the algebraically largest of all the roots, and has the same sign as that of k_0.

2. The asymptotic behavior of the reactor power $P(t)$ after a long time following the step change at $t = 0$ is governed by ω_0:

$$P(t) = \frac{k_0 P_0}{\omega_0}\left[\frac{l}{\beta} + \sum_{i=1}^{6}\frac{a_i\lambda_i}{(\lambda_i + \omega_0)^2}\right]^{-1} e^{\omega_0 t} \qquad (\text{for } t \text{ large}) \tag{15}$$

REACTOR PERIOD

The inverse of the logarithmic derivative of the instantaneous power is called the reactor period, and is denoted by T, i.e.,

$$1/T = d[\ln P(t)]/dt \tag{16}$$

It is time-dependent, and can be interpreted as the instantaneous e-folding time at time t. Equation (15) indicates that the reactor period asymptotically approaches a constant value $T_s = 1/\omega_0$ which is called the stable, or asymptotic, reactor period, after a sufficiently long time has elapsed. The term "reactor period" usually refers to the stable period. Hence, the algebraically largest root ω_0 of the inhour equation may be identified as the inverse reactor period.

3. The value of ω_0 for a given k_0 can be approximately evaluated in the limiting case by expanding the summation term in the inhour equation (5) in powers of ω_0 and $1/\omega_0$:

$$k_0 = \omega_0 \left[\frac{l}{\beta} + \sum_{i=1}^{6} \frac{a_i}{\lambda_i}\left(1 - \frac{\omega_0}{\lambda_i} + \frac{\omega_0^2}{\lambda_i^2} - \cdots\right)\right] \tag{17a}$$

and

$$k_0 = \omega_0 \left[\frac{l}{\beta} + \frac{1}{\omega_0} \sum_{i=1}^{6} a_i \left(1 - \frac{\lambda_i}{\omega_0} + \frac{\lambda_i^2}{\omega_0^2} - \cdots\right)\right] \tag{17b}$$

For small values of k_0, i.e., $| k_0 | \ll 1$, we can approximate the series in (17a) by retaining only the first term, and obtain

$$\omega_0 \approx k_0 \beta / l^* \tag{18a}$$

where

$$l^* \equiv l + \beta\bar{\tau} \tag{18b}$$

$$\bar{\tau} \equiv \sum_{i=1}^{6} (a_i/\lambda_i) \tag{18c}$$

Here, l^* is defined as the effective generation time, which includes the mean delay time $\bar{\tau}$ of all delayed neutrons. For ^{235}U, $\bar{\tau} = 12.8$ sec. The mean prompt generation time l is of the order of 10^{-4} sec. in thermal reactors, and 10^{-7} sec in fast reactors. The effective generation time is therefore essentially equal to $\beta\bar{\tau}$. Thus, the inverse period for small reactivity steps is predominantly determined by the mean delay time $\bar{\tau}$, which shows the importance of the delayed neutrons in the dynamic behavior of the reactor power.

For large, positive values of k_0, we use the expansion in (17b), and approximate it by keeping the first two terms. The resulting quadratic equation in ω_0 yields

$$\omega_0 = (\beta/2l)(k_0 - 1) + \{[(\beta/2l)(k_0 - 1)]^2 + (\lambda^* \beta/l)\}^{1/2} \tag{19}$$

where

$$\lambda^* \equiv \sum_{i=1}^{6} a_i \lambda_i \tag{19a}$$

and is the mean decay constant for all delayed neutrons, as introduced before in (3.2, Eq. 14). We note that $\lambda^* \neq 1/\bar{\tau}$. For ^{235}U, $\lambda^* = 0.46\ \text{sec}^{-1}$, whereas $1/\bar{\tau} = 0.08\ \text{sec}^{-1}$. When reactivity step k_0 is 1 \$, the inverse period is obtained from (19),

$$\omega_0 \approx (\beta\lambda^*/l)^{1/2} \tag{20}$$

When the net reactivity in a reactor is equal to 1 \$, the reactor is said to be prompt-critical. The importance of the concept of prompt-criticality in the dynamic behavior of a reactor will be discussed later. When the reactor is sufficiently above prompt-critical so that

$$(k_0 - 1)^2 \gg 4\lambda^* l/\beta \tag{21}$$

then (19) yields

$$\omega_0 \approx (k_0 - 1)\, \beta/l \tag{22}$$

which shows that the inverse period above prompt-critical is predominantly determined by the mean prompt generation time l (see Figure 3.3.1).

For large, negative values of k_0, neither of the expansions (17) is valid because ω_0 approaches $-\lambda_1$ and $1/(\lambda_1 + \omega_0)$ diverges. It is interesting to note that the inverse period, or the stable reactor period, becomes independent of the magnitude of k_0 in this limiting case, i.e., $T_s \approx -\lambda_1 \approx -80$ sec, which is the decay time of the most stable delayed neutron precursor.

4. The calculation of the roots ω_j of the inhour equation is facilitated by plotting the function

$$f(s) \equiv s \sum_{i=1}^{6} [a_i/(\lambda_i + s)]$$

which depends only on the dealyed neutron parameters, as a function of time. The intersections of this curve with the straight line $k_0 - s(l/\beta)$ gives the roots.

The coefficients of each term in the expression of $P(t)/P_0$ given by (8) may then be obtained by adding (l/β) to values of the following functions at $s = \omega_j$:

$$A(s) \equiv \sum_{i=1}^{6} [a_i\lambda_i/(\lambda_i + s)^2]$$

This function also contains only the delayed neutron parameters, and is insensitive to the type of reactor for a specific fuel (cf 3.1, Eq. 5).

Reactivity Determination by Means of Asymptotic Period

The foregoing analysis shows that the magnitude of the step reactivity can be determined by measuring the asymptotic reactor period T_s and then relating it to reactivity through the inhour equation

$$k_0 = \frac{1}{T_s}\left[\frac{l}{\beta} + \sum_{i=1}^{6} \frac{a_i T_s}{1 + \lambda_i T_s}\right] \tag{23}$$

In order to ensure that the reactor period attains its asymptotic value T_s, a sufficiently long time must elapse after the insertion of the step reactivity. A premature observation of the change in the reactor power may lead to considerable error in the period owing to the presence of transients [12]. The relative error $\epsilon = [T_s - T(t_w)]/T_s$ after a waiting time t_w can be estimated from (8) and (16) as

$$\epsilon = \left\{\sum_{i=1}^{6} \frac{\omega_i - \omega_0}{\omega_i} \frac{Y'(\omega_0)}{Y'(\omega_i)} e^{(\omega_i - \omega_0)t_w}\right\} \Big/ \left\{1 + \sum_{i=1}^{6} \frac{Y'(\omega_0)}{Y'(\omega_i)} e^{(\omega_i - \omega_0)t_w}\right\} \tag{24}$$

where $Y(\omega_i) = k_0$, $i = 1, 2, \ldots, 6$ (inhour equation). Considering only the slowest transient, corresponding to $i = 1$, ignoring the transient in the denominator compared to unity, and, finally, evaluating $Y'(\omega_0)/Y'(\omega_1)$, explicitly we obtain

$$\epsilon \approx [(\omega_1 - \omega_0)/\omega_1][1 + (\beta/l)\, A(\omega_0)][1 + (\beta/l)\, A(\omega_1)]^{-1} e^{(\omega_1 - \omega_0)t_w} \tag{25a}$$

where

$$A(\omega) \equiv \sum_{i=1}^{6} [a_i \lambda_i/(\lambda_i + \omega)^2] \tag{25b}$$

The waiting times to observe a period within an error of 1%, 5%, and 10% are computed by Toppel [12] for positive and negative reactivity steps [13].

Step Reactivity Insertion into a Subcritical Reactor

We shall now consider a step reactivity insertion into a subcritical reactor in which a stationary neutron population is maintained by a constant external source S_0. This analysis is motivated by the fact that a reactor is never free from an external neutron source due to the photoneutrons produced by the gamma field, and therefore is never truly critical. It is perhaps more realistic to retain the external source in the analysis, and to investigate the behavior of the reactor power following a step reactivity insertion in the presence of a constant external source. Toppel [12] showed that a significant increase in the waiting time for a specified accuracy can result even if the reactor is slightly subcritical before the step insertion of reactivity.

The stationary power level P_0 prior to $t = 0$, at which a step k_1 is inserted, is related to the external source S_0 by

$$P_0 = S_0 l / |k_s| \beta \tag{26}$$

where k_s is the negative reactivity measuring the degree of subcriticality. The net reactivity k_0 after the step change becomes $k_0 = k_1 + k_s$.

The response of the reactor power for $t > 0$ follows from (6) by substituting $\bar{S}(s) = (S_0/s)$ and observing that $\mu_i = 0$ and $\mathrm{Res}[\bar{S}(s)]_{s=0} = S_0$:

$$\frac{P(t)}{P_0} = k_1 \sum_{j=0}^{6} \frac{e^{\omega_j t}}{\omega_j Y'(\omega_j)} - \frac{|k_s|}{k_0} \tag{27}$$

In obtaining (27), we have used $Y(0) = 0$,

$$\lim_{s \to 0} [Y(s)/s] = l^*/\beta \tag{28}$$

and eliminated S_0 in favor of $|k_s|$.

The relative error ϵ in a period measurement at t_w can be found as

$$\epsilon = \frac{\omega_1 - \omega_0}{\omega_1} \frac{Y'(\omega_0)}{Y'(\omega_1)} e^{(\omega_1 - \omega_0) t_w} + \frac{|k_s| \omega_0}{k_0 k_1} Y'(\omega_0) e^{-\omega_0 t_w} \tag{29}$$

We assume that the step reactivity change k_1 is positive and sufficiently large to make the net reactivity k_0 positive. Then, $\omega_0 > 0$ and, after still longer times, the error reduces to the last term in (29). The waiting time for an error of 5% as a function of the positive asymptotic period has also been calculated [12, 13].

One Group of Delayed Neutrons

A more quantitative insight into the behavior of $P(t)$ in response to a step reactivity change can be obtained by using the so-called one-group delayed-neutron model, characterized by the group parameters $\beta^{(1)}$ and $\lambda^{(1)}$. These parameters can be determined by comparing the zero-power transfer function $Z(s)$, or its inverse $Y(s)$, in both six- and one-groups models in the asymptotic regions for large and small arguments. This procedure leads to $\lambda^{(1)} = \bar{\lambda}$ and $\beta^{(1)} = \beta$, where

$$1/\bar{\lambda} = \sum_{i=1}^{6} (a_i/\lambda_i) \tag{30}$$

The inverse reactor transfer function in the one-delayed-neutron-group model then becomes

$$Y(s) = s\{(l/\beta) + [1/(\bar{\lambda} + s)]\} \tag{31}$$

This particular choice of group parameters, however, does not represent the actual behavior of the reactor at prompt-critical. The inverse period at prompt-critical was obtained before in (20) as $\omega_0 \approx (\beta\lambda^*/l)^{1/2}$, which is replaced in the one-group model by $\omega_0 \approx (\beta\bar{\lambda}/l)^{1/2}$. It may be concluded that in the problems involving reactivities near critical a choice of $\beta^{(1)}$ and $\lambda^{(1)}$ as $\beta^{(1)} = \beta$ and $\lambda^{(1)} = \lambda^* \equiv \sum_i a_i\lambda_i$ (cf Eq. 19) may represent reactor behavior more accurately than the usual choice described above. We shall have occasion to come back to this point in the discussions of fast ramp reactivity insertion. For a discussion of the reduced delayed-neutron group representation and for other choices of $\beta^{(1)}$ and $\lambda^{(1)}$, the reader is referred to Skinner and Cohen [14] and Blokhintsev and Nikolaev [15].

The response of a critical reactor to a step change for the model of one group of delayed neutrons can be obtained from Eq. (4) as (Problem 7)

$$\frac{P(t)}{P_0} = \frac{1}{1 - k_0}\left\{\exp\left[\frac{\bar{\lambda}k_0}{1 - k_0}t\right] - k_0 \exp\left[-\frac{\beta}{l}(1 - k_0)t\right]\right\} \tag{32}$$

where we have ignored $\bar{\lambda}l$ as compared to $\beta(1 - k_0)$ on the basis that $\bar{\lambda} = 0.08\ \text{sec}^{-1}$, l is certainly less than 10^{-3} sec, and thus $\bar{\lambda}l < 8 \times 10^{-5}$, where as $\beta \mid 1 - k_0 \mid \approx 7.5 \times 10^{-3}$ for small reactivity insertion.

We observe from (32) that the first term represents the asymptotic behavior of the reactor power when $k_0 < 1$ \$, the second term being the transient solution, which vanishes exponentially. Hence, the asymptotic response is essentially governed by the delayed neutrons. The situation is reversed however, when $k_0 > 1$ \$, because then $1 - k_0 < 0$, and it is the second term which determines the asymptotic behavior of the reactor power, and the stable period is mainly determined by the mean prompt generation time l. It is clear that the response of the reactor drastically changes when k_0 exceeds 1 \$. When $k_0 > 1$, the reactor is said to be prompt-critical, implying that the chain reaction would be sustained by the prompt neutrons only. When $k_0 > 0$ but less than 1 \$, then the reactor is said to be delayed-critical. In this range, the chain reaction is controlled by the delayed neutrons.

B. Ramp Reactivity Insertion

The reactivity insertion into a subcritical reactor during startup can be characterized by a linear reactivity variation as

$$k(t) = k_0 + \gamma t \tag{33}$$

where k_0 is the initial reactivity in the shutdown state, and γ is the rate of insertion in dollars per second. We find it more convenient to use the standard form of the kinetic equations given by (3.1, Eqs. 1 and 2). Substituting (33) into these equations, we obtain

$$(l/\beta)\,\dot{P}(t) = [k_0 + \gamma t - 1]\,P(t) + \sum_{i=1}^{6} \lambda_i C_i(t) + (l/\beta)\,S_0 \tag{34a}$$

$$\dot{C}_i(t) = a_i P(t) - \lambda_i C_i(t), \qquad i = 1, 2, \ldots, 6 \tag{34b}$$

where S_0 is a constant source of neutrons.

Our task is to solve this set of seven linear differential equations exactly for $P(t)$. This problem was considered by Wallach [16], Garabedian *et al.* [17], Wilkins [18], and Smets [10]. The mathematical basis of the method used in the solution of this equation is described by Ince [19], and essentially consists in representing the solution by definite integrals. We shall closely follow Wilkins' analysis.

It was shown in our discussion of the inverse method in Section 3.2 that the reactor power is expected to behave as $\exp[\beta\alpha t^2/2l]$ when the reactivity insertion is a linear function of time. This indicates that the solution of (34) will not be of an exponential order, i.e., there is not a finite positive number m such that $\lim_{t\to\infty} P(t)\,e^{-mt} = 0$. Hence, we cannot use the conventional Laplace-transform technique to solve (34). However, the solution can still be expressed in a definite integral form as

$$P(t) = \int_{x_1}^{x_2} ds\,\bar{P}(s)\,e^{st} \tag{35a}$$

and

$$C_i(t) = \int_{x_1}^{x_2} ds\,\bar{C}_i(s)\,e^{st} \tag{35b}$$

where the limits x_1 and x_2 are real, and yet to be specified. In the case of the Laplace transform, the variable s is a complex number, and the path of integration is a vertical line parallel to the imaginary axis in the complex s-plane (Bromwich contour). In general, the path of integration in (35) may be a curve in the complex s–plane to be chosen depending on the nature of the problem under consideration. In the present problem, this path turns out to be portions of the real axis.

We assume that we can differentiate (35) under the integral sign, and observe

$$dP(t)/dt = \int_{x_1}^{x_2} ds\,[s\bar{P}(s)\,e^{st}] \tag{36a}$$

$$tP(t) = \bar{P}(s)\,e^{st}\Big|_{s=x_1}^{s=x_2} - \int_{x_1}^{x_2} ds\,[d\bar{P}(s)/ds]\,e^{st} \tag{36b}$$

substituting (35) into (34), using (36), and eliminating $\bar{C}_i(s)$ by choosing them as $\bar{C}_i(s) = a_i\bar{P}(s)/(\lambda_i + s)$, we obtain

$$\int_{x_1}^{x_2} ds\,e^{st}\left\{[Y(s) - k_0]\,\bar{P}(s) + \gamma\,\frac{d\bar{P}(s)}{ds}\right\} = \frac{l}{\beta}\,S_0 + \gamma\bar{P}(s)\,e^{st}\Big|_{s=x_1}^{s=x_2} \tag{37}$$

where $Y(s)$ is the inverse of the zero-power transfer function defined by (2).

The crucial point in this approach is that one can choose x_1 and x_2 so that

$$S_0(l/\beta) + \gamma\bar{P}(s)\,e^{st}\,|_{s=x_1}^{s=x_2} \equiv 0 \tag{38}$$

will hold for all t. We shall prove later that such a choice is indeed possible. With this condition, Eq. (37) is satisfied if we choose $\bar{P}(s)$ as the solution of

$$\gamma[d\bar{P}(s)/ds] + [Y(s) - k_0]\,\bar{P}(s) = 0 \tag{39}$$

i.e.,

$$\bar{P}(s) = B\prod_{i=1}^{6}\left(1 + \frac{s}{\lambda_i}\right)^{a_i\lambda_i/\gamma} \exp\left[-\frac{s^2}{2}\frac{l}{\beta\gamma} + \frac{k_0 - 1}{\gamma}\,s\right] \tag{40}$$

where B is the constant of integration (see Problem 8). We shall now determine x_1 and x_2 from the assumed condition (38), using the solution (40). We note that Eq. (34) is a linear equation with variable coefficients, and hence its general solution can be constructed as the superposition of the linearly independent solutions of the homogeneous equation, i.e., $S_0 = 0$, plus a particular solution of the inhomogeneous equation. The solutions of the homogeneous equation can be found by choosing x_1 and x_2 such that $\bar{P}(s)$ in (40) satisfies

$$\bar{P}(s)\, e^{st} \Big|_{s=x_1}^{s=x_2} = 0$$

To compress writing, we introduce

$$F(s, t) = \prod_{i=1}^{6} \left(1 + \frac{s}{\lambda_i}\right)^{a_i\lambda_i/\gamma} \exp\left[-\frac{s^2}{2\gamma}\frac{l}{\beta} + \left(\frac{k_0 + \gamma t - 1}{\gamma}\right) s\right] \tag{41}$$

The condition to be satisfied can now be written as $F(x_2, t) - F(x_1, t) = 0$. A moment's reflection will show that this identity is indeed satisfied using the following seven pairs: $(x_j, x_{j+1}) = (-\infty, -\lambda_6), (-\lambda_6, -\lambda_5), \ldots, (-\lambda_1, \infty)$. Each pair corresponds to a linearly independent solution of the homogeneous equation of the following form:

$$\int_{x_j}^{x_{j+1}} F(s, t)\, ds$$

Our next task is to find a particular solution of the inhomogeneous equation, which must satisfy $S_0 l/\beta + B[F(x_2, t) - F(x_1, t)] = 0$. Noting that $F(0, t) = 1$ and $F(\infty, t) = 0$, we find that this equation is satisfied if we choose the constant of integration as $B = S_0 l/\beta$. The desired general solution of the point kinetic equations for a ramp reactivity insertion can now be written as

$$P(t) = S_0 \frac{l}{\beta} \int_0^{\infty} ds\, F(s, t) + \sum_{j=1}^{7} B_j \int_{x_j}^{x_{j+1}} F(s, t)\, ds \tag{42}$$

where the B_j are the constants of integration to be determined from the seven intial conditions, $P(0)$ and $C_i(0)$, $i = 1, 2, \ldots, 6$, and where $x_1 = -\infty$, $x_2 = -\lambda_6$, $x_3 = -\lambda_5, \ldots, x_7 = -\lambda_1$, and $x_8 = \infty$. It is observed that the range of integration in the first six terms in the summation involves only negative values of s. Since $F(s, t) \to 0$ as $t \to \infty$ for $s < 0$ (cf Eq. 41), we conclude that these terms decrease rapidly for $t > 0$, and are important only at the very beginning of the transients. The asymptotic behavior of the reactor power is governed by the remaining terms in (42). In the case of a ramp reactivity insertion into a critical reactor, $S_0 = 0$ and the asymptotic behavior is simply given by

$$P(t) = B_7 \int_{-\lambda_1}^{\infty} F(s, t)\, ds \tag{43}$$

An asymptotic expansion for this integral has been worked out by Smets [10]. The result is complicated and will not be reproduced here. However, we shall discuss some of the properties of the exact solution given by (42) in the case of reactivity insertion into a critical reactor, i.e., $S_0 = 0$, $k_0 = 0$, using the model of one group of delayed neutrons, and obtain some practical conclusions.

In this special case, (42) reduces to

$$P(t) = B_1 \int_{-\infty}^{\lambda^*} ds\, F(s, t) + B_2 \int_{-\lambda^*}^{\infty} ds\, F(s, t) \tag{44}$$

where λ^* is the decay constant to be associated with the single delayed neutron group. We shall see that the proper value for λ^* in this problem is different than that given by (30). To proceed further, we introduce

$$\mu \equiv \lambda^*/l, \tag{45a}$$

$$y \equiv (l/\beta\gamma)^{1/2} (\lambda^* + s) \tag{45b}$$

$$z(t) \equiv (\beta/l\gamma)^{1/2} [\gamma t - 1 + (\lambda^* l/\beta)] \tag{45c}$$

and write (44) in a more explicit form:

$$P(t) = (\exp -\lambda^* t) \int_0^{\infty} dy\, (y^{\mu}[\exp(-y^2/2)]\{B_2 \exp[-yz(t)] + B_1 \exp[yz(t)]\}) \tag{46}$$

where B_1 and B_2 are redefined to absorb constants. Note that time dependence is implicit in $z(t)$. The constants B_2 and B_1 are to be determined by the initial conditions $P(0) = P_0$ and $\dot{P}(0) = 0$. We want to estimate the power level and the inverse reactor period at the instant the reactor is approximately prompt-critical, i.e., $\gamma t_p = 1$. It is expedient to choose t_p as the time at which $z(t_p) = 0$, i.e.,

$$t_p = (1/\gamma) - (\lambda^* l/\gamma\beta) \approx 1/\gamma \tag{47}$$

At $t = t_p$, the reactor power is obtained from (46) as

$$P(t_p) = (B_1 + B_2)\, 2^{(\mu-1)/2} [\exp(-\lambda^* t_p)]\, \Gamma\left(\frac{\mu + 1}{2}\right) \tag{48}$$

where $\Gamma(z)$ is the gamma function, defined by [20]

$$\Gamma\left(\frac{\mu+1}{2}\right) \equiv 2\int_0^\infty (\exp - y^2) y^\mu \, dy \tag{49}$$

The reciprocal reactor period $\omega(t_p) = \{d[\log P(t)]/dt\}|_{t=t_p}$ can be obtained, after some manipulation, as

$$\omega(t_p) = -\lambda^* + \left(\frac{2\beta\gamma}{l}\right)^{1/2} \left[\Gamma\left(\frac{\mu+2}{2}\right) \Big/ \Gamma\left(\frac{\mu+1}{2}\right)\right] \frac{B_1 - B_2}{B_1 + B_2} \tag{50}$$

The remaining task is to determine B_1 and B_2 from the initial conditions $P(0) = P_0$ and $\dot{P}(0) = 0$:

$$P_0 = \int_0^\infty dy \, (y^\mu (\exp - \tfrac{1}{2}y^2)\{B_2 \exp[-yz(0)] + B_1 \exp[yz(0)]\}) \tag{51a}$$

$$P_0 = (1/\lambda^*)(\beta\gamma/l)^{1/2}$$
$$\times \int_0^\infty dy \, (y^{\mu+1} (\exp - \tfrac{1}{2}y^2)\{-B_2 \exp[-yz(0)] + B_1 \exp[yz(0)]\}) \tag{51b}$$

where

$$z(0) = (\beta/l\gamma)^{1/2}[-1 + (\lambda^* l/\beta)] \approx -(\beta/l\gamma)^{1/2}$$

The exact calculation of B_1 and B_2 requires numerical integration. However, we can proceed analytically, following Wilkins, if we make use of the fact that $|z(0)| \to \infty$ as $l \to 0$. Indeed, $z(0)$ is a large number in practical applications. For example, let $\beta = 7.5 \times 10^{-3}$, $l \approx 10^{-5}$ sec, and $\gamma = 1$ \$/sec; then $|z(0)| \approx 25$. In a fast reactor for which $l \approx 10^{-7}$ sec, $|z(0)| \approx 250$. This number would be still larger if the reactivity insertion rate γ is less than 1 \$/sec. Hence, $z(0)$ is a large, negative number. This fact can be used in Eq. (51a) and (51b) with the following expansion:

$$\int_0^\infty dy \left\{ y^\mu \exp\left[-\frac{y^2}{2} - |z(0)|y\right] \right\}$$
$$= \frac{\Gamma(\mu+1)}{|z(0)|^{\mu+1}} - \frac{1}{2}\frac{\Gamma(\mu+3)}{|z(0)|^{\mu+3}} + \frac{1}{8}\frac{\Gamma(\mu+5)}{|z(0)|^{\mu+5}} + \cdots \tag{52}$$

which can be shown by integration by parts. Hence, the integral multiplying B_1 can be approximated by the first term in this expansion. The integrals multiplying B_2 in (51), however, diverge when $|z(0)| \to \infty$. Hence, in the limit $l \to 0$ for a fixed γ, (51) has exactly the following solution:

$$B_2 = 0$$

$$B_1 = P_0[1/\Gamma(\mu+1)](\beta/l\gamma)^{(\mu+1)/2} \tag{53}$$

Using these values in (48) and (50), we obtain the power level and the reciprocal period at near-prompt-criticality as

$$\frac{P(t_{\mathrm{p}})}{P_0} = \frac{1}{2}\left(\exp - \frac{\lambda^*}{\gamma}\right)\left(\frac{2\beta}{l\gamma}\right)^{(\mu+1)/2} \left[\Gamma\left(\frac{\mu+1}{2}\right)\Big/\Gamma(\mu+1)\right] \tag{54a}$$

and

$$\omega(t_{\mathrm{p}}) = -\lambda^* + \left(\frac{2\beta\gamma}{l}\right)^{1/2} \left[\Gamma\left(\frac{\mu+2}{2}\right)\Big/\Gamma\left(\frac{\mu+1}{2}\right)\right] \tag{54b}$$

where we recall $\mu = \lambda^*/\gamma$. These equations are valid when $(\beta/l\gamma)^{1/2} \gg 1$. We can obtain simpler results in two limiting cases:

1. Fast ramp insertion, $\mu = (\lambda^*/\gamma) \ll 1$:

$$P(t_{\mathrm{p}})/P_0 \approx (\beta\pi/2l\gamma)^{1/2} \tag{55a}$$

$$(\gamma l/\beta \ll 1, \quad \gamma \gg \lambda^*)$$

$$\omega(t_{\mathrm{p}}) \approx (2\beta\gamma/l\pi)^{1/2} \tag{55b}$$

2. Slow ramp insertion, $\mu = \lambda^*\gamma \gg 1$. Using the asymptotic value of the gamma functions for large arguments [20], i.e.,

$$\Gamma(\mu+1) \to e^{-\mu}\mu^{\mu}(2\pi\mu)^{1/2}, \qquad \mu \to \infty$$

we obtain

$$P(t_{\mathrm{p}})/P_0 \approx [\exp(-\lambda^*/2\gamma)](l\lambda^*)^{-\lambda^*/2\gamma} \tag{56a}$$

$$(\gamma l/\beta \ll 1, \quad \gamma \ll \lambda^*)$$

$$\omega(t_{\mathrm{p}}) \approx (\lambda^*\beta/l)^{1/2} \tag{56b}$$

We shall now discuss the proper value to be assigned to λ^*. Wilkins calculated the reciprocal period with six delayed neutron groups and found that if reduces to the form (56b) provided λ^* is evaluated as

$$\lambda^* = \sum_{i=1}^{6} a_i\lambda_i \tag{57}$$

in contrast with the mean decay constant $\bar{\lambda} = [\sum_{i=1}^{6}(a_i/\lambda_i)]^{-1}$ assigned to the single delayed neutron group in (30) on the basis of an asymptotic reactor period following a small step reactivity change [see the comments following (31)].

We reproduce in Figure 3.3.2 the variation of $\omega(t_p)$ as a function of the rate of reactivity insertion γ based on computed values with β_i and λ_i for ^{235}U, from Keepin *et al.* [9].

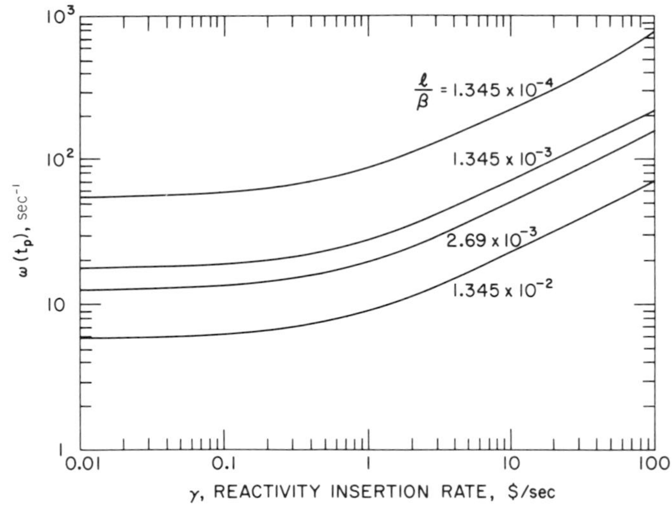

FIGURE 3.3.2. Variation of the reciprocal period at prompt-critical with rate of reactivity insertion.

PROBLEMS

1. Verify (3.2, Eq. 6).
   Hint: First, compute $P(t)\, k(t)$, noting that

$$\int_0^\infty D(u) \sin \omega(t-u)\, du = \text{Im}[e^{i\omega t} \bar{D}(i\omega)]$$

   where $\bar{D}(s) = \int_0^\infty du\, e^{-su} D(u) = \sum_i \lambda_i a_i/(\lambda_i + s)$. Use $Z(s) = (l/\beta)s + 1 - \bar{D}(s)$.

2. Verify (3.2, Eq. 8).
   Hint: First, prove and then use the following equality:

$$\int_0^{\pi/2} dx\, (1 - a \sin^2 x)^{-1} = \pi/2(1-a)^{1/2}$$

   (Gradshetyn and Ryzhik [21], p. 381, Formula 3.653-2).

3. Determine the reactivity insertion $k(t)$ for a power variation

$$\begin{aligned} P(t) &= P_0, & t &< 0 \\ &= P_0 e^{\alpha t}, & 0 &< t < T \\ &= P_0 e^{\alpha T}, & t &> T \end{aligned}$$

   and discuss the discontinuity in $k(t)$ at $t = T$ (Murray *et al.* [1]).

4. Using the inverse method, determine the second harmonic of the periodic reactivity insertion that gives rise to a pure sinuoisdal power variation, i.e., $P(t) = P_0 + p_1 \sin \omega t$, assuming $(p_1/P_0) \ll 1$.

5. Determine the reactivity insertion $k(t)$ required for a ramp power rise with cutoff, i.e.,

$$\begin{aligned} P(t) &= P_0\,, & t &< 0 \\ &= 1 + \alpha t, & 0 &< t < T \\ &= 1 + \alpha T, & t &> T \end{aligned}$$

(see Murray *et al.* [1].)

6. Using the sum rules given by (9), (11) and (13) of Section 3.3, prove the following identities:

$$\sum_{j=0}^{6} \frac{\omega_j}{\lambda_i + \omega_j} \left[1 + \sum_{k=1}^{6} \frac{(\beta_k/l)}{(\lambda_k + \omega_j)^2}\right]^{-1} = 1$$

and

$$\sum_{j=0}^{6} \frac{\omega_j}{(\omega_j + \lambda_i)(\omega_j + \lambda_l)} \left[1 + \sum_{k=1}^{6} \frac{(\beta_k/l)}{(\lambda_k + \omega_j)^2}\right]^{-1} = 0$$

for $\lambda_i \neq \lambda_l$.

7. Find the response of a critical reactor to a step reactivity insertion k_0 at $t = 0$ using the model of one group of delayed neutrons, and derive (3.3, Eq. 32) by ignoring $\bar{\lambda} l$ as compared to $\beta(1 - k_0)$.

8. Find the solution of

$$\gamma[d\bar{P}(s)/ds] + [Y(s) - k_0]\,\bar{P}(s) = 0$$

where

$$Y(s) \equiv s \left\{(l/\beta) + \sum_{i=1}^{6} [a_i/(\lambda_i + s)]\right\}$$

REFERENCES

1. R. L. Murray, C. R. Bingham, and C. F. Marten, *Nucl. Sci. Eng.* **18**, 481 (1964).
2. R. W. Miller and F. Schroeder, IDO-16317. 1957. Nuclear Reactor Testing, Station, Arco, Idaho.
3. W. E. Horning and H. C. Corben, IDO-16446. 1958. Nuclear Reactor Testing, Station, Arco, Idaho.

4. H. C. Corben, *Nucl. Sci. Eng.* **5**, 127 (1959).
5. H. B. Smets, *Nukleonik* **5**, 181 (1961).
6. G. Hardy, J. Littlewood, and G. Polya, "Inequalities," p. 261, Thm. 368. Cambridge Univ. Press, London and New York, 1952.
7. W. K. Ergen, *J. Appl. Phys.* **25**, 702 (1954).
8. H. Greenspan, ANL-5577, Appendix by R. O. Britton. Argonne Nat. Lab., Argonne, Illinois, 1966.
9. G. R. Keepin, T. F. Wimett, and R. K. Zeigler, *Phys. Rev.* **107**, 1044 (1957).
10. H. B. Smets, *Acad. Roy. Belg., Mem. Collect. Cl. Sci.* 5° **45**, No. 3 (1959).
11. A. F. Henry, *Nucl. Sci. Eng.* **20**, 338 (1964).
12. B. J. Toppel, *Nucl. Sci. Eng.* **5**, 88 (1959); see also ANL-5800, p. 452. Argonne Nat. Lab., Argonne, Illinois, 1963.
13. Reactor physics constants. ANL-5800. At. Energy Comm., Div. of Tech. Inform., Argonne Nat. Lab., Argonne, Illinois, July 1963.
14. R. E. Skinner and E. R. Cohen, *Nucl. Sci. Eng.* **5**, 291 (1959).
15. D. I. Blokhintsev and H. A. Nikolaev, *Proc. Int. Conf. Peaceful Uses At. Energy, Geneva*, 1955, **3**, p. 35. Columbia Univ. Press, New York, 1956.
16. S. Wallach, WAPD 13. 1950. Westinghouse Atomic Power Division, Pittsburgh, Pennsylvania.
17. H. L. Garabedian, R. S. Varga, and G. G. Bilodeau, *Nucl. Sci. Eng.* **3**, 548 (1958).
18. J. Wilkins, Jr., *Nucl. Sci. Eng.* **5**, 207 (1959).
19. E. L. Ince, "Ordinary Differential Equations," p. 186, Dover, New York, 1956.
20. "Handbook of Mathematical Functions with Formulas, Graphs, and Mathematical Tables" (Appl. Math. Ser. 55). U. S. Dept. of Comm., Nat. Bur. of Stand., Washington, D.C., 1965.
21. I. S. Gradshteyn and I. M. Ryzhik, "Tables of Integrals, Series and Products." Academic Press, New York, 1965.

CHAPTER 4

Approximate Solutions of the Point Kinetic Equations without Feedback

In this chapter, we shall discuss approximate methods for solving the point kinetic equations for an arbitrary reactivity insertion. Approximate solutions are needed because the exact solutions are known only for a few special reactivity insertions, as discussed in the previous chapter, and they are too complicated for practical calculations. Furthermore, simplified results which were obtained in Chapter 3 using purely mathematical approximations to the exact solutions will now be obtained directly from the original equations by physically meaningful simplifications. These physical assumptions lead to a better understanding of the problem.

4.1. Approximation of Constant Delayed-Neutron Production Rate

In certain problems, such as when the reactor is shut down by rapid insertion of the safety rods, we are interested in the response of the reactor power to a given reactivity insertion in short time intervals following a time t_0. In such short time intervals, we may ignore the change in the rate of production of delayed neutrons, and replace the term $\sum_{i=1}^{6} \lambda_i C_i(t)$ in the point kinetic equations, which accounts for the production of delayed neutrons, by its value at t_0. Choosing the time origin at $t_0 = 0$, we then obtain the following approximate kinetic equation from (3.1, Eq. 1):

$$(l/\beta)\, \dot{P} = [k(t) - 1]\, P(t) + Q(t)(l/\beta) \tag{1}$$

where the quantity

$$Q(t) = (\beta/l) \sum_{i=1}^{6} \lambda_i C_i(0) + S(t) \tag{2}$$

acts as a modified external source. The general solution of this equation can be obtained, by using the method of integrating factors, as

$$P(t) = [\exp\ A(t)] \left\{ P(0) + \int_0^t dt' \, [\exp -A(t')] \, Q(t') \right\} \tag{3}$$

where

$$A(t) = (\beta/l) \int_0^t [k(t') - 1] \, dt' \tag{4}$$

A. Fast Ramp Reactivity Insertion

As an application (see also Problem 1), we consider a ramp reactivity insertion into a delayed critical reactor, i.e., $k = 0$ at $t = 0$, in the absence of external sources. Since the reactor is critical for $t < 0$, $\sum_i \lambda_i C_i(0) = P(0)$, and hence $Q(t) = P(0)\, \beta/l$. Substituting $k(t) = \gamma t$ in (4), we obtain $A(t) = (\beta/l)[(\gamma t^2/2) - t]$. Equation (3) can be expressed in a compact way by introducing

$$T \equiv (1 - \gamma t)\, T_0 \tag{5a}$$

$$T_0 \equiv (\beta/l\gamma)^{1/2} \tag{5b}$$

as

$$P(t)/P(0) = \exp[(T^2 - T_0^2)/2] + T_0 \int_T^{T_0} \exp[-(T'^2 - T^2)/2] \, dT' \tag{6}$$

We shall be interested in the power level and the reciprocal period at prompt-criticality $k = 1$, i.e., $t = t_p = 1/\gamma$, assuming that $T_0 \gg 1$ (this point will be discussed below),

$$P(t_{\mathrm{p}})/P(0) \approx T_0(\pi/2)^{1/2} = (\beta\pi/2l\gamma)^{1/2} \tag{7}$$

and the reciprocal period is given as

$$\omega(t_{\mathrm{p}}) = (2\beta\gamma/\pi l)^{1/2} \tag{8}$$

In obtaining (7), we have ignored $\exp(-T_0^2/2)$ as compared to $T_0(\pi/2)^{1/2}$, and used $\int_0^\infty \exp(-\frac{1}{2}x^2)\, dx = (\pi/2)^{1/2}$. In order for the above results to be valid, the time interval t_p must be short compared to a characteristic decay time associated with delayed neutron, i.e., $\lambda^*/\gamma \ll 1$. If we interpret

λ^* as $\sum a_i\lambda_i$, then this condition, and the results (7) and (8), become identical to those obtained from the exact solutions in the previous chapter for a fast ramp input (cf 3.3, Eq. 55). The condition $T_0 \gg 1$ implies $(\beta/l\gamma)^{1/2} \gg 1$, which is also identical to that introduced (3.3, Eq. 55) in the calculation of the integration constants B_1 and B_2 in the limit of $l \to 0$. We shall use these results in the discussion of reactor startup.

B. Kinetics of Reactor Shutdown

In the case of an emergency, such as the loss of coolant flow, the reactor is shut down by a rapid insertion of the safety rods. Since the rod insertion takes a finite time, of the order of tenths of a second, due to the mechanical inertia, the insertion of reactivity cannot be treated as a step change. A realistic analysis of the power variation should include the finite speed of the safety rods. A reasonable model is to assume a negative ramp insertion, i.e., $k(t) = -\gamma t$. The quantity of interest is the power variation during the rod motion and after the end of the stroke, so that the total energy release can be computed by integrating the power. Since the reactor is not being cooled as a result of loss of flow, the energy released in the reactor increases the temperature of the core and may cause damage.

We can treat this problem by using the short-time behavior during the reactivity insertion. Assuming that the reactor was critical for $t < 0$, and substituting $k(t) = -\gamma t$ in (4), we obtain from (3), after some straightforward algebra,

$$P(t)/P(0) = \{\exp[-(T^2 - T_0^2)/2]\}[1 - \sqrt{2}T_0F(T_0/\sqrt{2})] + \sqrt{2}T_0F(T/\sqrt{2}) \quad (9)$$

where $T = T_0(1 + \gamma t)$ and

$$F(T_0) \equiv \int_0^{T_0} \exp(T^2 - T_0^2)\, dT \quad (10)$$

Equation (9) gives the reactor power during the insertion of the rods. At the end of the stroke, the reactivity becomes constant (shutdown reactivity). We can therefore obtain the variation of the reactor power at subsequent times by solving the kinetic equations for a constant reactivity (cf 3.3, Eq. 1) with the initial power evaluated from (9) at the end of the stroke. The initial values for the delayed neutron precursors can be approximated by $a_iP(0)/\lambda_i$, i.e., the densities of the precursors just before the safety rods are inserted.

The foregoing analysis is crude in the sense that it does not include the effect of feedback, but it provides a useful estimate of the total energy release.

4.2. The Prompt-Jump Approximation

If the relative rate of change of reactor power in a mean prompt generation time is sufficiently small so that

$$|(l/\beta)[\dot{P}(t)/P(t)]| \ll |1 - k(t)| \tag{1}$$

holds, then the term $(l/\beta)\,\dot{P}(t)$ in the kinetic equations can be neglected as compared to $(k - 1)\,P(t)$. The resulting approximate equations read

$$0 = [k(t) - 1]\,P(t) + \sum_{i=1}^{6} \lambda_i C_i(t) + (l/\beta)\,S(t) \tag{2a}$$

$$\dot{C}_i(t) = a_i P(t) - \lambda_i C_i(t), \qquad i = 1, 2, \ldots, 6 \tag{2b}$$

This approximation is usually called the prompt-jump approximation [1–4] because it predicts a sudden change in the reactor power following a sudden change in reactivity [1, 2]. The delayed neutron precursor density and the external source remain continuous at the instant of the step change. The power level P_2 after the jump from k_1 to k_2 in reactivity is related to the initial power P_1 by

$$P_2/P_1 = (k_1 - 1)/(k_2 - 1) \tag{3}$$

which is obtained from (2a).

The prompt-jump approximation reduces the number of differential equations from seven to six. In the absence of external sources, it is equivalent to replacing the mean prompt generation time l by zero in the kinetic equations, and in their solutions. This approximation is particularly useful in the case of one delayed-neutron group, because the precursor density $C_1(t)$ can be eliminated between (2a) and (2b):

$$(1 - k)\,\dot{P} = P[\dot{k} + \bar{\lambda} k] + (l/\beta)(\dot{S} + \bar{\lambda} S) \tag{4}$$

This equation can be solved easily if $k(t)$ and $S(t)$ are given. The general solution is given by (4.1, Eq. 3) with

$$A(t) = \int_0^t \{[\dot{k}(t') + \bar{\lambda} k(t')]/[1 - k(t')]\}\,dt' \tag{5a}$$

$$Q(t) = (l/\beta)(\dot{S} + \bar{\lambda} S)/(1 - k) \tag{5b}$$

As an example, we consider a ramp reactivity insertion into a critical reactor in the absence of external sources. The result is (Problem 2)

$$P(t)/P_0 = e^{-\bar{\lambda}t}/(1 - \gamma t)^{1+(\bar{\lambda}/\gamma)} \tag{6}$$

It is clear that the prompt-jump approximation fails near prompt-critical, as expected from the condition (1), in which the right-hand side vanishes when $\gamma t = 1$. Combining (4) and (1), we can obtain a condition for the validity of the prompt-jump approximation in terms of $k(t)$ only. In the absence of external sources, (4) yields the instantaneous inverse period as

$$\omega(t) = \dot{P}/P = (\dot{k} + \bar{\lambda}k)/(1 - k) \tag{7}$$

Using this result in (1), we find

$$(1 - k)^2 \gg (l/\beta) \mid \dot{k} + \bar{\lambda}k \mid \tag{8a}$$

In the case of slow reactivity changes, so that $\mid \dot{k} \mid \ll \bar{\lambda}k$, (8a) reduces to $(1 - k) \gg (lk\bar{\lambda}/\beta)^{1/2}$. This can be further simplified to [1]

$$(1 - k) \gg (\bar{\lambda}l/\beta)^{1/2} \tag{8b}$$

which is a practical condition for the validity of the prompt-jump approximation. Using the values $\bar{\lambda} \approx 0.1\ \text{sec}^{-1}$, $l \approx 10^{-3}$ sec, and $\beta \approx 10^{-2}$, we find that (8) will be valid until the reactivity reaches roughly 90% of prompt-criticality.

The accuracy of the prompt-jump approximation as compared with exact solutions for a thermal ($l/\beta = 10^{-2}$ sec) and a fast ($l/\beta = 5 \times 10^{-5}$ sec) reactor has been calculated by Goldstein and Shotkin [4] using a step input of reactivity $k_0 = 0.5$ \$. They found that the relative error, i.e., (exact − PJA)/exact, is approximately 2 and 0.008%, respectively, for 0.1–10 sec after the insertion of the step.

Corrections to the Prompt-Jump Approximation

The mathematical basis of the prompt-jump approximation as the first term in an asymptotic expansion in powers of l has been discussed by Birkhoff [3], and by Goldstein and Shotkin [4]. Here, we present Birkhoff's technique to construct the first-oder correction to the prompt-jump approximation.

In the absence of external sources, the power equation can be cast in the form

$$\gamma(t)\, P(t) = Q(t) - (l/\beta)\, \dot{P}(t) \tag{9}$$

where, following Birkhoff, we have introduced

$$\gamma(t) \equiv 1 - k(t) \tag{10a}$$

$$Q(t) \equiv \sum_{i=1}^{6} \lambda_i C_i(t) \tag{10b}$$

The conventional prompt-jump approximation is obtained by ignoring $(l/\beta)\dot{P}$ in (9) as explained above, which leads to the set (2). A more refined approximation is obtained by computing $\dot{P}(t)$ from the asymptotic approximation $P(t) \sim Q(t)/\gamma(t)$. This leads to the following set of equations:

$$P(t) = \frac{Q(t)}{\gamma(t)} + \frac{l}{\beta}\left[\frac{1}{\gamma(t)}\right]^2 \left[\frac{\dot{\gamma}(t)\,Q(t)}{\gamma(t)} - \dot{Q}(t)\right] + O(l^2) \tag{11}$$

and (2b). If we substitute back from (11) into (2b), we obtain

$$\dot{C}_i + \lambda_i C_i = \frac{a_i}{\gamma(t)}\left[\left(1 + \tau\frac{\dot{\gamma}}{\gamma}\right) Q(t) - \tau\dot{Q}(t)\right] \tag{12a}$$

where

$$\tau \equiv l/\beta\gamma(t) \tag{12b}$$

These equations yield very accurate results if

$$\tau \ll 1 \text{ [i.e., } (l/\beta) \ll 1 - k(t)] \qquad \text{and} \qquad (l/\beta)\,\dot{\gamma}(t) \ll [1 - k(t)]^2$$

In the approximation of one group of delayed neutrons, (12a) reduces to

$$\dot{C}(t)\left[1 + \frac{l\bar{\lambda}}{\beta}\frac{1}{\gamma^2(t)}\right] + \bar{\lambda}\left\{1 - \frac{1}{\gamma(t)}\left[1 + \frac{l}{\beta}\frac{\dot{\gamma}(t)}{\gamma^2(t)}\right]\right\} C(t) = 0 \tag{13}$$

which can be integrated exactly using the general solution (4.1, Eq. 3). The lowest-order prompt-jump approximation is obtained in (13) by letting $l = 0$. [See Problems 4-6 for applications of (4) and (13).] It is clear from this remark that the prompt-jump approximation is more accurate for fast reactors ($l/\beta \approx 10^{-4}$ sec) than for thermal reactors ($l/\beta \approx 10^{-2}$ sec).

4.3. Gradual Reactivity Changes

An approximate solution of the point kinetic equation for an arbitrary but slowly varying reactivity insertion was obtained by Hurwitz [5] in 1949.

A. Hurwitz's Method

In this approach, one looks for a solution of the form

$$P(t) = \eta(t) \exp\left[\int_0^t \omega_0(t')\, dt'\right] \tag{1a}$$

$$C_i(t) = [f_i(t)\, \eta(t) + \epsilon_i(t)] \exp\left[\int_0^t \omega_0(t')\, dt'\right] \tag{1b}$$

where $\eta(t)$ and $\epsilon_i(t)$ are the new unknown functions to be determined, $f_i(t)$ is defined by

$$f_i(t) \equiv a_i/[\lambda_i + \omega_0(t)] \tag{2a}$$

and $\omega_0(t)$ is the algebraically largest root of the inhour equation with the instantaneous value of the reactivity insertion at t, i.e.,

$$k(t) = \omega_0 \left[\frac{l}{\beta} + \sum_{i=1}^{6} \frac{a_i}{\lambda_i + \omega_0}\right] \tag{2b}$$

If the reactivity were constant, then the new unknown functions $\eta(t)$, $f_i(t)$, $\epsilon_i(t)$, and $\omega_0(t)$ would be constant, and equations (1a) and (1b) would represent the exact asymptotic solution for a step reactivity change. It may therefore be expected that these functions would vary slowly in time when the reactivity is a slowly varying function of time. The equations for the unknown functions $\eta(t)$ and $\epsilon_i(t)$ can be obtained by substituting $P(t)$ and $C_i(t)$ from (1) into the standard form of the point kinetic equations (cf 3.1, Eqs. 1 and 2) in the absence of external sources. The results are

$$(l/\beta)\, d\eta/dt = \sum_{i=1}^{6} \lambda_i \epsilon_i \tag{3a}$$

$$d(\eta f_i)/dt + d\epsilon_i/dt = -(\omega_0 + \lambda_i)\, \epsilon_i \tag{3b}$$

It is worth mentioning that no approximation has yet been introduced; if we could solve equations (3) exactly, we would obtain the rigorous solution to the kinetic equations for an arbitrary input using (1). The approximation to be made to solve (3) is based on the following observation: Since $\omega_0(t)$ was chosen as the algebraically largest root of the inhour equation, $(\omega_0 + \lambda_i)$ is always positive. Hence, we may expect that $\epsilon_i(t)$ in (3b) will always tend toward the value

$$-(\lambda_i + \omega_0)^{-1}\, d(\eta f_i)/dt$$

with a relaxation time $(\omega_0 + \lambda_i)^{-1}$. If the reactivity variations are small in time intervals comparable to $(\omega_0 + \lambda_i)^{-1}$, then we can approximate the solution of (3b) by

$$\epsilon_i(t) = -[1/(\omega_0 + \lambda_i)]\, d(\eta f_i)/dt \tag{4}$$

We may note that the relaxation time becomes small for large, positive reactivities, whereas it may tend toward infinity for $i = 1$ when large, negative reactivities are involved. Hence, the rate of reactivity change must be smaller when the reactivity assumes large, negative values. Substituting (4) into (3a), we obtain

$$\frac{l}{\beta}\frac{d\eta}{dt} = -\sum_{i=1}^{6} \frac{\lambda_i}{\lambda_i + \omega_0} \frac{d(\eta f_i)}{dt} \tag{5}$$

which can also be written as

$$\frac{d(\ln \eta)}{dt} = -\sum_{i=1}^{6} \frac{df_i}{dt} \frac{\lambda_i}{\lambda_i + \omega_0} \Big/ \left[\frac{l}{\beta} + \sum_{i=1}^{6} \frac{\lambda_i f_i}{\lambda_i + \omega_0}\right] \tag{6}$$

Using the explicit form of $f_i(t)$ from (2a), we find

$$d(\ln \eta)/dt = -\tfrac{1}{2}(d/dt)[\ln Y'(\omega_0)] \tag{7}$$

where $Y(\omega_0)$ is the inverse of the zero-power transfer function defined in (3.3, Eq. 2). We can integrate this equation using the initial conditions $\eta = P_0$ and $\omega_0 = 0$, i.e., $k = 0$, at $t = 0$:

$$P(t)/P_0 = [Y'(0)/Y'(\omega_0)]^{1/2} \exp\left[\int_0^t \omega_0(t')\, dt'\right] \tag{8}$$

where $\omega_0(t)$, we recall, is the algebraically largest root of $k(t) = Y(\omega_0)$. Physically, ω_0 is the inverse stable period corresponding to the value of the reactivity at time t.

Discussion

In order to understand the physical implications of the assumptions inherent in (8), we shall consider a few applications. First, we calculate the instantaneous inverse period $\omega(t) \equiv d(\log P)/dt$ using (8):

$$\omega(t) = \omega_0 - \tfrac{1}{2}\{Y''(\omega_0)/[Y'(\omega_0)]^2\}(dk/dt) \tag{9}$$

where we have used $dk/dt = Y'(\omega_0)\, d\omega_0/dt$. We observe that (9) gives the correct inverse stable period in the case of constant reactivity. The time dependence of k, when it is not constant, is accounted for both by the presence of the second term, and by the time dependence of $\omega_0(t)$ through the inhour equation. Secondly, we consider the inverse period at prompt-critical for a ramp reactivity insertion $k(t) = \gamma t$. To facilitate the comparison with the exact results obtained in Section 3.3B, we must also assume here one group of delayed neutrons and use the limit of $l \to 0$. The first term in (9) at prompt-critical is obtained from (3.3, Eq. 20) as $\omega_0 = (\lambda^* \beta / l)^{1/2}$. The second term can be calculated by substituting $Y(s) = s/(\lambda^* + s)$ and using $\omega_0 \gg \lambda^*$ in the limit of $l \to 0$, as $-\omega_0/\lambda^*$. Combining these two results, we find from (9)

$$\omega(t_{\mathrm{p}}) = (\lambda^* \beta / l)^{1/2} [1 + (\gamma/\lambda^*)]$$

The exact result in the case of slow reactivity insertion, i.e., $\gamma \ll \lambda^*$, was obtained in (3.3, Eq. 56b) as $(\lambda^* \beta / l)^{1/2}$. It thus follows that (9) predicts the correct inverse period for a slow ramp insertion of the order of a few cents per second or less. This example provides a quantitative assesment of the basic assumption of gradual reactivity change. A more detailed comparison between the exact solutions for a ramp input and those obtained approximately from (8) was worked out by Wilkins [6], and essentially the same conclusion as mentioned above was drawn.

As a final remark, we may mention that the solution (8) represents the asymptotic behavior of the reactor power in the case of a step reactivity change. The approximation inherent in (8) is not sufficiently accurate to investigate transients, because the crucial assumption that the reactivity must be slowly varying in time intervals comparable to $(\lambda_i + s)^{-1}$ is violated initially when the reactivity undergoes a step change (see Goldstein and Shotkin [4] for a comparison of this approximation with the prompt-jump approximation discussed in the previous chapter).

B. Analysis of Startup Incidents

As an application of the previous approximate methods to an actual problem in reactor operation, we shall present an approximate analysis of reactor startup incidents. Our discussions will be based essentially on a paper by Hurwitz [7]. We consider a startup procedure in which the control rods are withdrawn continuously, and the reactivity is assumed to be inserted at a constant rate of γ dollars per second. In normal operation, the reactivity insertion continues until a preassigned reactivity period is reached [8]. This value of the reactor period is maintained up to the desired power level. We shall assume that, in the incident to be analyzed, the reactivity insertion continues at a constant rate as a result of malfunctioning of the period meter beyond the preassigned period until a power-level trip shuts the reactor down by inserting the safety elements. Initially, the reactor is assumed to be far below critical with a shutdown reactivity of k_0 dollars. The initial power level P_s is maintained by an external source S_0 of neutrons. The mathematical problem we want to solve is the determination of the power transient caused by the linear reactivity insertion $k(t) = -k_0 + \gamma t$ into a reactor at steady state. We are particularly interested in the power level P_d at delayed-criticality, which occurs at $t_d = k_0/\gamma$, and in the inverse period ω_T at the time t_T when the reactor power reaches the value P_T corresponding to the trip level. Following Hurwitz [7], we divide the power transient into three time intervals as indicated in Figure 4.3.1.

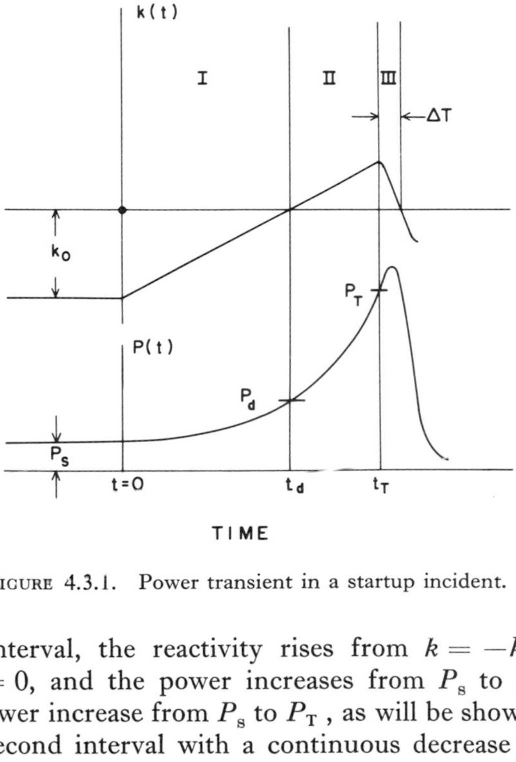

FIGURE 4.3.1. Power transient in a startup incident.

In the first interval, the reactivity rises from $k = -k_0$ to delayed-criticality, $k = 0$, and the power increases from P_s to P_d. However, most of the power increase from P_s to P_T, as will be shown below, takes place in the second interval with a continuous decrease in the reactor period. This is intuitively expected because the reactor is above delayed critical through out this interval. The third interval models the transients after the safety mechanism has been tripped at t_T. Most of the energy release occurs in this interval. An accurate description of the transients in this interval, which would enable us to calculate the total energy release, requires a reliable knowledge of the mechanical motion of the safety mechanism and the time dependence of the resulting reactivity insertion. As Hurwitz [7] states, "The uncertainty in the total energy release due to possible fluctuations in the speed of the mechanical operation of the safety system is greater than the uncertainty existing because of an approximate theoretical treatment of phase III." More importantly, the presence of feedback effects in this region may further obscure the analysis. We may therefore argue that an accurate mathematical description of phase III is usually difficult to obtain, and is not warranted in this illustrative analysis. An order-of-magnitude estimate of the total energy release in this interval may still be computed as the product of the trip level and the time ΔT required to reduce the reactivity to delayed-critical. This is justified because the safety system is so designed [7] that ΔT is short compared to the reactor period at t_T, and hence the power rise, after the safety mechanism begins to reduce the reactivity, cannot exceed the trip level P_T in order of magnitude. In these considerations, we assume that the reactor is not prompt-critical at t_T, and that t_T denotes the time at which the reactivity actually begins to decrease as a result of the action of the safety mechanism.

If the reactor is above prompt-critical at the instant t_T, the total energy release will be more nearly equal to the product of the peak power and the time required for the reactivity to fall back to prompt-criticality, because the power will begin to decrease after the reactivity has been reduced below prompt-criticality [7], because of the emission of delayed neutrons.

We see from the foregoing considerations that the crucial quantity to be calculated is the inverse period at the time the trip level is reached, in terms of the reactivity insertion rate γ. From this information, we can determine the maximum safe value of γ such that no damage to the core will result in case of an incident as described above [8].

Having disposed of the third interval by the arguments presented above, we now return to the analysis of the first and second phases.

The power variation in the first interval can be analyzed by using the prompt-jump approximation and the one-delayed-neutron-group model with the group parameters $\beta^{(1)} = \beta$ and $\lambda^{(1)} = \bar{\lambda}$, for the reactivities are far below prompt-critical (cf 4.2, Eq. 1).

The prompt-jump approximation is justified because the reactor is always below delayed-critical in this interval, and the power constantly tends toward $(l/\beta)\, S_0/|\, k(t)\,|$ at all times. Hence, the rate of change of reactor power is essentially determined by the rate of change in reactivity.

As a more quantitative justification, let us consider (4.2, Eq. 8b), which states, for the present problem,

$$1 + k_0 - \gamma t \gg (\bar{\lambda} l/\beta)^{1/2}$$

Since $(\bar{\lambda} l/\beta)^{1/2}$ is of the order of 10^{-2} for $\bar{\lambda} \approx 0.1\ \text{sec}^{-1}$, $l = 10^{-3}\,\beta$, whereas the left-hand side is always greater than unity in the first interval.

We want to compute $P(t_d)$ in terms of γ, by solving

$$\dot{P} = P\left[\frac{\gamma - k_0\bar{\lambda} + \bar{\lambda}\gamma t}{1 + k_0 - \gamma t}\right] + \frac{l\bar{\lambda}}{\beta}\,\frac{S_0}{1 + k_0 - \gamma t} \tag{10a}$$

with the initial condition $P = P_s$ at $t = 0$, where P_s is the source level, i.e.,

$$P_s = (l/\beta)\,S_0/k_0 \tag{10b}$$

Evaluating the solution of this equation (Problem 12) at $t_d = k_0/\gamma$, we obtain

$$P(t_d) = [\exp(-\mu k_0)](1 + k_0)^{\mu}\left[P_0 + \frac{l\bar{\lambda}}{\beta}\,\frac{S_0}{1 + k_0}\right.$$

$$\left.\times \int_0^{k_0/\gamma} dt'[\exp(\bar{\lambda} t')]\left(1 - \frac{\gamma t'}{1 + k_0}\right)^{\mu - 1}\right] + \frac{l}{\beta}\,S_0 \tag{11}$$

where $\mu = \bar{\lambda}/\gamma$. At this stage, we introduce a delayed-neutron buildup factor ξ, defined by

$$\xi = P(t_d)/S_0(l/\beta) \tag{12}$$

The denominator is the source level at delayed-criticality if the effects of delayed neutrons are neglected. This can be seen from (10b) with $k_0 = 1\,\$$. The latter follows from the fact that the reactor at delayed-criticality is 1 \$ below prompt-critical. The reason for introducing the buildup factor is that it can be expressed as a function of μ in most practical cases, and the computation of $P(t_d)$ can be simplified considerably by plotting ξ versus μ. Combining (11) and (12) and letting $x = \bar{\lambda} t'$, we obtain

$$\xi = 1 + [(1 + k_0)^{\mu}/k_0]\,e^{-\mu k_0} + e^{-\mu k_0}\int_0^{\mu k_0} dx\{e^{x}[1 + k_0 - (x/\mu)]^{\mu - 1}\} \tag{13}$$

In this expression, we can neglect unity as compared to k_0 because the magnitude of the shutdown reactivity k_0 is of the order of 10 \$ or more. For slow rates of reactivity insertion such that $\mu k_0 = \bar{\lambda} k_0/\gamma \gg 1$ or $\gamma \ll \bar{\lambda} k_0$ (typically, $\gamma \ll 80$ ¢/sec), we can neglect the second term in (13) and replace the upper limit of the integration by infinity. After some straightforward manipulations, we obtain ξ as

$$\xi = \int_0^\infty dx\{e^{-x}[1 + (x/\mu)]^\mu\} \tag{14a}$$

which can be approximated by (see Problem 8)

$$\xi = \tfrac{2}{3} + (\tfrac{1}{2}\mu\pi)^{1/2} \tag{14b}$$

for large values of μ. The variation of ξ as a function of μ is shown in Figure 4.3.2.

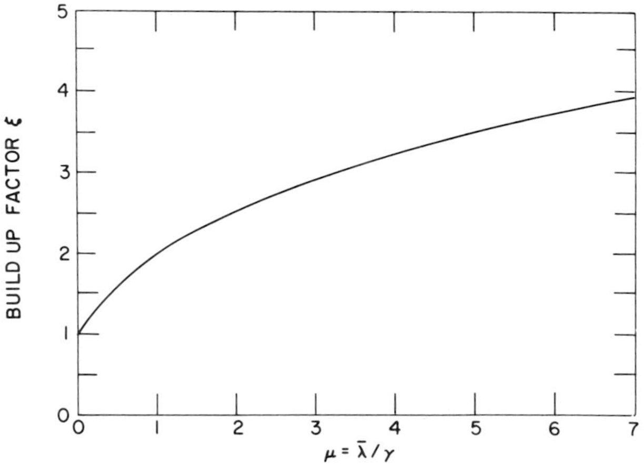

FIGURE 4.3.2. Variation of the delayed-neutron buildup factor ξ, Eq. (14a), with reactivity insertion rate γ.

The power level at the end of the first interval can now be easily calculated by using this curve and (12) as $P(t_d) = \xi k_0 P_s$. For example, $P(t_d) = 35P_s$ for $k_0 = 10\,\$$, $\bar{\lambda} \approx 0.10\ \text{sec}^{-1}$, and $\gamma = 2$ ¢/sec ($\mu = 5$ and $\xi = 3.5$).

In the case of a fast reactivity insertion, such that $\mu k_0 \ll 1$ ($\gamma \gg 0.80\,\$$), we can use the limit of ξ as μk_0 approaches zero, which follows from (13) as $\xi = 1 + (1/k_0)$, i.e., approximately unity for large k_0. It follows that the asymptotic form of ξ given by (14b) can be used as an approximation, for all values of γ, with an increasing accuracy for slow reactivity insertions.

The foregoing discussion indicates that most of the power buildup from P_s to P_T, which exceeds P_s by many decades, must take place in the second interval, because the ratio $P(t_d)/P_s$ in the first phase of the transient is less than 100.

We can analyze the power transient in the second interval in the two limiting cases of rapid and slow reactivity insertions. In the first case, in which γ is several dollars per second or more, we can use the approximation of constant delayed-neutron production rate, with the initial values $P(0) = P(t_d)$ and $\sum_i C_i(0)\,\lambda_i = P(t_d) - (l/\beta)\,S_0$. The latter follows from the equation $(k_0 + \gamma t - 1)\,P(t) + (l/\beta)\,S_0 + \sum \lambda_i C_i(t) = 0$, which was used in the first interval to describe the transients, with $t = t_d$. Substituting these initial conditions into (1), we obtain the following equation:

$$\dot{P} = (\beta/l)(\gamma t - 1)\,P + (\beta/l)\,k_0 \xi P_s \tag{15}$$

whose solution was already obtained in (4.1, Eq. 6) and discussed. More will be said about the solution of (15) presently.

The second limiting case comprises the gradual insertion of reactivity, i.e., γ is several cents per second or less, and can be investigated by means of (4.3, Eq. 8) with $P_0 = k_0 \xi P_s$:

$$P(t) = \xi k_0 P_s F(k) \exp\left[(1/\gamma)\int_0^k \omega_0(k')\,dk'\right] \tag{16}$$

where $F(k) \equiv [Y'(0)/Y'(\omega_0)]^{1/2}$, or, more explicitly,

$$F(k) = \left[\left(\frac{l}{\beta} + \sum_{i=1}^{6} \frac{a_i}{\lambda_i}\right)\Big/\left(\frac{l}{\beta} + \sum_{i=1}^{6} \frac{a_i \lambda_i}{(\lambda_i + \omega_0)^2}\right)\right]^{1/2} \tag{17}$$

Note that we have changed the variable t in the exponent in (16) to $k = \gamma t$. The calculations are facilitated by plotting $\omega_0(k)$ and $F(k)$ as functions of k as shown in Figures 4.3.3 and 4.3.4. The curves in

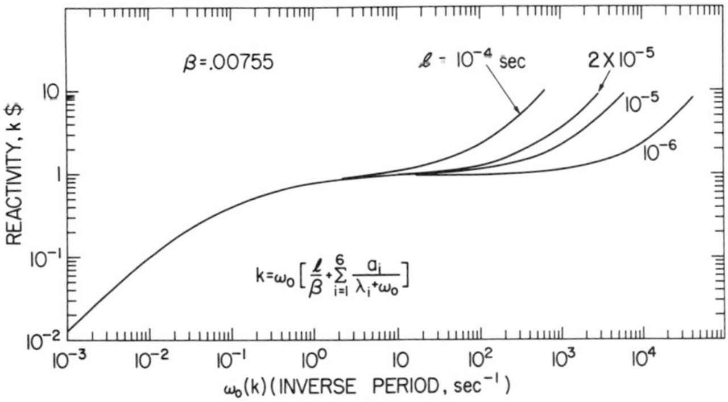

FIGURE 4.3.3. Six-group inverse period as a function of reactivity (^{235}U).

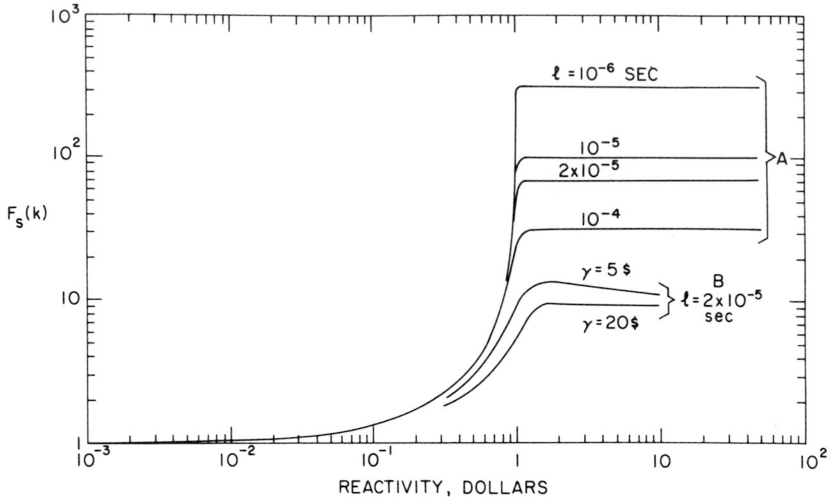

FIGURE 4.3.4. Variation of $F(k)$, Eq. (17), with reactivity; ^{235}U, six groups, $\beta = 0.00755$.

Figure 4.3.3 are obtained from the inhour equation for various values of l using ^{235}U as fuel. The curves A in Figure 4.3.4 are a plot of $F(k)$ from (17) with the values of $\omega_0(k)$ read from Figure 4.3.3 for the same values of l, and for ^{235}U. The asymptotic value of $F(k)$ for $k > 1$ is obtained from (17) as $[1 + (\beta\bar{\tau}/l)]^{1/2} \approx (\beta\bar{\tau}/l)^{1/2}$, which is about 70 for $l = 2 \times 10^{-5}$ sec. We observe that $F(k)$ varies about two decades when k increases from zero to above critical.

The integral in the exponent in (16) can be evaluated exactly, if we notice that

$$\int_0^k \omega_0(k')\, dk' = \int_0^{\omega_0} d\omega[\omega\, dk/d\omega] = \omega_0 k - \int_0^{\omega_0} d\omega\, k(\omega) \tag{18}$$

where $k(\omega)$ is given by the inhour equation, i.e.,

$$k(\omega) = \omega \left\{(l/\beta) + \sum_i [a_i/(\lambda_i + \omega)]\right\}$$

The last integral in (18) is readily performed using the explicit form of $k(\omega)$. The result is

$$\int_0^k \omega_0(k')\, dk' = \frac{\omega_0{}^2 l}{2\beta} + \sum_{i=1}^{6} a_i\lambda_i \left[\ln\left(1 + \frac{\omega_0}{\lambda_i}\right) - \frac{\omega_0}{\lambda_i + \omega_0}\right] \tag{19}$$

The variation of this integral with k is plotted in Figure 4.3.5 for various values of l, and for ^{235}U as fuel. With the aid of the curves in Figures 4.3.4 and 4.3.5, we can compute $P(t)$ as a function of time (or k, because

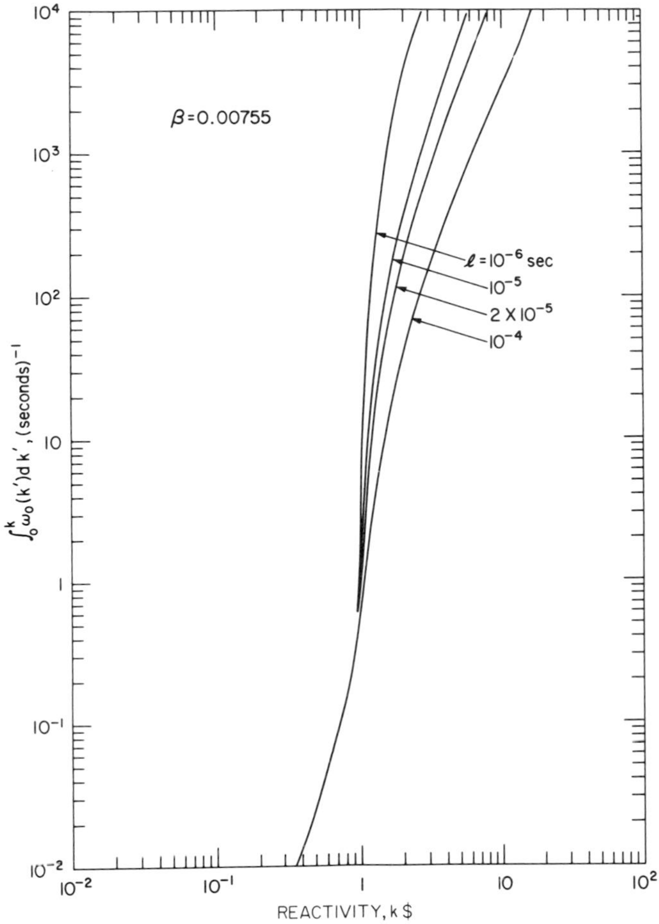

FIGURE 4.3.5. An evaluation of the integral $\int_0^h \omega_0(k')\,dk'$ for ^{235}U.

$k = \gamma t$) using (16) by reading $F(k)$ from Figure 4.3.4 and calculating the exponential factor with the aid of the curve in Figure 4.3.5. Since $F(k)$ varies only by about two orders of magnitude, most of the power increase is accounted for by the exponential factor when the final reactivity $k > 1$.

We now come to the final phase of our approximate analysis of a startup incident, i.e., the estimation of the inverse period $\omega_0(t_T)$ when the preset trip level P_T has been reached at $t = t_T$. The latter can be obtained by solving the following equation for k_T:

$$\int_0^{k_T} \omega_0(k')\, dk' = \gamma \ln[P_T/P_S k_0 \xi F(k_T)] \tag{20}$$

which is the logarithm of (16) with $k = k_T$. [For $k_T > 1$, this equation is easily solved because $F(k)$ is independent of k for $k > 1$; cf Figure 4.3.4.] The value of $\omega_0(t_T)$ is then obtained from Figure 4.3.3 with $k = k_T$. When the final reactivity is far above prompt-critical, so that $\omega_0(k) \approx (\beta/l)(k - 1)$ holds, the left-hand side of (20) can be approximated by $(\omega_0{}^2 l/2\beta)$ as can be seen from its expression in (19). Then, we obtain the following relation, yielding $\omega_0(t_T)$ directly:

$$\omega_0(t_T) \cong \{\ln[P_T/P_S \xi k_0 F(k_T)]\}^{1/2}(2\beta\gamma/l)^{1/2} \tag{21}$$

This result indicates that the inverse period when the trip level P_T is reached is essentially proportional to $(2\beta\gamma/l)^{1/2}$. Since the factors $\xi(\gamma)$ and $F(k_T)$ enter (21) logarithmically under the square root, their variations with γ and k_T, respectively, can be ignored. They can be approximated by $\xi = 1$ and $F(k_T) = 10$ as a crude conservative estimate of $\omega_0(t_T)$.

We now return to the case of rapid reactivity insertion, which is described by (15). The solution of the latter was already obtained in (4.1, Eq. 6). We can also express this solution in the form of (16), i.e., as a product of a slowly varying factor $F(k)$ and the rapidly varying factor $\exp[(1/\gamma) \int_0^k \omega_0(k')\, dk']$. The factor $F(k)$ in this case can be found as

$$F(k) = \left[\exp(-T_0^2/2) + T_0 \int_T^{T_0} \exp(-T'^2/2)\, dT'\right]$$
$$\times \exp\left[(T^2/2) - (1/\gamma) \int_0^k \omega_0(k')\, dk'\right] \tag{22}$$

where T and T_0 are defined by (4.1, Eq. 5), and $k = \gamma t$.

The variation of $F(k)$ with reactivity is also shown in Figure 4.3.4 for two values of rate of reactivity insertion, $\gamma = 5$ \$/sec and $\gamma = 20$ \$/sec (curve B). It is observed that $F(k)$ is unity at $k = 0$, and gradually increases, with k attaining values of the order of 10 above prompt-critical. For large values of γ (e.g., $\gamma \geqslant 15$ \$), $F(k)$ has a plateau for $k > 1$ at a value which can be approximated (Problem 9) by

$$F(k) \approx (2\pi\beta/l\gamma)^{1/2} \tag{23}$$

It follows from these observations that $F(k)$ varies above prompt-critical, from about 10 to 70 (for $l = 2 \times 10^{-5}$), when γ changes from 20 \$ (curve B) to very small values (curve A). Since the values of $F(k)$ for slow and fast reactivity insertions are of the same order of magnitude, we can also use (21) in the case of rapid reactivity insertions for crude estimates of $\omega_0(t_T)$.

C. The BWK Approximation

The BWK (Brillouin, Wentzel, and Kramers) approximation [9, 10], developed originally in quantum mechanics, was used first by Smets [11], and later by Tan [12] in solving the point kinetic equations. The BWK procedure is very similar to Hurwitz's method described above, but is applicable only to the model of one group of delayed neutrons.

The standard form (3.1, Eq. 1) of the kinetic equations in the case of one group of delayed neutrons can be written as

$$\ddot{P}(t) + 2Q(t)\,\dot{P}(t) + R^2(t)\,P(t) = 0 \tag{24a}$$

where

$$2Q(t) \equiv (1/l)[\bar{\beta} - \rho(t)] \tag{24b}$$

$$R^2(t) \equiv -(1/l)[\bar{\lambda}\rho(t) + \dot{\rho}(t)] \tag{24c}$$

$$\bar{\beta} \equiv \beta + \bar{\lambda} l \tag{24d}$$

Here, $\rho(t) = \beta k(t)$ and denotes the reactivity. Substituting

$$P(t) = N(t) \exp\left[-\int Q(t')\,dt'\right] \tag{25}$$

in (24a), we obtain for the new unknown function $N(t)$

$$\ddot{N}(t) + G^2(t)\,N(t) = 0 \tag{26}$$

where

$$G^2(t) = R^2 - Q^2 - \dot{Q} \tag{27a}$$

or

$$G^2(t) \equiv -(1/l)[\lambda\rho(t) + \tfrac{1}{2}\dot{\rho}(t)] - (1/4l^2)[\bar{\beta} - \rho(t)]^2 \tag{27b}$$

We can verify very easily that

$$u(t) = [G(t)]^{-1/2} \exp\left[-i\int G(t')\,dt'\right] \tag{28}$$

is a solution of the following differential equation:

$$\ddot{u} + u[G^2 - \tfrac{3}{4}(\dot{G}/G)^2 + \tfrac{1}{2}(\ddot{G}/G)] = 0 \tag{29}$$

Comparing (29) to (26), we conclude that $u(t)$ is an approximation to $N(t)$ if

$$|\,G(t)^2| \gg |\,-\tfrac{3}{4}(\dot{G}/G)^2 + \tfrac{1}{2}(\ddot{G}/G)| \tag{30}$$

is satisfied at all times. Substituting (27) into (30), we can obtain a condition directly on $\rho(t)$ for the validity of the above approximation, which is the BWK approximation.

It is interesting to note that the additional terms vanish exactly if $G(t)$ satisfies

$$3(\dot{G}/G) = 2(\ddot{G}/\dot{G}) \tag{31}$$

which can easily be integrated to yield

$$G(t) = 1/(C_1 t + C_2)^2 \tag{32}$$

where C_1 and C_2 are two arbitrary constants. Thus, substituting $G(t)$ from (27) into (32), we can obtain a first-order nonlinear differential equation for $\rho(t)$ for which the BWK solution is exact.

Substituting $G(t)$ from (27b) into (28) and introducing a new variable $S(t)$ as

$$S(t) \equiv [1 + (1/lQ^2)(\tfrac{1}{2}\dot{\rho} + \bar{\lambda}\rho)]^{1/2} \tag{33}$$

we cast $u(t)$ in the form

$$u(t) = (QS)^{-1/2} \exp\left[\int QS\, dt\right] \tag{34}$$

In order to construct the general solution, we need a second, independent solution. Using the general relation [13]

$$u\dot{v} - \dot{u}v = C \tag{35}$$

between two independent solutions, we can construct $v(t)$. Choosing the constant C in (35) as -2 for convenient normalization, following Tan [12], we solve (35) for $v(t)$ as

$$v(t) = -2u \int u^{-2}\, dt'$$

which, upon the substitution of $u(t)$ from (34), reduces to

$$v(t) = (QS)^{-1/2} \exp\left[-\int QS\, dt\right] \tag{36}$$

The general solution of (26) is obtained as $N = Au(t) + Bv(t)$, where

A and B are the constants of integration to be determined from the initial conditions. Using (25), we find the BWK solution for the reactor power as

$$P(t) = (QS)^{-1/2} \left\{ A \exp\left[- \int Q(1 - S)\, dt\right] + B \exp\left[- \int Q(1 + S)\, dt\right]\right\} \quad (37)$$

In the case of a step insertion of reactivity, this solution becomes exact. Tan [12] has shown that the validity of (37) is restricted to finite time intervals for a ramp insertion of positive reactivity. No such restriction exists for negative ramp reactivity insertions.

The power of the BWK method manifests itself for sinusoidal reactivity insertions, which have been investigated in detail by Tan [12, 14]. She showed that the condition (30) can be satisfied at all times for a reactivity input

$$\rho(t) = \rho_0 \sin \omega t \qquad \text{if} \quad |\rho_0|\,|\tfrac{1}{2}\omega + \bar{\lambda}| \ll (1/4l)(\bar{\beta} - |\rho_0|)^2 \quad (38)$$

is satisfied. The solution can be obtained from (37) approximating $(1 + x)^{1/2}$ by $1 + x/2$ as

$$P(t) = \frac{A}{Q\sqrt{S}} \exp\left[\int_0^t \frac{\bar{\lambda}\rho_0 \sin \omega\tau}{\bar{\beta} - \rho_0 \sin \omega\tau}\, d\tau\right] + \frac{B}{\sqrt{S}} \exp\left[-\frac{\bar{\beta}}{l}t - \frac{\rho_0}{l\omega}(\cos \omega t - 1) - \int_0^t \frac{\bar{\lambda}\rho_0 \sin \omega\tau}{\bar{\beta} - \rho_0 \sin \omega\tau}\, d\tau\right] \quad (39)$$

where Q and S are defined by (24b) and (33), respectively, with $\rho(t) = \rho_0 \sin \omega t$.

Figure 4.3.6 show a comparison of this result with the exact numerical solution, as well as with the perturbation solution (cf 4.6, Eq. 41) by Akcasu [15] for an input $\rho(t) = 0.004 \sin t$ (i.e., $\omega = 1$ rad/sec) with $\bar{\beta} = 0.008$, $\bar{\lambda} = 0.1\ \text{sec}^{-1}$, and $l = 10^{-5}$ sec. The agreement is remarkable for such a large reactivity variation of ± 0.5 \$.

4.4. Small-Amplitude Approximation (Linearization)

The response of the reactor power to arbitrary but small changes in reactivity is of particular importance in determining the stability of reactor systems for small perturbations. The crucial assumption here is that the small reactivity variations will produce small changes in reactor power from its equilibrium value P_0. We already know that this assumption is not true for a critical reactor because even a slight positive

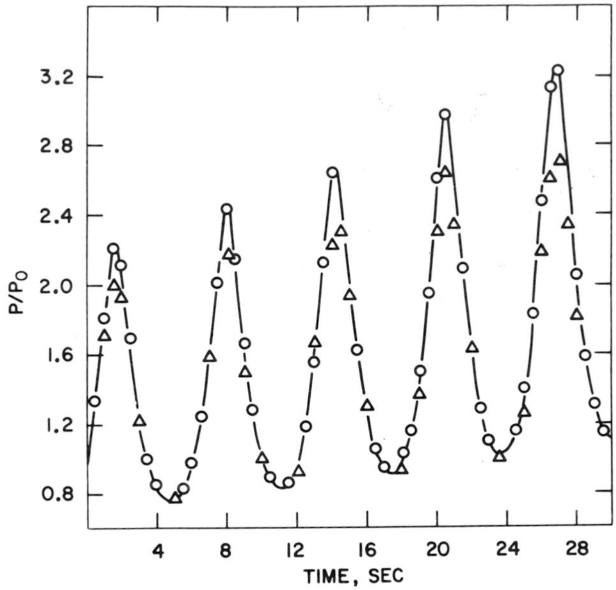

FIGURE 4.3.6. Response of a reactor to a sinusoidal reactivity insertion; $k(t) = (0.004/\beta) \sin t$. (——) Exact numerical results; (○) WKB approximation [14]; (△) after Akcasu [15]. $\beta = 0.00755$, $l = 10^{-5}$ sec, $\bar{\lambda} = 0.1$ sec^{-1}.

step change in reactivity gives rise to an exponentially increasing power response which eventually grows beyond any bound. However, the assumption will still be true if we consider only short times following the step insertion. Furthermore, for a certain class of reactivity changes, such as a periodic reactivity insertion with an appropriate negative bias, the resulting power variations remain small at all times, as we shall show later. In such cases, the point kinetic equations can be approximated by a set of linear differential equations with constant coefficients that can be solved exactly for any arbitrary reactivity insertion. This approximation is commonly referred to as the "linearization" of the kinetic equations.

Let $p(t)$ denote the variations in power about a reference level P_r, which will be specified later, and which depends on the problem under consideration. Substituting $P(t) = P_r + p(t)$ in the integrodifferential form of the point kinetic equations (cf 3.1, Eq. 9), we obtain

$$(l/\beta)\,\dot{p}(t) = k(t)[P_r + p(t)] + \int_0^\infty D(u)[p(t-u) - p(t)]\,du \tag{1}$$

This equation is a linear inhomogeneous equation in $p(t)$ with a time-dependent coefficient $k(t)$. The inhomogeneous term is $k(t)\,P_{\mathrm{r}}$. Because of the product term $p(t)\,k(t)$, the functional relationship between $p(t)$ and $k(t)$ characterized by (1) is not linear, i.e., the response to $k_1(t) + k_2(t)$ is not equal to the sum of the responses $p_1(t)$ and $p_2(t)$ to $k_1(t)$ and $k_2(t)$, separately, with the same initial conditions. However, if we ignore $p(t)\,k(t)$, then the resulting equation,

$$(l/\beta)\,\dot{p}(t) = P_{\mathrm{r}}k(t) + \int_0^\infty D(u)[p(t-u) - p(t)]\,du \tag{2}$$

becomes an inhomogeneous linear equation with constant coefficients, and represents a linear functional relationship between $p(t)$ and $k(t)$. In view of this observation, equation (2) is called the linearized point kinetic equation (in the absence of feedback effects), even though the original equation, (1), is already linear in $p(t)$. We shall always use the term "linearization" to imply linearization of the functional relationship between $p(t)$ and $k(t)$.

Equation (2) can be solved immediately by Laplace transforms. Let us assume that the reactor is operated at a constant power level P_0 prior to $t = 0$. It is convenient to choose the reference power level P_{r} as the equilibrium power level P_0 in this problem because then $p(t) \equiv 0$ for $t < 0$, and (2) reduces to (cf Section 4.4C)

$$(l/\beta)\,\dot{p}(t) = P_0k(t) + \int_0^t D(u)\,p(t-u)\,du - p(t) \tag{3}$$

Taking the Laplace transform of (3), we obtain

$$\bar{p}(s)/P_0 = Z(s)\,\bar{k}(s) \tag{4}$$

where $Z(s)$ is the zero-power transfer function introduced earlier (cf 3.2, Eq. 7; 3.3, Eq. 2), and $\bar{p}(s)$ and $\bar{k}(s)$ are the transforms of the power and reactivity variations, respectively. It is clear from (4) that the power response of the reactor to any arbitrary reactivity insertion can be obtained from (4) by finding the inverse Laplace transform of $Z(s)\,\bar{k}(s)$ in terms of the poles of $Z(s)$ and $\bar{k}(s)$, provided the incremental power is always small compared to P_0 (see Problem 10). The linear approximation is particularly suited to the investigation of the reactor response for small, periodic reactivity changes. Because of its importance in reactor stability analysis, we shall discuss the special case of a sinusoidal reactivity insertion in some detail. However, before proceeding to this problem, we shall examine briefly some basic concepts associated with linear systems as a digression in order to gain a deeper insight into the properties of the linear description of a reactor.

A. Reactor as a Linear System

The analysis of most physical systems can be reduced to the investigation of the relationship between two functions of time which constitute the "output" or "response" of the system to an "input" or "perturbation" (Figure 4.4.1). The input and output may be interpreted as cause and effect, respectively, in a broader sense. In a reactor, the input and output are usually chosen as the reactivity insertion and the resulting power response, although other pairs of dynamical variables, such as the inlet coolant temperature and pressure, are also considered as the input and output in some stability problems. A physical system may be pictured by a black box (Figure 4.4.1) which produces an output function for

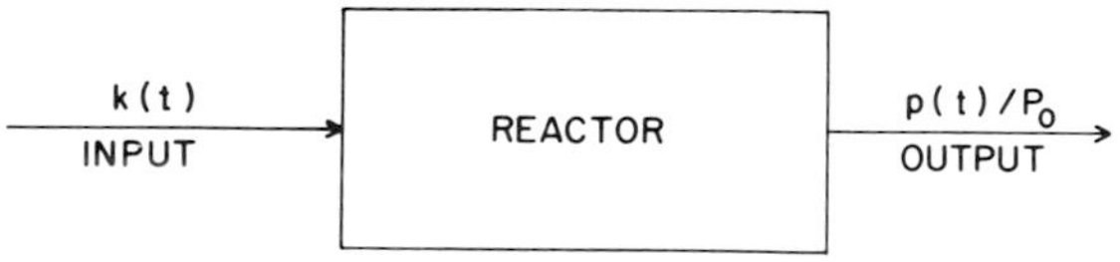

FIGURE 4.4.1. Input and output of a reactor.

a given input function [16]. The relationship between the input and output can be expressed mathematically by a functional, or transformation, as

$$p(t)/P_0 = F[k(t)] \tag{5}$$

A functional is essentially a prescription by which we can assign a value $p(t)$ to the output at a given time t when the input function is specified at all times. It is important to realize that the value of $p(t)$ at a time t may depend on the values of the input prior, or even subsequent, to t. In this sense, a functional may be viewed as an operator which produces the output function $p(t)$ when it acts on an input function $k(t)$. A physical system is thus characterized, from mathematical point of view, by a certain functional (or operator). The latter may be a set of ordinary or partial differential equations which determine uniquely, with the initial conditions, the output for a given forcing term $k(t)$. The point kinetic equation (1) may be viewed as a functional which relates the reactor power $p(t)$ to the reactivity insertion $k(t)$. In some cases, a functional describing a physical system may satisfy the following relation for two arbitrary numbers a_1 and a_2 :

$$a_1 p_1(t) + a_2 p_2(t) = P_0 F[a_1 k_1(t) + a_2 k_2(t)] \tag{6}$$

where p_1 and p_2 are the outputs corresponding to the inputs k_1 and k_2 .

Then, the physical system described by such a functional is said to be a linear system. The condition (6) is usually referred to as the principle of superposition, which states in more general physical terms that the combined effect of several causes is equal to the sum of the effects of each individual cause taken separately. Unfortunately, most physical systems are not linear in this sense, and a reactor is not an exception. Indeed, we can see from (1) that the response to $k_1 + k_2$ is not equal to $(p_1 + p_2)/P_0$, because of the product term $p(t)\,k(t)$, and hence the superposition principle is not obeyed. However, we obtain an approximate linear description of the reactor by ignoring this product term, since the resulting approximate equation (2) does obey the superposition principle.

We shall confine our discussion in the subsequent exposition to linear systems. When the parameters characterizing a linear system, such as l, β_i, and λ_i in a reactor, are independent of time, the linear operator is invariant under time translation, i.e.,

$$p(t - t_0) = P_0 F[k(t - t_0)] \tag{7}$$

where t_0 is an arbitrary constant. We shall always consider linear systems that are time-invariant. The physical implication of (7) is that the same cause always produces the same effect at any time.

A linear system with constant parameters can be described completely by its response $z(t)$ to a unit impulse input, i.e., $k(t) = \delta(t)$, where $\delta(t)$ is a Dirac delta function. The function $z(t)$, which is formally defined by

$$z(t) = F[\delta(t)] \tag{8}$$

is called the unit impulse response of the linear system. Any input $k(t)$ satisfies the identity

$$k(t) = \int_{-\infty}^{+\infty} dt' \, [\delta(t - t') \, k(t')] \tag{9}$$

because of the definition of $\delta(t)$. Substituting (9) into $F[k(t)]$, using the linearity and time invariance of $F[\cdots]$, we obtain

$$\begin{aligned} p(t) &= P_0 F \left\{ \int_{-\infty}^{+\infty} dt' \, [\delta(t - t') \, k(t')] \right\} \\ &= P_0 \int_{-\infty}^{+\infty} dt' \, \{k(t') \, F[\delta(t - t)']\} \end{aligned}$$

or

$$p(t) = P_0 \int_{-\infty}^{+\infty} dt' \, [k(t') \, z(t - t')] \tag{10}$$

The function $z(t - t')$ is the output of the system at t due to an impulse input at t'. Since the effect cannot precede its cause in time in a physical system,

$$z(t - t') \equiv 0 \qquad \text{for} \quad t < t' \tag{11}$$

must hold. This is known as the causality condition. Then, (10) reduces to

$$p(t) = P_0 \int_{-\infty}^{t} dt' \, [k(t') \, z(t - t')]$$

or

$$p(t) = P_0 \int_{0}^{\infty} dt' \, [k(t - t') \, z(t')] \tag{12}$$

Let us consider a linear system in which $k(t) = p(t) = 0$ for $t < 0$, e.g., an initially critical reactor. The response of this system to an arbitrary input $k(t)$ which is introduced at $t = 0$ is

$$p(t) = P_0 \int_{0}^{t} dt' \, [k(t - t') \, z(t')], \qquad t \geqslant 0 \tag{13}$$

Taking the Laplace transform of both sides,

$$\bar{p}(s) = P_0 Z(s) \, \bar{k}(s) \tag{14}$$

where

$$Z(s) \equiv \int_{0}^{\infty} e^{-st} z(t) \, dt \tag{15}$$

and is called the transfer function of the linear system. By comparing (14) to (4), we see that $Z(s)$ is the zero-power transfer function. It is clear that a linear system is completely described either by its unit impulse response or by its transfer function. For a reactor (cf Eq. 4),

$$Z(s) = \left[s \left(\frac{l}{\beta} + \sum_{i=1}^{6} \frac{a_i}{\lambda_i + s} \right) \right]^{-1} \tag{16}$$

and

$$z(t) = \frac{\beta}{l^*} + \sum_{j=1}^{6} \left\{ e^{\omega_j t} / \omega_j \left[\frac{l}{\beta} + \sum_{i=1}^{6} \frac{a_i \lambda_i}{(\lambda_i + \omega_j)^2} \right] \right\} \tag{17}$$

where ω_j are negative roots of $Y(\omega_j) \equiv [1/Z(s)] = 0$. The variation of $z(t)$ with t is plotted in Fig. 4.4.2. It is interesting to note that

$$\lim_{t \to \infty} z(t) = \beta / l^* \tag{18}$$

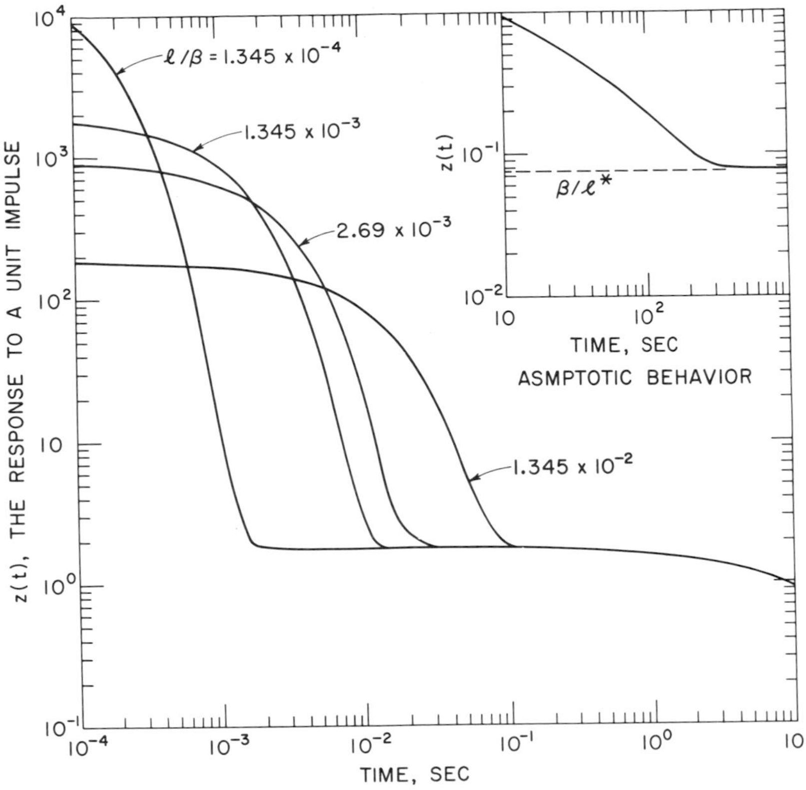

FIGURE 4.4.2. The unit impulse response function of a reactor.

which implies that the effect of a momentary cause in a linear system is remembered even after a long time. This memory effect is due to the pole of $Z(s)$ at $s = 0$. To better understand the implication of this pole, we shall now introduce the concept of the *stability* of linear systems. A linear system is said to be stable if its response to any bounded input is also bounded. The necessary and sufficient condition for stability is

$$\int_0^\infty | z(t)| \, dt < \infty \tag{19}$$

The sufficiency of the proof follows immediately, since

$$| p(t)| < P_0 \int_0^\infty dt' \, [| k(t - t')| \, | z(t')|] < MP_0 \int_0^\infty | z(t)| \, dt$$

where M is the bound of the input, i.e., $|k(t)| < M$. Its necessity can be shown, following Papoulis [16], by constructing a bounded input for which the output is unbounded if (19) does not hold. Let the input be

$$k(-t) = z(t)/|z(t)|$$

so that $|k(t)| = 1$. The corresponding response at $t = 0$ is

$$[p(0)/P_0] = \int_0^\infty dt' \, [k(-t')\, z(t')] = \int_0^\infty dt \, [z^2(t)/|z(t)|]$$
$$= \int_0^\infty dt \, |z(t)| \tag{20}$$

and thus is unbounded if (19) does not hold. Equation (17) indicates, that $z(t)$ is not absolutely integrable, because of the constant term β/l^*, and hence a critical reactor without feedback is unstable with respect to bounded inputs.

B. Relationship between Frequency Characteristics of a Reactor

We shall now discuss the relations between the real and imaginary parts, and also between the amplitude and phase of the transfer function $Z(s)$ for pure imaginary arguments. These relations follow from the analyticity of $Z(s)$ for Re $s \geqslant 0$ except for $s = 0$, where it has a simple pole.

We shall first determine the equations relating the real and imaginary parts, denoted by $R(\omega)$ and $X(\omega)$, respectively, of

$$Z(i\omega) = \left[i\omega \left(\frac{l}{\beta} + \sum_{j=1}^{6} \frac{a_j}{\lambda_j + i\omega} \right) \right]^{-1} \tag{21}$$

We consider the contour integral of $Z(s)/(s - i\omega_0)$ on the closed path C shown in Figure 4.4.3a. This function is analytic on the contour, and in its interior. Hence,

$$\int_C [Z(s)/(s - i\omega_0)] \, ds = 0$$

Since $Z(s)$ behaves as $1/s$ when $s \to \infty$, the integration along the large semi circle vanishes as $R \to \infty$. The integration on the small semi circles r_1 and r_2 and on the portions of the imaginary axis can be calculated easily as

$$\int_{-\infty}^{+\infty} [Z(i\omega)/(\omega - \omega_0)] \, d\omega + i\pi[Z(i\omega_0) + (\beta/l^*)(i/\omega_0)] = 0 \tag{22}$$

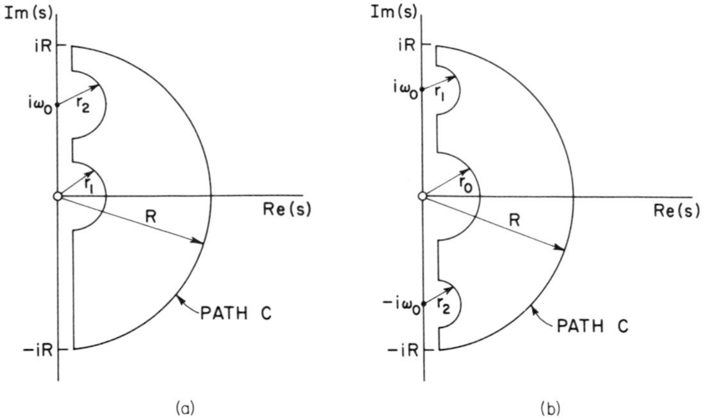

FIGURE 4.4.3. Figures used to prove dispersion relations.

where the term $(\beta/l^*)(i/\omega_0)$ stems from the integration around the pole at the origin, and where the improper integral is to be interpreted as the Cauchy principal part.† Equating real and imaginary parts to zero in (22), we obtain

$$R(\omega_0) = (1/\pi) \int_{-\infty}^{+\infty} [X(\omega)/(\omega_0 - \omega)]\, d\omega \tag{23}$$

and

$$X(\omega_0) = -(\beta/l^*\omega_0) - (1/\pi) \int_{-\infty}^{+\infty} [R(\omega)/(\omega_0 - \omega)]\, d\omega \tag{24}$$

These relations are known as Hilbert transforms. We observe that it is sufficient to know only either $R(\omega)$ or $X(\omega)$ to determine the reactor transfer function $Z(i\omega)$.

In order to determine the relation between the amplitude, $G(\omega) \equiv |Z(i\omega)|$, and the phase, $\phi(\omega) \equiv \arg Z(i\omega)$, we consider a function

$$Q(s) = [\ln Z(s)]/(s^2 + \omega^2) \tag{25}$$

which is analytic for Re $s \geqslant 0$ except for the simple poles at $s = \pm i\omega_0$ and the logarithmic singularity at $s = 0$. We note that $\ln Z(s)$ would have

† For example,

$$\int_{-\infty}^{+\infty} \frac{R(\omega)}{\omega_0 - \omega}\, d\omega = \lim_{\epsilon \to 0} \left[\int_{-\infty}^{\omega_0 - \epsilon} \frac{R(\omega)\, d\omega}{\omega_0 - \omega} + \int_{\omega_0 + \epsilon}^{\infty} \frac{R(\omega)\, d\omega}{\omega_0 - \omega} \right]$$

singularities in the right half s-plane, had $Z(s)$ possessed zeros with positive real parts.† For a reactor, all the zeros of $Z(s)$ are negative, and, in fact, are equal to $-\lambda_1, ..., -\lambda_6$, as can be seen from (16). We consider again the integral of $Q(s)$ on the closed path C shown in Fig. 4.4.3b. The integral on the large semi-circle vanishes because $Z(s)$ behaves as $1/s$ for large s, and $(s \ln s)/(s^2 + \omega_0^2) \to 0$ as $s \to \infty$.‡ The reason for choosing $Q(s)$ as in (25) is now clear. If we had chosen it as $[\ln Z(s)]/(s - i\omega_0)$, as one would have expected, then the integral on the large circle would diverge logarithmically. The integral around the singularity at the origin vanishes because $r_0 \ln r_0 \to 0$ as $r_0 \to 0$. The integration on the portions of the imaginary axis and on the semicircles r_1 and r_2 are evaluated as

$$\int_{-\infty}^{+\infty} \{[\ln Z(i\omega)]/(\omega_0^2 - \omega^2\} \, d\omega + (\pi/2\omega_0)[\ln Z(i\omega_0) - \ln Z(-i\omega_0)] = 0 \tag{26}$$

Using $\ln Z(i\omega_0) = [\ln G(\omega_0)] + i\phi(\omega_0)$, noting that $G(\omega_0) = G(-\omega_0)$ and $\phi(\omega_0) = -\phi(-\omega_0)$ [because $Z(-i\omega) = Z^*(i\omega)$], and equating the imaginary terms in (26) to zero, we obtain the desired relation:

$$\phi(\omega_0) = (2\omega_0/\pi) \int_0^{\infty} \{[\ln G(\omega)]/(\omega^2 - \omega_0^2)\} \, d\omega \tag{27}$$

which expresses the phase of the reactor transfer function on the imaginary axis, in terms of its logarithmic amplitude. As an application of (27), let us assume that $G(\omega) = 1/\omega^\nu$, where ν is a positive constant [17], and compute the corresponding phase angle:

$$\phi(\omega_0) = -(2\nu/\pi)\, \omega_0 \int_0^{\infty} [(\ln \omega)/(\omega^2 - \omega_0^2)] \, d\omega$$

which can be evaluated (see Problem 11) as

$$\phi(\omega_0) = -(\pi/2)\, \nu \tag{28}$$

This result is of course expected from the fact that the phase angle of $1/(i\omega)^\nu = (1/\omega^\nu) \exp[-i(\pi/2)\nu]$ is $-\pi\nu/2$.

† Systems whose transfer function has no zeros with positive real parts are referred to as being of minimum phase-shift type.

‡ The integration on the large semicircular arc is

$$\int_{+\pi/2}^{-\pi/2} Q(Re^{i\theta}) Rie^{i\theta} \, d\theta \to i \int_{\pi/2}^{-\pi/2} \{R[(\ln R) + i\theta]/R^2\} e^{-i\theta} \, d\theta = (2/R)(i - \ln R)$$

which vanishes as $R \to \infty$.

The foregoing proof indicates that the relation between $\phi(\omega)$ and $\ln G(\omega)$ given by (27) is not the only one; other relations similar to (27) can be derived by choosing different forms for $Q(s)$ in (25). However, for a given amplitude characteristic, they all predict the same phase characteristic, provided the system is of minimum phase-shift type. We shall not, however, dwell on this subject any longer; the relations derived above should be sufficient to illustrate the concepts and the methods involved in the analysis of linear systems (see Problem 12).

C. Response to a Small Sinusoidal Input

We discuss power oscillations induced by a reactivity variation of the form

$$k(t) = \delta k \sin \omega t$$

Substituting

$$\bar{k}(s) = [\omega/(\omega^2 + s^2)]\, \delta k$$

in (4) and finding the inverse Laplace transform of the resulting expression, we obtain

$$\frac{p(t)}{P_0} = \delta k \left[\frac{Z(i\omega)}{2i}\, e^{i\omega t} - \frac{Z(-i\omega)}{2i}\, e^{-i\omega t}\right] + \delta k\, \omega \sum_{j=0}^{6} \frac{e^{\omega_j t}}{(\omega^2 + \omega_j{}^2)\, Y'(i\omega_j)} \tag{29}$$

where the first two terms arise from the poles of $\bar{k}(s)$ on the imaginary axis at $s = i\omega$ and $s = -i\omega$, and the remaining terms are due to the poles of $Z(s)$ that are the roots ω_j of the inhour equation $Y(\omega_j) = 0$. Six of these terms for $j = 1, 2, \ldots, 6$ are associated with the negative roots of the inhour equation, and vanish as $t \to \infty$. However, the term corresponding to $\omega_0 = 0$ is a constant. The asymptotic behavior of the power oscillations as $t \to \infty$ therefore becomes (Figure 4.4.4)

$$p(t)/P_0 = (\delta k)\, G(\omega)[\sin(\omega t + \phi)] + (\delta k/\omega)\, (\beta/l^*) \tag{30}$$

where

$$G(\omega) = |\, Z(i\omega)| \tag{31a}$$

$$\phi(\omega) = \tan^{-1}[\operatorname{Im} Z(i\omega)/\operatorname{Re} Z(i\omega)] \tag{31b}$$

In obtaining (30), we have used $\sin x = (e^{ix} - e^{-ix})/2$, and , in (29), replaced $Z(i\omega)$ by $|\, Z(i\omega)|\, e^{i\phi}$. The constant term in (30) represents a shift of the average power level, in the presence of the power oscillations (Figure 4.4.4), from P_0 to $P_0 + \delta P_0$, where

$$\delta P_0 = P_0 \lim_{s \to 0} [s\bar{k}(s)\, Z(s)] \tag{32a}$$

or

$$\delta P_0 = P_0 (\delta k)\, \beta/l^* \omega \tag{32b}$$

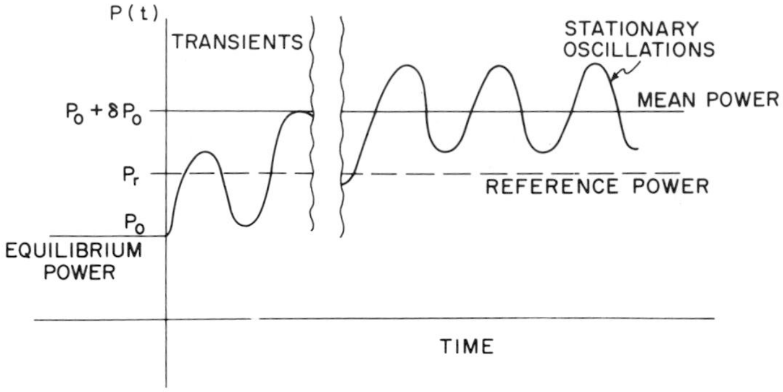

FIGURE 4.4.4. Power response of an initially critical reactor to a pure sinusoidal reactivity insertion (linearized response); $k(t) = \delta k \sin \omega t$; $\delta P_0 = P_0(\delta k)\, \beta/\omega l^*$.

For a sinusoidal reactivity insertion with an arbitrary initial phase, i.e., $k(t) = \delta k \sin(\omega t + \theta)$, for which $\bar{k}(s) = [\omega \cos \theta + s \sin \theta]/(s^2 + \omega^2)$, (32a) yields

$$\delta P_0 = P_0(\delta k)\, (\beta \cos \theta)/\omega l^* \tag{33a}$$

The change in the mean power from P_0 to $P_0 + \delta P_0$ affects the accuracy of the pile-oscillator experiment which will be discussed below.

The gain of a reactor is usually defined as the ratio of the relative amplitude (with respect to the mean power) of the sinusoidal power oscillations to the amplitude of the sinusoidal reactivity variations inducing the power oscillations. This definition is valid even in the presence of the nonlinearities which are discussed in the subsequent sections of this chapter. According to (30) (see also Figure 4.4.4), however, the gain as defined above is equal to

$$G(\omega)/\{1 + [(\delta k)\, \beta/l^* \omega)]\} \tag{33b}$$

which is slightly different than $G(\omega)$. Since the linear approximation implies the limit of $\delta k \to 0$ in the definition of the gain, the additional term in the denominator is in fact of higher order than retained in the linearization.

We can clarify this point also by remembering that the choice of the reference power level P_r used in the linearization procedure (cf Eqs. 1 and 2) is arbitrary. In (30), it is taken as the equilibrium power level P_0 prior to $t = 0$. If we solve (2) with an arbitrary P_r, we obtain (Problem 13), for an arbitrary input,

$$\bar{p}(s)/P_r = Z(s)\, \bar{k}(s) + (1/s)[(P_0/P_r) - 1] \tag{33c}$$

The last term vanishes if $P_r = P_0$, reducing (33c) to (4). In the case of a pure sinusoidal input, (33c) yields, after a long time,

$$p(t)/P_r = (\delta k)\, G(\omega) \sin(\omega t + \phi) + (\delta k/\omega)(\beta/l^*) + [(P_0/P_r) - 1] \quad (33d)$$

which is a generalization of (30). Figure 4.4.4 suggests that it is more appropriate to choose P_r as the mean power level in the presence of the power oscillations, because, then, the deviations are smallest at all times, as required by the linearization. The value of P_r in this case is obtained by equating the sum of the last two terms to zero:

$$P_r = P_0[1 - (\delta k\, \beta/\omega l^*)] \quad (33e)$$

which is consistent with $|\,\delta P_0\,| = (\delta k)\beta/\omega l^*$. With this choice of P_r, (33d) reduces to

$$p(t)/P_r = (\delta k)\, G(\omega) \sin(\omega t + \phi) \quad (33f)$$

which justifies the above definition of the reactor gain in the linearized treatment.

D. Variation of the Zero-Power Transfer Function with Frequency

We observe from (33f) that the relative amplitude of the sinusoidal power oscillations is proportional to $|\,Z(i\omega)|$, the magnitude of the zero-power transfer function $Z(s)$ at $s = i\omega$. Equation (31b) indicates that the relative phase of the reactivity oscillations and the resulting power oscillations is equal to the phase angle of $Z(i\omega)$. Hence, we can obtain the values of $Z(s)$ on the imaginary axis in the complex s-plane by measuring experimentally the amplitude and relative phase, as functions of frequency, of power oscillations induced by a sinusoidal reactivity insertion with a constant amplitude. This is the basis of the so-called zero-power pile-oscillator experiment. The values of $Z(s)$ for arbitrary complex arguments with a positive real part can be related to $R(\omega)$ by

$$Z(s) = (\beta/l^* s) + (2/\pi)\, s \int_0^\infty [R(\omega)/(s^2 + \omega^2)]\, d\omega, \qquad \operatorname{Re} s > 0 \quad (34)$$

(see Problem 14). The values of $Z(s)$ for $\operatorname{Re} s < 0$ can be obtained from (34) by analytical continuation. It follows from (34) that we can determine the reactor transfer function completely on the entire complex plane by measuring its real values.

However, we can extract information concerning the macroscopic properties of the reactor directly from the observed values of the amplitude and phase of $Z(i\omega)$, avoiding the need for constructing $Z(s)$ for all s. The functions $G(\omega) = |Z(i\omega)|$ and $\phi(\omega) = \arg Z(i\omega)$ are called, respectively, the amplitude and phase characteristics of a reactor at zero power. Figures 4.4.5 and 4.4.6 show the amplitude and phase characteristics

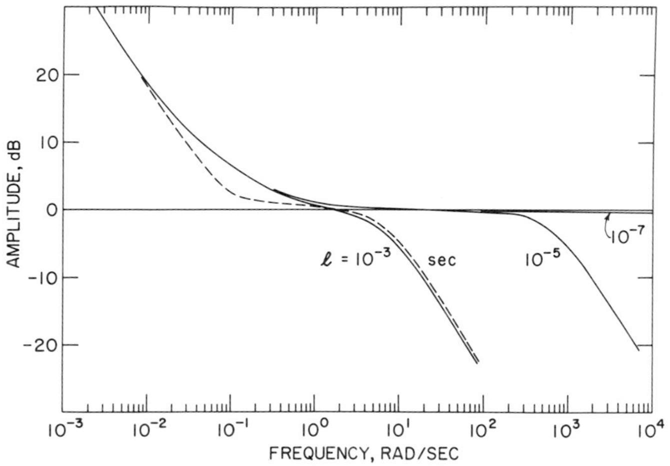

FIGURE 4.4.5. The amplitude of the zero-power transfer function (^{235}U).

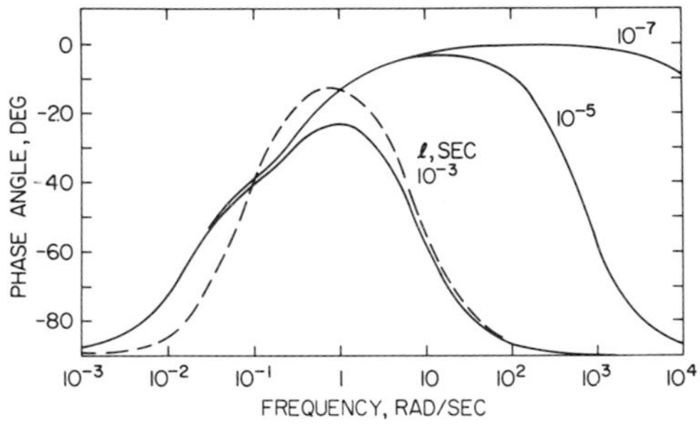

FIGURE 4.4.6. The phase of the zero-power transfer function (^{235}U).

for a reactor with ^{235}U as fuel and mean prompt generation times of $l = 10^{-3}$, 10^{-5}, and 10^{-7} sec. Both curves are plotted in the logarithmic frequency scale. The amplitude characteristic is expressed in decibels, as is customary in communications engineering; conversion of the amplitude to decibels is governed by

$$G_{\text{dB}}(\omega) = 20 \log_{10} G(\omega) \tag{35}$$

The behavior of these curves can be investigated by expressing $Z(i\omega)$ as

$$Z(i\omega) = \frac{\beta}{l} \frac{(i\omega + \lambda_1) \cdots (i\omega + \lambda_6)}{i\omega(i\omega - \omega_1) \cdots (i\omega - \omega_j)} \tag{36}$$

Where the ω_j are the roots of $Y(s) = 0$. The numerical values of ω_j for ^{235}U with $l = 10^{-5}$ sec are presented in Table 1. Equation (36) can

TABLE 1

ROOTS OF $Y(s) = 0$ FOR ^{235}U AND $l = 10^{-5}$ sec

$\omega_1 = -0.014$	$\omega_4 = -1.02$
$\omega_2 = -0.068$	$\omega_5 = -2.90$
$\omega_3 = -0.195$	$\omega_6 = -641.4$

be cast into more conventional [17–19] form as

$$Z(i\omega) = K \frac{(1 + i\omega\tau_1) \cdots (1 + i\omega\tau_6)}{i\omega(1 + i\omega t_1) \cdots (1 + i\omega t_0)} \tag{37a}$$

where

$$\tau_j = 1/\lambda_j \tag{37b}$$

$$t_j = 1/|\,\omega_j\,| \tag{37c}$$

$$K = (\beta/l)\, \lambda_1\lambda_2 \cdots \lambda_6/\omega_1\omega_2 \cdots \omega_6 \tag{37d}$$

The amplitude and phase characteristics can now be expressed as

$$G_{\mathrm{dB}}(\omega) = 20 \log_{10} K + 10[\log_{10}(1 + \omega^2\tau_1^2) + \cdots + \log_{10}(1 + \omega^2\tau_6^2) - \log_{10} \omega^2 - \log_{10}(1 + \omega^2 t_1^2) \cdots \log_{10}(1 + \omega^2 t_6^2)] \tag{38}$$

and

$$\phi(\omega) = \tan^{-1} \omega\tau_1 + \cdots + \tan^{-1} \omega\tau_6 - 90^\circ - \tan^{-1} \omega t_1 - \cdots - \tan^{-1} \omega t_6 \tag{39}$$

These equations indicate that both the amplitude and phase characteristics in Figures 4.4.5 and 4.4.6 are sums of several simpler curves which represent the frequency characteristcs of the individual terms in (38) and (39). It is therefore sufficient to investigate the magnitude and phase of each term as a function of frequency. These curves can be graphed approximately using their asymptotes for large and small frequencies, and then computing the errors from these asymptotes. There are two kinds of terms in (38) and (39): $(1 + i\omega\tau_j)$ and $(1 + i\omega t_j)^{-1}$. The frequency characteristics of these functions are shown in Figures 4.4.7

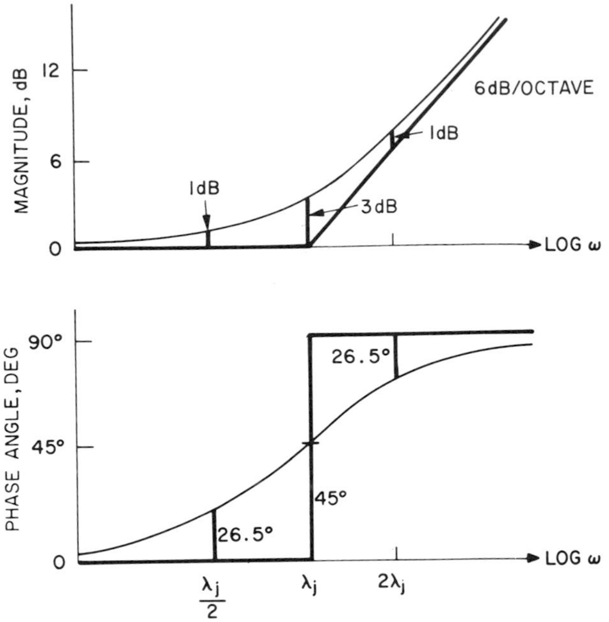

FIGURE 4.4.7. Frequency characteristics of $1 + i\omega\tau_j$.

and 4.4.8. The frequencies $\lambda_j = 1/\tau_j$ and $|\omega_j| = 1/t_j$ are called the "corner" frequencies. The heavy straight-line segments are the asymptotes of the curves away from the corner frequencies. The asymptotes in the magnitude plots for frequencies larger than the corner frequency have a slope 6 dB/octave, which is positive for terms like $(1 + i\omega\tau_j)$ and negative for $(1 + i\omega t_j)^{-1}$ (i.e., it increases or decreases 6 dB when the frequency is doubled). If we had terms like $(1 + i\omega\tau_j)^N$, then the slope would be $6N$ dB/octave (such terms may occur in feedback transfer functions, which we discuss in Chapter 5). The errors in the magnitude plot are calculated from $10 \log 2 = 3.01$ at the corner frequency, $10 \log(5/4) = 0.969$ at one-half of the corner frequency, and $[10 \log 5] - 6 = 0.989$ at double the corner frequencies.

In the phase plots, the asymptotes are horizontal straight-line segments at 0 deg for small frequencies, and at ∓ 90 deg, for large frequencies, depending on the type of term: positive for $(1 + i\omega\tau_j)$ (lead) and negative for $(1 + i\omega t_j)^{-1}$ (lag). The errors are calculated as $\tan^{-1}(1 + i) = 45$ deg at the corner frequency, $\tan^{-1}[1 + (i/2)] \simeq 26.5$ deg at one-half this frequency, and $\tan^{-1}[1 + (i/2)] \approx 26.5$ deg at double the corner frequency.

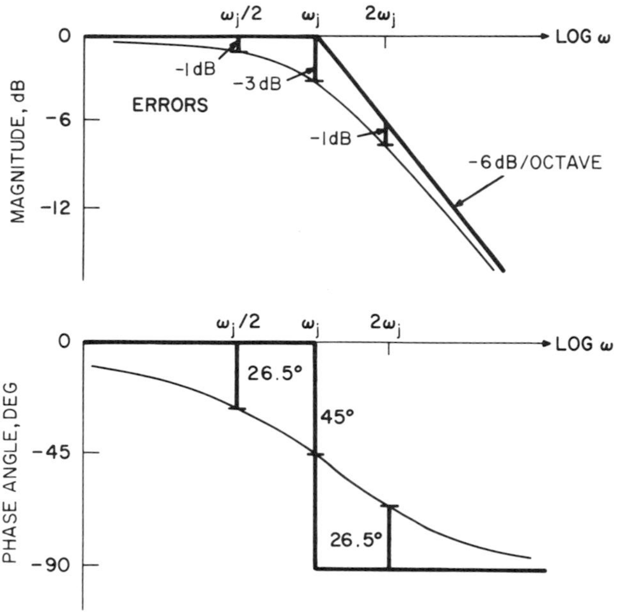

FIGURE 4.4.8. Frequency characteristics of $(1 + i\omega t_j)^{-1}$.

As an example, we present the model of one group of delayed neutrons in Figure 4.4.9, for which

$$Z(i\omega) = K(i\omega + \omega_L)/i\omega(i\omega + \omega_H)$$

where $K = \beta/l$, $\omega_L = \bar{\lambda}$, and $\omega_H = (\beta/l) + \bar{\lambda} \approx \beta/l$, with $l = 10^{-4}$ sec, $\bar{\lambda} = 0.08\ \text{sec}^{-1}$, and $\beta = 0.0075$.

The behavior of these curves is self-explanatory. We just note that the phase angle is always negative, equal to -45 deg at the break frequencies, and tends toward -90 deg in the low- and high-frequency regions, where the amplitude, expressed in decibels, has a constant slope. We recall at this point the result in (28), which was obtained by using the general relation between amplitude and phase characteristics. Finally, the phase angle tends toward zero in frequency regions where the amplitude is approximately constant, which is also a consequence of (27).

The corner frequencies ω_L and ω_H can be determined experimentally by fitting the straight-line-segment approximation to the measured amplitude and phase characteristics. From the knowledge of ω_H, we infer β/l, which is an important macroscopic nuclear property of a given

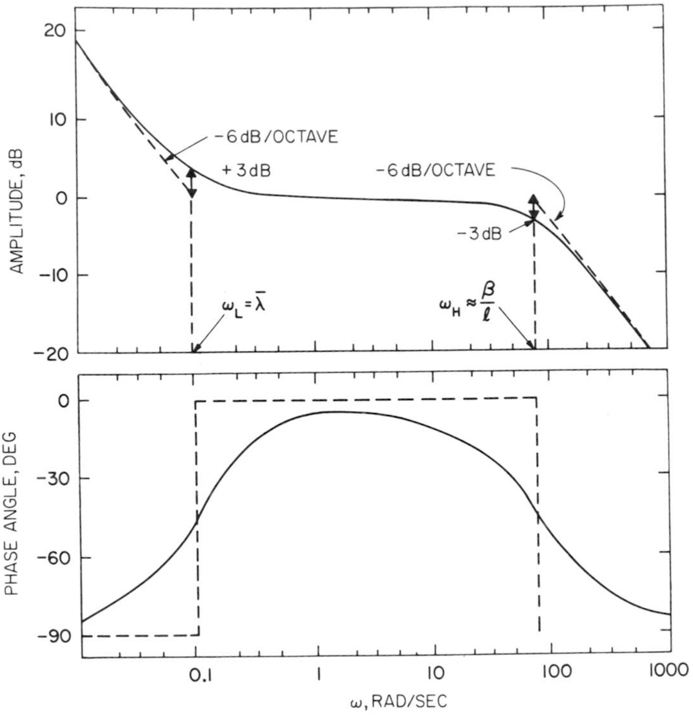

FIGURE 4.4.9. Amplitude and phase characteristics of a one-group delayed neutron model.

reactor. The pile-oscillator technique, as described above, is one of the most precise methods for measuring β/l.

It is concluded from Figures 4.4.5 and 4.4.9 that there are three more or less distinct frequency regions in which the phase characteristics have simple forms. At very low frequencies, $\omega \ll \bar{\lambda}$, the transfer function

$$Z(s) = \left[s\left(\frac{l}{\beta} + \sum_i \frac{a_i}{\lambda_i + s}\right)\right]$$

behaves as

$$Z(s) = \beta/l^*s, \qquad l^* = l + \sum_i (\beta_i/\lambda_i) \qquad \text{(low frequencies)} \tag{40a}$$

At intermediate frequencies, between $\omega = \bar{\lambda}$ and $\omega = \beta/l$, provided β/l is sufficiently larger than $\bar{\lambda}$, $Z(s)$ is constant:

$$Z(s) = 1 \qquad \text{(intermediate frequencies)} \tag{40b}$$

At high frequencies,

$$Z(s) = \beta/ls \qquad \text{(high frequencies)} \tag{40c}$$

The plateau observed in Figures 4.4.5 and 4.4.9 in the magnitude plots corresponds to (40b), where $G(\omega) = 1$ and $\phi(\omega) = 0$ (Problem 15).

4.5. Logarithmic Linearization

The linear approximation discussed in the preceding section requires the incremental power variations to be small compared to the equilibrium power level throughout the time interval of interest. In this section, we shall relax this requirement by linearizing [20] the functional relationship between the logarithm of the power and $k(t)$, rather than that between $p(t)$ and $k(t)$. Our starting point is again the integrodifferential form of the point kinetic equations (cf 3.1, Eq. 9) without an external source. Substituting

$$y(t) = \ln[P(t)/P_r] \tag{1}$$

in this equation, we obtain

$$(l/\beta)\, \dot{y}(t) - \int_0^\infty D(u)[e^{y(t-u)-y(t)} - 1]\, du = k(t) \tag{2}$$

where P_r is, for now, an arbitrary reference power level, although we shall ultimately choose it as the equilibrium power level before the insertion of the reactivity perturbation. Equation (2) describes a physical system whose input–output functional relation is nonlinear. We obtain an approximate description of this system by assuming that

$$|\, y(t-u) - y(t)| \ll 1 \tag{3a}$$

or

$$|\, \ln[P(t-u)/P(t)]| \ll 1 \tag{3b}$$

holds for all u and t, and by then linearizing the exponential term:

$$(l/\beta)\, \dot{y} - \int_0^\infty D(u)[\, y(t-u) - y(t)]\, du = k(t) \tag{4}$$

The implication of condition (3) is that it requires the variation of the logarithmic power in a time interval $(t - u, t)$ to be small compared to unity. Although the length of this integral varies from zero to infinity with u, the contribution arising from the large values of u to the integral in (2) is weighted by $D(u)$, which vanishes exponentially as $u \to \infty$. Hence, we may choose this interval as $(t - \bar{\tau}, t)$ where $\bar{\tau}$ is the mean decay time of all the delayed neutron groups. Condition (3) is much weaker than the $|\, P(t) - P_0 \,|/P_0 \ll 1$ which is the basis of linearization in the conventional sense, because it allows a gradual change in the reactor power, provided that the relative change of power in the interval

t and $t - \bar{\tau}$ is always small. The following examples will clarify this qualitative discussion.

Let us consider again a critical reactor operating at a constant power level P_0 prior to $t = 0$, and assume that $k(t)$ is varied arbitrarily for $t \geqslant 0$. Noting that $y(t) = 0$ for $t < 0$, if we choose $P_r = P_0$ (see Problem 16), we obtain from (4)

$$(l/\beta)\,\dot{y} + y(t) - \int_0^t D(u)\,y(t-u)\,du = k(t) \tag{5}$$

which is identical to (4.4, Eq. 3) with $p(t)/P_0$ replaced by $y(t)$. The Laplace transform of (5) yields

$$\bar{y}(s) = Z(s)\,\bar{k}(s)$$

which indicates that the linear system obtained by logarithmic linearization is also described by the zero-power transfer function $Z(s)$.

The response to a step input can be easily found (see Problem 17) as

$$y(t) = \delta k\,\frac{\beta}{l^*}\left(t + \frac{\beta}{l^*}\langle\tau^2\rangle_{\text{av}}\right) + \sum_{j=1}^{6} \frac{(\delta k)\,e^{\omega_j t}}{\omega_j Y'(\omega_j)} \tag{6}$$

where

$$\langle\tau^2\rangle_{\text{av}} = \sum_{i=1}^{6} (a_i/\lambda_i^2) \tag{7}$$

The asymptotic behavior of the reactor power is

$$P(t) = P_0 \exp[+\delta k(\beta/l^*)^2\langle\tau^2\rangle_{\text{av}} + \delta k(\beta/l^*)\,t] \tag{8}$$

which is identical to the results obtained from the exact solution (3.3, Eq. 15) in the limit of $\delta k \to 0$. We may point out that the usual linear approximation predicts the asymptotic response (see Problem 10) as

$$P(t) = P_0 + P_0(\delta k)(\beta/l^*)[t + (\beta/l^*)\,\langle\tau^2\rangle_{\text{av}}]$$

which is linear in time, and valid only for short times, in contrast to (8), which is valid at all times. This example justifies the remarks we have made about the validity of the logarithmic linearization (cf Eq. 3).

As a second example, we consider the response of the reactor to a sinusoidal reactivity change $k(t) = \delta k \sin \omega t$ after the transient effects have disappeared. Using the solution given in Problem 16, which was obtained for an arbitrary reference power P_r, we obtain

$$P(t) = P_r \exp[(\delta k)\,G(\omega)\sin(\omega t + \phi) + \delta k(\beta/\omega l^*) + y(0)] \tag{9a}$$

where $y(0) = \ln(P_0/P_r)$. [This result can be obtained from (4.4, Eq. (33d) by replacing $P(t)/P_r$ by $y(t)$.] We can choose P_r such that $[(\delta k)\beta/\omega l^*] + y(0) = 0$, or

$$P_r = P_0 \exp[(\delta k)\,\beta/\omega l^*] \tag{9b}$$

Then, (9a) reduces to

$$P(t) = P_r \exp[(\delta k)\,G(\omega)\sin(\omega t + \phi)] \tag{10}$$

which agrees with the result (4.4, Eq. (33f) obtained using linearization, as when δk is small. We note that (10) implies a periodic solution for a pure sinusoidal reactivity insertion with zero average. This contradicts the conclusions reached in Section 3.2A using the method that the average reactivity over a period must be negative when the power oscillations are periodic. Although (10) is extremely accurate in finite time intervals, as will be shown later, it is not sufficiently accurate to describe the asymptotic nature of the point kinetic equations in the case of pure sinusoidal reactivity insertion.

The Transfer Function at Period-Equilibrium

A last important application of the logarithmic linearization is the derivation of the reactor transfer function when the power is rising or falling with a constant period ω_0. This state of the reactor may be called [21] "period-equilibrium." The response of the reactor at this state to a small sinusoidal reactivity insertion is described by a modified transfer function which is sufficiently different than the conventional transfer function, and may be of considerable importance in the design of high-performance control systems.

We assume that $y(t) = \omega_0 t$ for $t \leq 0$, which is maintained by a constant reactivity k_0. Here ω_0 is the constant inverse period. At $t = 0$, we introduce an arbitrary but small reactivity $k_1(t)$. The resulting power variation is denoted by $\epsilon(t)$. Then, $y(t) = \omega_0 t + \epsilon(t)$ for $t > 0$. We desire to find the ratio of the Laplace transforms of $\epsilon(t)$ and $k_1(t)$, which defines the period-equilibrium transfer function. We then substitute

$$\begin{aligned} y(t) &\equiv \omega_0 t, & k(t) &\equiv k_0\,, & t &\leqslant 0 \\ y(t) &= \omega_0 t + \epsilon(t), & k(t) &= k_0 + k_1(t), & t &> 0 \end{aligned}$$

in (2) and obtain, for $t > 0$,

$$\begin{aligned} &(l/\beta)[\omega_0 + \dot{\epsilon}(t)] + 1 - e^{-\epsilon(t)} \left[\int_0^t du\, D(u)\, e^{-\omega_0 u}(e^{\epsilon(t-u)} - 1) + \bar{D}(\omega_0)\right] \\ &\quad = k_0 + k_1(t) \end{aligned} \tag{11}$$

This equation can be linearized in $\epsilon(t)$ by using $e^{\epsilon(t)} \approx 1 + \epsilon(t)$ and ignoring terms of ϵ^2 or higher:

$$(l/\beta)\,\dot{\epsilon}(t) + \bar{D}(\omega_0)\,\epsilon(t) - \int_0^t du\, D(u)\, e^{-\omega_0 u}\epsilon(t-u) = k_1(t) \tag{12}$$

where we have used the fact that ω_0 and k_0 satisfy the inhour equation. Since $\epsilon(0) = 0$, the Laplace transform of (12) immediately yields the desired transfer function $Z(\omega_0, s) = \bar{\epsilon}(s)/\bar{k}_1(s)$,

$$Z(\omega_0, s) = \left[s\left(\frac{l}{\beta} + \sum_{i=1}^{6} \frac{a_i\lambda_i}{(\lambda_i + \omega_0)(\lambda_i + s + \omega_0)}\right)\right]^{-1} \tag{13}$$

which, of course, reduces to the conventional zero-power transfer function when $\omega_0 = 0$. The presence of ω_0 affects only the low and intermediate-frequency responses:

$$Z(\omega_0, s) = \beta/sl^*(\omega_0) \qquad \text{(low frequencies)} \tag{14a}$$

$$l^*(\omega_0) \equiv l + \sum_{i=1}^{6} [\beta_i\lambda_i/(\lambda_i + \omega_0)^2] \tag{14b}$$

and

$$Z(\omega_0, s) = \left\{\sum_{i=1}^{6} [a_i\lambda_i/(\lambda_i + \omega_0)]\right\}^{-1} \qquad \text{(intermediate frequencies)} \tag{15}$$

In both cases, the gain of the reactor increases at a given frequency with increasing ω_0. The high-frequency behavior is still described by β/ls (Problem 18).

The foregoing examples (see also Problem 17) indicate that we can construct more accurate solutions to the reactor kinetic equations for an arbitrary reactivity insertion by using logarithmic linearization without any more mathematical complexity than is involved in the linear approximation. Such solutions, however, are still not sufficiently accurate to describe the asymptotic behavior, as we already pointed out.

4.6. Perturbation Analysis

In this section, we shall present a perturbation technique [15] for constructing in a systematic way approximate solutions of the point kinetic equations for an arbitrary reactivity insertion. In particular, we shall consider the power response to a sinusoidal reactivity input and compare it to the result obtained in the previous section by logarithmic linearization. We shall show that the power oscillations are not periodic and, in fact, increase exponentially in time.

A. Construction of the General Solution

The analysis starts with the standard form of the kinetic equation,

$$(l/\beta)\,\dot{P} = [k(t) - 1]\,P + \sum_{i=1}^{6} \lambda_i C_i \tag{1a}$$

$$\dot{C}_i = a_i P - \lambda_i C_i\,, \qquad i = 1, 2,..., 6 \tag{1b}$$

We assume that the reactivity insertion is of the form

$$k(t) = k_0 + \epsilon y(t) \tag{2}$$

where k_0 is the constant part of the reactivity, and $y(t)$ is a known function of time describing the time dependence of the reactivity. The parameter ϵ is a positive number which measures the magnitude of the reactivity variations and is assumed to be sufficiently small to allow a perturbation analysis. No restriction is imposed on the magnitude of k_0.

Our approach consists in finding linearly independent particular solutions of (1), and then obtaining the general solution as a linear combination of these solutions. We have to find as many particular solutions as the number of unknown functions, i.e., $P(t)$ and $C_i(t)$. This approach is permitted because the equations (1) are linear with time-dependent coefficients.

We look for a particular solution of the form

$$P(t) = \exp\Big[\int_0^t \omega(t')\,dt'\Big] \tag{3a}$$

$$C_i(t) = \exp\Big[\int_0^t \omega(t')\,dt'\Big]\,f_i(t) \tag{3b}$$

where $\omega(t)$ and $f_i(t)$ are new unknown functions of time to be determined. It is clear that $\omega(t)$ is the instantaneous inverse period, i.e., $\omega(t) = \dot{P}(t)/P(t)$. Substituting (3a) and (3b) into (1), we obtain

$$(l/\beta)\,\omega(t) = (k_0 - 1) + \epsilon y(t) + \sum_{i=1}^{6} \lambda_i f_i(t) \tag{4a}$$

$$\dot{f}_i(t) + f_i(t)\,\omega(t) = a_i - \lambda_i f_i(t), \qquad i = 1, 2,..., 6 \tag{4b}$$

This set of equations containing the unknown functions $\omega(t)$ and $f_1(t), f_2(t),..., f_6(t)$ is still exact, and is mathematically equivalent to the original point kinetic equations. However, as will be apparent below, this set lends itself better to a perturbation analysis in terms of the small parameter ϵ. We may point out that the equations (4) are nonlinear because of the product term $f_i(t)\,\omega(t)$ in (4b).

The perturbation analysis starts with the expansion of $\omega(t)$ and $f_i(t)$ into a power series in ϵ:

$$\omega(t) = \omega_0 + \epsilon\omega_1(t) + \epsilon^2\omega_2(t) + \cdots \tag{5a}$$

$$f_i(t) = f_{i0} + \epsilon f_{i1}(t) + \epsilon^2 f_{i2}(t) + \cdots \tag{5b}$$

where ω_0 and f_{i0} are constants, and $\omega_j(t)$ and $f_{ij}(t)$ are functions of time. We assume that ϵ is sufficiently small to ensure the convergence of these series. Substituting (5) into (4), and equating the coefficients of equal powers of ϵ, we get

$$(l/\beta)\,\omega_0 = (k_0 - 1) + \sum_{i=1}^{6} \lambda_i f_{i0} \tag{6a}$$

$$f_{i0} = a_i/(\lambda_i + \omega_0) \tag{6b}$$

$$\dot{f}_{i1}(t) + (\lambda_i + \omega_0)\,f_{i1}(t) = -\omega_1(t)\,f_{i0} \tag{7a}$$

$$(l/\beta)\,\omega_1(t) = y(t) + \sum_{i=1}^{6} \lambda_i f_{i1}(t) \tag{7b}$$

$$\dot{f}_{i2}(t) + (\lambda_i + \omega_0)\,f_{i2}(t) = -\omega_1(t)\,f_{i1}(t) - \omega_2(t)\,f_{i0} \tag{8a}$$

$$(l/\beta)\,\omega_2(t) = \sum_{i=1}^{6} \lambda_i f_{i2}(t) \tag{8b}$$

$$\dot{f}_{i3}(t) + (\lambda_i + \omega_0)\,f_{i3}(t) = -\omega_2(t)\,f_{i1} - \omega_1(t)\,f_{i2}(t) - \omega_3(t)\,f_{i0} \tag{9a}$$

$$(l/\beta)\,\omega_3(t) = \sum_{i=1}^{6} \lambda_i f_{i3}(t) \tag{9b}$$

$$\vdots$$

which can be solved by successive integration, starting from (6). Eliminating f_{i0} between (6a) and (6b) we obtain ω_0 as one of the roots of the inhour equation $1 = k_0 Z(\omega_0)$. There are seven roots of the inhour equation. Hence, we obtain seven particular solutions, starting each time with a different root of the inhour equation. These particular solutions to (3) are linearly independent, and thus the general solution can be formed as their linear combination.

We now consider (7) in order to determine a particular solution for $\omega_1(t)$ and $f_{i1}(t)$ using the values of ω_0 and f_{i0} already obtained from (6a) and (6b). Since we are looking for a particular solution of a linear inhomogeneous equation, we are free to choose the initial values of $\omega_1(t)$ and $f_{i1}(t)$ to obtain the particular solution in the simplest form. Taking the Laplace transform of (7), and eliminating $f_{i1}(s)$, we obtain a particular solution $\omega_1(t)$ as the inverse transform of

$$\bar{\omega}_1(s) = K(\omega_0\,, s)\,\bar{y}(s) \tag{10}$$

Since the initial values can be adjusted to eliminate the terms arising from the singularities of $K(\omega_0\,, s)$, it is sufficient to consider the singularities s_j of $\bar{y}(s)$ in finding the inverse Laplace transform of (10), i.e.,

$$\omega_1(t) = \sum_j K(\omega_0\,, s_j)\,\mathrm{Res}[\bar{y}(s)]_{s=s_j}\, e^{s_j t} \tag{11}$$

One can verify that (11) is indeed a solution of the set of equations (7). In (10) and (11), we have introduced

$$K(\omega_0\,, s) \equiv \left[\frac{l}{\beta} + \sum_{i=1}^{6} \frac{a_i\lambda_i}{(\omega_0 + \lambda_i)(\omega_0 + \lambda_i + s)}\right]^{-1} \tag{12}$$

Having determined $\omega_1(t)$ in terms of $y(t)$, we now proceed to solve (8) for $\omega_2(t)$. Using similar arguments as above, we obtain

$$\bar{\omega}_2(s) = -K(\omega_0\,, s) \sum_{i=1}^{6} \frac{\lambda_i}{\lambda_i + s + \omega_0}\,[f_{i1}(t)\,\omega_1(t)]_{\mathrm{L}} \tag{13}$$

where $[f_{i1}(t)\,\omega_1(t)]_{\mathrm{L}}$ denotes the Laplace transform of the product in the bracket. Here, also, we take into account the poles of $[f_{i1}(t)\,\omega_1(t)]_{\mathrm{L}}$ only to construct the particular solution, because the singularities of $K(\omega_0\,, s)$ can be eliminated by a suitable choice of the initial values of $f_{i2}(t)$. Furthermore, there are no additional poles arising from $(\lambda_i + s + \omega_0)$ under the summation, because $K(\omega_0\,, s)$ also vanishes when $(\lambda_i + s + \omega_0)$ is zero.

The functions $\omega_3(t)$, $\omega_4(t)$, etc. in (5a) can be calculated in a similar fashion. However, we shall consider only the first two terms in this expansion as a second-order approximation in ϵ. A few examples will illustrate the application of the foregoing results.

B. Step Input

If we set $\epsilon = 0$ in (2), we obtain the special case of a step reactivity insertion, $k(t) = k_0$ for $t > 0$. The general solution in this case is trivially obtained from (3) and (6) as

$$P(t) = \sum_{j=0}^{6} A_j \exp[\omega_{0j} t]$$

where ω_{0j} are the roots of the inhour equation corresponding to k_0, and A_j are the constants of integration. This is the exact result for a step input.

If we set $k_0 = 0$ and choose $y(t)$ as the step function, we obtain an approximate solution for small step reactivity insertions. Since $\bar{y}(s) = 1/s$ in this case, we obtain $\omega_1(t)$ from (11) as

$$\omega_1(t) = K(\omega_0, 0) = \left[\frac{l}{\beta} + \sum_{i=1}^{6} \frac{a_i \lambda_i}{(\omega_0 + \lambda_i)^2}\right]^{-1} \tag{14}$$

where ω_0 is one of the roots of the inhour equation corresponding to $k_0 = 0$. If we denote the seven roots by ω_{0j} and note that $\omega_{00} = 0$, the general solution reads as follows:

$$P(t) = A_0\{\exp[(\delta k)\, \beta t/l^*]\} + \sum_{j=1}^{6} A_j \exp\{[\omega_{0j} + (\delta k)\, K(\omega_{0j}, 0)]\, t\} \tag{15}$$

where we have used $K(0, 0) = \beta/l^*$, and replaced ϵ by δk. This is equivalent to (4.5, Eq. 8), which was obtained by using the logarithmic linearization. In the present approach, however, we can calculate the second approximation and obtain the stable period up to $(\delta k)^2$ using (13).

C. Ramp Input

As a second example, we consider $k(t) = k_0 + (\delta k)t$, and calculate the reactor response in the first approximation, ignoring $\omega_2(t)$, $\omega_3(t)$, etc. Substituting $\bar{y}(s) = 1/s^2$ in (10), we find $\omega_1(t)$ as

$$\omega_1(t) = K(\omega_0, 0)\, t + K'(\omega_0, 0) \tag{16}$$

where $K'(\omega_0, s) = dK(\omega_0, s)/ds$. The general solution follows from (3a) as

$$P(t) = \sum_{j=0}^{6} A_j \exp\{[\omega_{0j} + (\delta k)\, K'(\omega_{0j}, 0)]\, t + \tfrac{1}{2}(\delta k)\, K(\omega_{0j}, 0)\, t^2\} \tag{17}$$

where ω_{0j} are the roots of $1 = k_0 Z(\omega_{0j})$.

It is interesting to note that (17) is exact if the delayed neutrons are not present. Thus, $K(\omega_0, s) = \beta/l$, and $K'(\omega_0, s) \equiv 0$. Furthermore, the inhour equation reduces to $k_0 = \omega_0(l/\beta)$, which has a single root. With these simplifications, (17) becomes

$$P(t) = P(0) \exp\{(\beta/l)[k_0 t + \tfrac{1}{2}(\delta k)\, t^2]\} \tag{18}$$

which is the exact solution of the kinetic equation, $(l/\beta)\dot{P} = [k_0 + (\delta k)t]P(t)$, in the absence of delayed neutrons.

We point out that the solution in (17) is valid only in limited time intervals because the functions $\omega_1(t)$, $\omega_2(t)$, etc. are not bounded at all times, and hence the convergence of (5a) and (5b) becomes questionable when $(\delta k)t$ approaches or exceeds unity. However, (17) is exact in the absence of delayed neutrons, as pointed out before, because then the series (5a) terminates after the first two terms, and the question of convergence does not arise.

D. Sinusoidal Reactivity Insertion

The real advantage of the perturbation approach presented above is manifested in the case of a periodic reactivity insertion. Because of its importance in reactor stability analysis, we investigate this case in some detail. First, we consider a pure sinusoidal reactivity change, i.e., $k(t) = \delta k \sin \omega t$. The case of an arbitrary periodic input will be presented later. The small parameter ϵ is again replaced by δk.

It is convenient to write $k(t)$ as

$$k(t) = (\delta k)[(1/2i)\, e^{i\omega t} + \text{C.C.}] \tag{19}$$

where C.C. denotes the complex conjugate of the first term. Noting that $y(t) = \sin \omega t$ and using (19), we find $\omega_1(t)$ from (10) as

$$\omega_1(t) = [K(\omega_0, i\omega)/2i]\, e^{i\omega t} + \text{C.C.} \tag{20}$$

The integral of $\omega_1(t)$ which appears in the solution is obtained from (20) as

$$\int \omega_1(t)\, dt = [Z(\omega_0, i\omega)/2i]\, e^{i\omega t} + \text{C.C.} \tag{21}$$

where†

$$Z(\omega_0, i\omega) \equiv K(\omega_0, i\omega)/i\omega \tag{22}$$

† Note that $Z(\omega_0, i\omega)$ is the period-equilibrium transfer function introduced in (4.5, Eq. 13).

In these equations, ω_0 is a root of the inhour equation with $k_0 = 0$. Considering the particular solutions corresponding to the different roots ω_{0j}, we obtain the general solution in the first approximation as

$$P(t) = \sum_{j=0}^{6} A_j \exp\{\omega_{0j}t + (\delta k) \mid Z(\omega_{0j}, i\omega)| \sin(\omega t + \phi_j)\} \tag{23}$$

where ϕ_j is the argument of $Z(\omega_{0j}, i\omega)$. Since all ω_{0j} are negative except for $\omega_{00} = 0$, the asymptotic response of the reactor is given by the first term in (23):

$$P(t) = A_0 \exp[(\delta k) \mid Z(i\omega)| \sin(\omega t + \phi_0)] \tag{24}$$

where we have used $Z(0, i\omega) \equiv Z(i\omega)$, which is the zero-power transfer function. It is interesting to note that this result is identical to that in (4.5, Eq. 10) which was obtained by the method of logarithmic linearization. However, perturbation analysis enables one to compute the second-order approximation, which we shall present next. Furthermore, we can find the response of the reactor to an input of the form $k(t) = k_0 + \delta k \sin \omega t$ without restricting the magnitude of the constant reactivity k_0 to small values. This is not possible with logarithmic linearization, because the presence of a large k_0 violates the condition (4.5, Eq. 3).

We shall now calculate the power response of a reactor to a periodic reactivity insertion correct to the second order in the perturbation expansion given by (5). The reactivity input can be expressed in this case as

$$k(t) = \delta k \sum_{m=-\infty}^{+\infty} X_m e^{i\omega m t} \tag{25a}$$

where the X_m are complex numbers with the following property:

$$X_m{}^* = X_{-m} \tag{25b}$$

We assume that $X_0 = 0$, implying that the mean value of the periodic reactivity is zero. We also assume that $k_0 = 0$, and consider the particular solution corresponding to the zero root, i.e., $\omega_0 = 0$ of the inhour equation. This solution describes the power oscillations after the transients have died away. Substituting (25a) into (10), we obtain $\omega_1(t)$ as

$$\omega_1(t) = \sum_{m=-\infty}^{+\infty} X_m K_m e^{i\omega m t} \tag{26}$$

where

$$K_m \equiv K(0, i\omega m) \tag{27a}$$

or, more explicitly (cf Eq. 12),

$$K_m = \left[\frac{l}{\beta} + \sum_{j=1}^{6} \frac{a_j}{\lambda_j + im\omega}\right]^{-1} \tag{27b}$$

The function $f_{i1}(t)$ can be obtained from (7a) using (26):

$$f_{j1}(t) = -f_{j0} \sum_{m=-\infty}^{+\infty} X_m \frac{K_m}{\lambda_j + i\omega m} e^{i\omega mt} \tag{28}$$

In order to solve (13) for $\omega_2(t)$, we need the product $\omega_1(t) f_{j1}(t)$, which is readily obtained from (26) and (28) as

$$\omega_1(t) f_{j1}(t) = -f_{j0} \sum_{m,n=-\infty}^{+\infty} X_m X_n \frac{K_m K_n}{\lambda_j + i\omega n} e^{i\omega(m+n)t} \tag{29}$$

We note that this product contains a constant term, which is determined by setting $m + n = 0$ in (29) and nothing that $f_{j0} = a_j/\lambda_j$ (cf Eq. 6b):

$$-2a_j \sum_{m=1}^{\infty} \frac{|X_m|^2 |K_m|^2}{\lambda_j^2 + (m\omega)^2} \tag{30}$$

The contribution of this constant term to $\omega_2(t)$ in (13) is

$$\omega_{20} = 2K(0, 0) \sum_{m=1}^{\infty} |X_m|^2 |K_m|^2 \sum_{j=1}^{6} \frac{a_j}{\lambda_j^2 + (m\omega)^2} \tag{31}$$

or, using

$$\operatorname{Re}[1/Z(i\omega)] = \{\operatorname{Re}[Z(i\omega)]\}/|Z(i\omega)|^2 \tag{32}$$

and $K(0, 0) = \beta/l^*$, we obtain

$$\omega_{20} = 2(\beta/l^*) \sum_{m=1}^{\infty} |X_m|^2 \operatorname{Re}[Z_m] \tag{33}$$

In these equations, $Z_m = Z(i\omega m)$.

The periodic terms in $\omega_2(t)$ are obtained by inserting (29) into (13), and replacing the running index m by $m = n - l$:

$$\omega_2(t) = \omega_{20} + \sum_{l \neq n=-\infty}^{+\infty} K_l K_n K_{l-n} X_n X_{l-n} e^{i\omega lt} \sum_{j=1}^{6} \frac{a_j}{(\lambda_j + i\omega l)(\lambda_j + i\omega n)} \tag{34}$$

This expression can be simplified by expanding the last factor into simple fractions, i.e.,

$$\frac{1}{(\lambda_j + i\omega l)(\lambda_j + i\omega n)} = \frac{1}{i\omega(n - l)} \left[\frac{1}{\lambda_j + i\omega l} - \frac{1}{\lambda_j + i\omega n}\right] \tag{35}$$

and observing

$$\sum_{j=1}^{6} \left[\frac{a_j}{\lambda_j + i\omega l} - \frac{a_j}{\lambda_j + i\omega n}\right] = \frac{1}{K_l} - \frac{1}{K_n} \tag{36}$$

The final result is

$$\omega_2(t) = \omega_{20} + \sum_{l \neq n = -\infty}^{+\infty} Z_{l-n}(K_l - K_n)\, X_n X_{l-n} e^{i\omega l t} \tag{37}$$

The expression for $\omega(t)$ correct up to the second power of δk is obtained as $\omega(t) = (\delta k)\, \omega_1(t) + (\delta k)^2\, \omega_2(t)$, where $\omega_1(t)$ and $\omega_2(t)$are given by (26) and (37), respectively. The desired particular solution follows from (3a) by substituting $\omega(t)$ into $\int \omega(t)\, dt$. We shall present the result only in the case of a pure sinusoidal reactivity variation, which is a special case of the foregoing analysis with $X_1 = 1/2i$, $X_{-1} = -1/2i$, and $X_n = 0$ for $n = \mp 2, \mp 3, \ldots$. In this special case, (37) reduces to

$$\omega_2(t) = \omega_{20} - \tfrac{1}{4}[Z_1(K_2 - K_1)\, e^{2i\omega t} + \text{C.C.}] \tag{38}$$

The integral of $\omega_2(t)$ is

$$\int \omega_2(t)\, dt = \omega_{20} t - \tfrac{1}{4} \mid Z_1(2Z_2 - Z_1| \cos(2\omega t + \phi_2) \tag{39a}$$

where

$$\phi_2 = \arg[Z_1(2Z_2 - Z_1)] \tag{39b}$$

and

$$\omega_{20} = \tfrac{1}{2}(\beta/l^*)\, \text{Re}[Z_1] \tag{40}$$

The response of the reactor to a pure sinusoidal reactivity change after the transients have disappeared can now be obtained from the foregoing results as

$$P(t) = A_0 \exp\{(\delta k) \mid Z_1 \mid \sin(\omega t + \phi_1) - \tfrac{1}{4}(\delta k)^2 \mid Z_1(2Z_2 - Z_1)| \cos(2\omega t + \phi_2) + \tfrac{1}{2}(\delta k)^2(\beta/l^*)\, t\, \text{Re}[Z_1]\} \tag{41}$$

This result, which is due to Akcasu [15], was compared to the exact numerical solution by Tan [14] in Figure 4.3.6 of Section 4.3C for an input $k(t) = (0.004/\beta) \sin t$, using one group of delayed neutrons. It can be improved by extending the perturbation analysis to include terms of $O(k^3)$ in the expansion of the inverse reactor period $\omega(t)$. We shall present the final result after the following discussion. We observe in (41) that the factor $\exp[(\delta k)^2 \omega_{20} t]$ causes the average value, as well as the amplitude, of the power oscillations to increase exponentially, implying that the output is not periodic when the input is a pure sine wave. This result was obtained by Bethe [22] using the prompt-jump approximation (see Problem 4). The amount of negative reactivity bias which has to be introduced to keep the power oscillations steady can be obtained very easily in the framework of the present analysis. Let us assume that the reactivity input has the following form:

$$k(t) = \delta k(\sin \omega t) + (\delta k)^2 \eta \tag{42}$$

where η is a constant to be determined. Since this additional term is of the second order, it first appears in (8b), and gives rise to an additional constant term $(\beta/l^*)\eta$ in the expression of $\omega_2(t)$ given by (37). Thus, the total constant term becomes $\omega_{20} + (\beta/l^*)\eta$. The latter vanishes if we choose η as

$$\eta = -2 \sum_{m=1}^{\infty} | X_m |^2 \operatorname{Re}[Z_m] \tag{43}$$

in the case of an arbitrary reactivity input, and as

$$\eta = -\tfrac{1}{2} \operatorname{Re}[Z_1]$$

or

$$\eta = -\tfrac{1}{2} | Z(i\omega)|^2 \sum_{j=1}^{6} [\omega^2 a_j/(\lambda_j^2 + \omega^2)] \tag{44}$$

when the input is a sinusoidal function. It is interesting to compare this result to (3.2, Eq. 9), which was obtained by the inverse method. In (3.2, Eq. 9), k_{av} denotes $(\delta k)^2\eta$. If we use the linear description to relate p_1/p_0 to δk, i.e., $p_1/p_0 = (\delta k)\, Z(i\omega)$, then (43) and (3.2, Eq. 9) become identical. We conclude that the response of the reactor to a sinusoidal input $k(t) = \eta(\delta k)^2 + \delta k \sin \omega t$ is periodic up to the fourth order in δk if η is chosen as in (43). The explicit form of these periodic power oscillations is obtained from (41) by discarding the secular term linear in t. We can improve this result by extending the foregoing perturbation analysis to the third order in δk. For this, we need

$$f_{j2}(t) = \frac{a_j}{2\lambda_j} \frac{| K_1 |^2}{(\lambda_j^2 + \omega^2)} + \left\{ \frac{a_j}{4\lambda_j} \frac{1}{\lambda_j + 2i\omega} \left[Z_1(K_2 - K_1) - \frac{K_1^2}{\lambda_j + i\omega} \right] e^{2i\omega t} + \text{C.C.} \right\} \tag{45}$$

which is obtained from (8a) substituting $\omega_2(t)$ and $f_{j1}(t)\,\omega_1(t)$ from (37) and (29), respectively. If we use (26), (28), (38), and (45) in (9a) and (9b), we obtain, after lengthy manipulations,

$$\omega_3(t) = (1/8i)\,K_1[Z_2(Z_1 - Z_1^*) - Z_1^2]\,e^{i\omega t} + \text{C.C.} + \text{3rd harmonic} \tag{46}$$

The periodic power response is obtained, by substituting (46), (38), and (20) into (5a) and (3a), as

$$P(t) = A_0 \exp[(\delta k)\,N_1 \sin(\omega t + \phi_1) - (\delta k)^2\,N_2 \cos(2\omega t + \phi_2) + (\delta k)^3(\text{3rd harmonic})] \tag{47}$$

where

$$N_1 = |\,Z_1\,|\,|\,1 + \tfrac{1}{4}(\delta k)^2\,[Z_2(Z_1 - Z_1^*) - Z_1^2]| \tag{47a}$$

$$\phi_1 = \arg Z_1\{1 + \tfrac{1}{4}(\delta k)^2[Z_2(Z_1 - Z_1^*) - Z_1^2]\} \tag{47b}$$

$$N_2 = \tfrac{1}{4}\,|\,Z_1\,|\,|\,2Z_2 - Z_1\,| \tag{48a}$$

$$\phi_2 = \arg Z_1(2Z_2 - Z_1) \tag{48b}$$

Since we are going to ignore the third harmonic in the subsequent analysis, we do not write down the explicit form of the last term in (47). We note that the higher-order terms in N_1 and N_2 are proportional to $(\delta k)^4$ and $(\delta k)^2$, respectively. Hence, the exponent in (47) represents the solution correctly up to the third power of (δk) if we disregard the third harmonic.

We shall now discuss some aspects of the periodic power oscillations using (47). The first important observation is that this expression becomes exact when the delayed neutrons are not present, because then $Z(i\omega) = 1/li\omega$, $Z_1 = -Z_1^*$ and $2Z_2 = Z_1$, hence $N_1 = |\,Z_1\,|$ and $N_2 = 0$ (cf the discussion following 4.6, Eq. 17). The second observation is that the frequency-dependent factor $[Z_2(Z_1 - Z_1^*) - Z_1^2]$ in (47a) is bounded at all frequencies, approaching -20 (Problem 9) at very low and intermediate frequencies, and vanishing as $1/\omega^3$ at high frequencies. Hence, the magnitude of the correction term in (47a) is always of the order of $5(\delta k)^2$ and negligible as compared to unity at all frequencies for values of δk sufficiently less than 1 \$. The third observation concerns the ratio of the amplitude of the second harmonic, $(\delta k)^2\,N_2$, to that of the first harmonic, $\delta k N_1$. This ratio is approximately equal to

$$|\,2Z_2 - Z_1\,|\,(\delta k)/4 \tag{49}$$

which has a maximum [23] at $\omega = 0$, where it is equal to $(\delta k/4)\,\text{Re}\,Z(0) \approx (\langle\tau^2\rangle_{\text{av}}/\langle\tau\rangle_{\text{av}}^2)\,\delta k/4$. In conclusion, we find that the exponent in (47) varies essentially as $(\delta k) \mid Z_1 \mid \sin(\omega t + \phi_1)$, which may be very large at low frequencies even for small reactivity changes, because of the factor $\mid Z_1 \mid$. It is interesting to point out that the power oscillations may be highly distorted when $(\delta k) \mid Z_1 \mid$ is close to unity, but its logarithm will still be a pure sine wave. This conclusion was borne out by a computer solution, shown in Figure 4.6.1, in which the reactivity input was $k(t) = 0.7 \sin t$.

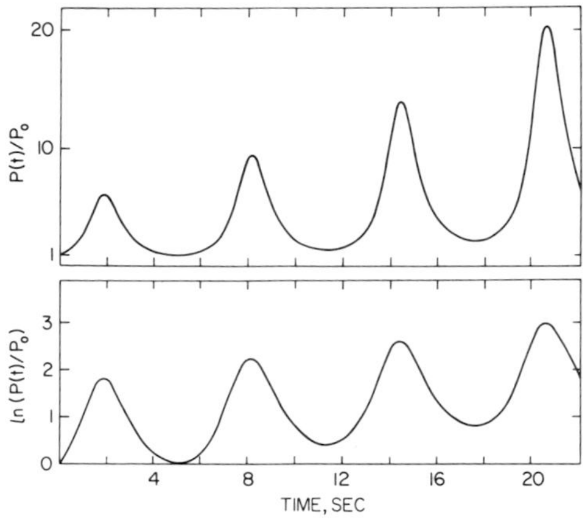

FIGURE 4.6.1. Power response to an oscillatory reactivity input $k(t) = 0.7 \sin t$.

We can investigate the harmonic content of the power oscillations rather than its logarithm by expanding the exponential function in (47) in a Fourier series. When δk is sufficiently smaller than unity, we can ignore the second harmonic N_2 compared to the fundamental at all frequencies, and approximate (47) by

$$P(t) = A_0 \exp[(\delta k) N_1 \sin(\omega t + \phi_1)] \tag{50}$$

The Fourier expansion of $P(t)$ can easily be obtained using the following expansion [24]:

$$\exp[x \cos \theta] = I_0(x) + 2I_1(x) \cos \theta + 2I_2(x) \cos 2\theta + \cdots \tag{51}$$

where the $I_n(x)$ are the modified Bessel functions of the first kind. Thus,

$$P(t) = A_0\{I_0([\delta k] N_1) + 2I_1([\delta k] N_1) \sin(\omega t + \phi_1) - 2I_2([\delta k] N_1) \cos[2(\omega t + \phi_1)] + \cdots\} \tag{52}$$

We observe from (52) that the relative magnitudes of the harmonics are independent of the constant of integration A_0, which depends on the initial conditions. The ratio of the amplitude of the first harmonic (fundamental) relative to the average power level to the amplitude of the sinusoidal reactivity input is called the gain of the reactor:

$$G(\delta k, \omega) = 2I_1(N_1[\delta k])/(\delta k)\, I_0([\delta k]\, N_1) \tag{53a}$$

or, approximating N_1 by $| Z_1 |$,

$$G(\delta k, \omega) = 2I_1([\delta k] \mid Z_1 \mid)/(\delta k)\, I_0([\delta k] \mid Z_1 \mid) \tag{53b}$$

which depends on the amplitude of the input as well as on the frequency. This amplitude dependence is a characteristic of systems that have a nonlinear input–output relation. When $(\delta k)\, N_1 \ll 1$, which is essentially the same as $(\delta k) \mid Z(i\omega)| \ll 1$ (cf 47a), the modified Bessel functions in (53) can be approximated by their asymptotic behavior for small arguments:

$$G(\delta k, \omega) = N_1[1 + \tfrac{1}{8}(N_1\, \delta k)^2]/[1 + \tfrac{1}{4}(N_1\, \delta k)^2] \tag{54}$$

It is important to note that (54) is valid only when $| Z(i\omega)|\, \delta k \ll 1$, whereas (53) is valid at all frequencies provided $\delta k \ll 1$. Since $Z(i\omega)$ is unbounded as $\omega \to 0$, there is always a frequency below which (54) is not valid no matter how small δk may be. We observe from (54) that the gain of the reactor approaches $| Z(i\omega)|$ and becomes independent of the amplitude of the reactivity input in the limit of $\delta k \to 0$. This limit corresponds to the linear description. Since $2I_1(x)/xI_0(x)$ is always less than unity and a decreasing function of x, we find from (53) that the nonlinear gain is always smaller than N_1 at all frequencies and decreases with increasing δk in the range of $\delta k < 1$.

We can obtain a more accurate expression for $G(\delta k, \omega)$ than that in (54) by expanding (47) in a power series in δk and neglecting the terms proportional to $(\delta k)^4$ or higher powers of δk. After some lengthy manipulations, we find

$$G(\delta k, \omega) = | Z_1 |\, | 1 + \tfrac{1}{4}(\delta k)^2(Z_2 - Z_1)\, Z_1 |/| 1 + \tfrac{1}{4}(\delta k)^2 \mid Z_1 |^2| \tag{55a}$$

or

$$G(\delta k, \omega) = | Z_1 |\, | 1 + \tfrac{1}{4}(\delta k)^2(Z_2 Z_1 - 2Z_1 \operatorname{Re} Z_1)| \tag{55b}$$

which was first obtained Akcasu [15]. Although this result corresponds to a consistent power-series expansion of $G(\delta k, \omega)$ which is correct up to $(\delta k)^2$, the convergence of this series is not uniform in frequency, because $(Z_2 Z_1 - 2Z_1 \operatorname{Re} Z_1)$ diverges as $1/\omega^2$ as $\omega \to 0$, and hence the correction term exceeds unity at some frequency for any fixed value of δk, no matter how small it may be. Hence, the appoximation inherent in (55) is valid in a restricted frequency range for a given δk, as discussed by Wasserman [25]. We shall return to this point in the next section in connection with the concept of describing functions.

4.7. The Response of a Reactor to Large Periodic Reactivity Variations

It was shown in the previous section that the response of a reactor to a periodic reactivity insertion is also periodic if the average reactivity is adjusted to a certain negative value which is approximately given by (4.6, Eq. 43). The purpose of this section is to present a Fourier analysis [25–28] of these periodic power oscillations, and to determine the amplitudes and phases of the various Fourier components in terms of the Fourier components of the periodic reactivity input. No restriction will be imposed on the magnitude of the reactivity changes in this analysis, except for the average reactivity, which will always be assumed to be equal to the correct negative bias. The latter will be determined exactly in terms of the Fourier components of the input and output. The case of a pure sinusoidal reactivity variation will be investigated in more detail in order to define the nonlinear gain and the concept of a describing function of a reactor.

A. Fourier Analysis of Power Oscillations

Let the periodic reactivity insertion be represented by

$$k(t) = k_0 + k_1 \sin(\omega t + \phi_1) + k_2 \sin(2\omega t + \phi_2) + \cdots \tag{1}$$

which can also be written as

$$k(t) = \sum_{n=-\infty}^{n=+\infty} x_n e^{i\omega n t} \tag{2a}$$

where

$$x_n = (k_n/2i)\, e^{i\phi_n} \tag{2b}$$

$$x_{-n} = x_n^* \tag{2c}$$

The power response is represented by

$$P(t) = P_0 + p_1 \sin(\omega t + \theta_1) + p_2 \sin(2\omega t + \theta_2) + \cdots \tag{3}$$

or

$$P(t) = P_0 \sum_{-\infty}^{+\infty} y_n e^{i\omega nt} \tag{4a}$$

with

$$y_n = (p_n/2P_0 i)\, e^{i\theta_n} \tag{4b}$$

$$y_{-n} = y_n^* \tag{4c}$$

We note that $y_0 = 1$. Our task is to determine $y_1, y_2, \ldots$ in terms of $x_1, x_2, \ldots$ using the point kinetic equation,

$$(l/\beta)\, \dot{P}(t) = [k(t) - 1]\, P(t) + \int_0^\infty D(u)\, P(t-u)\, du \tag{5}$$

Substituting (2a) and (4a) into (5) and equating the equal powers of $e^{i\omega t}$ on both sides of (5), we obtain

$$y_m = Z_m \sum_{n=-\infty}^{+\infty} x_n y_{m-n}, \qquad m = 0, \pm 1, \pm 2, \ldots \tag{6}$$

where we define $Z_m \equiv Z(i\omega m)$ as in the previous section. The infinite set of equations represented by (6) can be solved for y_m by an iteration procedure which we shall discuss presently. First, we would like to point out that the equations are homogeneous in y_m. We can determine y_m only within an arbitrary factor. We have removed this arbitrariness by normalizing y_0 to unity so that $|y_n| = p_n/2P_0$, i.e., y_n is proportional to the relative amplitude of the nth harmonic with respect to the average power P_0. The value of P_0 depends on the initial conditions, and cannot be determined from the set of homogeneous equations (6).

The equation corresponding to $m = 0$ is of particular interest. Since $Z_0 = Z(0)$ is infinite, we obtain from (6) with $m = 0$ the following relation:

$$\sum_{n=-\infty}^{+\infty} x_n y_n^* = 0 \tag{7}$$

Using the normalization condition $y_0 = 1$ and $x_0 = k_0$, we find

$$k_0 = -\sum_{\substack{n=-\infty \\ n \neq 0}}^{+\infty} x_n y_n^* \tag{8a}$$

or

$$k_0 = -2 \sum_{n=1}^{\infty} \mathrm{Re}[x_n y_n^*] \tag{8b}$$

This equation yields the critical value of the negative reactivity bias k_0 in terms of x_n and y_n, which is required to make the reactor response periodic, as assumed in (3). Although (8b) is an exact relation which is valid regardless of the magnitude of the periodic reactivity insertion, its usefulness is restricted by the fact that the right-hand side still contains the unknown amplitudes y_n. We may point out at this point that (8b) is a generalization of (4.6, Eq. 43) which was obtained for a pure sinusoidal reactivity change using the second-order perturbation analysis. We shall derive the latter as a special case of (8b) by approximating y_n by $Z_n x_n$.

In order to solve (6) for y_n by a perturbation technique, we first separate the terms corresponding to $n = 0$ and $n = m$ in the summation in (6) as

$$y_m = Z_m \left[x_0 y_m + y_0 x_m + \sum_{\substack{n=-\infty \\ (n \neq 0, m)}}^{+\infty} x_n y_{m-n} \right] \tag{9}$$

and then use $y_0 = 1$ and substitute x_0 from (8a):

$$y_m = Z_m \left[x_m + \sum_{\substack{-\infty \\ (n \neq 0, m)}}^{+\infty} x_n y_{m-n} - y_m \sum_{\substack{-\infty \\ (n \neq 0)}}^{+\infty} x_{-n} y_n \right], \qquad m \neq 0 \tag{10}$$

The reason for separating x_0 is that the latter is not an independent variable as $x_{\pm 1}$, $x_{\pm 2}$,... are, but rather, is an unknown quantity to be determined in terms of $x_{\pm 1}$, $x_{\pm 2}$,... in order to ensure the assumed periodicity of the power variations. Hence, it has to be eliminated in favor of x_n and y_n with $n \neq 0$. The reason for singling out y_0 is that it is not an unknown quantity, by virtue of our normalization as $y_0 = 1$, and hence the set of equations (6) is not homogeneous, as explicitly indicated by (10). To solve the latter systematically, we assume that the x_n are of the order of ϵ, and the y_n have a power-series expansion as

$$y_n = \sum_{\alpha=1}^{\infty} y_n^{(\alpha)} \epsilon^\alpha, \qquad n = \pm 1, \pm 2, \ldots \tag{11}$$

Substituting (11) into (10), we obtain

$$\frac{1}{Z_m} \sum_{\alpha=1}^{\infty} \epsilon^\alpha y_m^{(\alpha)} = \epsilon x_m + \sum_{\alpha=1}^{\infty} \epsilon^{\alpha+1} \sum_{\substack{n=-\infty \\ (n \neq 0, m)}}^{+\infty} x_n y_{m-n}^{(\alpha)} - \sum_{\alpha,\beta=1}^{\infty} \epsilon^{\alpha+\beta+1} y_m^{(\alpha)} \sum_{\substack{n=-\infty \\ (n \neq 0)}}^{+\infty} y_n^{(\beta)} x_{-n} \tag{12}$$

in which we equate the equal powers of ϵ on both sides:

$$y_m^{(1)} = Z_m x_m \tag{13a}$$

$$y_m^{(2)} = Z_m \sum_{\substack{n=-\infty \\ (n\neq 0,m)}}^{+\infty} x_n y_{m-n}^{(1)} \tag{13b}$$

$$y_m^{(3)} = Z_m \Bigg[\sum_{\substack{n=-\infty \\ (n\neq 0,m)}}^{+\infty} x_n y_{m-n}^{(2)} - y_m^{(1)} \sum_{\substack{n=-\infty \\ (n\neq 0)}}^{+\infty} x_{-n} y_n^{(1)} \Bigg] \tag{13c}$$

The equations (13) can be solved successively for $y_m^{(\alpha)}$ and all α and $m \neq 0$. We shall discuss only the special case of a pure sinusoidal reactivity insertion in detail. In this case, $k(t) = \delta k \sin \omega t$ (we have used δk instead of k_1 for comparison purposes), and hence $x_n = 0$ for all n except for $x_0 = k_0$ and $x_{\pm 1} = \pm(\delta k/2i)$. It immediately follows from (13) that

$$y_{\pm 1}^{(1)} = Z_{\pm 1} x_{\pm 1} \qquad y_n^{(1)} = 0 \qquad \text{for} \quad n = \pm 2, \pm 3, \ldots \tag{14}$$

$$y_{\pm 1}^{(3)} = Z_{\pm 1}[x_{\mp 1} y_{\pm 2}^{(2)} - y_{\pm 1}^{(1)}(x_{-1} y_1^{(1)} + x_1 y_{-1}^{(1)})] \tag{15}$$

$$y_{\pm 1}^{(2)} = 0, \qquad y_{\pm 2}^{(2)} = Z_{\pm 2}\, x_{\pm 1}\, y_{\pm 1}^{(1)}, \; y_n^{(2)} = 0, \qquad \text{for} \quad n = \pm 3, \pm 4, \ldots \tag{16a}$$

$$y_{\pm 2}^{(3)} = 0, \qquad y_{\pm 3}^{(3)} = Z_{\pm 3} x_{\pm 1} y_{\pm 2}^{(2)}, \qquad y_n^{(3)} = 0 \qquad \text{for} \quad n = \pm 4, \pm 5, \ldots \tag{16b}$$

Substituting these results into (11) with $\epsilon = 1$, we obtain

$$y_1 = Z_1 x_1 [1 + Z_1 \,|\, x_1 \,|^2 (Z_2 - 2 \operatorname{Re} Z_1)] + O(|\, x_1 \,|^5) \tag{17}$$

$$y_2 = Z_1 Z_2 x_1{}^2 + O(|\, x_1 \,|^4) \tag{18}$$

$$y_3 = Z_1 Z_2 Z_3 x_1{}^3 + O(|\, x_1 \,|^5) \tag{19}$$

where $O(|\, x_1{}^m \,|)$ means that the magnitude of the subsequent term is proportional to $|\, x_1 \,|^m$. Combining these results with (3) and (4), we can construct the periodic power response to a sinusoidal reactivity insertion. The following remarks are of interest at this point.

1. We observe from (17)–(19) that none of the y_n are zero, which implies that the power oscillations contain all the harmonics with decreasing amplitudes. The amplitude of the nth harmonic in the lowest approximation is

$$p_n = P_0 \,|\, Z_1 Z_2 \cdots Z_n \,|\, (\delta k/2)^n \tag{20}$$

The ratio of the amplitudes of the consecutive harmonics follows from (20) as

$$p_n/p_{n-1} = |Z_n| \, \delta k/2 \tag{21}$$

where we recall that $Z_n = Z(i\omega m)$. Since the amplitude of the zero-power transfer function is a monotonically decreasing function of its argument (cf Section 4.4, Fig. 4.4.5), the ratio p_n/p_{n-1} decreases rapidly with increasing n, being proportional to $1/n$ for large n.

2. The negative bias k_0 can be calculated to any desired accuracy by substituting y_n from (13) into (8b).

For a pure sine input, the appoximation equivalent to neglecting terms proportional to $(\delta k)^6$ and higher yields k_0, after substituting y_1 from (17), is

$$k_0 = -\tfrac{1}{2}(\delta k)^2 \operatorname{Re}[Z_1 + \tfrac{1}{4}(\delta k)^2 Z_1^2(Z_2 - 2 \operatorname{Re} Z_1)] \tag{22}$$

This is a higher-order approximation to k_0 than is (4.6, Eq. 43).

3. In the case of a pure sinusoidal reactivity input, (9) can be written as

$$y_m U_m = x_1^* y_{m+1} + x_1 y_{m-1} \tag{23a}$$

where

$$U_m = (1/Z_m) - k_0 \tag{23b}$$

Evaluating (23a) for $m = 1$ and solving the resulting equation for U_1, we obtain

$$U_1(i\omega) = \frac{[y_2 + (x_1/x_1^*)]x_1}{y_1} \tag{24}$$

This result suggests, as pointed out by Babala [29], a method for measuring the zero-power transfer function $Z(i\omega)$ by a pile-oscillator technique using large amplitudes. Since $Z(i\omega)$ describes the linear response of a reactor to a sinusoidal reactivity change, one may think that its measurement must be performed using small power oscillations to avoid nonlinear effects. However, at least in principle, we can still extract $Z(i\omega)$ from (24) even in the presence of large amplitudes. If we measure the complex amplitudes of the first and second harmonics relative to the average power level, i.e., y_1 and y_2, we can determine $U(i\omega)$ experimentally from (24). The zero-power transfer function is then obtained as

$$Z(i\omega) = [k_0 + U(i\omega)]^{-1} \tag{25a}$$

where k_0 is to be computed from (8b), i.e.,

$$k_0 = -2 \operatorname{Re}[x_1 y_1^*] \tag{25b}$$

in terms of the measured values y_1 and x_1.

3. Another interesting application of (23a) is that it enables one to express the ratio y_1/x_1, whose physical meaning will be discussed below, in the form of an infinite fraction. From (23a), we get

$$y_m/x_1 y_{m-1} = [U_m - x_1{}^*(y_{m+1}/y_m)]^{-1} \tag{26}$$

which yields the ratio y_m/y_{m-1} in terms of y_{m+1}/y_m, and can be solved by an interation procedure. For example, y_1/y_0x_1 can be found as

$$\frac{y_1}{y_0 x_1} = \cfrac{1}{U_1 - |x_1|^2 \cfrac{1}{U_2 - |x_1|^2 \cfrac{1}{U_3 - |x_1|^2 \cfrac{1}{U_4 - \cdots}}}} \tag{27}$$

which was obtained by Lauber [28]. In general, we have

$$\frac{y_m}{x_1 y_{m-1}} = \cfrac{1}{U_m - |x_1|^2 \cfrac{1}{U_{m+1} - |x_1|^2 \cfrac{1}{U_{m+2} - |x_1|^2 \cdots}}} \tag{28}$$

B. The Concept of the Describing Function

It was shown in Section 4.4 that a reactor is completely described in the linear approximation by the zero-power transfer function $Z(i\omega)$. The latter gives the phase and the amplitude of the sinusoidal power response at a given frequency ω in terms of the sinusoidal reactivity input. As we shall see in Chapter 6, the stability of a reactor system against small perturbations can be analyzed once the transfer function of each element in the system is determined from frequency response measurements.

We have seen in the previous sections that the reactor can be treated as a linear system only approximately. When large perturbations are involved, the nonlinearity of the input–output relation of the point reactor must be taken into account. At this point, the question arises whether it is possible to extend the concept of transfer functions in such a way that the stability of a reactor system for large disturbances can also be analyzed in terms of the sinusoidal response of the reactor. It has been found [30] that such an extension is possible for certain systems in which the harmonics generated by the nonlinearities are attenuated by the linear components of the system, by introducing the concept of the "describing" function. In general, the describing function of a nonlinear network is defined as the complex ratio of the amplitude of the fundamental Fourier component of the response to the amplitude of the sinusoidal input. The complex amplitude of a sinusoidal input $k(t) = \delta k \sin \omega t$ was denoted by $x_1 = \delta k/2i$, and the complex amplitude of the fundamental component was denoted by $y_1 = (p_1/2iP_0) \exp(\theta_1)$. The describing function, which we denote by $D(\delta k, \omega)$, is obtained from (17) as y_1/x_1 :

$$D(\delta k, \omega) = Z_1[1 + \tfrac{1}{4}(\delta k)^2 Z_1(Z_2 - 2 \operatorname{Re} Z_1)] \tag{29}$$

This expression for $D(\delta k, \omega)$ corresponds to the first two terms in the power-series expression of $D(\delta k, \omega)$ in δk. It was pointed out at the end of the last section that the convergence of this power series is not uniform [25] in ω, because $Z_1(Z_2 - 2 \operatorname{Re} Z_1)$ approaches infinity as ω tends to zero. An approximate expression for $D(\delta k, \omega)$ free from this drawback can readily be obtained [31] from (4.6, Eq. 52) as

$$D(\delta k, \omega) = [2I_1(N_1 \, \delta k)/(\delta k) \, I_0(N_1 \, \delta k)] \, e^{i\phi_1} \tag{30a}$$

where we recall (cf 4.6, Eq. 47a)

$$\begin{aligned} N_1 &\equiv | Z_1 | \, | 1 + \tfrac{1}{4}(\delta k)^2[Z_2(Z_1 - Z_1^*) - Z_1^2]| \\ \phi_1 &\equiv \arg[Z_1\{1 + \tfrac{1}{4}(\delta k)^2[Z_2(Z_1 - Z_1^*) - Z_1^2]\}] \end{aligned} \tag{30b}$$

These expressions are valid for all frequencies provided $\delta k < 1$. It is slightly less accurate than (29) for high frequencies and δk satisfying $\delta k \, | Z_1 | \ll 1$. The variation of the magnitude of the describing function, which was referred to in the previous section as the nonlinear gain of the reactor (cf 4.6, Eq. 53a), i.e., $G(\delta k, \omega) \equiv | D(\delta k, \omega)|$, is plotted in Figure 4.7.1 as a function of ω for various values of δk. The dashed curves in these figures represent the nonlinear gain calculated using (29). The various remarks made in the above discussion can be observed in these curves. Figure 4.7.2 shows the variation of $G(\delta k, \omega)$ with δk at a fixed ω.

Expanding the expression (4.6, Eq. 47) in a Fourier series without ignoring the second harmonic term in the exponent, Vaurio and Pulkkis [32] have recently obtained a more accurate describing function than (30). However, they found that the amplitude and phase characteristics of the describing function calculated from (30) agreed well with those calculated from their expression. The difference in amplitudes was less

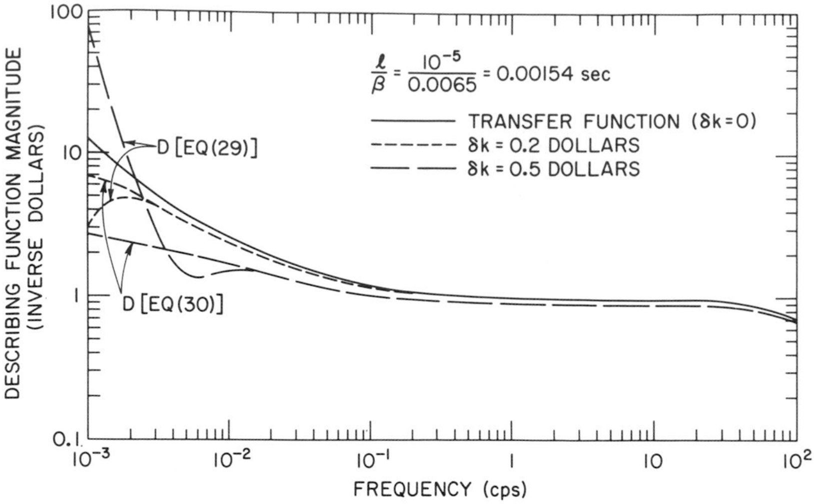

FIGURE 4.7.1. Describing-function magnitude versus frequency for various δk. (——) $\delta k = 0$; (— —) $\delta k = 0.5$; (- - -) $\delta k = 0.2$. $l/\beta = 10^{-5}/0.0065 = 0.00154$ sec.

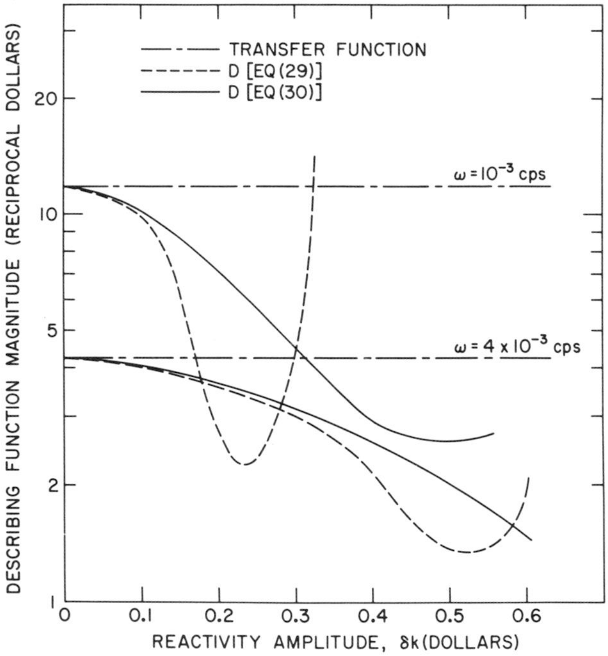

FIGURE 4.7.2. Describing-function magnitude versus δk for various frequencies. (— - —) Transfer function; (- - -) Eq. (29); (——) Eq. (30).

than 1% for $\delta k = 0.5$ \$ if $f > 10^{-2}$ Hz and for $\delta k = 0.20$ \$ if $f > 10^{-3}$ Hz. The improvement is not significant in view of the other approximations inherent in the starting equation (4.6, Eq. 47), and does not warrant the complication of the refined expression which involves a ratio of infinite series in modified Bessel functions. The agreement mentioned above justifies the neglect of the second harmonic in the derivation of (30).

The utility of the zero-power describing function lies in the fact that it extends the interpretation of pile-oscillator experiments to larger power variations. The zero-power transfer function is the limit of $D(\delta k, \omega)$ as $\delta k \to 0$. The main advantage of introducing the concept of describing function is that one can investigate nonlinear stability and the response to large perturbations of power reactors, in which the feedback effects are important, by the conventional methods of feedback control theory once an appropriate zero-power describing function has been found. We shall discuss this point further when we introduce the high-power describing function in Chapter 5. The appropriate zero-power describing function for the purpose of deriving a high-power describing function turns out to be slightly more general than that given by (29).

C. Multy-Input Zero-Power Describing Function

In order to extend the concept of describing function to multiple input, we consider a reactivity input containing both first and second harmonics with proper negative bias and determine the complex amplitude of the first harmonic in the resulting power oscillations. The input in complex notation reads

$$k(t) = x_0 + [x_1 \exp(i\omega t) + x_2 \exp(2i\omega t) + \text{C.C.}]$$

The complex amplitude of the fundamental of the power output can be obtained from (13) as $y_1 = y_1^{(1)} + y_1^{(2)} + y_1^{(3)}$ if we ignore the fourth and higher-order terms. The result can be expressed directly as the ratio (y_1/x_1) which defines the desired two-input (x_1 and x_2 are the inputs) zero-power describing function $D_z(x_1, x_2, \omega)$:

$$D_Z(x_1, x_2, \omega) = Z_1\{1 + [| x_1 |^2 Z_1(Z_2 - 2 \operatorname{Re} Z_1) + x_2(x_1^*/x_1)(Z_2 + Z_1^*)] + | x_2 |^2[(Z_1 + Z_2) Z_3 + Z_1^*(Z_2^* + Z_1) - 2Z_1 \operatorname{Re} Z_2]\} \quad (31)$$

We note that (31) reduces to (29) when $x_2 = 0$. It yields the complex amplitude of the fundamental of the power in the presence of the second harmonic in the reactivity input. Therefore, if the reactivity input in a pile-oscillator experiment is not a pure sine function as a result of distortion, (31) should be used in the interpretation of the ratio (y_1/x_1). Since the magnitude of the second harmonic is small in such cases as compared to the fundamental amplitude we may simplify (31) by ignoring terms proportional to $|x_2|^2$ as

$$D_Z(x_1, x_2, \omega) = Z_1[1 + |x_1|^2 Z_1(Z_2 - \mathrm{Re}\, Z_1) + x_2(x_1^*/x_1)(Z_2 + Z_1)] \quad (32)$$

We shall use this two-input zero-power describing function in Chapter 5 to derive a high-power describing function which will include feedback effects.

PROBLEMS

1. Determine the short-time behavior of the reactor power following a step reactivity insertion into a critical reactor using the general expression (4.1, Eq. 3), and compare it to the exact solution. Discuss the physical origin of the difference between these two results.

2. Find the response of an initially critical reactor to a ramp reactivity insertion in the prompt-jump approximation assuming one group of delayed neutrons. (Answer 4.2, Eq. 6.)

3. Write down the kinetic equation in the prompt-jump approximation assuming that the changes in the delayed neutron production rate are small (short-time behavior), and determine the reactor power for a ramp reactivity insertion into a critical reactor. Compare this result to that of Problem 2.

4. Find the response of a critical reactor to a sinusoidal reactivity input, i.e., $k(t) = k_0 \sin \omega t$, using the prompt-jump approximation and the model of one group of delayed neutrons.

   Answer:

$$p(t) = \frac{p(0)}{1 - k(t)} \exp\left[\bar{\lambda} \int_0^t \frac{k(t')}{1 - k(t')}\, dt'\right]$$

5. Find the response of a critical reactor to a step reactivity input using the refined prompt-jump approximation and one group of delayed neutrons, and compare the result to the exact solution [4].

6. Find the response of a critical reactor to a ramp reactivity input $k(t) = \gamma t$ using the refined prompt-jump approximation and one group of delayed neutrons [4].

7. Show that the solution of $(1 + k_0 - \gamma t)\dot{P} = (\gamma - \bar{\lambda}k_0 + \bar{\lambda}\gamma t)P + (\bar{\lambda}lS_0/\beta)$ for $P(t)$ with the initial condition $P(0) = P_0$ is

$$P(t) = \left\{\left[\exp -\bar{\lambda}t\right]\left(1 - \frac{\gamma t}{1 + k_0}\right)^{-\mu-1}\right\}\frac{1}{1 + k_0}$$

$$\times \left[P_0 + \frac{l\bar{\lambda}}{\beta}\frac{S_0}{1 + k_0}\int_0^t dt'(\exp \bar{\lambda}t')\left(1 - \frac{\gamma t'}{1 + k_0}\right)^{\mu-1}\right]$$

$$+ \frac{l}{\beta}\frac{S_0}{1 + k_0 - \gamma t}$$

where $\mu = \bar{\lambda}/\gamma$.

8. Show that

$$\int_0^\infty dx\, e^{-x}[1 + (x/\mu)]^\mu$$

behaves asymptotically for large μ as $(2/3) + (\mu\pi/2)^{1/2}$.
Hint: Write the second factor as $\exp\{\mu \ln[1 + (x/\mu)]\}$ and expand $\ln[1 + (x/\mu)]$ keeping the first three terms; use $\exp(x^3/3\mu^2) \approx 1 + (x^3/3\mu^2)$.

9. (a) Show that (4.3, Eq. 22) can be approximated above prompt-critical by

$$F(k) \cong \left(\frac{2\pi\beta}{l\gamma}\right)^{1/2} \exp\left[\frac{\beta}{2l\gamma}(k - 1)^2 - \frac{1}{\gamma}\int_0^k \omega_0(k')\, dk'\right]$$

Hint: Use $T_0 = (\beta/l\gamma)^{1/2} \gg 1$ (small l and large γ), ignore the first term in (22). The second term in (22) is approximated by $T_0(2\pi)^{1/2}$ because $T_0 \gg 1$ and $T = (1 - k)\, T_0$ is a large negative number (cf 4.1, Eq. 5) for $k > 1$.

(b) Using (4.3, Eq. 19) and $\omega_0 \approx (\beta/l)(k - 1)$ (cf 3.3, Eq. 22), show that $F(k)$ can further be reduced to

$$F(k) \cong (2\pi\beta/l\gamma)^{1/2}\, Q(k, \gamma)$$

where

$$Q(k, \gamma) = \prod_{i=1}^{6}\left(1 + \frac{\beta k}{\lambda_i l}\right)^{-a_i\lambda_i/\gamma}, \qquad Q(k, \gamma_1)^{\gamma_1} = Q(k, \gamma_2)^{\gamma_2}$$

The variation of $Q(k, \gamma)$ is shown in Fig. P9.
(c) Varify (4.3, Eq. 23) using Fig. P9.

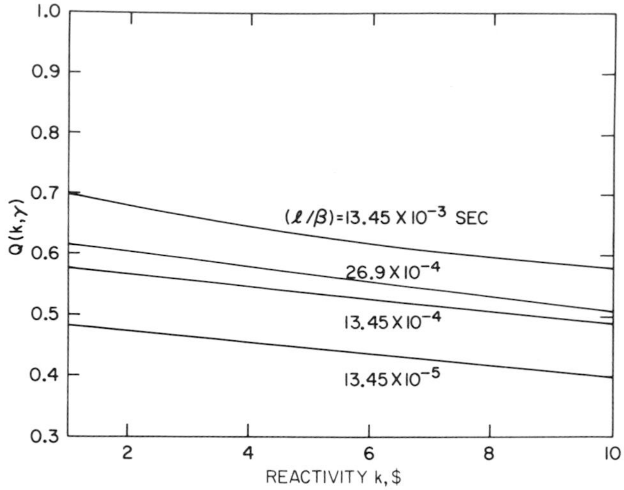

FIGURE P9. Variation of $Q(k, \gamma)$ with k for $\gamma = 5$ \$/sec.

10. Find the response of a reactor to a small step reactivity change using the linearized kinetic equations and compare the result to the exact response given by (3.3, Eq. 8) and discuss the discrepancies.

11. Show that

$$\omega_0 \int_0^\infty [(\ln \omega)/(\omega^2 - \omega_0^2)]\, d\omega = \pi^2/4$$

Hint: Let $x = \omega/\omega_0$ use $\int_0^\infty dx\,(x^2 - 1)^{-1} = 0$ and $\int_0^1 (\ln x)(x^2 - 1)^{-1}\, dx = \pi^2/8$.

12. Show that the phase of a linear system (of minimum-phase type) with a gain $G(\omega)$ given by

$$G(\omega) = 1, \qquad \text{for} \quad 0 \leqslant \omega < \omega_0$$

$$= 1/\omega, \qquad \text{for} \quad \omega > \omega_0$$

is $-\pi/4$ at $\omega = \omega_0$.
Hint: Use (4.4, Eq. 27) and Problem 11.

13. Show that the solution of

$$(l/\beta)\,\dot{p}(t) = P_{\mathrm{r}}k(t) + \int_0^\infty D(u)[p(t - u) - p(t)]\, du$$

for $P(t) \equiv P_0$ for $t \leqslant 0$ is

$$\bar{p}(s)/P_r = Z(s)\,\bar{k}(s) + (1/s)[(P_0/P_r) - 1]$$

Compare this result with (4.4, Eq. 4).

14. a. Show that

$$Z(s) = (\beta/l^*s) + (2/\pi)\,s \int_0^\infty [R(\omega)/(s^2 + \omega^2)]\,d\omega, \qquad \operatorname{Re}(s) > 0$$

where $R(\omega) = \operatorname{Re}[Z(i\omega)]$.
Hint: Use

$$Z(s)/2s = (1/2\pi i) \oint_\Gamma dz\, Z(z)/(z^2 - s^2)$$

where the path Γ is shown in Fig. P14.

b. Verify the above equality for

$$Z^{-1}(s) = s\{(l/\beta) + [1/(s + \bar{\lambda})]\}$$

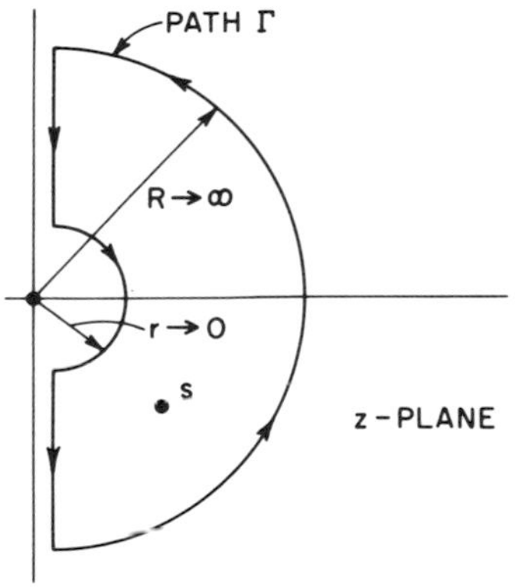

FIGURE P14. Path of integration.

15. Find the external source-reactor power transfer function in a subcritical reactor using point kinetics and one group of delayed neutrons. Plot the amplitude and phase characteristics using the straight-line-segment approximation, and compare them to those for the reactivity power transfer function (cf 4.4, Figure 4.4.9). Compare the low-frequency behavior in both cases.

16. Show that the solution of (4.5, Eq. 4) for a reactor operated at a constant power level P_0 prior to $t = 0$ is

$$\bar{y}(s) = Z(s)\,\bar{k}(s) + (1/s)\,y(0)$$

where $y(0) = \ln(P_0/P_r)$.

17. Find the response of an initially critical reactor to a step and ramp reactivity insertion using the logarithmic linearization technique (cf Section 4.5). Compare the results with those obtained using the conventional linearized kinetic equations (see Problem 10), and the prompt-jump approximation (see Problems 2, 5, and 6). In the case of the step input compare the result with the exact solution.

18. Plot the amplitude and phase characteristics of the period-equilibrium transfer function $Z(\omega_0, s)$ (cf 4.5, Eq. 13) using one group of delayed neutrons for $\omega_0 = 10$, 1, and 0.1 sec^{-1}, and compare the resulting curves with that for $\omega_0 = 0$ (cf 4.4, Figure 4.4.9).

19. (a) Show that $[Z_2(Z_1 - Z_1^*) - Z_1^2]$ approaches

$$(\beta/l^*)^3[2\tau_2^2(\beta/l^*) - 3\tau_3]$$

in the limit of $\omega \to 0$, where τ_n is defined as

$$\tau_n = \sum_{i=1}^{6} [a_i/(\lambda_i)^n]$$

(b) Show that $[2Z_2 - Z_1]$ approaches $(\beta/l^*)^2\,\tau_2$ as $\omega \to 0$. (Hint: First show that

$$Z_1(s) = \left\{s\left[(l/\beta) + \sum_{i=1}^{6} \frac{a_i}{s + \lambda_i}\right]\right\}^{-1}$$

$$= \frac{\beta}{l^* s}\left[1 + s\frac{\beta}{l^*}\tau_2 + s^2\frac{\beta}{l^*}\left(\frac{\beta}{l^*}\tau_2^2 - \tau_3\right) + O(s^3)\right]$$

then verify $(l^*/\beta) \to \tau_1$ when $l = 0$.)

REFERENCES

1. E. R. Cohen, *Proc. U. N. Int. Conf. Peaceful Uses At. Energy*, *2nd*, *Geneva*, 1958 p. 629. United Nations, New York, 1958.
2. H. Soodak, *Nucl. Reactor Theory Proc. Symp. Appl. Math.*, 11*th*, *New York*, 1959 (G. Birkhoff and E. P. Wigner, eds.). Am. Math. Soc., Providence, Rhode Island, 1961.

3. G. Birkhoff, *in* "Numerical Solutions of Nonlinear Differential Equations" (D. Greenspan, ed.), p. 3. Wiley, New York, 1966.
4. R. Goldstein and L. M. Shotkin, *Nucl. Sci. Eng.* **38**, 94 (1969).
5. H. Hurwitz, Jr., *Nucleonics* **5**, 61 (1949).
6. J. Wilkins, Jr., *Nucl. Sci. Eng.* **5**, 207 (1957).
7. H. Hurwitz, Jr., *Nucl. Sci. Eng.* **6**, 11 (1959).
8. "Reactor Handbook" (H. Soodak, ed.), 2nd ed., Vol. III, Part A. Wiley (Interscience), New York, 1962.
9. L. Brillouin, *Quart. Appl. Math.* **6**, 167 (1948).
10. L. Brillouin, *Quart. Appl. Math.* **7**, 363 (1949).
11. H. B. Smets, Low and high power nuclear reactor kinetics. D.Sc. Thesis, Nucl. Eng. Dept., M.I.T., Cambridge, Massachusetts, 1958.
12. S. Tan, *Nukleonik* **8**, 480 (1966).
13. E. L. Ince, "Ordinary Differential Equations." Dover, New York, 1956.
14. S. Tan, *Nucl. Sci. Eng.* **30**, 436 (1967).
15. A. Z. Akcasu, *Nucl. Sci. Eng.* **3**, 456 (1958).
16. A. Papoulis, "The Fourier Integral and Its Applications." McGraw-Hill, New York, 1962.
17. V. V. Solodovnikov, "Introduction to the Statistical Dynamics of Automatic Control Systems." Dover, New York, 1960.
18. H. W. Bode, "Network Analysis and Feedback Amplifier Design. Van Nostrand, Princeton, New Jersey, 1945.
19. M. A. Shultz, "The Control of Nuclear Reactors and Power Plants." McGraw-Hill, New York, 1955.
20. H. B. Smets, The describing function of nuclear reactors. *IRE Trans. Nucl. Sci.* NS-6, No. 4, **8** (1959).
21. S. Singer, LA-2654 (1962), Los Alamos Scientific Laboratory.
22. H. Bethe, APDA-117. 1956, Atomic Power Development Associates.
23. H. B. Smets, *Nukleonik* **7** (7), 399 (1965).
24. "Handbook of Mathematical Functions" (Appl. Math. Ser. 55), p. 375. Nat. Bur. Stand., Washington, D. C., 1964.
25. A. A. Wasserman, IDO-16755. 1962, Phillips Petroleum Co., Idaho Falls, Idaho.
26. M. Nelkin, Appendix to the report by H. Hurwitz, KAPL-1138. 1955, Knolls Atomic Power Lab. Schenectady, New York.
27. H. A. Sandmeier, *Nucl. Sci. Eng.* **11**, 85 (1959).
28. R. Lauber, *Atomkenenergic* **7**, 95 (1962).
29. D. Babala, Letters to the editor. *Nucl. Sci. Eng.* **17** (1963).
30. J. G. Truxal, "Automatic Feedback Control Synthesis." McGraw-Hill, New York, 1955.
31. A. Z. Akcasu and L. M. Shotkin, Technical note. *Nucl. Sci. Eng.* **32**, 262 (1968).
32. J. K. Vaurio and G. Pulkkis, Technical Note, *Nucl. Sci. Eng.* **32**, 283 (1970).

CHAPTER 5

Mathematical Description of Feedback

It was shown in Chapter 2 (see Section 2.6) that feedback effects can be described mathematically by a functional of the reactor power $P(t)$, which was denoted by $\delta\rho_f[P]$. The computation of this functional for a given reactor involves the solution of a set of coupled nonlinear integro-partial differential equations (see Section 1.3) describing the heat transfer in the reactor with a distributed heat source as well as the creation and destruction of nuclear species by nuclear reactions. Except for simple cases, some of which will be discussed in the subsequent section, a realistic evaluation of $\delta\rho_f[P]$ is extremely difficult. Therefore, one usually avoids this difficulty by introducing various crude and mostly intuitive simplifying approximations in the description of the feedback mechanism, and thereby obtains simple analytical expressions of $\delta\rho_f[P]$. Because of the lack of precise justification for these approximations, the resulting feedback functionals are usually referred to as "feedback models." The validity of a given model for a certain reactor type is verified experimentally by comparing the measured and predicted temporal behavior. We shall present the derivation of such models in the following sections and discuss their validity in certain cases in the light of experimental data.

5.1. Feedback Kernels

We begin our discussions by casting the point kinetic equations (2.6, Eqs. 18) in the presence of feedback into the standard form (cf 3.1, Eqs. 1 and 2):

$$(l/\beta)\,\dot{P} = \{\delta k_{\mathrm{ext}}(t) + \delta k_{\mathrm{f}}[P] - 1\}\,P + \sum_{i=1}^{6} \lambda_i C_i + (l/\beta)\,S \tag{1a}$$

$$\dot{C}_i = a_i P - \lambda_i C_i\,, \qquad i = 1,\ldots, 6 \tag{1b}$$

We also reproduce the definitions of the symbols here to facilitate reference in this chapter:

$$(l/\beta) \equiv (1/\mathscr{L})\langle N_0^+ \mid N_0\rangle \tag{2}$$

$$S \equiv [\langle N_0^+ \mid S\rangle/\langle N_0^+ \mid N_0\rangle]\,P_0 \tag{3}$$

$$a_i \equiv (1/\mathscr{L})\langle N_0^+ \mid M_i[N_0] \mid N_0\rangle \tag{4}$$

$$\delta k_{\mathrm{ext}}(t) = (1/\mathscr{L})\left\langle N_0^+ \,\middle|\, \sum_i \delta N_i^{\mathrm{ext}}(\partial\mathscr{H}_0/\partial N_{i0}) \,\middle|\, N_0\right\rangle \tag{5}$$

$$\delta k_{\mathrm{f}}[P] \equiv \delta k_{\mathrm{c}}[P] + \delta k_{\mathrm{T}}[P] \tag{6}$$

$$\delta k_{\mathrm{T}}[P] \equiv (1/\mathscr{L})\langle N_0^+ \mid \delta T[(P/P_0)\,N_0](\partial\mathscr{H}_0/\partial T_0) \mid N_0\rangle \tag{7}$$

$$\delta k_{\mathrm{c}}[P] \equiv (1/\mathscr{L})\left\langle N_0^+ \,\middle|\, \sum_i \delta N_i^{c}[(P/P_0)\,N_0](\partial\mathscr{H}_0/\partial N_{i0}) \,\middle|\, N_0\right\rangle \tag{8}$$

$$\mathscr{L} \equiv \left\langle N_0^+ \,\middle|\, \sum_{i=1}^{6} M_i[N_0] \,\middle|\, N_0\right\rangle \tag{9}$$

In these equations, reactivities are measured in dollars. Furthermore, the angular density is approximated (cf 2.6, Eq. 17) as $n(\mathbf{r}, u, \mathbf{\Omega}, t) = (P/P_0)\,N_0(\mathbf{r}, u, \mathbf{\Omega}, t)$, where $N_0(\mathbf{r}, u, \mathbf{\Omega})$ is the equilibrium distribution. When $P(t) \equiv P_0$, (1a) implies

$$\delta k_{\mathrm{f}}(P_0) = 0 \tag{10}$$

where $\delta k_{\mathrm{f}}(P_0)$ is an ordinary function of P_0. Thus, $\delta k_{\mathrm{f}}[P]$ measures the feedback reactivity from its value at equilibrium, i.e., the incremental feedback reactivity, as the symbol suggests. Similarly, $\delta k_{\mathrm{ext}}(t)$ measures the incremental reactivity from its constant positive value k_0 at equilibrium. The latter just compensates the equilibrium feedback reactivity, which we denote by $k_f(P_0)$. Hence,

$$k_0 + k_f(P_0) = 0 \tag{11}$$

determines P_0 which was introduced in (2.4, Eq. 21) as an arbitrary normalization constant in the absence of feedback.

A reactor with feedback can be represented, in the absence of external sources, by a block diagram as shown by Figure 5.1.1. Here, $p(t)$ is

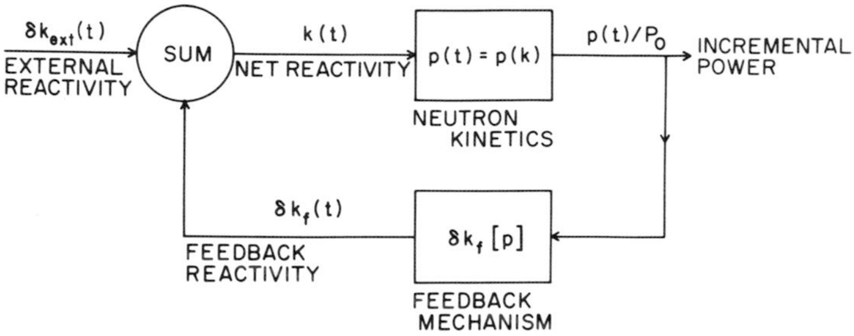

FIGURE 5.1.1. Block diagram of a reactor with feedback.

the incremental power, i.e., $p(t) = P(t) - P_0$, and $\delta k_f[p]$ is the feedback functional expressed in terms of the incremental power, so that $\delta k_f(0) = 0$ $(\delta k_f[p] \equiv \delta k_f[P_0 + p])$.

The black box corresponding to the neutron kinetics is completely described by the functional $p(t) = p[k]$, where the input is the net reactivity and the output is the incremental power. The functional $p[k]$ is determined by solving the point kinetic equations, i.e.,

$$(l/\beta)\,\dot{p} = k(t)(P_0 + p) + \int_0^\infty D(u)[p(t-u) - p(t)]\,du \tag{12}$$

Since the exact solution of (12) is not available in general, we cannot express $p(t)$ explicitly in terms of known operations on $k(t)$, except for the case of no delayed neutrons. In the absence of delayed neutrons, $p(t)$ is given by

$$p(t) = P_0 \left\{ \exp\left[(1/l) \int_0^t \rho(u)\,du\right] - 1 \right\} \tag{13}$$

where ρ is the net reactivity. In the case of delayed neutrons, we obtain explicit forms such as (13) only approximately (as we discussed in detail in Chapter 4). For example, if the power variations are sufficiently small, we can use a linear approximation, or, more accurately, logarithmic linearization, to express $p(t)$ in terms of $k(t)$. When the power variations are slow, we may use the prompt-jump approximation combined with the approximation of one group of delayed neutrons.

The input reactivity is given by

$$k(t) = \delta k_{\text{ext}}(t) + \delta k_{\text{f}}[p] \tag{14}$$

Description of the feedback functional $\delta k_{\text{f}}[p]$ in terms of the nuclear characteristics and the geometry of the reactor will be presented in the following sections when we consider specific reactor types. In this section, we discuss some aspects of $\delta k_{\text{f}}[p]$ from a mathematical point of view.

We first note that $\delta k_{\text{f}}[p]$ is invariant under a time translation as defined in (4.4, Eq. 7) when the feedback parameters are not explicit functions of time. The time invariance of $\delta k_{\text{f}}[p]$ is expressed mathematically as

$$\delta k_{\text{f}}(t - t_0) = \delta k_{\text{f}}[p(t - t_0)] \tag{15}$$

where t_0 is an arbitrary constant. Thus, if $\delta k_{\text{f}}(t)$ is the response of the feedback mechanism to $p(t)$, then the response to $p(t - t_0)$ is $\delta k_{\text{f}}(t - t_0)$.

Secondly, the feedback reactivity $\delta k_{\text{f}}(t)$ at a time t is uniquely determined if $p(t)$ is known only in the interval $(-\infty, t)$ (causality).

Thirdly, the feedback reactivity is bounded for any bounded input (stability). In other words, if $|p(t)| < M$ for all t, then $|\delta k_{\text{f}}(t)| < W(M)$, where $W(M)$ is a positive function independent of the input, and vanishes at $M = 0$ as M.

Functionals in general enjoy properties similar to those of ordinary functions. For example, we can define continuity, derivatives of any order, power-series expansion, and so on, for functionals as well as for ordinary functions (the interested reader is referred to the book by Voltera [1] for a clear discussion of functionals). For example, a time-invariant, causal functional may be represented by a power series as follows [2]:

$$\delta k_{\text{f}}[p] = \sum_{n=1}^{\infty} \int_{-\infty}^{t} du_1 \int_{-\infty}^{t} du_2 \cdots \int_{-\infty}^{t} du_n \, G_n(t - u_1, \ldots, t - u_n)\, p(u_1) \cdots p(u_n) \tag{16}$$

where the constant term is not included because $\delta k_{\text{f}}[p]$ denotes the incremental feedback reactivity from the steady-state value, and hence $\delta k_{\text{f}}[0] = 0$. The series (16) is called a "functional power series." Not all functionals, however, can be represented by a power series of the form (16). For this reason, functionals which have a power-series expansion are called "analytic functionals" [1]. We shall assume that the feedback functionals we shall be concerned with are analytic.

The functions $G_n(u_1, u_2, ..., u_n)$ appearing in (16) may be called feedback kernels of the nth order. We can always suppose that $G_n(u_1, u_2, ..., u_n)$ is symmetric with respect to its n variables, for, if not, we could symmetrize it. For example, let $G_2(u_1, u_2) \neq G_2(u_2, u_1)$. Since

$$\int_{-\infty}^{t} du_1 \int_{-\infty}^{t} du_2\, G_2(u_1, u_2)\, p(u_1)\, p(u_2) = \int_{-\infty}^{t} du_2 \int_{-\infty}^{t} du_1\, G_2(u_2, u_1)\, p(u_1)\, p(u_2) \tag{17}$$

by an interchange of the dummy variables u_1, u_2, we can replace $G_2(u_1, u_2)$ by $[G_2(u_1, u_2) + G_2(u_2, u_1)]/2$ without changing the value of the double integral in (17). This symmetrization procedure can be repeated for any n by considering all the permutations of $(u_1, u_2, ..., u_n)$, summing the functions G_n corresponding to each arrangement of $u_1, u_2, ..., u_n$, and dividing the sum by $n!$.

The time invariance of (16) can be verified as follows. The response to $p(t - t_0)$ at a time t_1 is

$$\delta k_f[p(t - t_0)]$$
$$= \sum_{n=1}^{\infty} \int_{-\infty}^{t_1} du_1 \cdots \int_{-\infty}^{t_1} du_n\, G_n(t_1 - u_1, ..., t_1 - u_n)\, p(u_1 - t_0) \cdots p(u_n - t_0)$$

If we let $u_1 - t_0 = v_1, ..., u_n - t_0 = v_n$, we find

$$\delta k_f[p(t - t_0)]$$
$$= \sum_{n=1}^{\infty} \int_{-\infty}^{t_1 - t_0} dv_1 \cdots \int_{-\infty}^{t_1 - t_0} dv_n\, G_n(t_1 - t_0 - v_1, ..., t_1 - t_0 - v_n)\, p(v_1) \cdots p(v_n)$$

which is identical to the response of the system for $p(t)$ at $(t_1 - t_0)$, as required by (15).

The stability of the functional is guaranteed if the feedback kernels are absolute integrable, i.e.,

$$\gamma_n \equiv \int_{0}^{\infty} du_1 \cdots \int_{0}^{\infty} du_n \mid G_n(u_1, ..., u_n) \mid < \infty \tag{18}$$

and if, in addition, the series $\sum_{n=1}^{\infty} \gamma_n M^{n-1}$ is convergent for all M. Then,

$$|\,\delta k_f[p]| \leqslant M \sum_{n=1}^{\infty} \gamma_n M^{n-1} \tag{19}$$

holds whenever $|\,p\,| \leqslant M$ for all t.

The causality of $\delta k_f[p]$ can also be expressed by requiring $G_n(u_1, u_2, ..., u_n)$ to vanish whenever any one of its variables is negative. The functional power-series expansion (16) involves the values of G_n for only positive arguments. With the above extension of the definition of G_n to negative arguments, we can replace (16) equivalently by

$$\delta k_f[p] = \sum_{n=1}^{\infty} \int_{-\infty}^{+\infty} du_1 \cdots \int_{-\infty}^{+\infty} du_n\, G_n(t - u_1, ..., t - u_n)\, p(u_1) \cdots p(u_n) \tag{20}$$

When the power variations are sufficiently small, the functional power-series expansion can be terminated after the first term, i.e.,

$$\delta k_f[p] = \int_{-\infty}^{t} du\, G(t - u)\, p(u) = \int_{0}^{\infty} du\, G(u)\, p(t - u) \tag{21}$$

where we have dropped the subscript of $G_1(u)$. As pointed out earlier (cf 4.4, Eqs. 6 and 10), (21) defines a linear functional. When the feedback functional is of this form, the corresponding feedback mechanism is called linear, and the function $G(t)$ is referred to as the "linear" feedback kernel. Physically, and quite generally, $G(t)$ is the reactivity at $t > 0$ due to a unit energy released at $t = 0$, when the feedback is linear. We can see this easily by substituting $p(u) = Q\,\delta(u)$ in (21). Since $\int_{-\infty}^{+\infty} du\, \delta(u) = 1$, and the integral of power is energy, Q must be equal to the unit of energy. Mathematically, $G(t)$ is the first derivative† of the functional $\delta k_f[p]$ at $p = 0$, i.e.,

$$G(t) = \lim_{\epsilon \to 0} (1/\epsilon)\, \delta k_f[\epsilon\, \delta(u)] \tag{22}$$

as can be seen from (20) with $p(u) = \epsilon\, \delta(u)$.

† The derivative of a functional $F[y(t)]$ at $u(t)$ is defined by [1]

$$F'[y(t), u] = \lim_{\epsilon \to 0} (1/\epsilon)\{F[y(t) + \epsilon\, \delta(u)] - F[y(t)]\}$$

In the present case, $y(t) \equiv 0$ and $F[0] = 0$.

The causality and stability conditions reduce, in the case of a linear feedback, to

$$G(t) = 0 \qquad \text{for} \quad t < 0 \tag{23a}$$

$$\int_0^\infty |G(t)|\, dt < \infty \tag{23b}$$

Although a linear feedback model may be sufficient for the stability analysis of most of the reactor types, there are cases in which the feedback is manifestly nonlinear. The combined effect of temperature and xenon feedback, as will be shown in a later section, is a quadratic functional of the incremental power, and can be represented by

$$\begin{aligned} \delta k_{\mathrm{f}}[p] = & \int_{-\infty}^{t} du\, G(t-u)\, p(u) \\ & + \int_{-\infty}^{t} du_1 \int_{-\infty}^{t} du_2\, G_2(t-u_1, t-u_2)\, p(u_1)\, p(u_2) \end{aligned} \tag{24}$$

In the case of a linear feedback, the point kinetic equations in integro-differential form are given by

$$\begin{aligned} (l/\beta)\, \dot{P} = & \left\{\delta k_{\mathrm{ext}} + \int_0^\infty du\, G(u)[P(t-u) - P_0]\right\} P \\ & + \int_0^\infty du[P(t-u) - P(t)]\, D(u) + S(l/\beta) \end{aligned} \tag{25}$$

In order to discuss an immediate physical implication of $G(u)$, let us suppose that we operate the reactor at a constant power level P_0' (different from P_0) in the absence of external sources by introducing a constant reactivity $\delta k_{\mathrm{ext}}(t) \equiv \delta k_0$. It readily follows from (25) with $\dot{P} = 0$ and $P(t) \equiv P_0'$ that

$$\delta k_0 = -\gamma(P_0' - P_0) \tag{26}$$

where we have introduced

$$\gamma \equiv \int_0^\infty G(u)\, du \tag{27}$$

Thus, the incremental change in the steady-state power level is proportional to the incremental change in the external reactivity. The proportionality constant γ is called the "power coefficient" of reactivity [the latter is measured in dollars in (25)]. Stability considerations to be discussed in subsequent chapters require γ to be negative. Indeed, if γ

were positive, a step increase in the power level due to a disturbance would produce a positive feedback reactivity in (25). Since the external reactivity remains unchanged, this additional positive reactivity would cause the power level to increase further. As a result of this cumulative effect, the reactor power would rise at a rate faster than an exponential unless corrective measures were taken by moving the control rods. This inherent instability against slow changes as described above is avoided by designing the reactor so that it has a negative power coefficient.

5.2. Temperature Feedback

In this and in the following sections, we shall illustrate the application of the general results just obtained by considering simple reactor types, and introduce several feedback models which are often used in reactor stability analysis.

In this section, we shall consider the temporal changes in the reactor power in time intervals of the order of several minutes or less (short-time behavior). In these time intervals, the variation in the atomic concentrations due to production and burnup of nuclei can be ignored. Then, the "short"-time behavior of a reactor is determined primarily by temperature variations and the characteristics of delayed neutrons.

The kinetic equations (see Section 1.3) in the diffusion approximation describing the short-time behavior are obtained from (1.4, Eq. 3) and (1.3, Eq. 5) as

$$\begin{aligned}[1/v(u)]\,\partial\phi(\mathbf{r},u,t)/\partial t = {} & \nabla\cdot D(\mathbf{r},u,t)\,\nabla\phi(\mathbf{r},u,t) \\ & - \Sigma(\mathbf{r},u,t)\,\phi(\mathbf{r},u,t) \\ & + \int_0^\infty du'\,\Big[\sum_j f_0{}^j(u)\,\nu^j(u')(1-\beta^j)\,\Sigma_f{}^j(\mathbf{r},u,t) \\ & + \Sigma_s(\mathbf{r},u'\to u,t)\Big]\,\phi(\mathbf{r},u',t) \\ & + \sum_{i=1}^{6}\lambda_i f_i(u) C_i(\mathbf{r},t) + S(\mathbf{r},u,t) \end{aligned} \tag{1}$$

$$[\partial C_i(\mathbf{r},t)/\partial t] + \lambda_i C_i(\mathbf{r},t) = \int_0^\infty du \sum_j \beta_i{}^j \nu^j(u)\,\Sigma_f{}^j(\mathbf{r},u,t)\,\phi(\mathbf{r},u,t) \tag{2}$$

$$\mu\{[\partial T(\mathbf{r},t)/\partial t] + \mathbf{V}(\mathbf{r},t)\cdot\nabla T(\mathbf{r},t)\} - \nabla\cdot K\nabla T(\mathbf{r},t) = H(\mathbf{r},t) \tag{3}$$

$$H(\mathbf{r},t) = W_f \int_0^\infty du\,\Sigma_f(\mathbf{r},u,t)\,\phi(\mathbf{r},u,t) \tag{4}$$

The cross sections in these equations are functions of the local temperature $T(\mathbf{r}, t)$, which is the origin of the temperature feedback. In order to compute the kinetic parameters defined in (5.1, Eqs. 2–9), we first have to determine the steady-state flux $\phi_0(\mathbf{r}, u)$ and $T_0(\mathbf{r})$ by solving the following coupled nonlinear integro-partial differential equations obtained from (5.1, Eqs. 1–9):

$$\nabla \cdot D_0(\mathbf{r}, u)\, \nabla\phi_0(\mathbf{r}, u) - \Sigma(\mathbf{r}, u)\, \phi_0(\mathbf{r}, u)$$

$$+ \int_0^\infty du' \left[\sum_j f^j(u)\, \nu^j(u')\, \Sigma_{f_0}^j(\mathbf{r}, u') + \Sigma_{s_0}(\mathbf{r}, u' \to u)\right] \phi_0(\mathbf{r}, u') = 0 \tag{5}$$

$$\mu\mathbf{V}_0(\mathbf{r}) \cdot \nabla T_0(\mathbf{r}) - \nabla \cdot K\, \nabla T_0(\mathbf{r}) = H_0(\mathbf{r}) \tag{6}$$

$$H_0(\mathbf{r}) = W_f \int_0^\infty du\, \Sigma_{f_0}(\mathbf{r}, u)\, \phi_0(\mathbf{r}, u) \tag{7}$$

The solution of this set requires, in general, numerical integration, and is one of the fundamental problems of reactor "statics." In reactor dynamics, we assume that the steady-state distribution of the dynamical variables is known. In the present problem, these are $T_0(\mathbf{r})$, $\mathbf{V}_0(\mathbf{r})$, and $\phi_0(\mathbf{r}, u) = v(u)\, N_0(\mathbf{r}, u)$.

The calculation of the kinetic parameters l/β, a_i, and S appearing in the point kinetic equations (5.1, Eq. 1) can be performed using their definitions in (5.1, Eqs. 2–9). This calculation reduces to a numerical integration once the steady-state distribution is known. In the diffusion approximation, the angular density $N_0(\mathbf{r}, u, \mathbf{\Omega})$ and its adjoint $N_0^+(\mathbf{r}, u, \mathbf{\Omega})$ entering these definitions are replaced by the scalar density $N_0(\mathbf{r}, u)$, and the definition of the scalar product is modified to mean integration over u and $\mathbf{r}$ only (cf Section 2.7).

The calculation of the feedback functional $\delta k_f[p]$, which contains the temperature feedback $\delta k_T[p]$ only when one is interested in the short-time behavior of the reactor, requires the solution of the time-dependent equations (1)–(4). This will be discussed next considering the lumped- and distributed-parameter descriptions of the feedback mechanism.

A. "Modal" Representation of the Temperature Feedback Functional

The temperature feedback functional $\delta k_T[p]$ can be expressed, in the diffusion approximation, using (2.7, Eq. 8) in (5.1, Eq. 7), as

$$\delta k_T[p] = \int_R d^3r\, \delta T(\mathbf{r}, t)\, \alpha(\mathbf{r}) \tag{8}$$

where $\alpha(\mathbf{r})$ can be interpreted as the "local temperature coefficient" of reactivity, and is defined by

$$\alpha(r) \equiv \frac{1}{\mathscr{L}} \int_0^\infty du \left[-\frac{3}{v} \frac{\partial \Sigma_{\text{tr}}}{\partial T_0} \mathbf{J}_0^+ \cdot \mathbf{J}_0 + \frac{1}{v} \phi_0^+ \frac{\partial \Gamma_0}{\partial T_0} \phi_0 \right] \tag{9}$$

The operator Γ_0 was defined in (2.7, Eq. 9). Since $\alpha(\mathbf{r})$ is calculable in terms of the equilibrium fluxes $\phi_0(\mathbf{r}, u)$ and $\phi_0^+(\mathbf{r}, u)$, our main task here is to establish the functional relationship between $\delta T(\mathbf{r}, t)$ and $p(t)$. Since we have already neglected terms proportional to $(\delta T)^2$ in obtaining (8) (cf 2.6, Eq. 7), we may linearize the temperature equations about the equilibrium values. Replacing $\phi(\mathbf{r}, u, t)$ in (4) by $[P(t)/P_0]\, \phi_0(\mathbf{r}, u)$ and linearizing, we obtain from (3)

$$\mu_0\{[\partial(\delta T)/\partial t] + \mathbf{V}_0 \cdot \nabla \delta T + \delta \mathbf{V} \cdot \nabla T_0\} - \nabla \cdot K_0 \nabla \delta T = [p(t)/P_0]\, H_0(\mathbf{r}) \tag{10a}$$

where

$$p(t) = P(t) - P_0 \tag{10b}$$

The variations $\delta \mathbf{V}$ in the fluid velocity can be related to the temperature variations δT again by a set of linearized equations obtained from the equations representing conservation of momentum. In the present discussion, we shall assume that the coolant velocity is not affected by temperature variations, so that $\delta \mathbf{V} = 0$. This assumption is realistic, for example, in the case of a pressurized water reactor with forced circulation. The solution of (10a) can then be exhibited in terms of a Green's function as

$$\delta T(\mathbf{r}, t) = (1/P_0) \int_0^\infty du\, p(t-u) \int_R d^3r'\, \mathscr{G}(\mathbf{r}, \mathbf{r}'; u)\, H_0(\mathbf{r}') \tag{11}$$

where $\mathscr{G}(\mathbf{r}, \mathbf{r}'; u)$ satisfies

$$\mu_0[(\partial \mathscr{G}/\partial t) + \mathbf{V}_0 \cdot \nabla \mathscr{G}] - \nabla \cdot K_0\, \nabla \mathscr{G} = \delta(\mathbf{r} - \mathbf{r}')\, \delta(t) \tag{12}$$

with the condition that $\mathscr{G}(\mathbf{r}, \mathbf{r}'; t) = 0$ for $t < 0$. Substituting (11) into (8), we obtain

$$\delta k_{\text{T}}[p] = \int_0^\infty du\, p(t-u)\, G(u) \tag{13}$$

where we have introduced

$$G(t) \equiv (1/P_0) \int_R d^3r\, \alpha(\mathbf{r}) \int_R d^3r'\, \mathscr{G}(\mathbf{r}, \mathbf{r}'; t)\, H_0(\mathbf{r}') \tag{14}$$

which is the linear feedback kernel defined by (5.1, Eq. 21).

Example. In order to illustrate the application of (14), we consider a bare thermal reactor which is cooled by keeping the temperature at the outer surface at a constant value, say $T_s = 0$. We assume that the core contains only solid regions, i.e., $\mathbf{V}_0 = 0$ everywhere. The heat capacity μ_0 and the heat conductivity K_0 are assumed to be independent of position. The nuclear properties of the core, which are characterized by $D_0(\mathbf{r}, T_0)$, $\Sigma_{a_0}(\mathbf{r}, T_0)$, and $\Sigma_{f_0}(\mathbf{r}, T_0)$ in the one-speed diffusion approximation, are also assumed to be uniform (in the presence of the temperature feedback) during steady state operation. The steady-state neutron density $\phi_0(\mathbf{r})$ and temperature $T_0(\mathbf{r})$ satisfy (cf Eq. 5)

$$D_0 \nabla^2 \phi_0 + (\nu\Sigma_{f_0} - \Sigma_{a_0})\,\phi_0 = 0 \tag{15a}$$

$$K_0 \nabla^2 T_0 + W_f \Sigma_{f_0} \phi_0 = 0 \tag{15b}$$

where $\phi_0(\mathbf{r})$ and $T_0(\mathbf{r})$ both vanish on the surface of the core. Clearly, ϕ_0 is the lowest eigenfunction of the ∇^2-operator appropriate to the geometry of the core, and

$$T_0(\mathbf{r}) = (W_f \Sigma_{f_0}/K_0 B_0^2)\,\phi_0(\mathbf{r}) \tag{16a}$$

$$B_0^2 = (\nu\Sigma_{f_0} - \Sigma_{a_0})/D_0 \tag{16b}$$

We now consider the Green's function $\mathscr{G}(\mathbf{r}, \mathbf{r}', t)$ defined by (12). This function satisfies the same boundary condition as the temperature, so that, for every $\mathbf{r}'$ in the core and for all $t > 0$, we have

$$\mathscr{G}(\mathbf{r}_s, \mathbf{r}'; t) = 0 \tag{17}$$

where $\mathbf{r}_s$ is a point on the surface. In order to solve (12), we may substitute

$$\mathscr{G}(\mathbf{r}, \mathbf{r}', t) = A(\mathbf{r}'t)\,\phi_0(\mathbf{r}) \tag{18}$$

and obtain

$$[\mu_0(\partial A/\partial t) + AK_0B_0^2]\,\phi_0(\mathbf{r}) = \delta(t)\,\delta(\mathbf{r} - \mathbf{r}') \tag{19}$$

Multiplying both sides by $\phi_0(\mathbf{r})$ and integrating over the reactor volume, we find

$$(\partial A/\partial t) + \eta_0 A = (1/\mu_0)\,\phi_0(\mathbf{r}')\,\delta(t) \tag{20a}$$

where

$$\eta_0 \equiv K_0 B_0^2/\mu_0 \tag{20b}$$

The solution of (20a) is readily obtained using the condition $A(\mathbf{r}', t) = 0$ for $t < 0$. Substituting this solution into (18), we obtain

$$\begin{aligned} \mathscr{G}(\mathbf{r}, \mathbf{r}', t) &= (1/\mu_0)\, \phi_0(\mathbf{r})\, \phi_0(\mathbf{r}')\, e^{-\eta_0 t}, \qquad & t > 0 \\ &= 0, & t < 0 \end{aligned} \tag{21}$$

The feedback kernel $G(t)$ follows from (14) and (21) as

$$G(t) = (\eta_0/P_0)\, e^{-\eta_0 t} \int_R d^3r\, \alpha(\mathbf{r})\, T_0(\mathbf{r}) \tag{22}$$

where we have used $H_0(\mathbf{r}) = W_{\mathrm{f}} \Sigma_{\mathrm{f}_0} \phi_0(\mathbf{r})$ and (16a). We may simplify (22) by defining an overall temperature coefficient of reactivity as

$$\alpha = (1/T_0) \int_R d^3r\, \alpha(\mathbf{r})\, T_0(\mathbf{r}) \tag{23}$$

where T_0 is an average temperature at equilibrium, e.g.,

$$T_0 \equiv (1/V) \int_R d^3r\, T_0(\mathbf{r}) \tag{24}$$

(V is the reactor volume). The result is

$$G(t) = \epsilon \alpha e^{-\eta_0 t}, \qquad \epsilon \equiv \eta_0 T_0/P_0 \tag{25}$$

An explicit form for the temperature coefficient can be obtained by substituting $\alpha(\mathbf{r})$ from (9) into (23):

$$\begin{aligned} \alpha = (V/\beta\nu\Sigma_{\mathrm{f}_0}) \Big\{ \int_R d^3r\, [3(\partial \Sigma_{\mathrm{tr}_0}/\partial T_0) \mid \mathrm{J}_0 \mid^2 \\ - (\partial/\partial T_0)(\Sigma_{\mathrm{a}_0} - \nu\Sigma_{\mathrm{f}_0})\, \phi_0{}^2]\, \phi_0 \Big\} \Big[\Big(\int_R d^3r\, \phi_0{}^2 \Big) \Big(\int_R d^3r\, \phi_0 \Big) \Big]^{-1} \end{aligned} \tag{26}$$

We recall that (26) does not include the feedback reactivity due to the change in the reactor size with temperature.

Remarks

The temperature feedback in the above example can be described equivalently by the following first-order differential equation:

$$\delta k_{\mathrm{T}}[p] = \alpha\, \delta T(t) \tag{27a}$$

$$[d(\delta T)/dt] + \eta_0\, \delta T = \epsilon p(t) \tag{27b}$$

This simple description is possible because, by virtue of the boundary condition imposed on the temperature (i.e., $T = 0$ on the outer boundary) and the assumption that μ_0 and K_0 are independent of position, the incremental temperature $\delta T(\mathbf{r}, t)$ is separable in time and space with the same spatial distribution as that of the equilibrium flux. In general, however, the temperature Green's function has a modal expansion as

$$\mathscr{G}(\mathbf{r}, \mathbf{r}'; t) = \sum_{n=0}^{\infty} \Phi_n(\mathbf{r}) \, \Phi_n(\mathbf{r}') \, e^{-\eta_n t} \tag{28}$$

where the $\Phi_n(\mathbf{r})$ are the eigenfunctions of

$$\nabla \cdot K_0(\mathbf{r}) \, \nabla \Phi_n(\mathbf{r}) + \mu_0(\mathbf{r}) \, \eta_n \Phi_n(\mathbf{r}) = 0 \tag{29}$$

with the proper boundary conditions on temperature. The orthogonality relation for $\Phi_n(\mathbf{r})$ is $\langle \Phi_n \mid \mu_0 \Phi_m \rangle = \delta_{nm}$. Substituting (28) into (14), we obtain

$$G(t) = (1/P_0) \sum_{n=0}^{\infty} \int_R d^3r \, \alpha(\mathbf{r}) \, \Phi_n(\mathbf{r}) \int_R d^3r' \, \Phi_n(\mathbf{r}') \, H_0(\mathbf{r}') \, e^{-\eta_n t} \tag{30}$$

The steady-state temperature $T_0(\mathbf{r})$ satisfies $-\nabla \cdot K_0 \, \nabla T_0 = H_0(\mathbf{r})$. It can be expanded in $\Phi_n(\mathbf{r})$ as

$$T_0(\mathbf{r}) = \sum_{n=0}^{\infty} (H_{0n}/\eta_n) \, \Phi_n(\mathbf{r}) \tag{31}$$

where

$$H_{0n} = \int d^3r \, H_0(\mathbf{r}) \, \Phi_n(\mathbf{r}) \tag{32}$$

We can define a temperature coefficient of reactivity associated with the nth mode, by substituting (31) into (23), as

$$\alpha_n = (1/T_0)(H_{0n}/\eta_n) \int_R d^3r \, \alpha(\mathbf{r}) \, \Phi_n(\mathbf{r}) \tag{33}$$

Using (32) and (33) in (30), we obtain a generalization of (25) as

$$G(t) = \sum_{n=0}^{\infty} \epsilon_n \alpha_n e^{-\eta_n t} \tag{34}$$

where

$$\epsilon_n = (T_0/P_0) \, \eta_n \tag{35}$$

The temperature feedback functional can be described in the above general case by an infinite set of ordinary differential equations:

$$\delta k_{\mathrm{T}}[p] = \sum_{n=0}^{\infty} \alpha_n \, \delta T_n \tag{36}$$

$$[d(\delta T_n)/dt] + \eta_n \, \delta T_n = \epsilon_n p(t), \qquad n = 0, 1, 2, \ldots \tag{37}$$

The analysis presented above is called "modal analysis," and is a very convenient method for attacking space-dependent problems. However, it requires the knowledge of the eigenvalues and eigenfunctions of the temperature operator in (29), which is not readily available in an analytic form in many practical situations involving complicated geometries and heterogeneities in the core. In such cases, the lumped-parameter description which is described below proves to be more convenient.

B. "Nodal" Representation of the Temperature Feedback Functional

A lumped-parameter description of the temperature feedback may be obtained by dividing the reactor into N different regions, such as the fuel, moderator, reflector, etc. For the sake of discussion, we consider a reactor with solid regions only, i.e., $\mathbf{V}_0 = 0$ everywhere. The temperature of the reactor is then governed by (cf Eq. 10a)

$$\mu_0[\partial(\delta T)/\partial t] - \nabla \cdot K_0(\nabla \delta T) = [p(t)/P_0] \, H_0(\mathbf{r}) \tag{38}$$

where $\delta T(\mathbf{r}, t)$ denotes the incremental temperature. The heat capacity μ_0 and the heat conductivity K_0 are different in general in each region. We assume that the steady-state temperature distribution $T_0(\mathbf{r})$ is known. We then define the following average quantities in each region:

$$\mu_j = (1/T_{0j}) \int_{V_j} d^3r \, \mu_0(\mathbf{r}) \, T_0(\mathbf{r}) \tag{39a}$$

$$T_{0j} = (1/V_j) \int_{V_j} d^3r \, T_0(\mathbf{r}) \tag{39b}$$

$$\delta T_j = (1/V_j) \int_{V_j} d^3r \, \delta T(\mathbf{r}, t) \tag{39c}$$

$$H_{0j} = \int_{V_j} d^3r \, H_0(\mathbf{r}) \tag{39d}$$

where V_j is the volume of the jth region. Integrating (38) over the volume of the jth region, we obtain

$$\mu_j[d(\delta T_j)/dt] - \int_{S_j} ds\, \hat{\mathbf{n}} \cdot K_0(\nabla\delta T) = [p(t)/P_0]\, H_{0j} \tag{40}$$

where the integration over S_j represents the total heat flux across the surface of the jth region. We assume that

$$-\int_{S_j} ds\, \hat{\mathbf{n}} \cdot K_0(\nabla\delta T) = \sum_{i=1}^{N} X_{ji}(\delta T_j - \delta T_i) + X_{j0}\, \delta T_j \tag{41}$$

where X_{ji} is coefficient of the heat transfer from the jth into the ith region through their common interface. If the regions are not adjacent, then $X_{ji} = 0$. It is clear from its physical nature that

$$X_{ij} = X_{ji} \tag{42}$$

The term $X_{j0}\, \delta T_j$ in (41) accounts for the heat transfer from the jth region to the surrounding medium, whose temperature is assumed to be constant (ambient temperature). In the regions that are not adjacent to the outer boundary of the reactor, $X_{j0} = 0$, if the reactor is cooled by heat conduction through its outer surface only. However, one may also interpret this term as the heat removed directly from the jth region by a coolant provided the variations in the coolant temperature can be neglected. In a more accurate description of the temperature feedback, the variations in the coolant temperature can be taken into account by treating the coolant as a separate region and allowing convective heat transfer between the regions by keeping the fluid velocity in (10a). We shall discuss such a model in the following example.

Substituting (41) into (40), we obtain

$$\mu_j[d(\delta T_j)/dt] + \sum_{i=1}^{N} X_{ji}(\delta T_j - \delta T_i) + X_{j0}\, \delta T_j = H_{0j}\, p(t)/P_0 \tag{43}$$

We can associate with the jth region a temperature coefficient of reactivity by breaking up (8) as

$$\delta k_{\mathrm{T}}[p] = \sum_{j=1}^{N} \int_{V_j} d^3r\, \delta T(\mathbf{r}, t)\, \alpha(\mathbf{r}) \tag{44}$$

and defining

$$\bar{\alpha}_j = (1/T_{0j}) \int_{V_j} d^3r\, T_0(\mathbf{r})\, \alpha(\mathbf{r}) \tag{45}$$

Equation (44) then reduces approximately to

$$\delta k_{\mathrm{T}}[p] = \sum_{j=1}^{N} \bar{\alpha}_j \, \delta T_j \tag{46}$$

Equations (43) and (46) describe the functional relationship between δk_{T} and $p(t)$. In order to obtain the feedback kernel as a sum of exponentials, we shall first cast (43) into a diagonal form. For this purpose, we define the following matrices [3]:

$$\mathsf{A} = [a_{ij}], \qquad a_{ij} = -X_{ij} + \delta_{ij} \sum_{k=0}^{N} X_{ik} \tag{47}$$

$$\boldsymbol{\mu} = [\delta_{ij}\mu_j] \tag{48}$$

$$\boldsymbol{\delta}\mathbf{T} = \mathrm{col}(\delta T_1, \delta T_2, \ldots, \delta T_N) \tag{49}$$

$$\mathbf{H} = \mathrm{col}(H_{01}, H_{02}, \ldots, H_{0N}) \tag{50}$$

We note that A is real and symmetric because the X_{ij} are real and satisfy $X_{ij} = X_{ji}$. The matrix $\boldsymbol{\mu}$ is diagonal with nonvanishing elements because μ_j is the temperature-weighted heat content per unit volume in the *j*th region, and hence is always positive.

With these matrices, we write (43) as

$$\boldsymbol{\mu}\, \boldsymbol{\delta}\dot{\mathbf{T}} + \mathsf{A}\, \boldsymbol{\delta}\mathbf{T} = [p(t)/P_0]\mathbf{H} \tag{51}$$

We now diagonalize A by a transformation

$$\boldsymbol{\delta}\mathbf{T} = \mathsf{R}\mathbf{X} \tag{52}$$

where R is a constant, real, and nonsingular ($|\,\mathsf{R}\,| \neq 0$) matrix, and $\mathbf{X}$ is a new column matrix replacing $\boldsymbol{\delta}\mathbf{T}$. According to a theorem in matrix algebra,† one can determine R in such a way that [4]

$$\tilde{\mathsf{R}}\boldsymbol{\mu}\mathsf{R} = \mathsf{I} \tag{53a}$$

$$\tilde{\mathsf{R}}\mathsf{A}\mathsf{R} = \boldsymbol{\eta} \tag{53b}$$

† If $\boldsymbol{\mu}$ and A are two symmetric $n \times n$ matrics, and $\boldsymbol{\mu}$ is positive-definite, then there is a nonsingular, real matrix R such that

$$\tilde{\mathsf{R}}\boldsymbol{\mu}\mathsf{R} = \mathsf{I}, \qquad \mathsf{RAR} = \mathrm{col}(\eta_1, \eta_2, \ldots, \eta_n)$$

The quantities $\eta_1, \eta_2, \ldots, \eta_n$ are necessarily the roots of the polynomial equation $|\, x\boldsymbol{\mu} - \mathsf{A}\,| = 0$ for any choice of the matrix R (Perlis [4], theorem 9-13).

where $\tilde{\mathsf{R}}$ is the transported matrix, i.e., $\tilde{R}_{ij} = R_{ji}$, I is the identity matrix, i.e., $\mathsf{I} = [\delta_{ij}]$, and η is a diagonal matrix, i.e., $\eta = [\eta_i \delta_{ij}]$. Substituting (52) into (51), multiplying the resulting equation by $\tilde{\mathsf{R}}$ from the left, and using the fact that $\tilde{\mathsf{R}}$ is a constant matrix, we get

$$\dot{\mathbf{X}} + \eta \mathbf{X} = p(t) \boldsymbol{\epsilon} \tag{54}$$

where the column matrix $\boldsymbol{\epsilon}$ is defined by

$$\boldsymbol{\epsilon} = \tilde{\mathsf{R}} \mathbf{H} / P_0 \tag{55}$$

In terms of the components, (54) reads

$$\dot{X}_j + \eta_j X_j = p(t)\, \epsilon_j \tag{56}$$

where

$$\eta_j = \sum_{m,n=1}^{N} R_{mj} R_{nj} A_{nm} \tag{57a}$$

$$\epsilon_j = \sum_{m=1}^{N} R_{jm} H_{0m} / P_0 \tag{57b}$$

The solution of (56) is readily obtained as

$$X_j(t) = \epsilon_j \int_0^\infty du\, e^{-\eta_j u} p(t-u) \tag{58}$$

The feedback functional $\delta k_{\mathrm{T}}[p]$ follows from (49) and (52) as

$$\delta k_{\mathrm{T}}[p] = \sum_{j=1}^{N} \alpha_j X_j \tag{59}$$

where

$$\alpha_j = \sum_{i=1}^{N} \bar{\alpha}_j R_{ij} \tag{60}$$

Substituting X_j from (58) into (59) yields the linear feedback kernel as

$$G(t) = \sum_{j=1}^{N} \epsilon_j \alpha_j e^{-\eta_j t} \tag{61}$$

The canonical parameters η_j, ϵ_j, and α_j are related to the physical constants A_{jk}, H_{0j}, and $\bar{\alpha}_j$ by (57) and (59), respectively. These relations involve the elements of the transformation matrix R.

C. Distributed-Parameter Description of the Temperature Feedback

We shall discuss the distributed-parameter description, considering a reactor which consists of a solid-fuel region and a coolant. The latter flows in the positive z direction (Figure 5.2.1). We assume that heat

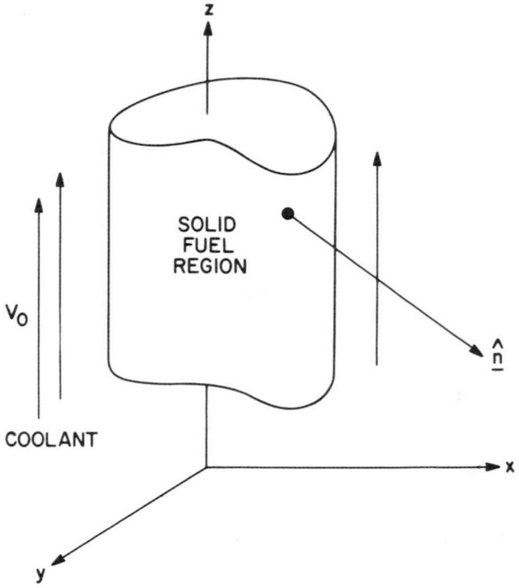

FIGURE 5.2.1. Geometry for a two-region temperature description.

is generated only in the fuel region, and that the nuclear and thermal properties of the core are independent of z. Denoting the incremental temperatures in the fuel and coolant by $T_f(\mathbf{r}, t)$ and $T_c(\mathbf{r}, t)$, respectively, we obtain the following equations from (10a):

$$\mu_f[\partial T_f(\mathbf{r}, t)/\partial t] - K_f \nabla^2 T_f = [p(t)/P_0] H_0(r) \qquad \text{(fuel)} \tag{62a}$$

$$\mu_c\{[\partial T_c(\mathbf{r}, t)/\partial t] + V_0[\partial T_c(\mathbf{r}, t)/\partial z]\} - K_c \nabla^2 T_c = 0 \qquad \text{(coolant)} \tag{62b}$$

We assume that the transverse distribution of $T_f(\mathbf{r}, t)$ and $T_c(\mathbf{r}, t)$ remains unaltered during transients. Integrating (62) in the xy plane, and defining

$$T_f(z, t) = (1/S_f) \int_{S_f} dx\, dy\, T_f(\mathbf{r}, t) \tag{63a}$$

(where S_f is the cross-sectional area of the fuel region), with a similar definition for $T_c(z, t)$, we obtain

$$\mu_f[\partial T_f(z, t)/\partial t] - K_f[\partial^2 T_f(z, t)/\partial z^2] - K_f \int_{S_{int}} ds\, \hat{\mathbf{n}} \cdot \nabla_2 T_f(\mathbf{r}, t)$$

$$= [p(t)/P_0]\, H_0(z) \tag{63b}$$

$$\mu_c\{[\partial T_c(z, t)/\partial t] + V_0[\partial T_c(z, t)/\partial z]\} - K_c[\partial^2 T_c(z, t)/\partial z^2]$$

$$= K_c \int_{S_{int}} ds\, \hat{\mathbf{n}} \cdot \nabla_2 T_c(\mathbf{r}, t) \tag{63c}$$

In (63b) and (63c), the heat capacities μ_f and μ_c are defined per unit length rather than per unit volume, as they are in (62), the vector $\nabla_2 T(\mathbf{r}, t)$ is the temperature gradient in the xy plane, and $\mathbf{n}$ is the unit reactor normal to the cylindrical interface S_{int}, between the fuel and coolant. Thus, the surface integrals denote the heat transfer from the fuel to the coolant per unit length in the z direction, which is assumed to be given by $h(T_f - T_c)$, where h is the heat transfer coefficient through the interface (cf Eq. 41). The heat conduction in the z direction both in the fuel and coolant (terms containing $\partial^2 T/\partial z^2$) can be neglected because the temperature gradient in the z direction is small. With these simplifying assumptions, Eqs. (63) reduce to

$$\mu_f(\partial T_f/\partial t) + h(T_f - T_c) = [p(t)/P_0]\, H_0(z) \tag{64a}$$

$$\mu_c[(\partial T_c/\partial t) + V_0(\partial T_c/\partial z)] = h(T_f - T_c) \tag{64b}$$

These equations can be solved with $T_c(0, t) = 0$ (i.e., constant inlet coolant temperature). Taking the Laplace transform of (64), and using the initial conditions $T_c(z, 0) = 0$ and $T_f(z, 0) = 0$, we obtain

$$(\mu_f s + h)\, \bar{T}_f - h\bar{T}_c = (\bar{p}/P_0)\, H_0(z) \tag{65}$$

$$(d\bar{T}_c/dz) + [(\mu_c s + h)/\mu_c V_0]\, \bar{T}_c = (h/V_0 \mu_c)\, \bar{T}_f \tag{66}$$

Eliminating $\bar{T}_f(s, z)$, we find the following equation for $\bar{T}_c(s, z)$:

$$(d\bar{T}_c/dz) + U(s)\, \bar{T}_c = V(z, s)\bar{p}(s)/P_0 \tag{67a}$$

where

$$U(s) \equiv (s/V_0)\{1 + (\mu_f/\mu_c)[h/(\mu_f s + h)]\}, \tag{67b}$$

$$V(z, s) \equiv (h/\mu_c V_0)[1/(\mu_f s + h)]\, H_0(z) \tag{67c}$$

The solution of (67a) is obtained, with the initial condition $\bar{T}_c(s, 0) = 0$, as

$$\bar{T}_c(s, z) = [\bar{p}(s)/P_0] \int_0^z du\, e^{-U(s)(z-u)} V(s, u) \tag{68}$$

The fuel temperature follows from (65) as

$$\bar{T}_f(s, z) = [\bar{p}(s)/P_0] \left[h \int_0^z du\, e^{-U(s)(z-u)} V(s, u) + H_0(z)\right] [1/(\mu_f s + h)] \tag{69}$$

The temperature feedback functional is obtained by substituting (68) and (69) into (8) and performing the integration in the xy plane,

$$\delta k_T[p] = \int_0^L dz\, [\alpha_f(z)\, T_f(z, t) + \alpha_c(z)\, T_c(z, t)] \tag{70}$$

where the fuel temperature coefficient is defined as the weighted average of $\alpha(\mathbf{r})$ in (9),

$$\alpha_f(z) = [1/T_{f_0}(z)] \int_{S_f} dx\, dy\, \alpha(\mathbf{r})\, T_{f_0}(\mathbf{r}) \tag{71}$$

where $T_{f_0}(z)$ is the steady-state fuel temperature distribution averaged in the xy plane according to (63a) using $T_{f_0}(\mathbf{r})$. The coolant temperature coefficient $\alpha_c(z)$ is defined in a similar way. In (70), L is the height of the channel.

It is easier, and also more useful from the stability point of view, to calculate the feedback transfer function $H(s)$, which is the Laplace transform of the feedback kernel $G(t)$. Its definition can also be written as

$$H(s) = (\delta k_T\, [p]_L)/\bar{p}(s) \tag{72}$$

Substituting $\bar{T}_c(z, s)$ and $\bar{T}_f(z, s)$ from (68) and (69) into the transform of (70), we obtain $H(s)$ as

$$P_0 H(s) = \int_0^L dz\, \{\alpha_c(z) + \alpha_f(z)[h/(\mu_f s + h)]\} \times \int_0^z du\, e^{-U(s)(z-u)} V(s, u) + \int_0^L dz\, [\alpha_f(z)\, H_0(z)/(\mu_f s + h)] \tag{73}$$

As an illustration, we shall simplify (73) by ignoring the fuel time constant μ_f/h. This restricts the validity of (73) to frequencies $\omega \ll h/\mu_f$.

The time constant μ_c/h of the coolant is retained. With this approximation, (73) reduces to

$$H(s) = (1/\mu_c V_0 P_0) \int_0^L dz\, [\alpha_c(z) + \alpha_f(z)] \times \int_0^z du\, \{\exp[-(s/V_0')(z-u)]\}\, H_0(u) + (1/P_0 h) \int_0^L dz\, \alpha_f(z)\, H_0(z) \tag{74a}$$

Where V_0' is the modified coolant velocity, defined as

$$V_0' \equiv V_0[1 + (\mu_f/\mu_c)]^{-1} \tag{74b}$$

The temperature coefficients $\alpha_f(z)$ and $\alpha_c(z)$ are very complicated functions of z (cf Eq. 71). The uncertainties and crude approximations in the physical description of the temperature feedback mechanism usually do not warrant an accurate evaluation of $\alpha_f(z)$ and $\alpha_c(z)$. Therefore, the effect of weighting is often included qualitatively by assuming $\alpha_f(z)$ and $\alpha_c(z)$ to be proportional to the first or second power of the steady-state axial distribution of the scalar neutron density $\phi_0(\mathbf{r})$. The simplest approximation, however, is to choose these functions as constants, namely $\alpha_f(z) \simeq \alpha_f/L$ and $\alpha_c(z) \simeq \alpha_c/L$, where α_f and α_c are the axially integrated reactivity coefficients.

In addition, we shall suppose that the heat production is axially uniform, i.e., $H_0(z) = H_0$, where H_0 is the heat energy per unit time per unit length. With these approximations, the integrations in (74) can be performed explicitly, and $H(s)$ reduces to

$$H(s) = A + B[(s\tau - 1 + e^{-s\tau})/(s\tau)^2] \tag{75}$$

where

$$A \equiv (\alpha_f/P_0 h)\, H_0 \tag{76a}$$

$$B \equiv [(\alpha_f + \alpha_c)/\mu_c P_0]\, H_0 \tau \tag{76b}$$

$$\tau \equiv L/V_0' \tag{77}$$

REMARKS

We observe that $H(s)$ is a transcendental function of s. This feature is a characteristic of physical systems with distributed parameters in general [the distributed parameter in our example was $T_c(z, t)$]. The transfer function characterizing a linear lumped-parameter system can always be expressed as a ratio of two polynomials in s with real coefficients (meromorphic function [5]), with the order of the numerator not exceeding the order of the denominator. The above difference between the distributed- and lumped-parameter descriptions of the feedback mechanism plays an important role in the stability analysis which we discuss in subsequent chapters.

Another interesting observation about (75) is that $H(s) = 0$ may have roots with real positive parts, indicating that the corresponding physical system is not of a minimum-phase type, and that the dispersion relation (cf Section 4.4) between the amplitude and phase of $H(i\omega)$ is not valid in general. Therefore, both the gain and phase must be measured in a pile-oscillator experiment that is designed to determine $H(i\omega)$.

At low frequencies $\omega < 2\pi/\tau$, where τ is the coolant transit time L/V_0' in the core, we may approximate $H(s)$ by a lumped-parameter system by replacing the exponential factor in (76a) by

$$e^{-\tau s} = [1 - (s\tau/2)]/[1 + (s\tau/2)] \tag{78a}$$

The result reads

$$H(s) \cong A + \tfrac{1}{2}B\{1/[1 + (\tau s/2)]\} \tag{78b}$$

which is a meromorphic function.

D. Temperature Feedback with a Transport Time Delay

In certain reactor types, the temperature feedback acts with a constant time delay τ. This delay may be due to the transport of the coolant from one region to the other within the reactor. Such a model was introduced by Bethe to investigate the stability of EBR-1 [6]. The time delay τ was visualized in his model as the transit time of the coolant from the top plate to the bottom. Mathematically, this model is described by

$$\delta k_{\mathrm{T}}[p] = \alpha\, \delta T(t - \tau) \tag{79a}$$

$$\mu[d(\delta T)/dt] + h\, \delta T = H_0 p(t)/P_0 \tag{79b}$$

or by a feedback kernel $G(t)$ as

$$G(t) = \alpha\epsilon U(t - \tau)\, e^{-\eta(t-\tau)} \tag{80a}$$

where $\epsilon = \eta T_0/P_0$, $\eta = h/\mu$ (cf Eqs. 27 and 28), and $U(t)$ is the unit step function. The associated transfer function $H(s)$ follows from (80a):

$$H(s) = [\alpha\epsilon/(s + \eta)]\, e^{-s\tau} \tag{80b}$$

which is again a transcendental function of s. Systems with time delays are often called "hereditary" or "systems with a past history." They require special attention in stability analysis, as we shall discuss later, and their future cannot be determined solely by the initial conditions of the dynamical variables; the future depends on the past history in the time interval $(-\tau, 0)$.

Transport time delays may also arise from the motion of the fuel in the core. The circulating fuel reactor is a typical example of a reactor with moving fuel.

E. Circulating Fuel Reactor

The dynamics of the circulating fuel reactor was discussed by Welton [7], Ergen and Weinberg [8], Ergen [9], and Fleck [10, 11]. Here, we shall follow Fleck's report [11] closely, in which he developed the temperature-dependent kinetics of circulating fuel reactors in the absence of delayed neutrons.

A circulating reactor consists of a bare, right cylindrical core of length L, and an external loop containing the heat exchanger. The circulating fuel passes through the core once before entering the heat exchanger. All particles of the fuel move with the same velocity V_0. Fission takes place only in the core.

We choose the z axis parallel to the axis of the core, and use the one-group diffusion model to describe the neutron flux. We ignore heat conduction in the fuel and the variations in the fuel velocity during transients. With these assumptions, (10a) reduces to

$$[\partial(\delta T)/\partial z] + (1/V_0)[\partial(\delta T)/\partial t] = [p(t)/V_0\mu P_0]\, H_0(z) \tag{81}$$

The steady-state flux distribution $\phi_0(r, z)$ in a bare cylindrical core in the absence of temperature feedback is

$$\phi_0(r, z) = A \sin[(\pi/L)z]\, J_0(\beta_r r) \tag{82}$$

where $J_0(x)$ is the zeroth-order Bessel function and $\beta_r = 2.405/R$, R being the radius of the core. We choose $\phi_0(\mathbf{r})$ as the shape function [this corresponds to choosing the reference reactor as the cooled reactor (cf Section 2.6)]. In (81), $H_0(z)$ is the power generated per unit length of the core at steady state, i.e.,

$$H_0(z) = 2W_f A\pi \int_0^R dr\, r\Sigma_{f_0} J_0(\beta_r r) \sin[(\pi/L)z] \tag{83a}$$

$$= H_0 \sin[(\pi/L)z] \tag{83b}$$

The incremental temperature $\delta T(z, t)$ in (81) is the radial average of $\delta T(r, z, t)$ (cf Eq. 63a), i.e.,

$$\delta T(z, t) = (2/R^2) \int_0^R dr\, r\, \delta T(r, z, t) \tag{84}$$

The solution of (81) with $\delta T(z, 0) = 0$ and $\delta T(0, t) = 0$ can be obtained as a special case of (68):

$$\delta \bar{T}(z, s) = [\bar{p}(s)/P_0 V_0 \mu] \int_0^z du\, e^{-(s/V_0)(z-u)} H_0(u) \tag{85}$$

The feedback functional is obtained from (70) as

$$\delta k_{\mathrm{T}}[p] = \int_0^L dz\, \alpha(z)\, \delta T(z, t) \tag{86}$$

from which we derive the feedback transfer function $H(s)$ using (72):

$$H(s) = (1/P_0 \mu V_0) \int_0^L dz\, \alpha(z) \int_0^z du\, e^{-(s/V_0)(z-u)} H_0(u) \tag{87}$$

In order to perform the integrations in (87), we have to express $\alpha(z)$ explicitly. It is obtained from (9) after substituting $\phi_0(r, z)$ from (82) and performing the radial integration. Here, we assume that

$$\alpha(z) = (\alpha/L) \sin^2[(\pi/L)z] \tag{87a}$$

This follows from (9) exactly if we ignore the change in the diffusion coefficient. Thus, (87) reduces to

$$H(s) = 2\pi[H(0)/\theta] \int_0^\theta dx\, \sin^2[(\pi/\theta)x] \int_0^x dy\, e^{-sy} \sin[(\pi/\theta)(x - y)] \tag{88}$$

where all the constants are lumped in $H(0)$. Here, θ is the core transit time, L/V_0. Upon performing the indicated integrations, we find

$$H(s) = \frac{H(0)}{\theta} \left\{ (1 - e^{-\theta s}) \left[\frac{1}{s} + \frac{s}{3[s^2 + 4(\pi^2/\theta^2)]} \right] + \frac{4}{3} \frac{s(1 + e^{-\theta s})}{s^2 + (\pi^2/\theta^2)} \right\} \tag{89}$$

which was obtained by Fleck [11]. If, instead of (87a), we assume $\alpha(z) = \alpha/L$, then (89) acquires a simpler form:

$$H(s) = [2H(0)/\theta](s\theta - 1 + e^{-\theta s})/s^2\theta \tag{90}$$

which was used by Welton [7] and by Ergen and Weinberg [8] in their studies of the stability of the circulating fuel reactor.

As pointed out by Fleck [11], the neglect of the delayed neutrons is a good approximation for large circulation velocities because, then, a large fraction of the delayed neutrons are lost to the external loop.†

The dynamics of a circulating fuel reactor based on (90) provides an excellent example for explaining several interesting features of nonlinear reactor dynamics, such as the appearance of semistable limit cycles. We shall consider these points in Chapter 7.

F. Concluding Remarks

In this section, we have illustrated the method of obtaining the temperature feedback transfer function $H(s)$ in terms of the thermal and nuclear properties of a reactor. We have shown in Sections 5.2A and 5.2B that $H(s)$ can be written as

$$H(s) = \sum_{j=1}^{N} [\alpha_j \epsilon_j/(s + \eta_j)] \tag{91}$$

if a lumped-parameter description (e.g., modal or nodal description) of the feedback mechanism is adapted. In the case of a distributed-parameter description, as well as in the presence of feedback with pure transport time delays, $H(s)$ is a transcendental function of s, involving exponential terms. We reproduce here the transfer function for temperature feedback with a delay and for the circulating reactor model discussed in Sections 5.2D and 5.2E:

$$H(s) = H(0)[\eta/(s + \eta)]\, e^{-s\tau} \tag{92}$$

$$H(s) = H(0)(2/\theta)(\theta s - 1 + e^{-s\theta})/s^2\theta \tag{93}$$

The feedback kernel $G(t)$ associated with these feedback models can be found by taking the inverse transform of $H(s)$:

$$G(t) = \sum_{j=1}^{N} \alpha_j \epsilon_j e^{-\eta_j t} \qquad \text{(lumped parameter)} \tag{94}$$

$$G(t) = \eta H(0)\, e^{-s(t-\tau)} U(t - \tau) \qquad \text{(time delay)} \tag{95}$$

$$\begin{aligned} G(t) &= (2/\theta)\, H(0)[1 - (t/\theta)] \qquad \text{for} \quad 0 \leqslant t \leqslant \theta \\ &= 0 \qquad \text{otherwise} \end{aligned} \qquad \begin{matrix}\text{(circulating} \\ \text{fuel reactor)}\end{matrix} \tag{96}$$

Figure 5.2.2 shows the variation of these kernels with time.

† For a discussion of the kinetics of circulating fuel reactors, see Fleck [10] and Meghreblian and Holmes ([12], p. 590).

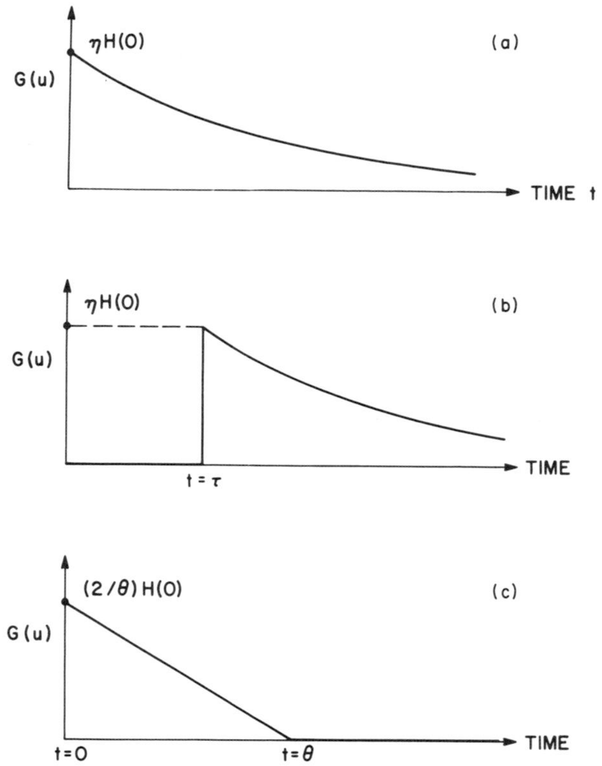

FIGURE 5.2.2. Feedback kernel variation for several different models. (a) Single temperature model; (b) temperature feedback with a delay; (c) circulating fuel reactor model.

We shall investigate reactor stability in Chapters 6 and 7 in terms of $H(s)$ [or $G(t)$] in general, and illustrate the application of the various stability criteria considering specific analytic forms for $H(s)$. This approach provides a unified treatment of the stability of various reactor types which are characterized by the particular form of their feedback transfer function.

5.3. Xenon Feedback

In the preceding discussion, we considered only the short-term behavior of a reactor, which involves changes in the power level in time intervals of the order of minutes and is determined primarily by temperature feedback. In time intervals of the order of hours, the behavior of a reactor is governed by both the temperature feedback and the buildup and burnup of higher cross-section fission-product poisons, e.g., ^{135}Xe. Since the thermal time constants η_j in (5.2, Eq. 34) are much less than those of ^{135}Xe and ^{135}I (9.2 and 6.7 hr, respectively), the temperature feedback can be treated in the quasistatic approximation.† The latter is equivalent to treating $p(t)$ in

$$\delta k_{\mathrm{T}}[p] = \int_0^\infty du\, p(t-u)\, G(u) \tag{1a}$$

as slowly varying, and approximating the integral as

$$\delta k_{\mathrm{T}}[p] \cong p(t) \int_0^\infty du\, G(u) = \gamma p(t) \tag{1b}$$

where γ is the power coefficient of reactivity, which was defined in (5.1, Eq. 27).

Having disposed of temperature feedback by the quasistatic approximation, we now focus our attention on the xenon feedback functional $\delta k_{\mathrm{Xe}}[p]$, which is obtained from (5.1, Eq. 8) by considering the term corresponding to changes in xenon concentration. In the diffusion approximation (cf 2.7, Eq. 8), $\delta k_{\mathrm{Xe}}[p]$ can be written as

$$\begin{aligned}\delta k_{\mathrm{Xe}}[p] = (1/\mathscr{L})[\langle\phi_0^+ \mid \delta N_{\mathrm{xe}}(1/v)(\partial\Gamma_0/\partial N_{\mathrm{xe}_0})\mid \phi_0\rangle \\ - \langle \mathbf{J}_0^+ \mid \delta N_{\mathrm{xe}}(3/v)(\partial\Sigma_{\mathrm{tr}}/\partial N_{\mathrm{xe}_0})\mid \mathbf{J}_0\rangle]\end{aligned} \tag{2}$$

where $\delta N_{\mathrm{xe}}(\mathbf{r}, t)$ is the change in the xenon concentration, and Γ_0 is defined in (2.7, Eq. 7). Introducing a weighting function $\alpha_{\mathrm{Xe}}(\mathbf{r})$ similar to $\alpha(\mathbf{r})$ in (5.2, Eqs. 8 and 9) as

$$\begin{aligned}\alpha_{\mathrm{Xe}}(\mathbf{r}) = (1/\mathscr{L}) \int_0^\infty du\, \{-[3/v(u)](\partial\Sigma_{\mathrm{tr}}/\partial N_{\mathrm{xe}_0})\, \mathbf{J}_0^+ \cdot \mathbf{J}_0 \\ + (1/v)\, \phi_0^+(\partial\Gamma_0/\partial N_{\mathrm{xe}_0})\, \phi_0\}\end{aligned} \tag{3}$$

we can express $\delta k_{\mathrm{Xe}}[p]$ in a simpler form:

$$\delta k_{\mathrm{Xe}}[p] = \int_R d^3r\, \delta N_{\mathrm{xe}}(\mathbf{r}, t)\, \alpha_{\mathrm{Xe}}(\mathbf{r}) \tag{4}$$

We observe that $\alpha_{\mathrm{Xe}}(\mathbf{r})$ is the local reactivity coefficient of xenon. It is calculated in terms of the equilibrium cross sections and fluxes.

† This approximation may not be valid in high-power reactors with long temperature-time constants, and hence will be relaxed in Section 6.3.

In order to establish the functional relationship between $\delta N_{\text{xe}}(\mathbf{r}, t)$ and $p(t)$, we use (1.3, Eqs. 6 and 7), which we reproduce here in a more conventional form:

$$\partial \mathrm{I}/\partial t = -\lambda_{\mathrm{I}}\mathrm{I} + y_{\mathrm{I}}\sigma_{\mathrm{f}}(\mathbf{r})\,\phi_0(\mathbf{r}, t) \tag{5}$$

$$\partial \mathrm{Xe}/\partial t = \lambda_{\mathrm{I}}\mathrm{I} + y_{\mathrm{Xe}}\sigma_{\mathrm{f}}(\mathbf{r})\,\phi_0(\mathbf{r}, t) - \lambda_{\mathrm{Xe}}\mathrm{Xe} - \mathrm{Xe}\sigma_{\mathrm{Xe}}(\mathbf{r})\,\phi_0(\mathbf{r}, t) \tag{6}$$

where we have introduced

$$\phi_0(\mathbf{r}, t) = \int_0^\infty du \int d\Omega\, vn(\mathbf{r}, u, \mathbf{\Omega}, t)$$

treated $^{135}\mathrm{I}$ as a direct fission product (cf the comments after 1.2, Eq. 34), and defined

$$\mathrm{I}(\mathbf{r}, t) = N_{\mathrm{I}}/N_{\mathrm{f}} \tag{7a}$$

$$\mathrm{Xe}(\mathbf{r}, t) = N_{\mathrm{Xe}}/N_{\mathrm{f}} \tag{7b}$$

In this notation, I and Xe denote the iodine and xenon concentrations per fuel atom. The cross sections $\sigma_{\mathrm{f}}(\mathbf{r})$ and $\sigma_{\mathrm{Xe}}(\mathbf{r})$ are defined by

$$\phi_0(\mathbf{r})\,\sigma_{\mathrm{f}}(\mathbf{r}) \equiv \int_0^\infty du\, \sigma_{\mathrm{f}}(u, T_0)\,\phi_0(\mathbf{r}, u) \tag{8a}$$

$$\phi_0(\mathbf{r})\,\sigma_{\mathrm{Xe}}(\mathbf{r}) \equiv \int_0^\infty du\, \sigma_{\mathrm{aXe}}(u, T_0)\,\phi_0(\mathbf{r}, u) \tag{8b}$$

and denote the flux-averaged effective microscopic cross sections at equilibrium.

Equations (5) and (6) can be solved exactly in terms of $\phi_0(\mathbf{r}, t)$:

$$\mathrm{Xe}(\mathbf{r}, t) = \int_0^\infty dt' \exp\left\{-\lambda_{\mathrm{Xe}}t' - \sigma_{\mathrm{Xe}}(\mathbf{r}) \int_0^{t'} dt''\, \phi_0(\mathbf{r}, t - t'')\right\} \times \{\lambda_{\mathrm{I}}\mathrm{I}(\mathbf{r}, t - t') + y_{\mathrm{Xe}}\sigma_{\mathrm{f}}(\mathbf{r})\,\phi_0(\mathbf{r}, t - t')\} \tag{9}$$

where

$$\mathrm{I}(\mathbf{r}, t) = y_{\mathrm{I}}\sigma_{\mathrm{f}}(\mathbf{r}) \int_0^\infty d\tau\, e^{-\lambda_{\mathrm{I}}\tau}\phi_0(\mathbf{r}, t - \tau) \tag{10}$$

The steady-state distributions of xenon and iodine follow from Eqs. (9) and (10) if we let $\phi_0(\mathbf{r}, t) = \phi_0(\mathbf{r})$:

$$\mathrm{Xe}_0(\mathbf{r}) = (y_{\mathrm{I}} + y_{\mathrm{Xe}})\,\sigma_{\mathrm{f}}(\mathbf{r})\,\phi_0(\mathbf{r})/[\lambda_{\mathrm{Xe}} + \sigma_{\mathrm{Xe}}(\mathbf{r})\,\phi_0(\mathbf{r})] \tag{11a}$$

$$\mathrm{I}_0(\mathbf{r}) = y_{\mathrm{I}}\sigma_{\mathrm{f}}(\mathbf{r})\,\phi_0(\mathbf{r})/\lambda_{\mathrm{I}} \tag{11b}$$

The desired xenon feedback functional is obtained by substituting $\delta N_{xe}(\mathbf{r}, t) = [\mathrm{Xe}(\mathbf{r}, t) - \mathrm{Xe}_0(\mathbf{r})]N_f$ from (9) and (11a) into (4) with $\phi_0(\mathbf{r}, t) \approx [P(t)/P_0]\,\phi_0(\mathbf{r})$ in (9). This leads to a nonlinear functional as a consequence of the exponential factor in the integrand of (9).

A simpler description of the xenon feedback can be obtained by using a space-independent model. For this purpose, we integrate the xenon and iodine equations (5) and (6) over the reactor volume, and introduce

$$\mathrm{Xe}(t) \equiv (1/V) \int_R d^3r\, \mathrm{Xe}(\mathbf{r}, t) \tag{12a}$$

$$\mathrm{I}(t) \equiv (1/V) \int_R d^3r\, \mathrm{I}(\mathbf{r}, t) \tag{12b}$$

$$\sigma_f \equiv \int_R d^3r\, \sigma_f(\mathbf{r})\, \phi_0(\mathbf{r}) \Big/ \int_R d^3r\, \phi_0(\mathbf{r}) \tag{12c}$$

$$\sigma_{\mathrm{Xe}} \equiv V \int_R d^3r\, \sigma_{\mathrm{Xe}}(\mathbf{r})\, \mathrm{Xe}_0(\mathbf{r})\, \phi_0(\mathbf{r}) \Big/ \left\{ \left[\int_R d^3r\, \mathrm{Xe}_0(\mathbf{r})\right] \left[\int_R d^3r\, \phi_0(\mathbf{r})\right] \right\} \tag{12d}$$

It proves convenient to choose P_0 in $\phi_0(\mathbf{r}, t) \approx [P(t)/P_0]\,\phi_0(\mathbf{r})$ as the average flux ϕ_0 defined by

$$\phi_0 \equiv (1/V) \int_R d^3r\, \phi_0(\mathbf{r}) \tag{13}$$

[With this choice, $P(t)$ has the dimension of flux.] Using (12) and (13), we obtain the following lumped-parameter description from (5) and (6):

$$d\mathrm{I}(t)/dt = -\lambda_\mathrm{I}\mathrm{I}(t) + y_\mathrm{I}\sigma_f P(t) \tag{14a}$$

$$d[\mathrm{Xe}(t)]/dt = \lambda_\mathrm{I}\mathrm{I}(t) + y_{\mathrm{Xe}}\sigma_f P(t) - \lambda_{\mathrm{Xe}}\mathrm{Xe}(t) - \mathrm{Xe}(t)\,\sigma_{\mathrm{Xe}}P(t) \tag{14b}$$

In obtaining the last term in (14b), we have used the approximation

$$(1/V) \int_R d^3r\, \mathrm{Xe}(\mathbf{r}, t)\, \sigma_{\mathrm{Xe}}(\mathbf{r})\, \phi_0(\mathbf{r}, t) \cong \sigma_{\mathrm{Xe}}\mathrm{Xe}(t)\, P(t)$$

which implies that $\mathrm{Xe}(\mathbf{r}, t)$, as well as $\phi_0(\mathbf{r}, t)$, is separable in time and space. Using the same approximation in (4), we obtain

$$\delta k_{\mathrm{Xe}}[p] = \alpha_{\mathrm{Xe}}\, \delta\mathrm{Xe}(t) \tag{15}$$

where α_{Xe} is the average xenon reactivity coefficent, defined by

$$\alpha_{\mathrm{Xe}} \equiv V \int_R d^3r\, \alpha_{\mathrm{Xe}}(\mathbf{r})\, N_f(\mathbf{r})\, \mathrm{Xe}_0(\mathbf{r}) \Big/ \int_R d^3r\, \mathrm{Xe}_0(\mathbf{r}) \tag{16}$$

In these equations, Xe and I denote the xenon and iodine concentrations per fuel atom (cf Eq. 7). The factor N_f in (16) is introduced to convert the relative xenon concentration $Xe_0(\mathbf{r})$ at equilibrium to the actual xenon concentration.

It has become conventional [13–15] to express α_{Xe} as

$$\alpha_{Xe} = -\sigma_{Xe}/\beta\sigma_f c \tag{17}$$

where c is a number whose definition follows from a comparison of (16) and (17):

$$1/c \equiv -\beta\langle\sigma_f \mid \phi_0\rangle\langle\alpha_{Xe}(\mathbf{r}) \mid N_f Xe_0\rangle/\langle\phi_0 \mid \sigma_{Xe} Xe_0\rangle \tag{18}$$

Here, the scalar product implies integration on $\mathbf{r}$ over the reactor volume. The explicit form of $\alpha_{Xe}(\mathbf{r})$ was given in Eq. (3). If we assume that the xenon feedback affects only the absorption cross section, and treat the energy dependence in the one-group approximation, then (3) reduces to

$$\alpha_{Xe}(\mathbf{r}) = -\sigma_{Xe}(\mathbf{r})\,\phi_0^2(\mathbf{r})/\beta\nu\langle\phi_0 \mid N_f\sigma_f \mid \phi_0\rangle \tag{19}$$

Substituting (19) into (18) yields

$$1/c = (1/\nu)\langle\sigma_f \mid \phi_0\rangle\langle\phi_0^2 \mid \sigma_{Xe} N_f Xe_0\rangle/\langle\phi_0 \mid N_f\sigma_f \mid \phi_0\rangle\langle\phi_0 \mid \sigma_{Xe} \mid Xe_0\rangle \tag{20}$$

In the case of constant cross sections and flat flux and xenon distributions, $c = \nu$, where ν is the mean number of neutrons per fission.

The equations describing the time behavior of a reactor in the presence of xenon feedback are compiled below for future reference (defining the power coefficient as $-\gamma$):

$$(l/\beta)\,\dot{P} = [\delta k_{ext}(t) - (\sigma_{Xe}/c\beta\sigma_f)\,\delta Xe - \gamma(P - P_0)]\,P - P + \sum_{i=1}^{6}\lambda_i C_i + S(l/\beta) \tag{21a}$$

$$\dot{C}_i = a_i P - \lambda_i C_i\,, \qquad i = 1,\ldots, 6 \tag{21b}$$

$$\dot{Xe} = y_{Xe}\sigma_f P - (\lambda_{Xe} + \sigma_{Xe} Xe)\,P + \lambda_I I \tag{21c}$$

$$\dot{I} = -\lambda_I I + y_I\sigma_f P \tag{21d}$$

The following transformation due to Smets [15, 16] casts these equations in a more compact form which is often prefered in the stability analysis of xenon-controlled reactors:

$$z(t) \equiv (l/\beta)\,p(t) - [\delta Xe(t)/c\beta\sigma_f] - [\lambda_I\,\delta I(t)/c\beta\sigma_f(\lambda_I - \lambda_{Xe})] \tag{22}$$

Eliminating $\delta\mathrm{Xe}(t)$ in favor of $z(t)$ in (21), we obtain, in the absence of external reactivity and source,

$$(l/\beta)\dot{p} = \delta k_{\mathrm{f}}[p](P_0 + p) + \int_0^\infty du\,[p(t-u) - p(t)]\,D(u) \tag{23}$$

where

$$\delta k_{\mathrm{f}}[p] = z\sigma_{\mathrm{Xe}} + \alpha_{\mathrm{I}}\,\delta\mathrm{I} - \alpha_{\mathrm{p}}p \tag{24a}$$

$$\dot{z} + \lambda_{\mathrm{Xe}}z = \alpha_2 p + \int_0^\infty du\,[p(t-u) - p(t)]\,D(u) - \gamma p^2 \tag{24b}$$

$$\delta\dot{\mathrm{I}} + \lambda_{\mathrm{I}}\,\delta\mathrm{I} = y_{\mathrm{I}}\sigma_{\mathrm{f}}p \tag{24c}$$

The parameters α_{I}, α_p, and α_2 are defined by

$$\alpha_{\mathrm{I}} \equiv \lambda_{\mathrm{I}}\sigma_{\mathrm{Xe}}/\beta c\sigma_{\mathrm{f}}(\lambda_{\mathrm{I}} - \lambda_{\mathrm{Xe}}) \tag{25a}$$

$$\alpha_{\mathrm{p}} \equiv \gamma + (\sigma_{\mathrm{Xe}}/\beta)l \tag{25b}$$

$$\alpha_2 \equiv \lambda_{\mathrm{Xe}}(l/\beta) - \gamma P_0 - [(y_{\mathrm{Xe}}\sigma_{\mathrm{f}} - \mathrm{Xe}_0\sigma_{\mathrm{f}})/\beta c\sigma_{\mathrm{f}}] - [\lambda_{\mathrm{I}}y_{\mathrm{I}}/\beta c(\lambda_{\mathrm{I}} - \lambda_{\mathrm{Xe}})] \tag{25c}$$

$$\mathrm{Xe}_0 \equiv [(y_{\mathrm{Xe}} + y_{\mathrm{I}})/(\lambda_{\mathrm{Xe}} + P_0\sigma_{\mathrm{Xe}})]\,\sigma_{\mathrm{f}}P_0 \tag{25d}$$

The derivation of (24) is left as an exercise. We observe that in (24b) the differential equations describing the feedback functional are not linear, because of the last term γp^2. However, we can still express $\delta k_{\mathrm{f}}[p]$ in terms of feedback kernels by solving (24b) and (24c) for $z(t)$ and $\delta\mathrm{I}(t)$ as

$$z(t) = \int_{-\infty}^{t} [\alpha_2 p(u) - \gamma p^2(u)]\,e^{-\lambda_{\mathrm{Xe}}(t-u)}\,du$$

$$+ \int_{-\infty}^{t} du\,e^{-\lambda_{\mathrm{Xe}}(t-u)} \int_{-\infty}^{u} dv\,[p(v) - p(u)]\,D(u-v) \tag{26a}$$

$$\delta\mathrm{I}(t) = \int_{-\infty}^{t} du\,e^{-\lambda_{\mathrm{I}}(t-u)} y_{\mathrm{I}}\sigma_{\mathrm{f}}p(u) \tag{26b}$$

and substituting them into (24a). The resulting expression is

$$\delta k_{\mathrm{f}}[p] = \int_{-\infty}^{t} du\,G(t-u)\,p(u) - \gamma\sigma_{\mathrm{Xe}} \int_{-\infty}^{t} du\,p^2(u) \tag{27}$$

where the kernel G is defined by

$$G(t) = -\alpha_p \, \delta(t) + K(t) \tag{28}$$

$$K(t) = \sigma_{Xe}\left(\alpha_2 + \sum_{i=1}^{6} \frac{\lambda_i a_i}{\lambda_i - \lambda_{Xe}}\right) e^{-\lambda_{Xe}t} - \sigma_{Xe} \sum_{i=1}^{6} \frac{a_i \lambda_i}{\lambda_i - \lambda_{Xe}} e^{-\lambda_i t} + \alpha_I y_I \sigma_f \, e^{-\lambda_I t} \tag{29}$$

The transfer function corresponding to (28) is

$$H(s) = -\alpha_p + \frac{A_0}{\lambda_{Xe} + s} - \sum_{i=1}^{6} \frac{A_i}{\lambda_i + s} + \frac{\alpha_I y_I \sigma_f}{\lambda_I + s} \tag{30}$$

where A_0 and A_i are the coefficients of $e^{-\lambda_{Xe}t}$ and $e^{-\lambda_i t}$ in (29).

It is interesting to observe in (27) that the feedback functional becomes linear in the absence of the temperature feedback, which is represented in the quasistatic approximation by γ. We also note that the nonlinearity is quadratic in $p(t)$, as was mentioned in Section 5.1 (cf the comments following Eq. 23).

The description of the "intermediate term" behavior of a reactor can be simplified if we also treat the delayed neutrons in (23) in the quasistatic approximation. Since the time constants of the delayed neutron precursors are also of the order of minutes or less as the temperature time constants, we may use the same technique to simplify the delayed neutron term in (23) as that used to obtain (1a) from (1b). For this purpose, we expand $p(t - u)$ about t in

$$\int_0^\infty du\, D(u)[p(t-u) - p(t)] = -\dot{p}(t) \int_0^\infty du\, [uD(u)] + \tfrac{1}{2}\ddot{p}(t) \int_0^\infty du\, [u^2 D(u)] + \cdots \tag{31}$$

Retaining only the first term in (31), and combining it with $\dot{p}(t)(l/\beta)$ on the left-hand side in (23), we obtain

$$(l^*/\beta)\,\dot{p}(t) = \delta k_f[p](P_0 + p) \tag{32}$$

where l^* is defined by (cf 3.3, Eq. 18b), namely $l^* = l + \beta \sum_i (a_i/\lambda_i)$. The feedback functional is still described by (24a), (24c), and

$$\dot{z} + \lambda_{Xe} z = \alpha_2 p - \gamma p^2 \tag{33}$$

provided the mean generation time l in z, α_p, and α_2 is replaced by l^*.

The transfer function $H(s)$ describing the linear portion of the feedback functional (the first term in Eq. 27) becomes, in simplified form,

$$H(s) = -\alpha_p + \alpha_2[1/(\lambda_{Xe} + s)] + \alpha_I[y_I\sigma_f/(\lambda_I + s)] \qquad (34)$$

which will be used in Chapter 6 to discuss the stability of a xenon-controlled reactor.

5.4. Transfer Function of a Reactor with Feedback

In discussing the response of a reactor to external reactivity insertions in Chapters 3 and 4, we ignored the effect of reactivity feedback, which represents the coupling between the neutron population and the nuclear properties of the medium in which the neutrons diffuse. In Chapters 1 and 2, we investigated the origin of this coupling and developed equations for a mathematical description of the feedback mechanism. In the previous sections of this chapter, we discussed the derivation of the feedback functionals for various reactor types. In this and following sections, we shall consider the effects of feedback on the reactor response to an external reactivity insertion.

There are three types of experimental excitation of a reactor. First, a small-amplitude reactivity variation, which can be accomplished by deliberately oscillating a control rod, although random variations due to reactor noise or arbitrary reactivity impulses can also be used. The purpose of this type of excitation is usually to measure the power-to-reactivity transfer function, from which valuable information about the feedback mechanism can be extracted. Furthermore, by measuring the power-to-reactivity transfer function at various power levels, one can predict the threshold of instability of a power reactor. The second type of excitation is a controlled large-amplitude sinusoidal reactivity variation in order to measure the reactor describing function in the presence of feedback, which is needed in nonlinear stability analysis. Third is a large and prompt reactivity insertion—either intentionally or accidentally—which leads to a power excursion. We shall consider the first and second types of excitation in this and following sections. The response to large prompt reactivity insertion will not be investigated in this book.

A typical series of reactivity–power transfer function measurements at several power levels is shown in Figure 5.4.1. These were obtained by inserting small-amplitude, periodic reactivity variations, through control-rod oscillations, into a boiling-water reactor (EBWR) [17], and

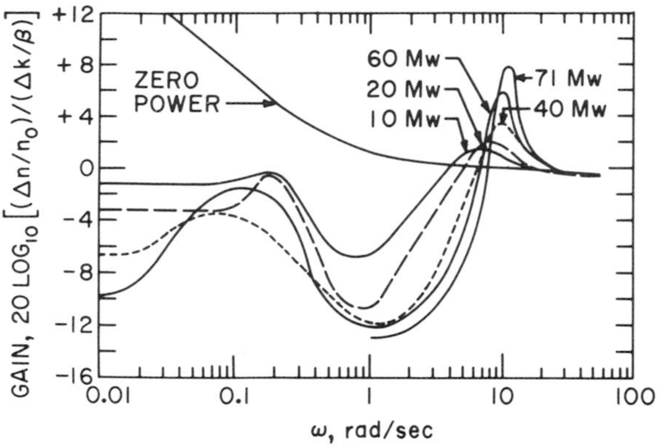

FIGURE 5.4.1. Transfer function measurements in EBWR at several power levels.

measuring the amplitude and phase of the resulting sinusoidal power oscillations at different frequencies. For our purpose in this section, it is sufficient to note that the transfer function at power differs considerably from the zero-power transfer function (cf Chapter 4), showing that the feedback effects are indeed important in determining the reactor behavior. We particularly note that the amplitude of the transfer function exhibits a marked resonance as the power level is increased. No such resonance occurs in the zero-power transfer function. The resonance peak becomes narrower and higher as the power level increases. It may be expected intuitively that the reactor will be unstable if the height of this peak can become infinite at a finite power level. This new behavior is clearly due to the presence of the feedback.

In order to derive a general expression for the transfer function that includes feedback effects, we start with (5.1, Eq. 12), which we reproduce here for convenience:

$$(l/\beta)\dot{p} = k(t)(p + P_0) + \int_0^\infty D(u)[p(t-u) - p(t)]\, du \tag{1}$$

The net reactivity $k(t)$ consists of the external reactivity $\delta k_{\text{ext}}(t)$ and the feedback reactivity $\delta k_{\text{f}}(t)$,

$$k(t) = \delta k_{\text{ext}}(t) + \delta k_{\text{f}}(t) \tag{2}$$

In general, δk_{f} is expressed as a functional power-series expansion as given by (5.1, Eq. 20). When small power oscillations about the equilibrium level are involved, as we consider in this section, we may represent the feedback by a linear functional of reactor power and use

(5.1, Eq. 21) to represent it mathematically. Substituting the latter into (1), we obtain

$$(l/\beta)\dot{p} = \left[\delta k_{\text{ext}}(t) + \int_0^\infty du\, G(u)\, p(t-u)\right](p + P_0)$$
$$+ \int_0^\infty du\, D(u)[p(t-u) - p(t)] \tag{3}$$

In general, the feedback kernel $G(t)$ depends on the equilibrium power level P_0 at which the feedback mechanism is linearized. This dependence will be put in evidence as $G(P_0, t)$ when we discuss the prediction of the instability threshold.

It is to be noted that (3) is a nonlinear integrodifferential equation as a result of the first term, even though the feedback is treated linearly. This equation with $\delta k_{\text{ex}}(t) \equiv 0$ is frequently used in the nonlinear stability analysis of reactors with the justification that the physical parameters entering the description of the feedback are usually insensitive to rapid power variations, and hence can be assumed to be independent of $p(t)$.

In the present discussion, we consider only small power variations, so that $p \ll P_0$. Hence, we can linearize (3) as

$$(l/\beta)\dot{p}(t) + p(t) = P_0\, \delta k_{\text{ext}}(t) + \int_0^\infty du\, [D(u) + P_0 G(u)]\, p(t-u) \tag{4}$$

which is the starting equation of this section. As in Chapters (3) and (4), we shall assume that the reactor is operated at its equilibrium power level P_0 prior to $t = 0$, so that $p(t) \equiv 0$ for $t < 0$, and we try to find its response to an arbitrary reactivity insertion. Taking the Laplace transform of (4) with $p(t) \equiv 0$ for $t < 0$, we obtain

$$\bar{p}(s)/[P_0\, \overline{\delta k}_{\text{ext}}(s)] = Z(s)/[1 - P_0 H(s)\, Z(s)] \tag{5}$$

where $Z(s)$ is the usual zero-power transfer function, and $H(s)$ [the Laplace transform of the feedback kernel $G(t)$] is the "feedback" transfer function. The right-hand side of (5) is called the "power-to-reactivity" or simply "power" transfer function (it is also called "closed-loop" transfer function). We shall denote it by $L(s)$, i.e.,

$$L(s) \equiv Z(s)/[1 - P_0 H(s)\, Z(s)] \tag{6}$$

It is noted that $L(s)$ reduces to $Z(s)$ as $P_0 \to 0$. This explains the reason for calling $Z(s)$ the zero-power transfer function.

The response of the reactor to any small reactivity insertion can now be obtained by finding the inverse transform of $L(s)\,\overline{\delta k}_{\text{ext}}(s)$, and can be written in general as

$$p(t) = P_0 \int_0^t du\,[l(t-u)\,\delta k_{\text{ext}}(u)] \tag{7}$$

where $l(t)$ is the inverse transform of $L(s)$, and is called the unit impulse response of the reactor with feedback (cf 4.4, Eq. 17 for the unit impulse response at zero power). Its behavior depends on the poles of $L(s)$ in the complex s plane. It is interesting to note that $L(s)$ has no simple pole at the origin, in contrast with the zero-power transfer function $Z(s)$, but rather, is analytic at $s = 0$ with a value

$$L(0) = 1/[-P_0 H(0)] = \int_0^\infty l(t)\,dt \tag{8}$$

The response of the reactor to a step reactivity insertion δk_0 follows from (7) as

$$p(t) = P_0\,\delta k_0) \int_0^t l(t')\,dt' \tag{9a}$$

which approaches

$$p(\infty) = P_0(\delta k_0)\,L(0) \tag{9b}$$

The new equilibrium power level is then obtained as

$$P_0' = P_0[1 + (\delta k_0)\,L(0)] \tag{10}$$

It may be recalled that the response of a zero-power reactor (cf 3.3, Eq. 15) grows exponentially for long times following a positive step reactivity insertion without reaching any finite value. In the presence of feedback, the positive external reactivity is decreased gradually by the negative feedback reactivity,

$$\delta k_f(t) = \int_0^t G(t-u)\,p(u)\,du \tag{11a}$$

which eventually approaches

$$\delta k_f(\infty) = p(\infty)\,H(0) \tag{11b}$$

Using (9b) and (8), we find $\delta k_f(\infty) = -\delta k_0$, which indicates that the feedback reactivity just compensates the positive external reactivity at the new equilibrium power level. For a stable reactor (cf Chapter 6), $H(0)$ must be negative, so that a positive power increase produces a negative feedback reactivity. $H(0)$ is called the "power coefficient" of reactivity.

The response of a reactor with feedback to a sinusoidal reactivity insertion

$$\delta k_{\text{ext}}(t) = \delta k \sin(\omega t + \theta) \tag{12}$$

can easily be obtained from (5), after all transients die out, as

$$p(t)/P_0 = |L(i\omega)| \sin(\omega t + \theta + \phi) \tag{13a}$$

where the relative phase angle of the output is given by

$$\phi(\omega) = \arg[L(i\omega)] \tag{13b}$$

As in the case of a zero-power reactor, the gain of the reactor is defined, in decibels, by

$$G(\omega) = 20 \log_{10} |L(i\omega)| \tag{14}$$

The presence of a resonance in the reactor gain can be seen from (13a) and (6) to occur when

$$1 - P_0 H(i\omega) Z(i\omega) \equiv 1 + P_0 |H(i\omega) Z(i\omega)| e^{i\Phi} \to 0 \tag{15}$$

where $\Phi(\omega) = \arg[-Z(i\omega) H(i\omega)]$. The two conditions that must be satisfied in order for (15) to hold are

$$P_0 |H(i\omega) Z(i\omega)| = 1 \tag{16a}$$

$$\Phi(\omega) = \pi \tag{16b}$$

These conditions yield a critical power level P_c and a resonance frequency ω_0. The latter is the root of (16b) and depends on the equilibrium power level P_0 through the dependence of $H(i\omega)$ on P_0. In a truly linear feedback, $H(i\omega)$, and thus ω_0, is independent of P_0. In general, however, ω_0 is a slowly varying function of P_0. Therefore, the resonance peak occurs approximately at ω_0. The height of the peak is given approximately by

$$|L(i\omega_0)| = |Z(i\omega_0)|/[1 - (P_0/P_c)] \tag{17}$$

which indicates that the gain approaches infinity when the critical power level is approached. At $P_0 = P_c$, the reactor is said to be "unstable." The question of stability will be dealt with more precisely in the next chapter.

Before discussing the utility of the measurement of $G(\omega)$ and $\phi(\omega)$, it is interesting to note that the average power level in the presence of oscillations is the same as the initial equilibrium power level P_0 when the feedback is taken into account. We recall that the average power level in the presence of power oscillations is different than the initial equilibrium level in a zero-power reactor by an amount given in (4.4, Eq. 32). The shift of the average power level is calculated by

$$\delta P_0 = P_0 \lim_{s\to 0} [s\, \overline{\delta k}_{\text{ext}}(s)\, L(s)] \tag{18}$$

which is obtained from (4.4, Eq. 32) by replacing $Z(s)$ by $L(s)$. Since $L(s)$ has no pole at the origin, the above limit is always zero regardless of the initial phase θ of the sinusoidal reactivity input, in contrast with the zero-power reactor, in which δP_0 depends on θ. This result indicates that a linearized reactor with feedback has no "memory" (cf discussion following 4.4, Eq. 18) for the initial conditions.

Utility of Transfer Function Measurements

We now discuss the type of information that can be obtained by measuring the power transfer function $L(P_0, i\omega_0)$ of a reactor with feedback at various power levels P_0 and as a function of frequency.

First of all, we can solve (6) for $H(P_0, i\omega)$ and obtain

$$P_0 H(P_0, i\omega) = [1/Z(i\omega)] - [1/L(P_0, i\omega)] \tag{19}$$

Consequently, since the zero-power transfer function is known, it is possible to determine $H(P_0, i\omega)$ from a measurement of $L(P_0, i\omega)$. Knowledge of the form of $H(P_0, i\omega)$ is the most information one can hope to obtain about the physics of feedback from a transfer function measurement. Clearly, this information does not in itself yield a unique description of the feedback processes. However, it can be used to check the validity of a given theoretical feedback model, and to determine numerical values of some feedback parameter by fitting the experimental and theoretical feedback transfer functions.

In the absence of any theoretical model, one can still use the knowledge of $H(P_0, i\omega)$ at lower power levels to predict the reactor behavior at a higher power level by using an appropriate extrapolation technique. For this purpose, one assumes the following form for $H(P_0, s)$:

$$H(P_0, s) = H(P_0, 0) \prod_j (1 + st_j) \Big/ \prod_i (1 + s\tau_i) \tag{20}$$

and determines the phenomenological time constants t_j and τ_i, and the power coefficient $H(P_0, 0)$ from the experimental feedback transfer function. The time constants are determined by resolving [18] the experimental amplitude and phase responses of the feedback into the Bode diagram [18, 19] (break-frequency method). The number of time constants to be included in (20) depends on the actual curves and on the desired accuracy in describing the experimental curve by (20). The time constants obtained in this manner do not have to correspond to actual time constants in the feedback, although some of them may be so identified if a crude model for the feedback is available. The feedback transfer function at a higher power level can then be obtained by extrapolating the time constants and the power coefficient using their values at lower power levels. The main point in this extrapolation technique is that the time constants and $H(P_0, 0)$ are usually slowly varying, monotonic functions of P_0. A combination of this method with a crude feedback model was successfully used to predict the 40-MW performance of the EBWR by Deshong and Lipinski [18, 20].

The critical power level P_c can be predicted by extrapolating the "gain" and "phase" margins from low-power measurements. The gain margin is equal to $(1 - |P_0ZH|)$ at a frequency for which the argument of $-ZH$ is 180 deg, and the phase margin measures the deviation from 180 deg of the phase angle of $-ZH$ when $|P_0ZH| = 1$. It is clear from Eqs. 16a,b that both quantities vanish at the instability threshold. Hence, the critical power level can be determined by extrapolating either the gain or phase margin to zero using their values at lower power levels. Both of these quantities require a complete knowledge of $H(P_0, i\omega)$ at various power levels as a function of frequency. This technique was used by Deshong and Lipinski [18] to predict the instability threshold of EBWR.

The knowledge of $H(P_0, i\omega)$ can in certain cases be used to determine even the nonlinear form of the feedback functional, as shown by Brooks [2], in spite of the fact that $H(P_0, i\omega)$ is measured using small-amplitude oscillations. Let us assume that the feedback reactivity is of the following functional form:

$$k_f[P] = \int_{-\infty}^{t} F[t - t', P(t')]\, dt' \tag{21}$$

where $F(t, P)$ is an ordinary function of its arguments. This form represents a causal, time-invariant functional of $P(t)$ which is a special case of the general form of such functionals described by (5.1, Eq. 16).

The connection between the latter and (21) can be established by expanding $F(t, P)$ in a Taylor series about P_0 as

$$F(t, P) = \sum_{n=0}^{\infty} (1/n!)\, F^{(n)}(t, P_0)\, p^n(t) \tag{22}$$

where $F^{(n)}(t, P_0)$ is the nth derivative of $F(t, P)$ with respect to P, evaluated at $P = P_0$. The constant term in (22) gives rise to the constant feedback reactivity at the equilibrium power level, i.e.,

$$k_f(P_0) = \int_0^{\infty} F(t, P_0)\, dt \tag{23}$$

Hence, the incremental feedback reactivity becomes

$$\delta k_f[p] = \sum_{n=1}^{\infty} \frac{1}{n!} \int_{-\infty}^{t} F^{(n)}(t - t', P_0)\, p^n(t')\, dt' \tag{24}$$

If we define

$$G_n(t_1, t_2, \ldots, t_n, P_0) = \frac{1}{n!} F^{(n)}(t_1, P_0)\, \delta(t_2 - t_1) \cdots \delta(t_n - t_{n-1}) \tag{25}$$

the desired connection between (5.1, Eq. 16) and (21) is immediately established.

Since the power transfer function measurement is performed with small amplitudes, we keep only the first term in (24), and obtain

$$\delta k_f[p] \equiv \int_{-\infty}^{t} [\partial F(t - t', P_0)/\partial P_0]\, p(t')\, dt' \tag{26}$$

Comparing this to (5.1, Eq. 21), we find the linear feedback kernel as

$$G(P_0, t) = \partial F(t, P_0)/\partial P_0 \tag{27}$$

Integrating both sides from $P_0 = 0$ to $P_0 = P$ and using $F(t, 0) \equiv 0$, we get

$$F(t, P) = \int_0^P dP_0\, G(P_0, t) \tag{28}$$

or

$$F(t, P) = \int_0^P dP_0\, \mathscr{L}^{-1}[H(P_0, s)] \tag{29}$$

It follows that the nonlinear feedback kernel $F(t, P)$ can be obtained in principle from the measurement of $H(P_0, s)$ at various power levels if the feedback can be described by the functional form given by (21). In practice, one usually does not have enough information to determine whether the above form is applicable to a given reactor. Nevertheless, the Brooks analysis does show the possibility of determining feedback kernels with rather complicated power dependences, using the knowledge of power transfer functions as a function of ω and P_0.

5.5. Response of a Reactor with Feedback to a Large Periodic Reactivity Insertion

In this section, we consider the response of a reactor with feedback to a large periodic reactivity insertion. The analysis presented here is an extension of that in Section 4.7 to include the effects of the feedback. The results of this analysis are used to define a "high-power" or simply "power" reactor describing function which is a logical generalization of the concept of the zero-power describing function introduced in Section 4.7B to high powers at which the feedback effects cannot be ignored.

The feedback will be assumed to be linear, but otherwise arbitrary. Then, the reactor power will be described by

$$(l/\beta)\dot{p} + P = \left[k_{\text{ext}}(t) + \int_0^\infty du\, G(u)\, P(t-u)\right] P(t) + \int_0^\infty du\, D(u)\, P(t-u) \quad (1)$$

This equation is identical to (5.4, Eq. 3), with the modification that $k_{\text{ext}}(t)$ here denotes the total external reactivity, including the positive reactivity, which is compensated by the feedback reactivity at the equilibrium, i.e., $k_{\text{ext}}(t) = \delta k_{\text{ext}}(t) + P_0 \mid H(0)|$. The present form of the kinetic equation is more convenient for the purpose of this section because the average power level in the presence of power oscillations, as will be clear below, depends on the magnitude of the reactivity oscillations as well as on the static positive reactivity.

We consider a periodic reactivity insertion given by

$$k_{\text{ext}}(t) = \sum_{n=-\infty}^{+\infty} k_n e^{in\omega t} \quad (2)$$

and look for a periodic solution of the following form:

$$P(t) = \sum_{n=-\infty}^{+\infty} P_n e^{in\omega t} \quad (3)$$

In (2) and (3), k_n and P_n denote the complex amplitudes of the nth Fourier component of the reactivity and power oscillation, respectively (cf Section 4.7). Our task is to determine P_0, $P_{\pm 1}$, $P_{\pm 2}$,... in terms of k_0, $k_{\pm 1}$,..., substituting (2) and (3) into (1) and equating the coefficients of exp($n\omega t$) on both sides of the equation. Verifying

$$\int_0^\infty du\, G(u)\, P(t-u) = \sum_{n=-\infty}^{+\infty} H_n P_n e^{in\omega t} \tag{4a}$$

and

$$\int_0^\infty du\, D(u)\, P(t-u) = \sum_{n=-\infty}^{+\infty} \bar{D}_n P_n e^{in\omega t} \tag{4b}$$

where $H_n = H(i\omega n)$ and $\bar{D}_n = D(i\omega n)$, we obtain

$$P_n/Z_n = P_0(k_n + P_n H_n) + \sum_{\substack{m=-\infty \\ m\neq 0}}^{+\infty} (k_{n-m} + H_{n-m}P_{n-m})\, P_m \qquad n = 0, \pm 1,... \tag{5}$$

where Z_n denotes the zero-power transfer function evaluated at $s = i\omega_n$.

For $n = 0$, $1/Z_n = 0$, and (5) yields

$$P_0(k_0 + P_0 H_0) = -2 \sum_{m=1}^{\infty} \text{Re}(k_m P_m{}^*) - 2 \sum_{m=1}^{\infty} |\, P_m\,|^2 \,\text{Re}\, H_m \tag{6}$$

In the absence of forced oscillations, the right-hand side vanishes, and (6) reduces to the equilibrium condition

$$k_e + P_e H_0 = 0 \tag{7}$$

which relates the positive static reactivity k_e to the equilibrium power level P_e. It is seen from (6) that the mean value P_0 of the forced power oscillations is not equal to the equilibrium power level corresponding to the static reactivity k_e. We can determine from (6) the constant reactivity bias, i.e.,

$$\begin{aligned} \delta k_0 &= k_0 - k_e \\ &= k_0 + P_0 H_0 \end{aligned} \tag{8}$$

which must be applied along with the reactivity oscillations in order to ensure $P_e = P_0$. Substituting (8) into (7), we obtain

$$\delta k_0 = -2 \sum_{m=1}^{\infty} \mathrm{Re}(k_m y_m{}^*) - 2P_0 \sum_{m=1}^{\infty} | y_m |^2 \mathrm{Re}\, H_m \tag{9a}$$

where y_n is the normalized Fourier amplitude, i.e.,

$$y_n = P_n/P_0 \tag{9b}$$

Alternatively, we can calculate the change in mean power, i.e.,

$$\begin{aligned} \delta P_0 &= P_0 - P_e \\ &= (1/H_0)(k_0 + P_0 H_0) \end{aligned} \tag{10}$$

due to the forced oscillations, assuming that $k_0 = k_e$. Substituting (10) into (6), we get

$$\delta z_0(1 + \delta z_0) = (2/k_0) \sum_{m=1}^{\infty} \mathrm{Re}(k_m z_m{}^*) - 2 \sum_{m=1}^{\infty} | z_m |^2 \mathrm{Re}(H_m/H_0) \tag{11a}$$

where z_n and δz_0 are defined as

$$z_n = P_n/P_e \tag{11b}$$

$$\delta z_0 = z_0 - 1 \tag{11c}$$

The foregoing discussion indicates that we can solve (5) assuming either a constant average power or a constant static reactivity. These two cases will be discussed separately for a cosinusoidal input, i.e., $k_{\mathrm{ext}}(t) = k_0 + \delta k \cos \omega t$, for which

$$\begin{aligned} k_m &= \delta k/2 \qquad &\text{for} \quad m = \pm 1 \\ &= 0 \qquad &\text{for} \quad m = \pm 2, \ldots \end{aligned}$$

A. High-Power Describing Function in the Case of Constant Mean Power

In this case, we use the normalization shown in (9b), and obtain from (5)

$$y_n/L_n = k_n + \sum_{\substack{m=-\infty \\ m \neq 0}}^{+\infty} (k_{n-m} + P_0 H_{n-m} y_{n-m})\, y_m \tag{12a}$$

where L_n denotes the power transfer function evaluated at $s = i\omega n$:

$$1/L_n = [1/Z(i\omega n)] - P_0 H(i\omega n) \tag{12b}$$

For $n = 0$, (12a) yields (9a), which reduces to

$$\delta k_0 = -\left[\delta k \operatorname{Re} y_1 + 2P_0 \sum_{n=1}^{\infty} |y_n|^2 \operatorname{Re} H_n\right] \tag{13}$$

in the special case of a pure cosinusoidal input.

For $n = 1$ and $n \geqslant 2$, (12a) becomes

$$y_1/L_1 = (\delta k/2)(1 + y_2) + y_1\, \delta k_0 + P_0 \sum_{\substack{m=-\infty \\ m\neq 0,1}}^{+\infty} H_m y_m y_{1-m} \tag{14}$$

$$y_n/L_n = (\delta k/2)(y_{n+1} + y_{n-1}) + y_n\, \delta k_0 + P_0 \sum_{\substack{m=-\infty \\ m\neq 0,n}}^{+\infty} H_m y_m y_{n-m}\,, \qquad n \geqslant 2 \tag{15}$$

A systematic way of solving these equations for y_1, y_2,... is to expand y_n in powers of δk as

$$y_n = \sum_{m=1}^{\infty} y_n^{(m)} (\delta k)^m, \qquad n = \pm 1, \pm 2, \ldots \tag{16}$$

and then equate the like powers of δk on both sides. However, δk_0 must first be eliminated in favor of y_n using (13). This method was used in Section 4.7 when we discussed the zero-power describing function. Here, we simply sketch the results.

We observe from (14) that $y_1^{(1)} = (\delta k) L_1/2$. This corresponds to the linearized treatment discussed in the previous section. Equation (13) indicates that δk_0 is of the order of $(\delta k)^2$. Anticipating that $y_n^{(m)} = 0$ for $m < n$, we in fact obtain δk_0 in the lowest approximation from (13) as

$$\delta k_0 = -\tfrac{1}{2}(\delta k)^2 [\operatorname{Re}(L_1) + P_0 |L_1|^2 \operatorname{Re} H_1] + O[(\delta k)^4] \tag{17a}$$

or, more compactly, as

$$\delta k_0 = -\tfrac{1}{2}(\delta k)^2 |L_1/Z_1|^2 \operatorname{Re} Z_1 + O[(\delta k)^4] \tag{17b}$$

which shows that $\delta k_0 < 0$. It is instructive to obtain this result directly from the expression of the negative bias (cf 4.7, Eq. 22) required for the stationarity of the forced oscillation in a reactor without feedback, which we reproduce here for convenience:

$$\delta k_0 = -\tfrac{1}{2}(\delta k_n)^2 \operatorname{Re} Z_1 \tag{17c}$$

In this expression, δk_n is the amplitude of the net reactivity oscillation, which is equal to that of the external reactivity insertion when feedback is not present. In the case of feedback, it is given by

$$|\delta k_n| = |L_1/Z_1|\,\delta k \tag{17d}$$

where L_1/Z_1 is the transfer function from external reactivity to net reactivity. Substitution of (17d) into (17c) yields (17b).

In order to obtain y_n in the lowest approximation, we use $y_n^{(n)} \sim (\delta k)^n$ and $y_n^{(m)} = 0$ for $m < n$ in (15), and get

$$y_n^{(n)}/L_n = (\delta k/2)\, y_{n-1}^{(n-1)} + P_0 \sum_{m=1}^{\infty} H_m y_m^{(m)} y_{n-m}^{(n-m)}, \qquad n \geqslant 2 \tag{18}$$

We present y_2 and y_3 explicitly as an illustration:

$$y_2 = (\delta k/2)^2 L_1 L_2 (1 + P_0 H_1 L_1) + O[(\delta k)^4] \tag{19a}$$

$$y_3 = (\delta k/2)^3 L_1 L_2 L_3 (1 + P_0 H_1 L_1)[1 + P_0(H_1 + H_2) L_1] + O[(\delta k)^5] \tag{19b}$$

We note that (19) reduces to (4.7, Eqs. 18 and 19) in the absence of feedback, i.e., when $H_n \to 0$, $L_n \to Z_n$.

The high-power describing function is defined as in the case of zero-power describing function, i.e.,

$$D(\delta k, \omega, P_0) \equiv y_1/k_1$$

Since $k_1 = \delta k/2$ in the case of a pure cosinusoidal input, $D = 2y_1/\delta k$. The next higher approximation to y_1 is obtained by substituting y_2 and k_0 from (19) and (17) into (14):

$$y_1/L_1 = (\delta k/2)(1 + y_2^{(2)}) + y_1^{(1)}\, \delta k_0^{(2)} + P_0[H_{-1} y_{-1}^{(1)} y_2^{(2)} + H_2 y_2^{(2)} y_{-1}^{(1)}] \tag{20}$$

from which we obtain

$$\begin{aligned} D(\delta k, \omega, P_0) = L_1\{1 &+ [(\delta k)^2/4][L_1 L_2 \,|\, 1 + P_0 H_1 L_1 |^2 \\ &+ |L_1|^2 P_0 H_2 L_2 (1 + P_0 H_1 L_1) \\ &- 2L_1 \operatorname{Re}(L_1 + P_0 H_1 |L_1|^2)]\} + O[(\delta k)^4] \end{aligned} \tag{21a}$$

or, more compactly,

$$D(\delta k, \omega, P_0) = L_1\{1 + [(\delta k)^2/4] \mid L_1/Z_1 \mid^2 \times L_1[(1 + Z_1{}^*P_0H_2)L_2 - 2 \operatorname{Re} Z_1]\} + O[(\delta k)^4] \quad (21b)$$

which is the desired high-power describing function.

Several attempts were made to obtain the high-power describing function in the past, with disagreement in the final results [21–24]. The result obtained here agrees with that obtained by Wasserman [24] in 1962. The same expression was also obtained more recently by Smets [25] with a slightly different approach. For a critical discussion of the high-power describing function, the reader is referred to the last two references.

It is interesting to obtain at this stage the amplitude ratio of the second harmonic to the fundamental component of the power oscillations using (14a) and (15):

$$y_2/y_1 = (\delta k/2)L_1L_2(1 + P_0H_1L_1)/D(\delta k\,, \omega, P_0) \quad (22)$$

which, in the lowest nonvanishing approximation, reduces to

$$\begin{aligned} y_2/y_1 &= (\delta k/2)L_2(1 + P_0H_1L_1) \\ &= (\delta k/2)L_1L_2/Z_1 \end{aligned} \quad (23)$$

Again, (22a) is in agreement with Wasserman's calculations [24].

It may be pointed out that (23) reduces to (4.7, Eq. 21) in the absence of feedback, i.e., when $H = 0$ and hence $L_2 \to Z_2$. Similarly, the zero-power describing function given by (4.7, Eq. 29) is obtained from (21) in the limit of $P_0 \to 0$.

Equation (21) is the power-series expansion of $D(\delta k, \omega, P_0)$ in δk. In contrast to the zero-power describing function (cf 4.7, Eq. 29), this series converges at all frequencies, because the coefficients of $(\delta k)^{2n}$ remain finite as $\omega \to 0$. The zero-frequency limit of (21) yields

$$D(\delta k, 0, P_0) = -1/P_0H(0) \quad (24)$$

This result becomes obvious if we recall the definition of D, i.e., $P_1/P_0 = D(\delta k, \omega, P_0)\,\delta k$. The latter reduces to $P_1 = -[\delta k/H(0)]$ for $\omega = 0$, which is equivalent to the equilibrium condition (10).

The value of the describing function in the intermediate-frequency range where $Z_2 = Z_1 = 1$ can easily be obtained with $L_1^{-1} = 1 - P_0 H_1$ and $L_2^{-1} = 1 - P_0 H_2$. The result is

$$D(\delta k, \omega, P_0) = L_1(1 + \tfrac{1}{4}(\delta k)^2 \mid L_1 \mid^2 L_1\{[(1 + P_0 H_2)/(1 - P_0 H_2)] - 2\}) + O[(\delta k)^4] \tag{25}$$

which was also obtained by Smets [26].

At high frequencies, $L_n \to Z_n$ and the describing function reduces to

$$D(\delta k, \omega, P_0) = Z_1\{1 + \tfrac{1}{4}(\delta k)^2[Z_1 Z_2 - 2Z_1 \operatorname{Re} Z_1]\} + O[(\delta k)^4] \tag{26}$$

which is the zero-power describing function (cf 4.7, Eq. 29).

Needless to say, $D(\delta k, \omega, P_0)$ approaches L_1 in the limit of $\delta k \to 0$.

We shall discuss an application and the utility of the high-power describing function in the next section.

B. High-Power Describing Function in the Case of Constant Static Reactivity

The analysis presented here applies to an experiment in which the reactor is initially critical at a power level P_e which is maintained by a positive static reactivity k_e, and the forced oscillations are produced by a periodic reactivity insertion without changing the mean position of the control rods, so that $k_0 = k_e$. Again, we consider a pure cosinusoidal reactivity variation, and use the normalization defined by (11) with respect to the initial equilibrium power level. Since the steps of the derivation are similar to those in the previous case, we shall simply present the results.

The power shift is obtained from (11a) as

$$\delta z_0 = -(\delta k^2/2H_0 P_e) \mid L_{e1}/Z_1 \mid^2 \operatorname{Re} Z_1 + O[(\delta k)^4] \tag{27a}$$

where L_{e1} is defined at equilibrium, i.e.,

$$L_{e1} = [(1/Z_1) - P_e H_1]^{-1} \tag{27b}$$

It is interesting to note that δP_0 and δk_0 in the previous case are related to each other in the lowest approximation by $\delta P_0 H_0 + \delta k_0 = 0$, as can be seen from (17a) and (27). We observe from (27) that the mean power is always larger than the equilibrium value because $H(0) < 0$ and hence $\delta z_0 > 0$. Although (27) is the lowest approximation to δP_0, the foregoing conclusion drawn from it is an exact statement. In Section 3.2A, it was proven, using the inverse method, that the mean value of periodic reactivity variations must be negative in order to maintain stationary forced power oscillations. Since the static part of the external reactivity is not adjusted in the present case, this negative reactivity bias must be provided by the feedback mechanism. Being a linear system, the feedback can produce this additional negative static reactivity only as a response to an increase in the mean value of the reactor power.

We now continue with the derivation of the describing function. Equations corresponding to (14) and (15) in the previous case are

$$z_1/L_{e1} = (\delta k/2)(1 + z_2) + (\delta z_0)[(\delta k/2) + z_1 P_e(H_0 + H_1)] + P_e \sum_{\substack{m=-\infty \\ (m \neq 0,1)}}^{+\infty} H_m z_m z_{1-m} \tag{28}$$

and

$$z_n/L_{en} = (\delta k/2)(z_{n+1} + z_{n-1}) + z_n(\delta z_0)\, P_e(H_0 + H_n) + P_e \sum_{\substack{m=-\infty \\ (m \neq 0,n)}}^{+\infty} H_m z_m z_{n-m} \tag{29}$$

In these equations, L_{en} is the power transfer function at equilibrium, i.e.,

$$1/L_{en} = (1/Z_n) - P_e H_n \tag{30}$$

Solving (28) and (29), we find

$$z_2 = (\delta k/2)^2 L_{e1} L_{e2}[1 + P_e H_1 L_{e1}] + O[(\delta k)]^4 \tag{31}$$

and

$$\bar{D}(\delta k, \omega, P_e) = L_{e1}\{1 + (\delta k/2)^2 \,|\, L_{e1}/Z_1 \,|^2 L_{e1} \times [(1 + Z_1{}^* P_e H_2) L_{e2} - 2(1 + 1/Z_1 H_0 P_e) \operatorname{Re} Z_1] + O[(\delta k)^4] \tag{32}$$

The last expression, which is the desired high-power describing function in the case of constant static reactivity, differs from its counterpart in (21b) in several respects due to the factor $[1 + (1/Z_1 H_0 P_e)]$ in the last term.

The most striking difference becomes apparent in the limit of weak feedback. Letting $H_n \to 0$, we obtain

$$\bar{D}(\delta k, \omega, P_e) \to Z_1\{1 + (\delta k/2)^2 Z_1[Z_2 - 2(1 + 1/Z_1 H_0 P_e) \operatorname{Re} Z_1]\} + O[(\delta k)^4] \tag{33a}$$

$$\to Z_1\{1 + \tfrac{1}{2}(\delta k)^2[(\operatorname{Re} Z_1)/|\, H_0 \,|\, P_e]\} \tag{33b}$$

which indicates that $\bar{D}$ tends to be large as the effect of the feedback is reduced. In the constant-mean-power case, D approaches the zero-power describing function. This becomes plausible if we recall that there can be no stationary forced oscillations in the absence of feedback when the mean reactivity is not adjusted to the proper negative bias, and that the amplitude of the oscillation grows beyond any bound.

Using (27) and (33b), we can show that the ratio P_1/P_0 approaches $Z_1\,\delta k$ as the feedback decreases, indicating that the linearized treatment remains valid even for larger reactivity variations if the static reactivity is not adjusted.

We also note in (33b) that there is a preferential increase in the amplitude of the describing function in the intermediate-frequency range, where $\text{Re}\, Z_1 = 1$. At very low and high frequencies, for which $Z_1 \sim 1/i\omega$ and thus $\text{Re}\, Z_1 = 0$, there is no shift in the mean power level, as can be seen from (27). Furthermore, the describing function in both cases is equal to $Z_1[1 - \frac{1}{8}(\delta k)^2 \mid Z_1 \mid]$. In the intermediate-frequency range, the constant-mean-power describing function is equal to $[1 - \frac{1}{4}(\delta k)^2]$ whereas, Eq. (33b) gives $[1 + \frac{1}{2}(\delta k)^2 \mid H_0 \mid P_e]$.

In the case of a strong feedback with large power coefficient $H(0)$ and/or at high power levels, the term $1/Z_1H_0P_e$ is negligible compared to unity, and the distinction between the above two cases becomes unnecessary.

C. Utility of the Concept of the Describing Function

The above analysis indicates that we can interpret the results of an oscillator test with large reactivity variations in a power reactor in terms of the proper describing function. The oscillator test can be performed by keeping either the average power or the average reactivity constant. In the former case, the static reactivity must be adjusted to the value calculated by (17b). In the latter case, the average power level in the presence of forced oscillations can be predicted in advance using (27a). If the predicted power increase is regarded as within safe limits, one can in fact try to correlate the measured power increase with the calculated value to measure the power reactivity coefficient $H(0)$ using (27a).

In either mode of operation, one can investigate the dependence of the amplitude and phase of the fundamental component of the forced oscillations on the test frequency and the magnitude of the sinusoidal reactivity changes. In the case of constant mean power, the complex amplitude P_1 of the fundamental is given by

$$P_1(k, \omega) = (\delta k)\, P_0 D(\delta k, \omega, P_0) \tag{34}$$

whereas in the case of constant reactivity, it is calculated from

$$P_1(k, \omega) = (\delta k)\, P_e \bar{D}(\delta k, \omega, P_e) \tag{35}$$

The expressions for $D(\delta k, \omega, P_0)$ and $\bar{D}(\delta k, \omega, P_e)$ are given by (21b) and (32), respectively. For small reactivity variations, both (34) and (35) reduce to the results of the linearized treatment, in which the high-power describing function is replaced by the high-power transfer function $L_1(\omega, P_0)$. In this case, there is no appreciable change in the power level if the equilibrium static reactivity is not too small.

We may ask what additional information one may extract from an oscillator test with large amplitudes over that obtainable from a conventional transfer function measurement. The answer is that one can in principle check the linearity of the feedback mechanism. Since the nonlinearity of the kinetic equation is taken into account in the derivation of the describing function, any disagreement with the experiment may be attributed to the nonlinearity of the feedback mechanism. In general, one can obtain a more stringent test of a proposed feedback model by adding the amplitude of the reactivity oscillations as a new variable both in the theory and the experiment.

D. Derivation of the High-Power Describing Function from the Zero-Power Describing Function

The high-power reactor describing function can also be derived by analyzing the closed feedback loop consisting of the zero-power reactor describing function and the reactivity-feedback transfer function. In addition to shedding light on the behavior of a reactor with feedback, this derivation also illustrates the usefulness of the zero-power describing function.

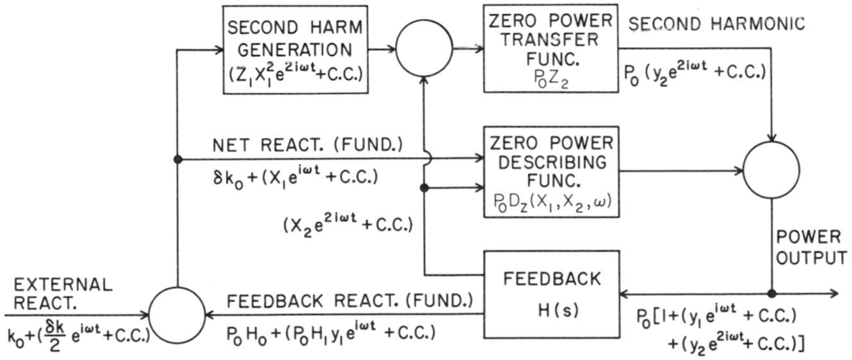

FIGURE 5.5.1. Block diagram of a reactor with feedback for large amplitudes.

Figure 5.5.1 shows the block diagram of the feedback loop under consideration. Here, $D_Z(x_1, x_2, \omega)$ denotes the zero-power describing function in the presence of the second harmonic, which was calculated in (4.7, Eq. 32). This describing function relates the variation of the net reactivity to the power oscillations. It was shown in Section 4.7 (cf 4.7, Eq. 18) that the complex amplitude of the second harmonic produced by the reactor in the power output is given by $P_0Z_1Z_2x_1^2$ where x_1 is the complex amplitude of the first harmonic of the reactivity input. This acts as an input into the loop as indicated in Figure 5.5.1. In the lowest approximation, the loop can be treated linearly with respect to the second harmonics. Hence, we can easily verify that

$$x_2 = Z_1Z_2x_1^2\, P_0H_2/(1 - P_0H_2Z_2)$$
$$= Z_1P_0H_2L_2x_1^2 \tag{36a}$$

and

$$y_2 = x_2/P_0H_2 = Z_1L_2x_1^2 \tag{36b}$$

The complex amplitudes x_1 and y_1 are of course related to each other by $y_1 = x_1D_Z(x_1, x_2, \omega)$. Hence, we can calculate x_1 in terms of $\delta k/2$ from

$$\tfrac{1}{2}\delta k + y_1H_1P_0 = x_1$$

as

$$x_1 = (\delta k/2)\{1/[1 - P_0H_1D_Z(x_1, x_2, \omega)]\} \tag{37}$$

The high-power describing function $D(\delta k, \omega, P_0)$, which was defined as $y_1/(\delta k/2)$, follows from above as

$$D(\delta k, \omega, P_0) = D_Z(x_1, x_2, \omega)/[1 - P_0H_1D_Z(x_1, x_2, \omega)] \tag{38}$$

where the zero-power transfer function is given by (4.7, Eq. 32), in which x_2 is substituted from (36a). After some manipulations, we obtain $D_Z(x_1, x_2, \omega)$ as

$$D_Z(x_1, x_2, \omega) = Z_1\{1 + | x_1 |^2Z_1[(1 + P_0H_2Z_1^*)L_2 - 2 \operatorname{Re} Z_1]\} \tag{39}$$

Equations (38) and (39) express the high-power describing function in terms of the zero-power describing function.† We observe that D_Z is

† Note that the expansion in (39) does not lack convergence as $\omega \to 0$, because $| x_1Z_1 |$ remains finite as $Z_1 \to \infty$. This can be seen from (37) by multiplying both sides by Z_1 and substituting D_z from (39). As $Z_1 \to \infty$, one finds the following asymptotic form of (37): $[1 - | x_1Z_1 |^2]\, | x_1Z_1 | = \delta k/2P_0 | H_0 |$. This means that $| x_1Z_1 | \to \delta k/2P_0 | H_0 |$ for sufficiently small values of $\delta k/P_0 | H_0 |$. In the absence of feedback, $| H_0 | \to 0$ and $| x_1Z_1 | \to \infty$, as was discussed in Section 4.7.

of the form $D_Z = Z_1 + Z_1 A \mid x_1 \mid^2$, where A denotes the coefficient of $\mid x_1 \mid^2$ in (39). Then, (38) becomes

$$D(\delta k, \omega, P_0) = L_1(1 + A \mid x_1 \mid^2)/(1 - L_1 P_0 H_1 A \mid x_1 \mid^2) \tag{40}$$

If $\mid L_1 P_0 H_1 A \mid \mid x_1 \mid^2 < 1$, we can expand (40) in a power series in $\mid x_1 \mid^2$, and express its right-hand side as

$$L_1[1 + \mid x_1 \mid^2 A(1 + L_1 P_0 H_1) + O(\mid x_1 \mid^4)]$$

Using $1 + L_1 P_0 H_1 = L_1/Z_1$, we obtain

$$D(\delta k, \omega, P_0) = L_1\{1 + \mid x_1 \mid^2 L_1[(1 + P_0 H_2 Z_1^*) L_2 - 2 \operatorname{Re} Z_1]\} \tag{41}$$

If we use the same expansion in (37), we can approximate x_1 by $(\delta k) L_1/2Z_1$. Substitution of the latter into (41) yields (21b) exactly, which was obtained by direct Fourier analysis. However, the original form (38) of the describing function is more accurate when $\mid L_1 P H_0 A \mid \mid x_1 \mid^2 \to 1$. Nothing has so far been said about P_0 in the foregoing derivation. It is implicitly assumed, however, that the net reactivity contains the proper negative bias. Hence, P_0 and k_0 must satisfy

$$k_0 + H_0 P_0 = -\tfrac{1}{2}(\delta k)^2 \mid L_1/Z_1 \mid^2 \operatorname{Re} Z_1 \tag{42}$$

This equation suggests that one can adjust either k_0 or P_0, or both, to satisfy (42). The describing function corresponding to any choice of k_0 and P_0 will have exactly the same form as in (38). However, the describing function will be numerically different in each case because of the different choice of P_0.

The foregoing observation indicates that the describing functions in the cases of constant mean power and mean reactivity must be connected to each other. In the former case, $P_0 = P_e = -k_0/H_0$, whereas in the latter case, $P_0 = P_e + \delta P_0$ (see Figure 5.5.2). Since the external

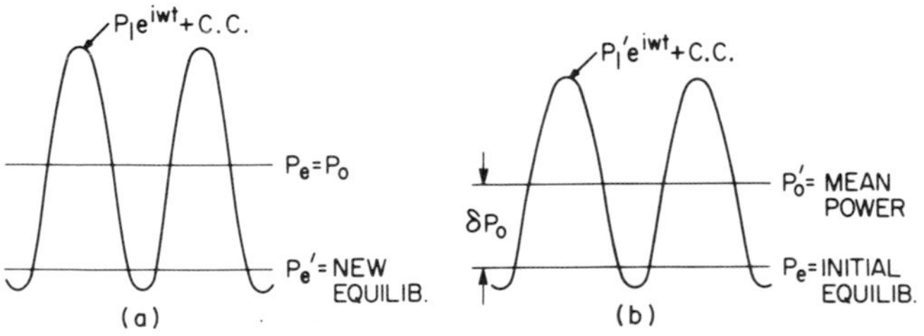

FIGURE 5.5.2. Two modes of operation in an oscillator test (a) constant mean power (b) constant mean reactivity.

reactivity is adjusted to a different value in case (a), the equilibrium power level associated with the static reactivity reduces to a new value P_e'. Then, the two cases actually refer to two different conditions. Since the only change is in the equilibrium power levels in the absence of forced oscillations, the describing functions in each case can be related to each other as follows: In case (b), we have $P_1' = P_e\bar{D}(P_e)\,\delta k$ (we suppress the irrelevant arguments in $\bar{D}$), which defines $\bar{D}$. The relative amplitude with respect to the new mean power is $P_1'/P_0' = \delta k(P_e/P_0')\,\bar{D}(P_e)$. But this is calculated in case (a) as $P_1/P_0 = D(P_0)$, in which the shifted equilibrium level does not appear explicitly. Hence, if we equate P_0 to P_0', then both formulas apply to the same condition, and P_1/P_0 must be equal to P_1'/P_0'. This leads to

$$\bar{D}(P_e) = (P_0'/P_e)\,D(P_0') \tag{43}$$

where $P_0' = P_e + \delta P_0$, or, more explicitly (cf 27a)

$$P_0' = P_e - [(\delta k)^2/2H_0]\,|\,L_{e1}/Z_1\,|^2\ \mathrm{Re}\ Z_1 \tag{44}$$

Using the expressions for $\bar{D}(P_e)$ and $D(P_0)$ and keeping terms proportional to $(\delta k)^2$, one can directly verify (43).

An interesting consequence of (43) is that it enables one to calculate the ratio of the absolute amplitudes in the two cases under consideration. Using $P_1 = P_0D(P_0)\delta k$ and $P_1' = P_e\bar{D}(P_e)\delta k$, we obtain

$$P_1'/P_1 = \bar{D}(P_e)/D(P_e) \tag{45a}$$

or, using (43), we get

$$P_1'/P_1 = [1 + (\delta P_0/P_e)][D(P_e + \delta P_0)/D(P_e)] \tag{45b}$$

This indicates that $P_1' = P_1$ when δP_0 is small, which is the case in the presence of a large power-reactivity coefficient $|\,H_0\,|$.

The foregoing discussion indicates that the results of oscillator tests with constant mean power and those with mean reactivity are connected in a simple way with each other. A choice between these two extreme cases is left to the experimenter. However, it is probably preferable to produce forced oscillations with a mean value nearly equal to the initial equilibrium power level. One does not have to aim at keeping the mean power level constant exactly, because intermediate cases between these two extremes can be analyzed equally well with (43) and (45b) provided the amount of power shift is also measured.

The high-power describing function given by (38) can also be used to investigate the amplitude and frequency of autonomous power oscillations. Autonomous oscillations occur when the denominator of (38) or of (37) vanishes. In this case, (37) yields a finite value for x_1 in the absence of any external reactivity, i.e., when $\delta k = 0$. Thus, the amplitude and frequency of the autonomous power oscillation are found by solving

$$1 = P_0 H_1 Z_1 \{1 + |x_1|^2 Z_1 [(1 + P_0 H_2 Z_1^*) L_2 - 2 \operatorname{Re} Z_1]\} \tag{46a}$$

where P_0 is also a function of $|x_1|^2$:

$$H_0(P_0 - P_e) = -2 |x_1|^2 \operatorname{Re} Z_1 \tag{46b}$$

In the absence of delayed neutrons, $Z = 1/il\omega$, so that $\operatorname{Re} Z_1 = 0$ and $Z_2 = Z_1/2$, and (46a) reduces to

$$1 = P_e H_1 Z_1 \{1 + \tfrac{1}{2} |x_1 Z_1|^2 [(L_{e2}/Z_2) - 2]\} \tag{47a}$$

which is often used to investigate autonomous oscillations in the absence of delayed neutrons [26, 27]. We recall that $|x_1|$ in (46) and (47a) is the complex amplitude of the sinusoidal net reactivity variations during autonomous oscillations. If we express the reactivity oscillations as $k(t) = \delta k \sin \omega t$, then $|x_1| = \delta k/2$. The corresponding power oscillations can be obtained by calculating the response of the zero-power reactor to $k(t)$. For example, if we use (4.6, Eq. 50) to relate $k(t)$ to the logarithm of the power oscillations, i.e.,

$$\log[P(t)/P_0] = N_0 + (\delta k) N_1 \sin(\omega t + \phi_1)$$

and approximate N_1 by $|Z_1|$ (cf 4.6, Eq. 47a), then $|x_1 Z_1|$ can be identified as $N_1/2$ and (47a) becomes

$$1 = P_e H_1 Z_1 \{1 + \tfrac{1}{8} N_1^2 [(L_{e2}/Z_2) - 2]\} \tag{47b}$$

which is the form used by Smets [26]. Note that (47b) gives directly the amplitude of the first harmonic of the logarithm of the power. If we recall that $|x_1| = |H_1| p_1/2$, where p_1 is the amplitude of the fundamental of the power oscillations rather than that of the logarithm, i.e., $P(t) = P_0 + p_1 \sin(\omega t + \theta_1) + \cdots$, then Eqs. (46) or (47a), in the absence of delayed neutrons, directly yield either N_1 or p_1, as well as the frequency of autonomous oscillations.

E. Logarithmic Response of a Reactor with Linear Feedback to Large Reactivity Insertions

Although the foregoing analysis may be sufficient to analyze any realistic oscillator test with large amplitudes or the autonomous oscillations in power reactors with feedback, it is still interesting to discuss some of the improvements one can obtain by using a different approach. Such an approach is to consider the logarithm of the power oscillations, rather than the power itself, as the response of the reactor. This approach is an extension of that used in Section 4.6, and was employed by Smets [26] and Akcasu and Shotkin [27] to investigate autonomous power oscillations. Here, we follow the derivation presented in the latter reference. We shall first discuss the logarithmic response in the absence of delayed neutrons, and comment on the extension of the analysis to include the delayed neutrons.

In the absence of delayed neutrons, the point kinetic equations can be written as

$$(l/\beta)\,\dot{N}(t) = k(t) + P_{\mathrm{e}} \int_0^\infty G(u)[e^{N(t-u)} - 1]\, du \tag{48}$$

where

$$N(t) \equiv \log[P(t)/P_{\mathrm{e}}] \tag{49}$$

and where the feedback is assumed to be linear. Here, P_{e} is the equilibrium power level in the absence of the external reactivity variations $k(t)$.

We consider a sinusoidal reactivity insertion, which we write as

$$\begin{aligned} k(t) &= k_1 e^{i\omega t} + \text{C.C.} \\ &= 2\mid k_1 \mid \cos(\omega t + \theta) \end{aligned} \tag{50}$$

for convenience. We look for a solution of the following form:

$$N(t) = N_0 + N_1 \cos(\omega t) + N_2 \cos(2\omega t + \phi_2) + \cdots \tag{51}$$

We need $e^{N(t-u)}$ in (48). We shall use the following expansion in terms of the modified Bessel functions $I_n(x)$ [28]:

$$\begin{aligned} \exp[N_1(\cos \omega t)] &= I_0(N_1) + 2I_1(N_1)(\cos \omega t) \\ &\quad + 2I_2(N_1)(\cos 2\omega t) + \cdots \end{aligned} \tag{52}$$

which was also used previously in Section 4.6 (cf 4.6, Eq. 51). We could have used the same expansion for $\exp[N_2 \cos(2t + \phi_2)]$ and thus allow for larger variations in N_2. But we prefer working with the approximation $N_2 \ll 1$, and use

$$\exp[N_2 \cos(2\omega t + \phi_2)] \simeq 1 + N_2 \cos(2\omega t + \phi_2) \tag{53}$$

It will be clear later that the condition $N_2 \ll 1$ is more readily satisfied than $N_1 \ll 1$. Combining (51), (52), and (53), and neglecting the higher harmonics we obtain,

$$\begin{aligned} e^{N(t)} \cong e^{N_0}\{&I_0(N_1) + I_2(N_1)\, N_2(\cos \phi_2) \\ &+ I_1(N_1)[2(\cos \omega t) + N_2 \cos(2\omega t + \phi_2)] \\ &+ 2I_2(N_1)(\cos 2\omega t) + I_0(N_1)\, N_2 \cos(2\omega t + \phi_2)\} \end{aligned} \tag{54}$$

Substituting (54) into (48), using

$$\int_0^\infty \cos[n\omega(t - u)]\, G(u)\, du = \tfrac{1}{2} H_n e^{in\omega t} + \text{C.C.}, \qquad n = 1, 2, \ldots \tag{55}$$

and equating the coefficients of $\exp(in\omega t)$ for various n, we obtain the following set of equations:

$$e^{-N_0} = I_0(N_1) + N_2 I_2(N_1) \cos \phi_2 \tag{56a}$$

$$N_1/Z_1 = 2 \mid k_1 \mid e^{i\theta} + P_e e^{N_0} I_1(N_1)[2 + N_2 e^{i\phi_2}]\, H_1 \tag{56b}$$

$$(N_2/Z_2)\, e^{i\phi_2} = P_e e^{N_0}[2I_2(N_1) + I_0(N_1)\, N_2 e^{i\phi_2}]\, H_2 \tag{56c}$$

where $Z_n = 1/in\omega l$. We observe from (54) that

$$P_{av}/P_e = e^{N_0}[I_0(N_1) + N_2 I_2(N_1) \cos \phi_1]$$

When combined with (56a), this equation shows that $P_{av} = P_e$, which is rigorously true in the absence of delayed neutrons.

Equations (56b) and (56) contain complex quantities, and thus actually provide a relation to solve for N_1, N_2, and ϕ_2 in terms of k_1. Eliminating N_0 between (56a) and (56c), we get

$$I_0(1 - F_2)\, N_2 e^{i\phi_2} + \tfrac{1}{2} I_2 N_2^2 [1 + e^{i2\phi_2}] = 2I_2 F_2 \tag{57a}$$

where we omitted the arguments of $I_0(N_1)$ and $I_2(N_1)$, and defined

$$F_2 \equiv P_e Z_2 H_2 \tag{57b}$$

which is the open-loop transfer function evaluated at 2ω in the linearized treatment. Since N_2 was already assumed to be less than unity, we can solve (57a) for $N_2 e^{i\phi_2}$ by ignoring the quadratic term,

$$N_2 e^{i\phi_2} = (2I_2/I_0)[F_2/(1 - F_2)]$$

$$= 2(1 - \eta)[F_2/(1 - F_2)] \tag{58}$$

where we introduced

$$\eta(N_1) \equiv 2I_1(N_1)/N_1 I_0(N_1) \tag{59}$$

and made use of the following identity:

$$I_2(N_1) = I_0(N_1) - [2I_1(N_1)/N_1] \tag{60}$$

It must be borne in mind that the neglect of the quadratic term in (57a) is not justified if F_2 is very close to unity. Henceforth, we shall assume that

$$2(1 - \eta)| F_2/(1 - F_2)| \ll 1 \tag{61}$$

so that the approximation $N_2 \ll 1$ can be justified. The origin of this restriction on $1 - F_2$ lies in the resonance phenomenon which might occur when 2ω and P_e approach the critical frequency ω_c and power P_c. The latter are given by

$$P_c Z(i\omega_c) H(i\omega_c) = 1 \tag{62}$$

If this condition is approximately satisfied, then the approximation $N_2 \ll 1$ as well as the neglect of all the higher harmonics are no longer valid. Consequently, (48) can no longer be linearized with respect to the second harmonic. With these remarks, we assume henceforth that $| 1 - F_2 |$ is sufficiently large so that (61) can be satisfied with sufficiently large amplitudes N_1. [Note that $\eta(N_1) \to 1$ as $N_1 \to 0$. Hence, (61) can always be satisfied if the amplitude of the forced oscillations is sufficiently small.]

We are now in a position to calculate N_1 and ϕ_2 from (56b) by substituting $N_2 e^{i\phi_2}$ from (58), and approximating e^{-N_0} by $I_0(N_1)$ consistently with the approximation inherent in (61). Using the identity in (60), we find

$$N_1/k_1 = 2Z_1/[1 - P_e H_1 D_Z(N_1, \omega, P_e)] \tag{63}$$

where $k_1 = | k_1 | e^{i\phi}$ and

$$D_Z(N_1, \omega, P_e) \equiv Z_1 \eta(N_1)\{1 + (1 - \eta)[F_2/(1 - F_2)]\} \tag{64}$$

We shall presently attach a physical significance to $D_Z(N_1, \omega, P_e)$.

The high-power describing function can be obtained by expanding $P(t)$ as

$$P(t) = \sum_{m=-\infty}^{+\infty} P_m e^{i\omega_m t} \qquad (P_0 = P_e) \tag{65}$$

and comparing it to the expansion in (54). The complex amplitude of the fundamental of $P(t)$ is found to be

$$P_1/P_e = (N_1/2Z_1)\, D_Z(N_1, \omega, P_e) \tag{66}$$

from which we obtain the desired describing function as $D = P_1/k_1P_e$:

$$D(N_1, \omega, P_e) = D_Z(N_1, \omega, P_e)/[1 - P_e H_1 D_Z(N_1, \omega, P_e)] \tag{67}$$

Here, N_1 is to be determined from (63) as

$$N_1 = (\delta k) \mid Z_1 \mid / \mid 1 - P_e H_1 D_Z(N_1, \omega, P_e) \mid \tag{68a}$$

where $\delta k \equiv 2 \mid k_1 \mid$ is the amplitude of the cosinusoidal input. For sufficiently small δk, we can approximate N_1 by replacing D_Z in (68) by Z_1 and obtaining

$$N_1 \cong \mid L_1 \mid \delta k \tag{68b}$$

where L_1 is the closed-loop transfer function. Substituting (68b) into (67), we get

$$D(\delta k, \omega, P_e) = D_Z(\delta k, \omega, P_e)/[1 - P_e H_1 D_Z(\delta k, \omega, P_e)] \tag{69a}$$

where

$$D_Z(\delta k, \omega, P_e) = Z_1 \eta(\mid L_1 \mid \delta k)\{1 + [1 - \eta(\mid L_1 \mid \delta k)]\, F_2/(1 - F_2)\} \tag{69b}$$

Comparing (69a) to (38), we conclude that $D_Z(\delta k, \omega, P_e)$ may be identified as the proper zero-power describing function in the presence of the second harmonic generated internally. Since δk is related to the amplitude of net reactivity δk_n by $\delta k \mid L_1 \mid = \delta k_n \mid Z_1 \mid$, we find that (69b) reduces, when the term $(1 - \eta)\, F_2/(1 - F_2)$ arising from the effect of the second harmonic is neglected, to (4.7, Eq. 30b) (use in the latter $Z_1 = -Z_1{}^*$ and $Z_2 = Z_1/2$, which hold when the delayed neutrons are not present, and observe that $N_1\, \delta k_n \to \mid Z_1 \mid \delta k_n$). Thus, we could have obtained (69a) and (69b) starting from the zero-power describing function given by (4.7, Eq. 30b) and using the feedback-loop analysis only. In order to include the effect of the second harmonic on the high-power transfer function in the latter approach, we must first generalize the zero-power describing function to an input containing a second harmonic as well as the fundamental. For example, solving the point kinetic equations without delayed neutrons for

$$k(t) = x_1 e^{i\omega t} + \text{C.C.} + x_2 e^{i2\omega t} + \text{C.C.}$$

Keeping track of the second harmonic linearly, and again using $x_2 = Z_1 P_e H_2 L_2 x_1^2$ (cf Eq. 36a), one can show that the zero-power transfer function is

$$D_Z(\delta k, \omega, P_e) = Z_1 \eta(\delta k \mid Z_1 \mid)[1/(1 - F_2)] \\ \times \{1 - F_2[1 - \tfrac{1}{8}(\delta k)^2 \mid Z_1 \mid]\} \tag{70}$$

If we use the expansion

$$\eta(x) \cong (1 + \tfrac{1}{8}x^2)/(1 + \tfrac{1}{4}x^2) \cong 1 - \tfrac{1}{8}x^2 \tag{71}$$

in the last term of (70), we establish the equivalence of (69b) and (70). This derivation illustrates once again how one can construct the high-power describing function through an appropriate zero-power transfer function.

It is interesting to show that (69b) reduces to (39), which was obtained by a power-series expansion. Using (71) in (69b), we have

$$D_Z(\delta k, \omega, P_e) \cong Z_1(1 - \tfrac{1}{8} \mid x_1 \mid^2)\{1 + \tfrac{1}{8}[(L_2/Z_2) - 1] \mid x_1 \mid^2\} \\ = Z_1\{1 + \tfrac{1}{8} \mid x_1 \mid^2[(L_2/Z_2) - 2]\}$$

which is identical to (39) in the absence of delayed neutrons.

Akcasu and Shotkin [27] used (69b) to investigate the autonomous oscillations in various reactor models, replacing (67b) by

$$1 = P_e H_1 \eta(N_1)[1 - \eta(N_1) F_2]/(1 - F_2) \tag{72}$$

The foregoing treatment may be generalized to include delayed neutrons. Since the method described in the previous section, which is based on a systematic power-series expansion, contains the effect of delayed neutrons, we shall not calculate the logarithmic response with delayed neutrons. If the effect of the second harmonic can be ignored, i.e., the feedback mechanism behaves as a sharp low-pass filter so that $H_2 \approx 0$, then we can use the zero-power describing function given by (4.7, Eq. 30a), which already includes the delayed neutrons.

When large-amplitude oscillations are involved, the describing function based on the logarithmic response of the reactor may be expected to be more accurate than that derived by a power-series expansion.

5.6. Equivalent Space-Independent Temperature Models

In Section 5.2, we considered the reduction of certain space-dependent temperature models to multipoint space-independent equivalents. We take this point up again with the idea of determining the conditions under which the continuous model and the space-independent equivalent yield essentially the same results. Obviously, if we break up the continuous model into many hundreds of very small regions, we could get better results than breaking it up into one or only a few regions. We expect, then, that if a transient is slow enough, the results will be the same, and we shall determine an explicit mathematical definition of the adjective "slow."

Because of its basic importance in cooling a power reactor, we consider a system containing a flowing coolant; secondly, to reduce the complexity, we consider only the question of axial (in the direction of coolant flow) lumping.†

A. A One-Temperature Model

Suppose the temperature distribution is accurately described by the equation in a simplified notation

$$(\partial T/\partial t) + v(\partial T/\partial x) = \phi(x)\, P(t), \quad 0 \leqslant x \leqslant h \tag{1}$$

where $\phi(x)$ is a static shape factor for the axial power distribution, $P(t)$ is the variable power level, and v is the coolant velocity. We reduce Eq. (1) to $N + 1$ point equations (the first point labeled 1) describing the average temperatures in N lumps. Define

$$\bar{T}^k(t) \equiv (1/\Delta x) \int_{k\Delta x}^{(k+1)\Delta x} T(x, t)\, dx, \qquad k = 1,\ldots, N + 1 \tag{2a}$$

$$T_k(t) = T(k\, \Delta x, t) \tag{2b}$$

Certainly, for small enough Δx, the temperature variation inside the increment Δx will not vary rapidly; hence, we may define the average temperature inside in terms of the endpoint values

$$2\bar{T}^k = T_{k+1} + T_k \tag{3}$$

The transit time through the region is given by

$$t_c \equiv h/v = (N\, \Delta x)/v \tag{4}$$

† We break the axial direction into a number of regions or "lumps;" hence, lumping.

With the above definitions, we may reduce Eq. (1) to

$$\dot{\bar{T}}^k + (2N/t_c)(\bar{T}^k - T_k) = G^k P(t), \qquad k = 1,\ldots, N+1 \tag{5a}$$

$$G^k \equiv (1/\Delta x) \int_{k\Delta x}^{(k+1)\Delta x} \phi(x)\, dx \tag{5b}$$

Subtracting out the equilibrium solution and taking a Laplace transform, we find

$$\bar{T}^k(s) = [G^k P(s)/s][1 - A(s)] + A(s)\, T_k(s) \tag{6a}$$

$$A(s) = [1 + (st_c/2N)]^{-1} \tag{6b}$$

If $T_1(s) = g(s)$, a given function, then, using the transform of Eq. (3), we solve Eq. (6a) for any point k to yield

$$\bar{T}^k(s) = A(2A-1)^{k-1} g(s) + [(1-A)\, P(s)/s] \left[2A \sum_{j=1}^{k-1} (2A-1)^{k-1-j}\, G^j + G^k\right] \tag{7a}$$

$$T_{k+1}(s) = (2A-1)^k g(s) + [2(1-A)\, P(s)/s] \sum_{j=1}^{k} (2A-1)^{k-j}\, G^j \tag{7b}$$

Consider now a reflector region in which $P(s) = 0$; then, the exit temperature from such a region is given by

$$T_{N+1} \equiv T_{\text{exit}} = \{[1 - (st_c/2N)/[1 + (st_c/2N)]\}^N g(s) \tag{8}$$

Obviously,

$$|\, st_c/2N \,| < 1 \tag{9}$$

In the limit $N \to \infty$, Eq. (8) reduces to

$$T_{\text{exit}} = e^{-st_c} g(s) \tag{10}$$

which is the analytic solution of Eq. (1). The general solution, Eqs. (7a,b), contains the term $2A - 1$; hence, the stipulation Eq. (9) is valid whether $P(t) \geqslant 0$.

The question reduces to a more explicit statement of Eq. (9). If we choose that our equation should be accurate to within 1 % of the correct answer, we find the more explicit criterion

$$|\{[2 - (st_c/N)]/[2 + (st_c/N)]\}^N - e^{-st_c}\,| < 0.01\, |\, e^{-st_c}\,| \tag{11a}$$

or $| st_c/N |$ must be less than x, where x satisfies

$$(2 - x)/(2 + x) = (0.99)^{1/N} e^{-x} \tag{11b}$$

Equation (11b) is the 1% criterion for the replacement of Eq. (1) by the $N + 1$ point equations (5). In particular, if $N = 1$, $x = 0.48$; hence, a one-point model is valid for frequencies less than $1/(2t_c)$ rad/sec.

B. The Two-Temperature Model

Suppose that, in place of Eq. (1), we had

$$\partial T_f/\partial t = \phi(x) P(t) - \eta_1(T_f - T_c) \tag{12a}$$

$$(\partial T_c/\partial t) + v(\partial T_c/\partial x) = \eta_2(T_f - T_c), \qquad 0 \leqslant x \leqslant h \tag{12b}$$

That is, suppose the reactor is well described axially by a two-temperature model in which we separate the fuel temperature from the coolant temperature. Introducing the new variable

$$\bar{T}_f^k(t) \equiv (1/\Delta x) \int_{k\Delta x}^{(k+1)\Delta x} T_f(x, t)\, dx \tag{13}$$

we may reduce Eq. (12) to an N-lump model. The procedure followed above is repeated, and we are led to the following definition for $\bar{T}_c^k$:

$$\bar{T}_c^k = \left[\frac{2N}{t_c} T_c^k + \frac{\eta_2 G^k P(s)}{s + \eta_1}\right] \Big/ \left[s\left(1 + \frac{\eta_2}{s + \eta_1}\right) + \frac{2N}{t_c}\right] \tag{14}$$

we define

$$A \equiv (1 + (st_c/2N)\{1 + [\eta_2/(s + \eta_1)]\})^{-1} \tag{15}$$

whence Eq. (14) reduces to

$$\bar{T}_c^k = AT_k + [(1 - A)\, \eta_2 P(s)\, G^k/s(s + \eta_2 + \eta_1)] \tag{16}$$

replacing $P(s)$ in Eq. (6a) by $\eta_2 P(s)/(s + \eta_1 + \eta_2)$, we are led again to 5.6, Eq. 11 except that, wherever $st_c/2N$ appears, we replace it by $(st_c/2N)\{1 + [\eta_2/(s + \eta_1)]\}$. Hence, a one-lump model is accurate to within 1% if, for $P(s) = 0$,

$$| st_c\{1 + [\eta_2/(s + \eta_1)]\}| < 0.48$$

This is exactly the expression obtained by Storrer (in a different way) for inlet coolant temperature oscillation [29]. The frequency criterion for power oscillation is more complicated since it involves an integration over the shape function $\phi(x)$. Let

$$f(s) = st_c\{1 + [\eta_2/(s + \eta_1)]\} \tag{17}$$

then the analytic solution to Eq. (12) is

$$T_c(h, s) = g(s)\, e^{-f(s)} + [\eta_2 t_c P(s)\, e^{-f(s)}/h(s + \eta_1)] \int_0^h e^{f(s)x/h}\, \phi(x)\, dx \tag{18}$$

if $y \equiv x/h$, the integral may be replaced by

$$\int_0^1 [1 + yf(s) + y^2 f^2(s)/2! + \cdots]\, \phi(y)\, dy$$

Define now

$$\langle y^k \rangle_{av} \equiv \int_0^1 y^k \phi(y)\, dy \Big/ \int_0^1 \phi(y)\, dy \tag{19}$$

then the analytic solution is replaced by

$$T_c(h, s) = g(s)\, e^{-f(s)} + [t_c \eta_2 P(s)/(s + \eta_1)] \langle \phi \rangle\, e^{-f(s)} \sum_{k=0}^{\infty} \langle y^k \rangle_{av}\, f^k(s)/k!] \tag{20}$$

The one-term solution for $T_c(h, s)$ in Eq. (14) is given by

$$T_{c,\text{exit}} = [(1 - \tfrac{1}{2}f)/(1 + \tfrac{1}{2}f)]\, g(s) + [\eta_2 t_c \langle \phi \rangle\, P(s)/(s + \eta_2)(1 + \tfrac{1}{2}f(s)] \tag{21}$$

Expanding Eq. (20) and Eq. (21) [i.e., $e^x \to 1 + x + \frac{1}{2}x^2 + \cdots$; $(1 + x)^{-1} = 1 - x + \frac{1}{2}x^2 - \cdots$] and desiring accuracy to within 1%, we require

$$\frac{\langle \phi \rangle\, |[\langle y \rangle_{av}(1 - \langle y \rangle_{av}) - \frac{1}{2} + \frac{1}{2}\langle y^2 \rangle_{av} \cdots] f^2 \,|}{|\, e^{-f} \int_0^1 e^{fy} \phi(y)\, dy \,|} < 0.01 \tag{22a}$$

For small values of $|f|$, the denominator can be reasonably replaced by $\langle \phi \rangle$; hence, we have

$$|\langle y \rangle_{av}(1 - \langle y \rangle_{av}) - \tfrac{1}{2} + \tfrac{1}{2}\langle y^2 \rangle_{av}\,|\,|f|^2 < 0.01 \tag{22b}$$

where, for all $\phi(y) \geqslant 0$,

$$0 \leqslant \langle y \rangle_{av} \leqslant \tfrac{1}{2}$$

$$0 \leqslant \langle y^2 \rangle_{av} \leqslant \tfrac{1}{3}$$

Hence, we may bound $|f|$ as follows: For no $\phi(y) \geqslant 0$ will the one-lump (or single-point) model yield an accuracy to within 1% if

$$|f| > 0.345 \tag{22c}$$

For any $\phi(y) \geqslant 0$, the one-lump model will always yield accuracy to within 1% if

$$|f| < 0.14 \tag{22d}$$

This latter result was also obtained by Storrer [29].

C. A Two-Region Model

We suppose now that the reactor contains two consecutive regions with differing nuclear characteristics. In the second, no heat production occurs; hence, it may be considered a moderating or reflecting region. For simplicity, we assume each region is well characterized by Eq. (1) but with v and $\phi(x)$ region-dependent. Proceeding as before, we find that the average temperature in the reflector region is given by

$$\bar{T}_{\mathrm{r}}(s) = B \sum_{k=1}^{M} (2B - 1)^{k-1}\, T_{\mathrm{c}}(h, s) \tag{23a}$$

$$T_{\mathrm{c}}(h, s) = T_{\mathrm{exit}} = 2(1 - A) \sum_{j=1}^{N+1} (2A - 1)^{N+1-j}\, G^{j}\, P(s)/s \tag{23b}$$

$$A = [1 + (st_{\mathrm{c}}/2N)]^{-1}, \qquad B = [1 + (st_{\mathrm{r}}/2M)]^{-1} \tag{23c}$$

Equations (23) assume that the core region has a constant inlet temperature [$g(s) = 0$] and is broken into N lumps, while the reflector is broken into M lumps and has no power production. The exact solutions are easily found and a comparison of the $N = M = 1$ model with the continuous one is most easily accomplished for the case $\phi(x) = 1$, i.e., $G_j = 1$. In this case, we find the 1% criterion to be given by $|st_{\mathrm{c}}| < x$, where, if $\alpha = t_{\mathrm{r}}/t_{\mathrm{c}}$,

$$|[4x^2\alpha/(2 + x)(2 + \alpha x)] - (1 - e^{-x})(1 - e^{-\alpha x})| \leqslant 0.01\, |(1 - e^{-x})(1 - e^{-\alpha x})| \tag{24}$$

for $\alpha \to 0$, $x \to 0.345$; for $\alpha \to \infty$, $x \to 0$; at $\alpha = 1$, $x = 0.255$. Generally, the limiting value of x decreases monatonically with increasing α.

The above results can be extended to the case of A given by Eq. (16), but the resulting equation replacing Eq. (24) is more complex.

It seems clear that the methods used above can be used generally to secure limiting criteria for any approximation.

REFERENCES

1. V. Voltera, "Theory of Functionals." Dover, New York, 1959.
2. H. Brooks, Temperature coefficients and stability. *Nucl. Reactor Theory Proc. Symp. Appl. Math. 11th, New York*, 1959, p. 91. Am. Math. Soc., Providence, Rhode Island, 1961.
3. W. K. Ergen, H. J. Lipkin, and H. A. Nohel, *J. Math. Phys. (Cambridge, Mass.)* **36**, 36 (1957).
4. S. Perlis, "Theory of Matrices." Addison-Wesley, Reading, Massachusetts, 1952.
5. E. A. Guillemin, "The Mathematics of Circuit Analysis." Wiley, New York, 1949.
6. H. Bethe, APDA-117. 1956.
7. T. A. Welton, Kinetics of stationary reactor systems. *Proc. Int. Conf. Peaceful Uses At. Energy, Geneva*, 1955, **5**, P/610, pp. 377–388. Columbia Univ. Press, New York, 1956.
8. W. K. Ergen and A. M. Weinberg, *Physica (Utrecht)* **20**, 413 (1954).
9. W. K. Ergen, *J. Appl. Phys.* **25**, 702 (1954).
10. J. A. Fleck, Jr., BNL 334. Brookhaven Nat. Lab., Upton, New York, 1955; also *Nucleonics* **12**, 11 (1954).
11. J. A. Fleck, Jr., BNL 357. Brookhaven Nat. Lab., Upton, New York, 1955.
12. R. V. Meghreblian and D. K. Holmes, "Reactor Analysis." McGraw-Hill, New York, 1960.
13. A. M. Weinberg and E. P. Wigner, "The Physical Theory of Neutron Chain Reactor." Univ. of Chicago Press, Chicago, Illinois, 1958.
14. J. Chernick, *Nucl. Sci. Eng.* **8**, 233 (1960).
15. H. B. Smets, *Bull. Cl. Sci. Acad. Roy. Belg.* **47**, 382 (1961).
16. H. B. Smets, *Nucl. Sci. Eng.* **11**, 133 (1961).
17. E. Wimuric, M. Petric, W. Lipinski, and H. Iskenderian, Performance characteristics of EBWR from 0-1000 MW. ANL-6775. Argonne Nat. Lab., Argonne, Illinois, 1963.
18. J. A. Deshong, Jr. and W. C. Lipinski, Analyses of experimental power reactivity feedback transfer functions for a natural circulation boiling water reactor. ANL-5850. Argonne Nat. Lab., Argonne, Illinois, 1958.
19. D. F. Tuttle, "Network Synthesis," Vol. 1. Wiley, New York, 1958.
20. Reactor physics constants. ANL-5800. Argonne Nat. Lab., Argonne, Illinois, 1963.
21. H. B. Smets, The describing function of nuclear reactors. *IRE Trans. Nucl. Sci.* **6**, 8–12 (1959).
22. H. A. Sandmeir, Nonlinear treatment of large perturbation in power reactor stability. *Nucl. Sci. Eng.* **3**, 85–92 (1959).
23. R. Lauber, Nichtlineare Stabilitatsuntersuchung von Reactoren und Reaktor-Reaktor-Regelsystemen. *Atomkernenergie* **1**, 95–101 (1962).
24. A. A. Wasserman, Contributions to two problems in space-independent, nuclear reactor dynamics. IDO 16755. 1962, Phillips Petroleum Co., Idaho Falls, Idaho.
25. H. B. Smets, *Nukleonik* **7**, 399 (1965).

26. H. B. Smets, Power oscillations in nuclear reactors. *Rev. A* **6**, 137 (1964).
27. A. Z. Akcasu and L. M. Shotkin, Power oscillations and the describing function in reactors with linear feedback. *Nucl. Sci. Eng.* **28**, 72–81 (1967).
28. "Handbook of Mathematical Functions" (Appl. Math. Ser. 55), p. 375. Nat. Bur. Stand., Washington, D. C., 1964.
29. F. Storrer, Temperature response to power, inlet coolant temperature and flow transients in solid fuel reactors. APDA-132. 1959.

CHAPTER 6

Linear Stability Analysis

The question of stability in reactors is concerned with the temporal behavior of the power and its distribution in the reactor following disturbances during operation. These disturbances may be fluctuations in the dynamical variables, such as the power, temperature, coolant flow, and so on. They may occur spontaneously as noise in the absence of an externally applied disturbance, or they may be programmed external perturbations in a kinetic experiment. After such a disturbance, the reactor may behave in essentially one of two ways. The dynamical variables may, after a transient period, return to their steady-state values, or some or all of them may tend to move continuously in either an oscillatory or a unidirectional manner away from their original steady-state values. The equilibrium state prevailing before the occurrence of the disturbance is said to be "stable" in the former case and "unstable" in the latter case. The concept of stability will be defined more precisely in mathematical terms in Chap. 7.

The possibility of an unstable response to a small disturbance within a system is a characteristic of physical systems with feedback, in which the output (or the effect) can influence the input of the system (the cause) through the feedback mechanism, forming a closed loop as indicated in Chap. 5, Figure 5.1.1. The origin of instabilities in feedback systems was known and understood by Maxwell ([1], see also Hammond [2]) when he referred in 1868 to the "dancing" of Jenkins' governor as due to the "possible (real) part of an impossible (complex) root of the equation of motion becoming positive." This chapter will essentially be an elaboration of this explanation.

The stability of any equilibrium state may depend on the magnitude of the disturbance. An equilibrium state may be unstable in the crude sense described above for large perturbations even though it may be stable for small disturbances. In the latter case, the transients of the dynamical variables involve small departures from the original steady-state values, and can be adequately described by the linearized kinetic equations. The stability of a reactor for small disturbances is therefore called "linear" stability. In the case of large perturbations, the nonlinearities of the kinetic equations cannot be ignored, because of the large variations of the dynamical variables during their transients, and thus the corresponding analysis is referred to as "nonlinear" stability analysis.

Since it is essential to have a firm understanding of the techniques of linear stability analysis before discussing the more complex problem of nonlinear stability, this chapter will be devoted entirely to a concise presentation of the stability theory of linear systems with feedback.

6.1. Characteristic Function and Linear Stability

The question of stability of a physical system is associated with an equilibrium (or steady state) of an autonomous system.

A physical system is said to be autonomous when the equations describing its temporal behavior are invariant under a translation of the origin of time. In an autonomous system, all the changes take place automatically as a response to the changes in the past, and none of the parameters characterizing the system can depend on time explicitly. Hence, in an autonomous point-reactor, the external reactivity and the external source are, by definition, constant in time. The temporal behavior of an autonomous point-reactor is described by

$$(l/\beta)\,\dot{P} = (\delta_0 + k_f[P])\,P + \int_0^\infty du[P(t-u) - P(t)]\,D(u) + S_0 \tag{1}$$

where δ_0 and S_0 are the constant values of the external reactivity and the source. We absorbed l/β in S_0 (cf 5.1, Eq. 25). Clearly, the remaining reactor parameters (l/β) and a_i and λ_i appearing in the delayed neutron kernel $D(u)$, as well as the feedback parameters implicit in the feedback functional $k_f[P]$ are all assumed to be independent of time.

An equilibrium state of an autonomous reactor is a solution of (1) that is independent of time. Substituting $P(t) \equiv P_0$ in (1), we obtain an algebraic equation for P_0:

$$[\delta_0 + k_f(P_0)]\,P_0 + S_0 = 0 \tag{2}$$

The feedback functional $k_f[P]$ becomes an ordinary function, denoted by $k_f(P_0)$ in (2), when $P(t)$ is independent of time. This function represents the feedback reactivity when the reactor is operated at a constant power level P_0, and in general, it is a nonlinear function of P_0 (see Figure 6.1.1). The equilibrium state is obtained as the intersection of the curves representing $-k_f(P)$ and $(S_0/P) + \delta_0$. In general

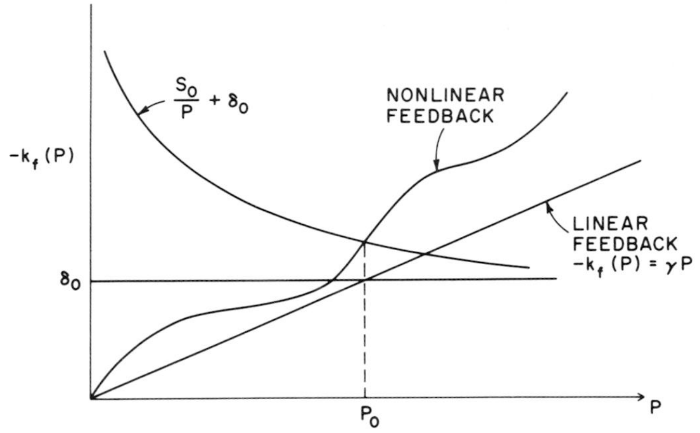

FIGURE 6.1.1. Determination of the equilibrium state.

there may be more than one equilibrium state. However, it is reasonable to assume, on physical grounds, that $-k_f(P)$ is a nondecreasing function of P, and that there is only one equilibrium state. (See 6.3, Eq. 3 for a case in which two equilibria exist.) For example, in the case of a linear feedback, we have

$$k_f[P] = \int_0^\infty G(u)\, P(t-u)\, du \tag{3a}$$

which, for a constant P, reduces to a linear function of P, i.e.,

$$k_f(P) = \gamma P \tag{3b}$$

where γ is called the power coefficient of reactivity. Thus, a reactor with linear feedback has only one equilibrium state for $S_0 \neq 0$.†

† Equation (2) may have more than one root. However, only one of these roots will be real and positive. For example, in the case of linear feedback, Eq. (2) reduces to $(\delta_0 + \gamma P_0)P_0 + S_0 = 0$, which has only one real, positive root when $\gamma < 0$ (negative power coefficient of reactivity) and no real, positive root when $\gamma > 0$.

Equation (2) indicates that $P_0 = 0$ is always an equilibrium state, irrespective of the form of the feedback functional, if the external source is not present. This equilibrium state represents the reactor in its shutdown, or zero-power, condition. In the shutdown state, the reactor is supercritical with a positive reactivity δ_0. Hence, the shutdown state is, of necessity, unstable because the reactor power tends to increase exponentially following a positive step-reactivity insertion if initial power ceases to be zero even momentarily as a result of a disturbance.

We shall be interested primarily in analyzing the stability of the operating equilibrium state which corresponds to the nonzero solution of (2) in the absence of external sources. Since we are interested in the linear stability of this equilibrium state, we must first express (1) in terms of the departures from P_0. Substituting $P(t) = P_0 + p(t)$ in (1) and defining the incremental feedback reactivity by

$$\delta k_{\mathrm{f}}[p] = k_{\mathrm{f}}[P] - k_{\mathrm{f}}(P_0) \tag{4}$$

we obtain

$$(l/\beta)\,\dot{p}(t) = \delta k_{\mathrm{f}}[p](p + P_0) + \int_0^\infty du\,[p(t-u) - p(t)]\,D(u) \tag{5}$$

which describes an autonomous reactor whose equilibrium power level is P_0. The linearized kinetic equation is obtained from (5) by ignoring p compared to P_0, and retaining the first term in the functional expansion of $\delta k_{\mathrm{f}}[p]$ (cf 5.1, Eq. 20):

$$(l/\beta)\,\dot{p} = P_0 \int_0^\infty p(t-u)\,G(u)\,du + \int_0^\infty du\,[p(t-u) - p(t)]\,D(u) \tag{6}$$

Let us suppose that the reactor is operated at the equilibrium state P_0 prior to $t = 0$, and assume that an initial perturbation $p(0)$ is introduced at $t = 0$. The temporal behavior of the reactor for $t > 0$ is governed by

$$(l/\beta)\,\dot{p} = P_0 \int_0^t p(t-u)\,G(u)\,du + \int_0^t p(t-u)\,D(u)\,du - p(t) \tag{7}$$

Taking the Laplace transform of (7), we get

$$\bar{p}(s) = [p(0)/(1/Z(s) - P_0 H(s)] \tag{8}$$

where $Z(s)$ is the zero-power transfer function and $H(s)$ is the Laplace transform of $G(t)$, i.e.,

$$H(s) = \int_0^\infty e^{-st} G(t)\, dt \tag{9}$$

The function $H(s)$ is called the feedback transfer function, and completely determines the linear feedback mechanism. Since the latter must be a stable linear system (cf 5.1, Eq. 23b), $G(t)$ is absolute-integrable, and the integral in (9) converges for all $\operatorname{Re} s \geqslant 0$. Thus, $H(s)$ does not have any poles with positive or zero real parts. Furthermore, $H(s)$ vanishes as $s \to \infty$ because the final-value theorem implies that $\lim_{s\to\infty} sH(s) = G(0) < \infty$. We may also mention that $H(0) = \gamma$, the power coefficient of reactivity.

Equation (8) indicates that the behavior of $p(t)$, $t > 0$, is determined by the singularities of $\bar{p}(s)$ on the complex s plane. These singularities occur at the zeros of

$$Q(s) = 1 - P_0 H(s)\, Z(s) \tag{10}$$

Equation (10) is called the "characteristic" equation.

Thus, the problem of linear stability of an equilibrium state, either shutdown or operating, is reduced to the problem of determining the sign of the real parts of the roots of the characteristic equation. If even one of these roots has a positive real part, then the reactor response $p(t)$ to an initial disturbance $p(0)$, will increase exponentially with time, and hence the equilibrium state P_0 will be unstable. We thus obtain: A reactor is linearly stable if the roots of the characteristic equation all have negative real parts. The response of the dynamical variables in the critical case, when any of these roots has real parts equal to zero, is not correctly described by linear analysis and depends on the nonlinearities of the kinetic equations.

In the following sections, we shall discuss the necessary and sufficient conditions for all the roots of the characteristic equation to have negative real parts. These conditions are referred to as "linear stability criteria," and enable one to investigate the question of stability of linear systems without actually solving the characteristic equation. In the following three sections, we shall present the algebraic stability criteria, and discuss the linear stability of a xenon, temperature-controlled reactor in detail. Section 6.5 will be devoted to the discussion of graphical stability criteria with an application to the two-temperature reactor model.

6.2. Routh–Hurwitz Stability Criterion

This criterion applies to those reactor systems in which the feedback mechanism can be characterized by a finite set of linear equations with constant coefficients, e.g., multitemperature feedback model (cf Sec. 5.2B). Then, the feedback transfer function $H(s)$ can be expressed as a ratio of two polynomials of s, and the characteristic equation can be written as a polynomial

$$\mathscr{P}(s) \equiv a_0 s^n + a_1 s^{n-1} + \cdots + a_{n-1} s + a_n = 0 \tag{1}$$

where we shall always assume a_0 to be positive without loss of generality. Before presenting the Routh–Hurwitz criterion in a general form, we consider a few special cases.

1. A necessary condition for all the roots of (1) to have negative real parts is that all the coefficients a_n are nonzero and positive [3].

Proof. Let us suppose that all the roots have negative real parts. Then, $s_1 = -|\alpha_1|$, $s_{2,3} = -|\alpha_2| \pm i\beta_2$,..., $s_n = -|\alpha_n|$, since $\mathscr{P}(s) = a_0(s - s_1),\ldots,(s - s_n)$, we immediately find

$$\mathscr{P}(s) = a_0(s + |\alpha_1|)[(s + |\alpha_2|)^2 + \beta^2] \cdots (s + |\alpha_n|)$$

which, upon multiplication, leads to a polynomial with strictly positive coefficients.

2. The foregoing necessary condition is also sufficient when $\mathscr{P}(s)$ is a first- or second-order polynomial.

3. In the general case $n \geqslant 3$, the positiveness of the coefficients ensures only the negativeness of the real roots, but does not yield information about the sign of the real parts of the complex roots. Indeed, suppose $\mathscr{P}(s) = a_0(s + |\alpha_1|)[(s - |\alpha_2|)^2 + \beta_2{}^2]$ so that $\mathscr{P}(s) = 0$ has one negative real root, and two complex roots with positive real parts. Then,

$$\mathscr{P}(s) = a_0[s^3 + (|\alpha_1| - 2|\alpha_2|)s^2 + (\alpha_2{}^2 + \beta_2{}^2 - 2|\alpha_1||\alpha_2|)s + (|\alpha_2|^2 + |\beta_2|^2)|\alpha_1|]$$

the coefficients of which may all be positive if $|\alpha_1|$, $|\alpha_2|$, and $|\beta_2|$ are properly adjusted.

4. The general conditions under which (1) can have only negative real parts were investigated by Routh [4] in 1877 and by Hurwitz [5] in 1895, who arrived, by different methods, at similar conclusions.

A. Routh–Hurwitz Criterion

Routh–Hurwitz conditions are expressed in terms of the Hurwitz determinants, which are formed from the coefficients of the characteristic polynomials of nth order as follows:

$$\Delta_n = \begin{vmatrix} a_1 & a_3 & a_5 & a_7 & \cdots & 0 & 0 & 0 \\ a_0 & a_2 & a_4 & a_6 & & 0 & 0 & 0 \\ 0 & a_1 & a_3 & a_5 & & 0 & 0 & 0 \\ 0 & a_0 & a_2 & a_4 & & 0 & 0 & 0 \\ \vdots & & & & & & & \\ 0 & 0 & 0 & 0 & & a_{n-2} & a_n & 0 \\ 0 & 0 & 0 & 0 & & a_{n-3} & a_{n-1} & 0 \\ 0 & 0 & 0 & 0 & & a_{n-4} & a_{n-2} & a_n \end{vmatrix} \tag{2}$$

The n elements of the first row are selected as the odd coefficients in the polynomial and a sufficient number of zeros in the indicated order in (2). The second row is obtained in a similar way with the even coefficients. The third and fourth rows are obtained by shifting the first and the second, respectively, by one step to the right and filling the preceding entries by zeros. Continuing in this manner, one obtains an $n \times n$ determinant denoted by Δ_n. We now state the Routh–Hurwitz stability criterion: The roots of $\mathscr{P}(s) = 0$ all have negative real parts if, and only if,

$$\begin{aligned} &a_0 > 0 \\ &\Delta_1 = a_1 > 0 \\ &\Delta_2 = \begin{vmatrix} a_1 & a_3 \\ a_0 & a_2 \end{vmatrix} > 0 \\ &\Delta_3 = \begin{vmatrix} a_1 & a_3 & a_5 \\ a_0 & a_2 & a_4 \\ 0 & a_1 & a_3 \end{vmatrix} > 0 \\ &\quad\vdots \\ &\Delta_n = a_n \Delta_{n-1} > 0 \end{aligned} \tag{3}$$

are satisfied.

The conditions (3) are not independent of each other. In the case of a third-order system, these conditions are equivalent to $a_1 > 0$, $a_2 > 0$, $a_3 > 0$, and $a_1 a_2 > a_0 a_3$. We observe that there is only one additional condition in addition to the positiveness of all the coefficients. It is interesting to verify that there is again only one additional condition, i.e., $a_3(a_1 a_2 - a_0 a_3) > a_4 a_1^2$, to the positiveness of all the coefficients in a fourth-order system, too. This observation is not true for higher-order systems. For example, in a fifth-order system, there are two additional conditions [3] (Problems 1 and 2).

B. Routh's Stability Criterion

An algorithm due to Routh [4] provides a simpler test for stability. It consists of arranging the coefficients in a triangular array as follows:

$$\begin{array}{ccccc} a_0 & a_2 & a_4 & a_6 & \cdots \\ a_1 & a_3 & a_5 & a_7 & \cdots \\ b_{11} & b_{13} & b_{15} & \cdot & \cdots \\ b_{21} & b_{23} & b_{25} & \cdot & \cdots \\ b_{31} & \cdot & \cdot & \cdot & \cdots \\ \cdot & \cdot & \cdot & \cdot & \cdots \end{array}$$

where

$$b_{11} = (a_1 a_2 - a_0 a_3)/a_1\,, \qquad b_{13} = (a_1 a_4 - a_0 a_5)/a_1$$

$$b_{15} = (a_1 a_6 - a_0 a_7)/a_1\,, \qquad b_{21} = (b_{11} a_3 - a_1 b_{13})/b_{11}$$

$$b_{23} = (b_{11} a_5 - a_1 b_{13})/b_{11}\,, \qquad b_{31} = (b_{21} b_{13} - b_{11} b_{23})/b_{21}$$

etc. The procedure for constructing the array should be clear from the above expression. The general term b_{ji} can be written as

$$b_{ji} = b_{j-2,i+2} - (b_{j-2,1}/b_{j-1,1})\, b_{j-1,i+2}\,, \qquad i = 1, 3, 5, \ldots; \quad j \geqslant 3 \tag{4}$$

Each succeeding row has fewer terms than the preceding row, and thus the array is triangular. The array terminates where there is only one term in the last row. There are $(n + 1)$ rows when the array is completed. Routh's criterion can now be stated:

The roots of $\mathscr{P}(s) = 0$ all have negative real parts if, and only if, all the terms in the first column in the Routh array are positive. If there are negative terms, then roots with positive real parts exist, and the system is unstable. The number of changes in sign gives the number of roots with positive real parts. For example, if the signs are positive up to a certain row, and all negative thereafter, then the number of sign changes is one, and there is only one root with a positive real part, which is of necessity real.

EXAMPLE. Consider $s^4 + 2s^3 + s^2 + 4s + 3 = 0$, and verify

$$\begin{array}{rrr} 1 & 1 & 3 \\ 2 & 4 & 0 \\ -1 & 3 & \\ 10 & 0 & \\ 3 & & \end{array}$$

There are two sign changes, and hence a complex pair with a positive real part exists. Note that there cannot be positive real roots, because all the coefficients are positive.

The one-to-one correspondence between the Routh–Hurwitz determinants and the terms in the first column can easily be demonstrated (see the literature [6–8] for further discussion of the Routh–Hurwitz criterion and Problem 3).

EXAMPLE. As an illustration of the use of the Routh–Hurwitz criterion, we consider the xenon stability problem. A detailed analysis of xenon-temperature stability appears in Sec. 6.3. Here, we wish only to consider the pure xenon feedback kernel. The model we use is that of (5.3, Eq. 21) with $k_{\text{ext}}(t) = \gamma = S = 0$; that is, without external source or reactivity variation and in the absence of temperature or power feedback. We reproduce the resulting equations here for convenience:

$$(l/\beta)\,\dot{P} = [\delta_0 - (\sigma_{\text{Xe}}\text{X}_\text{e}/\beta c\sigma_\text{f})]\,P + \sum_{j=1}^{6} \lambda_j C_j - P \tag{5a}$$

$$\dot{C}_j = a_j P - \lambda_j C_j \tag{5b}$$

$$\dot{\text{I}} = y_\text{I}\sigma_\text{f} P - \lambda_\text{I}\text{I} \tag{5c}$$

$$\dot{\text{X}}_\text{e} + \lambda_{\text{Xe}}\text{X}_\text{e} = y_{\text{Xe}}\sigma_\text{f} P + \lambda_\text{I}\text{I} - \sigma_{\text{Xe}}\text{X}_\text{e} P \tag{5d}$$

The equilibrium of Eqs. (5) is defined by

$$\begin{aligned} C_{j0} &= a_j P_0/\lambda_j\,; \qquad \text{I}_0 = y_\text{I}\sigma_\text{f} P_0/\lambda_\text{I} \\ \text{Xe}_0 &= \beta c\sigma_\text{f}\delta_0/\sigma_{\text{Xe}} = y\sigma_\text{f} P_0/(\lambda_{\text{Xe}} + \sigma_{\text{Xe}} P_0) \end{aligned} \tag{6}$$

where $y = y_\text{I} + y_{\text{Xe}}$.

In order to reduce the complexity of the system with its large number of parameters, we pass to the form where the dynamical variables are measured relative to their equilibrium values. Define

$$Y = \beta c\delta_0/y, \qquad Y_{\text{Xe}} = y_{\text{Xe}}/y \qquad Y_\text{I} = y_\text{I}/y \tag{7}$$

then the equilibrium values for Xe_0 and P_0 in Eq. (6) become

$$\text{Xe}_0 = y\sigma_\text{f} Y/\sigma_{\text{Xe}}\,, \qquad P_0 = \lambda_{\text{Xe}} Y/\sigma_{\text{Xe}}(1 - Y) \tag{8}$$

The new variables are defined by

$$\begin{aligned} p &= (P - P_0)/P_0\,, \qquad \delta\text{Xe} = (\text{Xe} - \text{Xe}_0)/\text{Xe}_0\,, \\ \delta C_j &= (C_j - C_{j0})/C_{j0}\,, \qquad \delta I = (I - I_0)/I_0 \end{aligned}$$

With these definitions, Eqs. (5) reduce to

$$(l/\beta)\,\dot{p} = -(yY/c\beta)\,\delta\mathrm{Xe}(1+p) + \sum_{j=1}^{\mathrm{I}} a_j(\delta C_j - p) \tag{9a}$$

$$\delta\dot{C}_j = \lambda_j(p - \delta C_j) \tag{9b}$$

$$\delta\dot{\mathrm{I}} = \lambda_{\mathrm{I}}(p - \delta\mathrm{I}) \tag{9c}$$

$$\delta\dot{\mathrm{X}}\mathrm{e} + \lambda_{\mathrm{Xe}}\,\delta\mathrm{Xe} = [\lambda_{\mathrm{Xe}}/(1-Y)][Y_{\mathrm{Xe}}p + Y_{\mathrm{I}}\,\delta\mathrm{I} - Y(p + \delta\mathrm{Xe} + p\,\delta\mathrm{Xe})] \tag{9d}$$

From Eq. (8), we observe that no physical equilibrium exists for P_0 if $Y > 1$. Hence, $\delta_0 \leqslant y/c\beta$. This result illustrates the saturating properties of xenon absorption, and is due basically to the phenomenon of xenon burnout. This can easily be seen in Eq. (6), where $\lim_{P_0\to\infty} \mathrm{Xe}_0 \to y\sigma_{\mathrm{f}}/\sigma_{\mathrm{Xe}}$ independent of P_0. In the limit of $\sigma_{\mathrm{Xe}} \to 0$, however (which implies no burnout), we see Xe_0 is continually proportional to P_0.

We proceed now to the question of stability. The linearized (see Sect. 6.1) form of Eq. (5) is reduced by a Laplace transform to the characteristic equation

$$s^3 + \lambda\left[\frac{1-(Y\lambda_{\mathrm{I}}/\lambda)}{1-Y}\right]s^2 + \lambda_{\mathrm{I}}\lambda_{\mathrm{Xe}}\left\{\frac{1+\omega_0(Y_{\mathrm{Xe}} - Y)/\lambda_{\mathrm{I}}]}{1-Y}\right\}s + \lambda_{\mathrm{I}}\lambda_{\mathrm{Xe}}\omega_0 = 0 \tag{10a}$$

where $\lambda \equiv \lambda_{\mathrm{I}} + \lambda_{\mathrm{Xe}}$, and

$$\omega_0 \equiv (yY/c\beta)/\{(l/\beta) + \sum_{j=1}^{6} [a_j/(s+\lambda_j)]\} \tag{10b}$$

Hence, Eq. (10) is not a third-order algebraic equation, but in fact a ninth-order one. We write Eq. (10b) in this particular form since we expect to show that in the stable region the natural oscillation frequencies are so small that $|s| \ll \lambda_j$, hence, that we can replace Eq. (10b) by

$$\omega_0 \simeq (yY/c\beta)\Big[(l/\beta) + \sum_{j=1}^{\mathrm{I}} (a_j/\lambda_j)\Big]^{-1} \equiv yY/cl^* \tag{10c}$$

Suppose Y is very small, so that $1 \gg Y_x \gg Y$; neglecting the prompt xenon yield, we find Eq. (10a) reduces to

$$s^3 + \lambda s^2 + \lambda_{\mathrm{I}}\lambda_{\mathrm{Xe}}(s + \omega_0) = 0 \tag{11}$$

The Routh–Hurwitz criterion reduces (since all coefficients are >0) to $\Delta_3 > 0$; hence,

$$\omega_0 < \lambda \tag{12}$$

Hence, the system is linearly stable if $Y < c\lambda l^*/y$. The numerical values for the physical parameters may be taken from Chernick [9]: $\lambda_{Xe} = 2.09 \times 10^{-5}\ \text{sec}^{-1}$, $\lambda_I = 2.87 \times 10^{-5}\ \text{sec}^{-1}$, $y_{Xe} = 0.003$, $y = 0.064$, $\beta = 0.0075$, $\sigma_{Xe} = 3 \times 10^6\ b = 3 \times 10^{-18}\ \text{cm}^2$, $c = 1.5$, $l^* = 0.1$ sec, and a_j, λ_j are taken from Table 1 in Sec. 1.1 for ^{235}U. The latter parameters are all typical of ^{235}U fueled reactors. Equation (12) implies $Y < 1.24 \times 10^{-4}$, hence that, for flux levels $P_0 > 8.64 \times 10^8$ neutrons $\text{cm}^{-2}\ \text{sec}^{-1}$, the system is unstable. On the boundary $\omega_0 = \lambda$, the value of the frequency is $s^2 = -\lambda_I \lambda_{Xe}$. The magnitude $|s|$ is $\sim 2 \times 10^{-5}\ \text{sec}^{-1}$; hence, we see Eq. (10c) is a very good approximation to Eq. (10b).

Making use of the Hurwitz criterion $\Delta_3 > 0$, we obtain from (10a) the following stability condition:

$$(\omega_0/\lambda_I)\, Y_{Xe} + 1 > (\omega_0/\lambda_I)\{Y + [(1 - Y)^2\, \lambda_I/(\lambda - Y\lambda_I)]\} \tag{13}$$

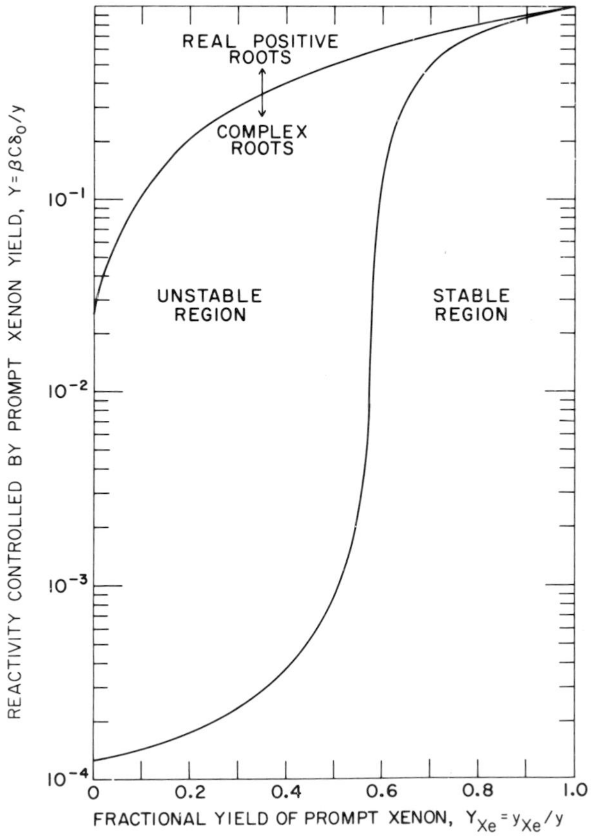

FIGURE 6.2.1. Stability regions for the linearized xenon problem.

In Figure 6.2.1, we plot the line defined by (13) as an equality in the form of $Y = f(Y_{Xe})$, which separates the stability and the instability regions. We observe that, up to $Y_{Xe} = 0.5$, Y is much smaller than unity so that (13) can be reduced to

$$\lambda/[1 - (\lambda/\lambda_I) Y_{Xe}] > \omega_0 \tag{14}$$

For values of $Y_{Xe} \ll 0.5$ $[(\lambda_I/\lambda) = 0.58]$, the stability limit is closely characterized by $\lambda > \omega_0$, which was used in (12). It also follows from (14) that unless the prompt xenon yield $y_{Xe} \to (\lambda_I/\lambda) = 0.58$, the reactor is stable only if its free period is of the order of hours, i.e., $P_0 < 8.64 \times 10^8$ neutrons cm^{-2} sec^{-1}. For values of $Y_{Xe} \geqslant 0.58$, the stability limit is determined by the condition (13) with all terms included.

We observe in (10a) that all the coefficients are always positive except for that of s, which becomes negative when

$$Y(Y - Y_{Xe}) > (cl^*/y)\,\lambda_I \approx 6.6 \times 10^{-5} \tag{15a}$$

or

$$Y \geqslant \{Y_{Xe} + [Y_{Xe}^2 + 4(cl^*\lambda_I/y)]^{1/2}\}/2 \tag{15b}$$

The line defined by (15b) as an equality represents the transition from complex roots to real, positive roots of (10a) (cf Figure 6.2.1).

6.3. Linear Stability Analysis of Xenon and Temperature Feedback

A linear stability analysis is useful in defining stable operating regions which are based on the physical parameters of reactor models. The Routh–Hurwitz criterion can be used to predict safe values of the equilibrium power level in reactors that are dominated by diverse feedback effects. Two important feedback effects in thermal reactors are xenon poisoning, which has a time scale for interaction with the flux of the order of hours, and temperature-dependent feedback, with time scales of the order of seconds or, at most, minutes. We shall now illustrate exactly what it is possible to learn of the physics of reactors in which these two feedback mechanisms are important, through a linear stability analysis.

A. Xenon with Prompt-Power Feedback

Xenon-135 is a fission product with a high absorption cross section for thermal neutrons. About 3% is produced directly in fission, while most is produced as a result of a long-lived decay from ^{135}I (cf 1.2, Eq. 31).

In the analysis of the stability of a reactor dominated by xenon feedback it is therefore reasonable, to a first approximation, to lump all temperature feedback effects in a prompt-reactivity constant γ and to consider the delayed neutron effects in an average generation time l^* (5.3, Eq. 32). In the former case, one assumes that, before enough xenon is produced through decay from fission products to materially affect stability, the power generated by fission has time to be completely transferred to the coolant and structural elements. Thus, in this approximation, the power is treated as acting "instantaneously" with respect to xenon production. The delayed neutrons have time decay constants of the order of minutes, while ^{135}I and ^{135}Xe have time decay constants of the order of hours. After fission by a given generation of neutrons, the delayed neutrons can also be considered to be produced "instantaneously" with respect to xenon. Also, for high flux levels ($P > 10^{12}$ neutrons cm^{-2} sec^{-1}), the quantity of xenon produced is much larger than the quantity β of delayed-neutron precursors. We may therefore consider the equations [10]

$$(l^*/\beta)\, dP(t)/dt = [\delta_0 - (\sigma_{\mathrm{Xe}}/\beta c \sigma_{\mathrm{f}})\, \mathrm{Xe}(t) - \gamma P(t)]\, P(t)$$

$$\equiv k(t, P, \mathrm{Xe})\, P(t) \tag{1a}$$

$$d\mathrm{Xe}(t)/dt = y_{\mathrm{Xe}}\sigma_{\mathrm{f}} P(t) - \lambda_{\mathrm{Xe}}\mathrm{Xe}(t) + \lambda_{\mathrm{I}}\mathrm{I}(t) - \sigma_{\mathrm{Xe}}\mathrm{Xe}(t)\, P(t) \tag{1b}$$

$$d\mathrm{I}(t)/dt = y_{\mathrm{I}}\sigma_{\mathrm{f}} P(t) - \lambda_{\mathrm{I}}\mathrm{I}(t) \tag{1c}$$

Note that $\gamma > 0$ defines a negative, or stabilizing, power coefficient.

In the reactivity term, $k(t, P, \mathrm{Xe})$ of Eq. (1a), δ_0 is the constant amount of reactivity necessary to maintain the steady-state in the face of xenon poison and power feedback. The xenon reactivity term, which is proportional to the macroscopic xenon cross section $\sigma_{\mathrm{Xe}}\mathrm{Xe}$, is a function of the power history, while the power feedback term γP is a function of the present instantaneous power. Thus, if we solve (1b) and (1c), we may obtain

$$\mathrm{Xe}(t) = \int_{-\infty}^{t} \exp[A(t') - A(t)] \Big[y_{\mathrm{Xe}}\sigma_{\mathrm{f}} P(t')$$

$$+ \lambda_{\mathrm{I}} \int_{-\infty}^{t'} y_{\mathrm{I}}\sigma_{\mathrm{f}} P(t'')[\exp - \lambda_{\mathrm{I}}(t' - t'')]\, dt'' \Big]\, dt' \tag{2}$$

$$A(t) = \int^{t} [\lambda_{\mathrm{Xe}} + \sigma_{\mathrm{Xe}} P(t')]\, dt'$$

which is obviously a function of the power history. The rest of the terminology in (1) has been defined in the Routh–Hurwitz example above.

Our first step in analyzing the stability of Eqs. (1) is to define the equilibrium states. These are found by setting the time derivatives in (1) equal to zero. The shutdown, or zero-power equilibrium state occurs when $P = \mathrm{Xe} = \mathrm{I} = 0$. Since xenon feedback is most important at high power levels, we shall not consider the shutdown state, but shall consider the operating equilibrium state defined by powers P_0 which are solutions of the equation

$$P_0^2 + \left[\frac{\lambda_{\mathrm{Xe}}}{\sigma_{\mathrm{Xe}}} + \frac{1}{\gamma}\left(\frac{y_{\mathrm{I}} + y_{\mathrm{Xe}}}{\beta c} - \delta_0\right)\right] P_0 - \frac{\lambda_{\mathrm{Xe}}\delta_0}{\gamma\sigma_{\mathrm{Xe}}} = 0 \tag{3}$$

Equation (3) was obtained from (1) by eliminating the steady-state xenon and iodine concentrations

$$\mathrm{Xe}_0 = (y_{\mathrm{I}} + y_{\mathrm{Xe}})\,\sigma_{\mathrm{f}} P_0/(\lambda_{\mathrm{Xe}} + \sigma_{\mathrm{Xe}} P_0), \qquad \mathrm{I}_0 = (y_{\mathrm{I}}\sigma_{\mathrm{f}}/\lambda_{\mathrm{I}})\, P_0 \tag{4}$$

from Eq. (1a). For $\gamma > 0$, only one equilibrium state exists, while two may exist for $\gamma < 0$. The operating equilibrium states are shown as the solid lines in Figure 6.3.1. We may visualize that γ is fixed by the

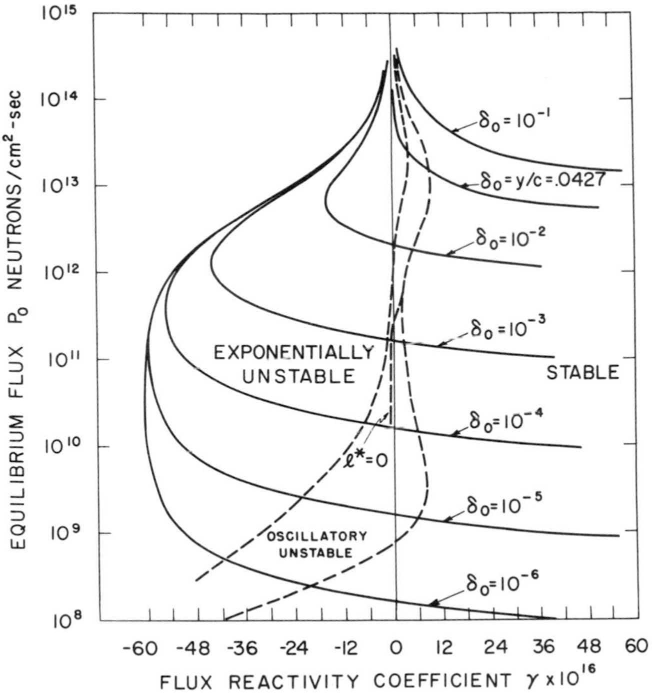

FIGURE 6.3.1. Regions of linear stability; equilibrium flux vs. prompt flux coefficient. Reactivities are measured in percent.

design of the reactor, but that, for a given γ, δ_0 may be varied (by control rod movement, say) to define different equilibrium states P_0.

Our next step is to investigate the stability of these equilibrium states using Routh–Hurwitz conditions. In order to derive the characteristic equation, we must first linearize Eqs. (1) using small departures from equilibrium,

$$P(t) = P_0(1 + p(t)) \tag{5a}$$

$$\mathrm{Xe}(t) = \mathrm{Xe}_0(1 + \delta_{\mathrm{Xe}}(t)) \tag{5b}$$

$$\mathrm{I}(t) = \mathrm{I}_0(1 + \delta_{\mathrm{I}}(t)) \tag{5c}$$

The stability problem we are investigating is such that, for fixed values of the reactor parameters (δ_0, γ), we perturb the power level and see whether or not it will return to its equilibrium value. We substitute (5) into (1) and neglect products of small quantities. The resulting linearized equations can be Laplace-transformed to give a set of equations in the transformed variables, $\bar{p}(s)$, $\overline{\delta\mathrm{Xe}}(s)$, and $\delta\bar{\mathrm{I}}(s)$. In matrix form, the transformed equations are

$$0 = \begin{pmatrix} (l^*/\beta)\, s + \gamma P_0 & \mathrm{Xe}_0(\sigma_{\mathrm{Xe}}/\beta c \sigma_{\mathrm{f}}) & 0 \\ \sigma_{\mathrm{Xe}} P_0 - (y_{\mathrm{Xe}}/y)(\lambda_{\mathrm{Xe}} + \sigma_{\mathrm{Xe}} P_0) & s + \lambda_{\mathrm{Xe}} + \sigma_{\mathrm{Xe}} P_0 & -(y_{\mathrm{I}}/y)(\lambda_{\mathrm{Xe}} + \sigma_{\mathrm{Xe}} P_0) \\ -\lambda_{\mathrm{I}} & 0 & s + \lambda_{\mathrm{I}} \end{pmatrix} \times \begin{pmatrix} \bar{p}(s) \\ \delta\bar{\mathrm{X}}\mathrm{e}(s) \\ \delta\bar{\mathrm{I}}(s) \end{pmatrix} \tag{6}$$

This homogeneous set of linear algebraic equations only has a solution if the determinant of the coefficient matrix is equal to zero. We shall write this determinant in the partially expanded form

$$\frac{l^*}{\beta} s + \gamma P_0 + \frac{\sigma_{\mathrm{Xe}} P_0 / c\beta}{s + \lambda_{\mathrm{Xe}} + \sigma_{\mathrm{Xe}} P_0} \left[\frac{y_{\mathrm{Xe}} \lambda_{\mathrm{Xe}} - y_{\mathrm{I}} \sigma_{\mathrm{Xe}} P_0}{\lambda_{\mathrm{Xe}} + \sigma_{\mathrm{Xe}} P_0} + \frac{y_{\mathrm{I}} \lambda_{\mathrm{I}}}{s + \lambda_{\mathrm{I}}} \right] = 0 \tag{7}$$

Equation (7) is the linearized and transformed form of Eq. (1a). The first term in (7) comes from the time-derivative of the power, the second term is the prompt feedback reactivity, while the last term is the transformed xenon reactivity feedback. It can be seen that $\sigma_{\mathrm{Xe}} P_0$ plays the role of a time constant associated with flux absorption by xenon. When $\sigma_{\mathrm{Xe}} P_0 \gtrsim \lambda_{\mathrm{Xe}}$, or $P_0 \gtrsim \lambda_{\mathrm{Xe}}/\sigma_{\mathrm{Xe}} = 7 \times 10^{12}$ neutrons cm^{-2} sec^{-1}, then it can be expected that flux absorption by xenon will play a significant role. Further, the expression (see Problem 4) [12]

$$K_{\mathrm{p}} \equiv (y_{\mathrm{Xe}} \lambda_{\mathrm{Xe}} - y_{\mathrm{I}} \sigma_{\mathrm{Xe}} P_0)/(\lambda_{\mathrm{Xe}} + \sigma_{\mathrm{Xe}} P_0) = (1/\sigma_{\mathrm{f}} P_0)[y_{\mathrm{Xe}} \sigma_{\mathrm{f}} P_0 - \sigma_{\mathrm{Xe}} \mathrm{Xe}_0 P_0] \tag{8}$$

is proportional to the net rate of prompt production ($K_p > 0$) or destruction ($K_p < 0$) of equilibrium xenon. We have $K_p = 0$ when $P_0 = 3.48 \times 10^{11}$ neutrons cm^{-2} sec^{-1} and is less than zero for higher values of the power (see Table 1). For very high power ($P_0 \gg 3.48 \times 10^{11}$),

TABLE 1

COEFFICIENT OF PROMPT XENON PRODUCTION VS. POWER

K_p	3.0×10^{-3} ($=y_{Xe}$)	2.91×10^{-3}	2.11×10^{-3}	0	-4.9×10^{-3}
P_0	0	10^{10}	10^{11}	3.48×10^{11}	10^{12}
K_p	-1.11×10^{-2}	-3.43×10^{-2}	-5.6×10^{-2}	-6.0×10^{-2} ($=-y_I$)	
P_0	2.0×10^{12}	10^{13}	10^{14}	∞	

prompt xenon production ($y_{Xe}\lambda_{Xe}$) may be effectively neglected in stability analyses.

Expanding Eq. (7) gives a cubic equation in the transform variable s which is the characteristic equation (Sec. 6.2) for the operating equilibrium states of Eq. (1):

$$a_0 s^3 + a_1 s^2 + a_2 s + a_3 = 0$$

with

$$a_0 = l^*/\beta, \qquad a_1 = \gamma P_0 + (l^*/\beta)(\lambda_I + \lambda_{Xe} + \sigma_{Xe} P_0)$$

$$a_2 = (\sigma_{Xe} P_0 / c\beta)\, K_p + (l^*/\beta)\,\lambda_I(\lambda_{Xe} + \sigma_{Xe} P_0) + \gamma P_0(\lambda_I + \lambda_{Xe} + \sigma_{Xe} P_0) \quad (9)$$

$$a_3 = \lambda_I \gamma P_0(\lambda_{Xe} + \sigma_{Xe} P_0) + (\sigma_{Xe} P_0 / c\beta)\,\lambda_I(y_I + K_p)$$

Application of the Routh–Hurwitz conditions gives

$$a_0 > 0, \qquad a_1 > 0, \qquad a_3 > 0 \quad (10a)$$

$$a_1 a_2 - a_0 a_3 > 0 \quad (10b)$$

For $\gamma > 0$, we can see that $a_0 > 0$ and $a_1 > 0$ are automatically satisfied. Since $-y_I \leqslant K_p \leqslant y_{Xe}$ (Table 1), then $a_3 > 0$ is satisfied (i.e., $y_I + K_p \geqslant 0$ always). It can also be shown (see Problem 5) that Eq. (10b) will be equal to zero before a_2 becomes less than zero. Thus, for a given P_0, we find $\gamma > 0$ to satisfy the equality in (10b). This gives the stability curve in Figure 6.3.1. This curve separates regions of oscillatory stability from regions of oscillatory instability. It is a characteristic of the xenon problem, as well as a characteristic of most reactor-kinetic stability problems, that the first instabilities encountered as the power level is raised are oscillatory rather than pure exponential.

This example, when Eq. (10b) equals zero to give oscillatory instability, is a special case of a more general result, shown by Duncan [7], that if the system is initially stable and some parameter (e.g., power level) is varied so as to approach instability, the onset of oscillatory instability is shown first by the next to highest Hurwitz determinant (see 6.2, Eq. 3).

The instability curve in Figure 6.3.1 crosses the $\gamma = 0$ line at $P_0 = 8.64 \times 10^8$ neutrons $\text{cm}^{-2}\ \text{sec}^{-1}$, which agrees with the result of Sec. 6.2.

Finally, we have plotted in Figure 6.3.1 a stability curve obtained from Eq. (7) with $l^* = 0$. Since l^* includes the delayed neutrons in this approximation, $l^* = 0$ means we have neglected the delayed neutrons. We thus can see that the delayed neutrons have little effect for $P_0 \gtrsim 10^{12}$ neutrons $\text{cm}^{-2}\ \text{sec}^{-1}$, but for low power levels, they exert a destabilizing effect. We have shown in Sec. 4.6 that delayed neutrons are destabilizing for forced oscillations of a reactor without feedback. The present example shows the destabilizing effect of delayed neutrons in an autonomous system with feedback.

Let us also note in this connection that setting $l^* = 0$ is equivalent to assuming the prompt-jump approximation in Eq. (1a). As already mentioned, the problem treated here is different from that in Sec. 4.2 in that we do not introduce reactivity into the system at $t = 0$, but, rather, perturb the power. Since we also neglect delayed neutrons, using the prompt-jump approximation in the present context requires that

$$(l/\beta)\ |\ \dot{P}/P\ | \ll |\ k(t)| \tag{11}$$

but does not place the restriction on the reactivity that $|\ k\ | < 1$.

B. Effect of Temperature Time Delay

We shall now illustrate the effect on stability of allowing the temperature feedback to act after a time delay [10], rather than instantaneously. To accomplish this, we replace the $-\gamma P(t)$ term in Eq. (1a) with a term $-\alpha T(t)$, and assume that the temperature is related to the power through a Newton's law of cooling model. Thus, Eqs. (1b) and (1c) remain unaffected, but Eq. (1a) is replaced by the two equations

$$(l^*/\beta)\, dP(t)/dt = [\delta_0 - (\sigma_{\text{Xe}}/\beta c \sigma_{\text{f}})\, \text{Xe}(t) - \alpha T(t)]\, P(t) \tag{12a}$$

$$dT(t)/dt = \lambda_{\text{T}}[(\gamma/\alpha)\, P(t) - T(t)] \tag{12b}$$

The "time delay constant" λ_T has been chosen in this specific form so that, in the limit $\lambda_T \to \infty$, we return to the instantaneous feedback model considered in the previous section. Also, since at equilibrium we have from (12a)

$$\gamma P_0 = \alpha T_0 \tag{13}$$

then the equilibrium states found in the previous section remain unaltered by the introduction of the temperature time delay. After linearizing and transforming (12b), we have

$$\delta \bar{T}(s) = \lambda_T \bar{p}(s)/(s + \lambda_T) \tag{14}$$

Linearizing and transforming (12a) then leads to

$$(l^*/\beta)\, s\bar{p}(s) = -(\sigma_{Xe}/\beta c\sigma_f)\, \mathrm{Xe}_0\, \delta\bar{\mathrm{X}}\mathrm{e}(s) - \alpha T_0\, \delta \bar{T}(s) \tag{15}$$

Using (13) and (14), the last term in (15) becomes

$$-\gamma P_0[\lambda_T/(s + \lambda_T)] \tag{16}$$

Thus, in the presence of a single temperature time delay, the characteristic equation (7) becomes

$$\begin{aligned}(l^*/\beta)\, s + \gamma P_0[\lambda_T/(s + \lambda_T)] + [(\sigma_{Xe} P_0/\beta c)/(s + \lambda_{Xe} + \sigma_{Xe} P_0)] \\ \times \{[(y_{Xe}\lambda_{Xe} - y_I \sigma_{Xe} P_0)/(\lambda_{Xe} + \sigma_{Xe} P_0)] + [y_I \lambda_I/(s + \lambda_I)]\} = 0\end{aligned} \tag{17}$$

Expansion of (17) gives a quartic in s:

$$a_0 s^4 + a_1 s^3 + a_2 s^2 + a_3 s + a_4 = 0 \tag{18}$$

and application of the Routh–Hurwitz criterion gives

$$a_0 > 0, \qquad a_1 > 0, \qquad a_3 > 0, \qquad a_4 > 0 \tag{19a}$$

$$a_3(a_1 a_2 - a_0 a_3) - a_4 a_1{}^2 > 0 \tag{19b}$$

For a given λ_T and P_0, we find γ to satisfy the equality in (19b). Following this procedure for several values of the variables leads to the solid curve shown in Figure 6.3.2. Also shown in Figure 6.3.2 are curves computed with one average group of delayed neutrons. In this case, with the prompt lifetime l equal to zero, we replace $l^* s/\beta$ in Eq. (17) with

$$s/(s + \lambda) \tag{20}$$

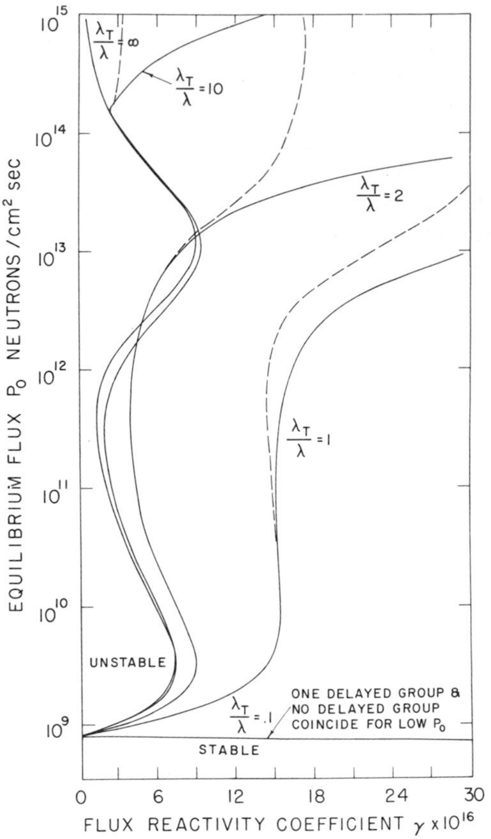

FIGURE 6.3.2. Regions of linear stability; Newton's law of cooling.

This still leads to a quartic of the form (18), only with different coefficients, and the result is used to construct the dotted curves shown in Figure 6.3.2.

The results shown in Figure 6.3.2 are not unexpected. They show that a time delay in a single-temperature feedback can have a destabilizing influence on a reactor dominated by xenon feedback. The longer the time delay, the more unstable the system. Also, the delayed neutrons exert a stabilizing influence at high power levels.

Since it is known [11] that the single-temperature system composed of Eq. (12) with Xe = 0 is always stable (for $\alpha > 0$, $\lambda_T > 0$), then the present analysis shows that the presence of xenon [Xe $\neq 0$ in Eq. (12)] has a destabilizing influence on a reactor dominated by temperature feedback.

6.4. Pontryagin's Stability Criteria for Transcendental Equations

The Routh–Hurwitz criterion is applicable only when the characteristic equation is a polynomial in s. This feature restricts its application to physical systems described by a set of linear equations with constant coefficients and no time delays. Most feedback models, however, have distributed parameters and are described by partial differential equations. Their transfer functions are nonrational functions of the complex variable s, and consequently, in the characteristic equation describing the linearized response of these reactor systems, are transendental. The test which we present here is applicable to those reactor systems whose characteristic equation can be expressed in the following general form:

$$Q(s) = h(z, e^z) = 0 \tag{1}$$

where z is related to s by a change of variable.

The location in the complex plane of the roots of an equation of this type has been investigated by Pontryagin ([13], see also Bellmann and Cooke [14]). We shall present his results without proof.

We assume that $h(z, e^z)$ is a polynominal of its arguments with real coefficients as

$$h(z, e^z) = \sum_{m,n} a_{mn} z^m e^{nz} \tag{2}$$

Definition 1. The term $a_{rt} z^r e^{tz}$ is called the principal term if, for each other term in Eq. (2), we have either $r > m$, $t > n$; $r = m$, $t > n$; or $r > m$, $t = n$. Clearly, not every polynomial h has a principle term; e.g., $h = z + e^z$.

A. Necessary Conditions for Stability

Theorem 1. If the polynomial h has no principal term, then $h(z, e^z) = 0$ has an unbounded number of zeros with arbitrarily large positive real part, and hence the system is unstable (Problem 6).

Theorem 2. Let the polynomial h have a principal term and define the real functions $F(y)$ and $G(y)$ by

$$h(iy, e^{iy}) \equiv F(y) + iG(y) \tag{3}$$

where y is a real number.

If all the zeros of the function $h(z, e^z)$ lie to the left of the imaginary axis, then the zeros of the functions $F(y)$ and $G(y)$ are real, alternating,† and, furthermore,

$$G'(y)F(y) - G(y)F'(y) > 0 \tag{4}$$

for each y.

It follows that the system associated with $h(z, e^z)$ is unstable if one of the above conditions is violated, e.g., if F or G has complex roots.

B. Sufficient Conditions for Stability

THEOREM 3. All the roots of the characteristic equation $h(s, e^s) = 0$ have negative real parts if h has a principal term, and if, in addition, one of the following conditions is satisfied:

(a) All the zeros of the functions $F(y)$ and $G(y)$ are real, alternating, and the inequality

$$G'(y)F(y) - F'(y)G(y) > 0 \tag{5}$$

holds at least for one value of y.

(b) All zeros y_0 of $F(y)$ are real and the inequality

$$F'(y_0)G(y_0) < 0 \tag{6}$$

holds for every zero y_0.

(c) All zeros y_0 of $G(y)$ are real and the inequality

$$G'(y_0)F(y_0) > 0 \tag{7}$$

holds for every zero y_0.

The application of the stability criteria of (b) and (c) requires an investigation of the zeros of functions $F(y)$ and $G(y)$ defined by Eq. (3). We can easily show from Eqs. (2) and (3) that these functions are of the form of $F(y) = f(y, \cos y, \sin y)$ and $G(y) = g(y, \cos y, \sin y)$, where $f(y, u, v)$ and $g(y, u, v)$ are polynomials with real coefficients. Indeed, from Eq. (2), we obtain for $F(y)$ [with a similar expression for $G(y)$] the following expansion

$$F(y) = \sum_{m,n} y^m \phi_m^{(n)}(\cos y, \sin y) \tag{8}$$

† The zeros of the function $F(y)$ and $G(y)$ "alternate" if (a) the zeros of these functions are simple, (b) they have no common zeros, and (c) between any two zeros of one of these functions there exists at least one zero of the other.

where

$$\phi_m^{(n)}(u, v) \equiv a_{mn} \operatorname{Re}[(i)^m (u + iv)^n] \tag{9}$$

which is a homogeneous polynomial of degree n with respect to the variables u and v, i.e., for any $\lambda \neq 0$

$$\phi_m^{(n)}(\lambda u, \lambda v) = \lambda^n \phi_m^{(n)}(u, v) \tag{10}$$

Our problem is therefore reduced to the investigations of the zeros of a function of $F(z) = f(z, \cos z, \sin z)$, where

$$f(z, u, v) \equiv \sum_{m,n} z^m \phi_m^{(n)}(u, v) \tag{11}$$

We define the "principal term" in the polynomial $f(z, u, v)$ as the term $z^r \phi_r^{(t)}(u, v)$ for which r and t simultaneously attain maximum values, i.e., $(r > m,\ t > n)$, or $(r = m,\ t > n)$, or $(r > m,\ t = n)$ for any other term $z^m \phi_m^{(n)}(u, v)$ in the expansion (11). We may again note that not every polynomial of the form of (11) has a principal term.

We can see from (9) that, if $h(s, e^s)$ has a principal term $a_{rt} s^r e^{ts}$, then $f(z, u, v)$ also has a principal term, which is $z^r \phi_r^{(t)}(u, v)$.

Let $\phi_*^{(t)}(u, v)$ denote the coefficient of z^r in (11), i.e.,

$$f(z, u, v) = z^r \phi_*^{(t)}(u, v) + \cdots \tag{12a}$$

where

$$\phi_*^{(t)}(u, v) \equiv \sum_{n \leq t} \phi_r^{(n)}(u, v) \tag{12b}$$

Clearly $\phi_*^{(t)}(u, v)$ is a polynomial of degree n with respect to u and v which is not homogeneous in general. Furthermore, $\phi_*^{(t)}(z) \equiv \phi_*^{(t)}(\cos z, \sin z)$ is a periodic function of z, and has exactly $2t$ zeros in each strip $0 \leqslant x < 2\pi$ $(z = x + iy)$ in the complex z plane. This can be seen by verifying that $\phi_*^{(t)}(z)$ is actually a polynomial of degree t in e^{iz}, and for a given one of the t roots of this polynomial, there are two values of z in $0 \leqslant x < 2\pi$.

Let ϵ be an arbitrary real number such that

$$\phi_*^{(t)}(\epsilon + iy) \neq 0 \tag{13}$$

for all y. Clearly, ϵ is any real number different from the real part of the roots of $\phi_*^{(t)}(z) = 0$.

We are now in a position to state the following theorem.

THEOREM 4. There exists a positive integer K such that the function $F(z)$ has exactly $4kt + r$ zeros (complex or real) in any strip $-2k\pi + \epsilon \leqslant \text{Re } z \leqslant 2k\pi + \epsilon$ in the complex z plane for all integers $k \geqslant K$.

COROLLARY. It follows from this theorem that a necessary and sufficient condition for the function $F(x)$ to have only real roots is that there must exist a positive integer K such that $F(x)$ has exactly $4kt + r$ real roots in the interval $-2k\pi + \epsilon \leqslant x \leqslant 2k\pi + \epsilon$ for all integers $k \geqslant K$, because then the real roots account for all the roots in the intervals under consideration.

This corollary is important because it provides a necessary and sufficient condition for all the zeros of $F(z)$ to be real. If we can show that the number of real roots is never equal to $4kt + r$ in any interval $-2k\pi + \epsilon \leqslant x \leqslant 2k\pi + \epsilon$ for any choice of ϵ and k, then $F(z)$ has nonreal roots and, by virtue of Theorem 2, the system associated with $h(z, e^z)$, which generates the functions $F(z)$ and $G(z)$, is unstable (Problem 7).

EXAMPLE [15]. We shall illustrate the application of the Pontryagin's stability criteria by considering a reactor model described by

$$\dot{p}(t) = -[A\,\Delta T(t-\tau) + Bp(t)][P_0 + p(t)] \tag{14a}$$

$$\Delta\dot{T}(t) = \epsilon[p(t) - \Delta T(t)] \tag{14b}$$

in which the delayed neutrons are neglected. This model contains a single-temperature feedback with a convection time delay τ, and a power feedback which may be interpreted as the contribution of all other fast-acting temperature feedback effects in the quasistatic approximation (cf Sec. 5.3).

The characteristic equation describing the linear stability of this reactor can be obtained as

$$Q(s) = [z^2 + pz + q]\, e^z + r \tag{15a}$$

where

$$p = [\epsilon + BP_0]\,\tau \tag{15b}$$

$$q = \tau^2 \epsilon BP_0 \tag{15c}$$

$$r = \tau^2 A\epsilon P_0/l \tag{15d}$$

$$z = \tau s \tag{15e}$$

This equation is transcendental, and a special case of (2). The principal term is $z^2 e^z$, so that $r = 2$ and $t = 1$. The functions $F(\omega)$ and $G(\omega)$ defined by (3) are

$$F(\omega) \equiv (q - \omega^2)(\cos \omega) - p\omega(\sin \omega) + r \tag{16}$$

$$G(\omega) \equiv (q - \omega^2)(\sin \omega) + p\omega(\cos \omega) \tag{17}$$

In order to investigate the stability of this reactor, we shall use the third stability criterion (c) discussed above. We first need to determine conditions under which all zeros of $G(z)$ are real. From (17) and (11), we obtain

$$f(z, u, v) = pzv + qu - z^2 u \tag{18}$$

The principal term in this polynomial is $z^2 u$, which corresponds to $r = 2, t = 1$, as it should. The function $\phi_*^{(l)}(u, v)$, which is the coefficient of z^2 (cf Eq. 12b), is

$$\phi_*^{(1)}(u, v) \equiv -u \tag{19}$$

Thus, $\phi_*^{(1)}(z) = -\sin z$ has exactly two zeros, i.e., $z = 0$ and $z = \pi$, in the interval $0 \leqslant x < 2\pi$. The number ϵ must be so chosen that

$$\phi_*^{(1)}(\epsilon + iy) = -\sin(\epsilon + iy) \neq 0$$

for all y. Since $\sin(\epsilon + iy) = (\sin \epsilon)(\cosh y) + i(\cos \epsilon) \sinh y$, any value of $\epsilon \neq 2\pi n$ $(n = 0, \pm 1, \ldots)$ can be chosen because $\cosh y$ is never zero. We shall choose $\epsilon = \pi/2$.

Since $r = 2$ and $t = 1$, it is sufficient to show that $G(\omega)$ has, for sufficiently large k, exactly $4k + 2$ real zeros in the interval $-2k\pi + (\pi/2) \leqslant \omega \leqslant 2k\pi + (\pi/2)$.

It is clear from (17) that $G(\omega)$ has a zero at $\omega = 0$. Bearing this in mind, we divide (17) by $\omega \sin \omega$, and obtain

$$\cot \omega = (\omega^2 - q)/\omega p \tag{20}$$

We shall first consider the case in which $p > 0$, $q \geqslant 0$ (or equivalently, $B > 0$ for ϵ and l always positive). The real zeros are shown in Figure 6.4.1. We see that the interval $-3\pi/2 \leqslant \omega \leqslant 5\pi/2$ contains exactly 6 roots, which is $4k + 2$ when $k = 1$. It is important to note that this equality holds for any value of $k \geqslant 1$. For example, the number of zeros in the interval corresponding to $k = 2$ is (from Figure 6.4.1) 10, which is just $4 \times 2 + 2$. Hence, in this particular example, the sufficiently large value of k is one. One can see from Figure 6.4.1 that there

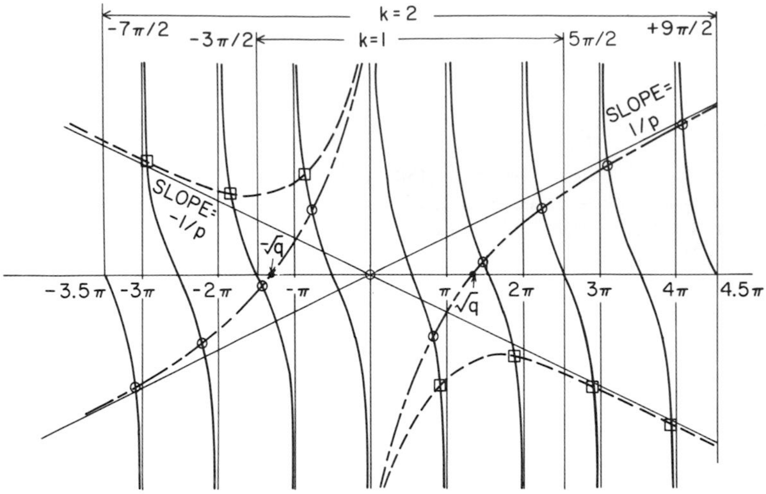

FIGURE 6.4.1. Real roots of $\cot \omega = (\omega^2 - q)/\omega p$ for $p > 0$, $q \geqslant 0$. (— - —) $(\omega^2 - q)/\omega p$, $p > 0$, $q > 0$; (——) $\cot \omega$; (- - -) $(\omega^2 - q)/\omega p$, $p < 0$, $q < 0$.

would be only 5 roots in the first interval ($k = 1$) had $q^{1/2}$ been greater than $3\pi/2$. In this case, k would have to be chosen as $k \geqslant 2$.

Having ascertained that $G(\omega)$ has only real roots when $p > 0$ and $q \geqslant 0$, we must now determine the conditions under which the inequality (7) holds for every zero of $G(\omega)$. A simple computation of $G'(\omega_0)\,F(\omega_0)$ from (16) and (17) shows that

$$G'(0)\,F(0) = (q + r)(q + p) \tag{21}$$

while, for $\omega_0 \neq 0$,

$$G'(\omega_0)\,F(\omega_0) = \{1 - [r(\sin \omega_0)/p\omega_0]\}[(\omega_0{}^2 - q)^2 + \omega_0{}^2(p^2 + p) + pq] \tag{22}$$

Since we are discussing the case of $p > 0$ and $q \geqslant 0$, we see that Eq. (21) is satisfied if

$$q + r \geqslant 0 \tag{23}$$

On the other hand, (22) holds for all $\omega_0 \neq 0$ if

$$1 > (r \sin \omega_0)/p\omega_0$$

Since $|(\sin \omega)/\omega\,|$ is a monotonically decreasing function of ω, this inequality is satisfied for every root if, for $r > 0$,

$$1 > (r \sin \omega_0)/p\omega_0 \tag{24a}$$

is true for the root in $(0, \pi)$, and if, for $r < 0$,

$$1 > (r \sin \omega_0)/p\omega_0 \tag{24b}$$

is true for the root in $(\pi, 2\pi)$.

We can now summarize the sufficient conditions for linear stability of the reactor model described by (14) as $B > 0$, $A + B > 0$,

$$(\tau A\epsilon \sin \omega_0)/(\epsilon + B)\, \omega_0 < 1 \qquad \begin{array}{ll} \text{for} \quad A > 0, & 0 < \omega_0 < \pi \\ \text{for} \quad A < 0, & \pi < \omega_0 < 2\pi \end{array}$$

The last condition is *a fortiori* satisfied if $\tau \mid A \mid \epsilon < \epsilon + B$.

As an illustration of a case in which one cannot find a finite integer k such that the number of real roots is exactly equal to $4k + 2$ in $-2\pi k + (\pi/2) \leqslant \omega \leqslant 2\pi k + (\pi/2)$, we now assume that $p < 0$ and hence $q < 0$. This condition implies a negative prompt power coefficient γ and a power level $P_0 > \epsilon/\mid B \mid$. One can see from Figure 6.4.1 that the roots (denoted by squares) in this case are located as follows: one root at the origin, $2k$ roots in $(0, 2k\pi)$, $2k$ roots in $(-2k\pi, 0)$, zero roots in $(2\pi k, 2\pi k + \frac{1}{2}\pi)$, one root in $(-2k\pi, -2k\pi + \frac{1}{2}\pi)$. Thus, $G(z)$ has only $4k$ roots in $-2k\pi + (\pi/2) \leqslant \omega \leqslant k\pi + (\pi/2)$ for all k, and hence the number of roots in this interval can never be made equal to $4k + 2$ by adjusting the value of ϵ, and increasing k. We therefore conclude that $G(z)$ must have complex roots in addition to the real roots, violating the necessary condition stated in Theorem 2, and thus implying instability.

In order to complete the stability analysis of the reactor model described by (14), we must also consider the case for which $p > 0$, a necessary condition for stability, and $q < 0$. When $q < -p$, which is equivalent to $P_0 > \epsilon/\mid B \mid(1 + \epsilon\tau)$, the reactor is unstable. When $q < 0$ but $p + q > 0$, then the reactor is stable if

$$[(\omega_0^2 - q)^2 + \omega_0^2(p^2 + p) + pq]\{1 - [r(\sin \omega_0)/p\omega_0]\} > 0 \tag{25}$$

holds for all $\omega_0 \neq 0$ and if $q + r > 0$ as well.

In summary, the region of linear stability of this reactor model in the plane of A vs. B for given values of ϵ and τ is determined by means of the Pontryagin criteria in the following manner. We first note that we need only consider $p > 0$, $q > 0$ (case I) and $p > 0$, $q < 0$ but $p + q > 0$ (case II), for only in these two cases are all the roots of $G(\omega) = 0$ real. We have $q + r > 0$ as a necessary condition for $F(0)\, G'(0) > 0$. We now determine, for any chosen value of B [and hence for p and q from (15)], the root ω_0 of (20) that lies in $(0, \pi)$. Then, we determine the value of r, and hence A by (15d), from (24) in case I, and from (25) in case II. The above procedure was used in computing the boundary curves shown in Figure 6.4.2 for $\epsilon = 1$ and various τ. See Problems 8 and 9 for other examples.

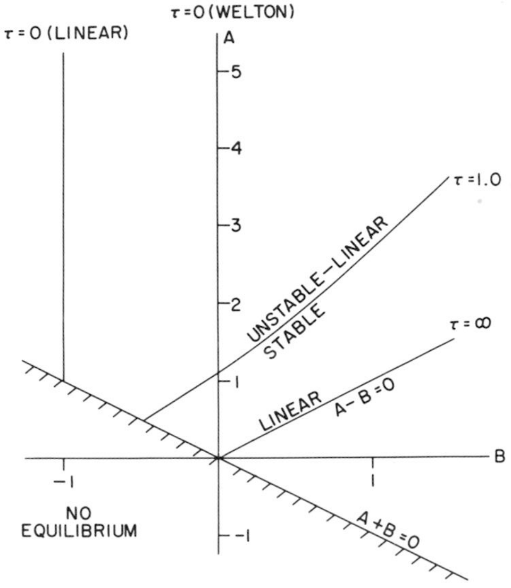

FIGURE 6.4.2. Regions of stability for a model with a convection time delay.

6.5. Nyquist Stability Criterion

In addition to the algebraic stability criteria presented in the preceding sections, there are also graphic stability tests which are widely used in the stability analysis of linear systems. These are the root-locus methods, the Bode plot, and the Nyquist stability criterion [6, 16]. We shall present here only the Nyquist criterion as a representive of graphic tests, because of its wide usage in reactor applications.

The Nyquist criterion is especially applicable to systems whose frequency response is available only graphically, and to systems whose characteristic equations are transcendental. The derivation of the Nyquist criterion is based on Cauchy's residue theorem.

Let $Q(s)$ be an analytic function of s, whose only singularities in any region of the complex s plane are poles. Let s_0 be a zero of $Q(s)$

with a multiplicity of n. Then, the Taylor series expansion about s_0 is

$$Q(s) = a_n(s - s_0)^n + a_{n+1}(s - s_0)^{n+1} + \cdots \tag{1}$$

where the a_n are constants. We want to show that the ratio $Q'(s)/Q(s)$ has a simple pole at $s = s_0$ with a residue equal to n. Differentiating $Q(s)$ with respect to s, we obtain

$$Q'(s) = a_n n(s - s_0)^{n-1} + a_{n+1}(n + 1)(s - s_0)^n + \cdots \tag{2}$$

and dividing the series in (1) and (2), we find

$$Q'(s)/Q(s) = [n/(s - s_0)] + c_0 + c_1(s - s_0) + \cdots \tag{3}$$

where c_0, c_1,... are some new constants depending on a_n. The first term in this expansion represents a simple pole with a residue which is equal to n.

We can show in a similar way, by replacing n by $-m$, that, if $Q(s)$ has a pole of multiplicity m at a point s_p, then $Q'(s)/Q(s)$ has again a simple pole at the same point with a residue equal to $-m$.

Another concept needed in the Nyquist criterion is the conformal mapping of contours in the complex plane. Let us assume that s is varied continuously on a closed path Γ, on which $Q(s)$ is analytic, as shown in Figure 6.5.1. It is clear that, for each value of s, there is a

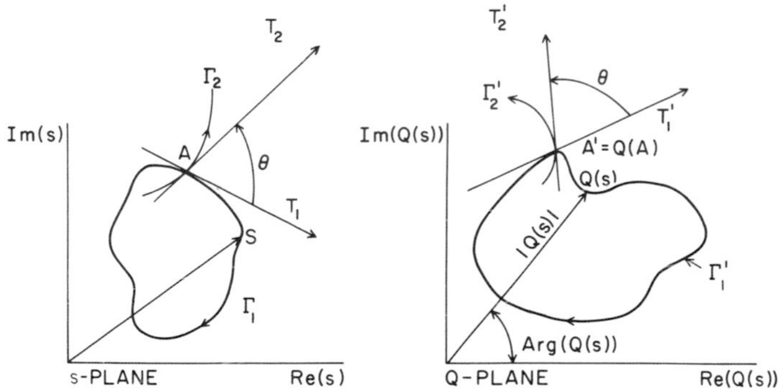

FIGURE 6.5.1. A mapping of the path Γ into the path Γ' by means of the analytic function $Q(s)$. Here, T_2 and T_1 are the tangents to Γ_2 and Γ_1, respectively, at A, and T_2' and T_1' are the tangents to Γ_2' and Γ_1', respectivity, at A'.

unique value of $Q(s)$, and thus a closed contour in the s plane is mapped into a closed path in the $Q(s)$ plane. The mapping (or transformation) induced by an analytic function is called a conformal mapping from the s plane into the $Q(s)$ plane. One can show that [6] the conformal mapping preserves the angles between the tangents of two intersecting curves (Γ_1 and Γ_2 in Figure 6.5.1 with tangents T_1 and T_2 at A), and the sense of rotation of the tangents ($T_1 \rightarrow T_2$ and $T_1' \rightarrow T_2'$).

We are now ready to derive the Nyquist criterion. Consider the following line integration on a closed path Γ on which $Q(s)$ is analytic:

$$\oint_{\Gamma} \{d[\ln Q(s)]/ds\}\, ds = \oint_{\Gamma} (\{d[\ln |Q(s)|]/ds\}\, ds + i\{d[\arg Q(s)/ds\}\, ds)$$

$$= \oint_{\Gamma} [Q'(s)/Q(s)]\, ds \tag{4}$$

Since $|Q(s)|$ has the same value at the beginning as at the end of the closed contour Γ', the integral $\int_{\Gamma} d(\ln |Q(s)|) = 0$. The integral of the second term in (4) is $i\, \Delta[\arg Q(s)]$, which is the net change in the angle of $Q(s)$ in the $Q(s)$ plane at the beginning and at the end when s describes the closed path Γ. Clearly, $\Delta[\arg Q(s)] = 0$ if the path Γ' does not encircle the origin, and is equal to 2π or -2π if Γ' encircles the origin in a clockwise or counterclockwise sense, respectively. We may conclude that $(1/2\pi)\, \Delta[\arg Q(s)]$ is the net number of encirclements of the origin when s describes a closed contour Γ.

The right-hand side of (4) is equal to $2\pi i$ times the sum of residues of the poles of $Q'(s)/Q(s)$ inside the closed contour Γ (Cauchy's residue theorem). But we have shown that the poles of $Q'(s)/Q(s)$ are all simple, and coincide with the zeros and poles of $Q(s)$. Hence, we establish the important relation

$$(1/2\pi)\, \Delta[\arg Q(s)] = Z - P \tag{5}$$

where Z is the sum of multiplicities of all the zeros, and P is the sum of multiplicities of all the poles of $Q(s)$ inside the closed contour Γ. This conclusion is stated as follows:

When s describes a closed contour in a given sense in the s plane, the corresponding closed contour in the $Q(s)$ plane encircles the origin $(Z - P)$ times in the same sense.

Application to Linear Stability Analysis

The above conclusion may be employed to ascertain whether the characteristic equation $Q(s) = 0$ has any zeros in the right half-plane. The contour enclosing the right half-plane may be chosen as the imaginary axis from $-iR$ to $+iR$ and a right semicircular arc (see Figure 6.52). as R approaches infinity. However, in feedback systems, it often happens that $Q(s)$ has poles on the imaginary axis in the s plane, usually at the origin. Therefore, the semicircular contour is chosen to avoid such points as illustrated in Figure 6.5.2. The transformation of this contour

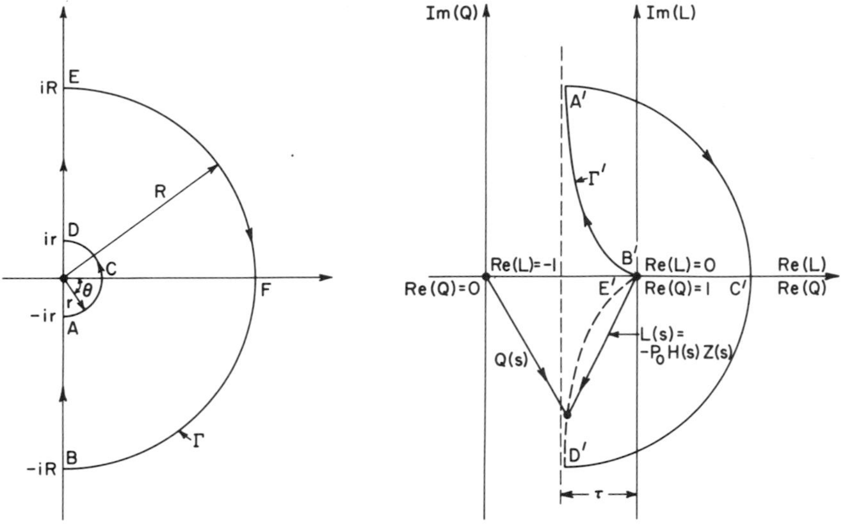

FIGURE 6.5.2. A semicircular contour enclosing the right half-plane.

into the $Q(s)$ plane is accomplished as follows. First, we observe that the product $H(s)\ Z(s)$ in $Q(s) \equiv 1 - P_0H(s)\ Z(s)$ behaves as $(1/s^n)$ for large values of s when $n \geqslant 1$. Then, $Q(s)$ maps into $+1$ along the large semicircular arc when $R \to \infty$. Secondly, in the neighborhood of the pole at the origin, $Q(s)$ can be approximated by

$$Q(s) \approx A/s^m \tag{6a}$$

where m is the order of the pole, and A is a constant. On the small semicircular arc, $s = re^{i\theta}$, where θ varies from $-\pi/2$ to $\pi/2$ when the point s goes from A to D counterclockwise, and hence

$$Q(s) = (A/r^m)\ e^{-i\theta m} \tag{6b}$$

Thus, a small semicircular arc around a pole of order m in the s-plane maps into a circular path of infinitely large radius over an angle $-\pi m$ in the Q plane. In other words, when the point s traverses a semicircular arc in a counterclockwise sense in the s plane, the image point $Q(s)$ in the Q plane undergoes m half rotations on a large circle in a clockwise sense.

It follows that it is sufficient to evaluate $Q(s)$ only on the imaginary axis and to plot $Q(i\omega)$ when ω is varied from $-\infty$ to $+\infty$. Since $Q(-i\omega)$ is the complex conjugate of $Q(i\omega)$, only one-half of the contour Γ' need be examined; the other half of the path is the mirror image of the first half with respect to the real axis. Figure 6.5.2 depicts the transformation of the closed contour Γ in the s plane in a specific case in which $Q(s)$ is given by

$$Q(s) = 1 + [1/s(1 + s\tau)]$$

Note that $Q(s)$ has a simple pole ($m = 1$) at the origin in the s plane, and thus the small semicircular arc maps into the large semicircular arc in the Q plane.

In Figure 6.5.2, we have introduced

$$L(s) = -P_0 H(s)\, Z(s) \tag{7}$$

which is called the open-loop transfer function. The characteristic function in terms of $L(s)$ is $Q(s) = 1 + L(s)$. Since an encirclement of the origin by $1 + L(s)$ is equivalent to an encirclement of the -1 point by $L(s)$, it is more convient to work with $L(s)$ and plot $L(i\omega)$ in the L plane. In the latter, the critical point is the -1 point rather than the origin, as illustrated in Figure 6.5.2.

In order to complete the derivation of the Nyquist criterion, we have also to consider the pole of $L(s) = P_0 H(s)\, Z(s)$. Since both $Z(s)$ and $H(s)$ have no poles on the right half-plane from physical considerations (cf Sec. 5.3), an encirclement by $L(s)$ of the -1 point can only be due to the zeros of $Q(s)$ in the right half-plane. We can now state the Nyquist stability criterion as follows:

If we can ascertain that the open-loop frequency response $L(i\omega)$ does not encircle the -1 point in the L plane when ω is varied from $-\infty$ to $+\infty$, then $Q(s) = 0$ cannot have roots with positive real parts, and the linear system described by the open-loop transfer function $-P_0 H(s)\, Z(s)$ is stable.

Example. We shall illustrate the application of the Nyquist stability criterion by considering a reactor model characterized by two-temperature feedback. This model may represent a reactor with two regions, e.g., fuel and moderator, with a Newton's law of heat removal in each. After the transformation described in Chap. 5 to diagonalize the temperature equations, we obtain the feedback transfer function as

$$H(s) = [H_{\mathrm{f}}/(1 + s\tau_{\mathrm{f}})] + [H_{\mathrm{S}}/(1 + s\tau_{\mathrm{S}})] \tag{8}$$

where H_f and H_S are the parameters related to the temperature coefficients of reactivity, and τ_f and τ_S are related to the time constants (cf Sec. 5.3). The parameters H_f and H_S may be positive or negative, whereas τ_f and τ_S, by virtue of their physical meaning, are always positive.

We shall use a one-group model of delayed neutrons to approximate the zero-power transfer function as

$$P_0 Z(s) = K(1 + s\tau_1)/s(1 + s\tau_2) \tag{9}$$

where

$$K = P_0\bar{\lambda}/[(\beta/l) + \bar{\lambda}](l/\beta) \tag{10a}$$

$$\tau_1 = 1/\bar{\lambda} \tag{10b}$$

$$\tau_2 = [(\beta/l) + \bar{\lambda}]^{-1} \tag{10c}$$

$$1/\bar{\lambda} = \sum_i a_i/\lambda_i \tag{10d}$$

The open-loop transfer function $L(s) = -P_0 H(s)\, Z(s)$ is obtained from (8) and (9) as

$$L(s) = -K(H_f + H_S)(1 + s\tau_0)(1 + s\tau_1)/s(1 + s\tau_2)(1 + s\tau_f)(1 + s\tau_S) \tag{11a}$$

where

$$\tau_0 = (H_f\tau_S + H_S\tau_f)/(H_f + H_S) \tag{11b}$$

We note that τ_0 may be positive or negative depending on the signs of H_f and H_S. In order to apply the Nyquist stability criterion to this reactor model, we must first plot the locus of $L(i\omega)$ when ω is varied from $-\infty$ to $+\infty$. The possible loci of $L(i\omega)$ were investigated by Miida and Suda [17], and the conditions for stability were determined in terms of feedback and nuclear parameters. We shall essentially reproduce their results in the sequel with a slightly different approach.

We first observe that the power coefficient of reactivity in this feedback model is $H(0) = H_f + H_S$, and hence a necessary condition for stability is

$$H_f + H_S < 0 \tag{12}$$

If (12) is not satisfied, the reactor is always unstable, as explained previously (cf Sect. 6.1). In order to determine necessary and sufficient conditions, we have to examine the locus of $L(i\omega)$ when $H_f + H_S < 0$.

Since $L(s)$ has a simple pole at $s = 0$, the locus of $L(i\omega)$ will describe the large semicircular arc in a clockwise sense when ω goes from 0^- to 0^+. For small ω, $L(i\omega)$ approaches a vertical line asymptotically, because

$$L(i\omega) \rightarrow K \mid H_f + H_S \mid (-i/\omega) \qquad \text{as} \quad \omega \rightarrow 0 \tag{13a}$$

which is pure imaginary. This does not mean, however, that the real part of $L(i\omega)$ necessarily vanishes in the limit of $\omega \rightarrow 0$, as one can verify from (11a) that

$$L_R(0) = K \mid H_f + H_S \mid [\tau_0 + \tau_1 - \tau_2 - \tau_f - \tau_S] \tag{13b}$$

The value of $L_R(0)$ determines the location of the vertical line mentioned above. Although $L_R(0)$ does not enter in the stability conditions, we mention it explicitly for the sake of preciseness. The relevant information obtained from (13a) is that $L(i\omega)$ describes the right, large semicircular arc when ω goes from 0^- to 0^+.

The behavior of the locus of $L(i\omega)$ in the vicinity of the origin is determined by the limiting values of $L(i\omega)$ as $\omega \rightarrow \pm\infty$. Since we know that $L(i\infty) = 0$, we need only consider the signs of the real and the imaginary parts of $L(i\omega)$ as $\omega \rightarrow \pm\infty$. We can easily show that the real part of $L(i\omega)$ behaves as

$$L_R(\omega) \rightarrow -K \mid H_S + H_f \mid (\tau_0\tau_1/\omega^2\tau_2\tau_S\tau_f), \qquad \omega \rightarrow \pm\infty \tag{14}$$

The asymptotic value of the imaginary part, $L_I(\omega)$, as $\omega \rightarrow \pm\infty$, can be obtained from

$$L_I(\omega) = K[a_1\omega^4 + a_2\omega^2 - a_3]/\omega[1 + (\omega\tau_f)^2][1 + (\omega\tau_S)^2][1 + (\omega\tau_2)^2] \tag{15a}$$

where

$$a_1 = \tau_f\tau_S(H_f\tau_S + H_S\tau_f)(\tau_1 - \tau_2) + (H_f\tau_S^2 + H_S\tau_f^2)\,\tau_1\tau_2 \tag{15b}$$

$$a_2 = (H_f\tau_S^2 + H_S\tau_f^2) + (H_f + H_S)\,\tau_1\tau_2 + (\tau_1 - \tau_2)(H_f\tau_f + H_S\tau_S) \tag{15c}$$

$$a_3 = \mid H_f + H_S \mid \tag{15d}$$

as

$$L_I(\omega) \rightarrow Ka_1/\omega^3\tau_f^2\tau_S^2\tau_2^2 \tag{16}$$

We observe from (14) and (16) that the locus will be tangent to the real axis as it approaches the origin, and the location of the curves will depend on the signs of a_1 and τ_0. We can have four cases: Case I: $a_1 < 0$, $\tau_0 > 0$; case II: $a_1 > 0$, $\tau_0 < 0$; case III: $a_1 > 0$, $\tau_0 > 0$; and case IV: $a_1 < 0$, $\tau_0 < 0$. The first three cases are depicted in Figure 6.5.3.

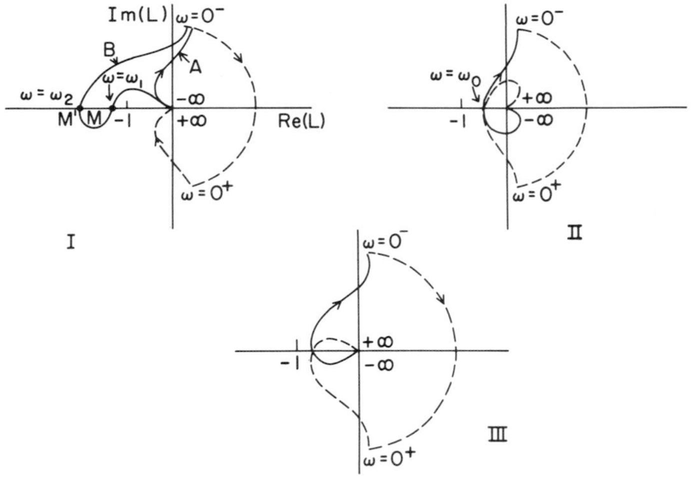

FIGURE 6.5.3. Polar loci of the open-loop transfer function $L(i\omega)$. Case I: $a_1 < 0$, $\tau_0 > 0$; case II: $a_1 > 0$, $\tau_0 < 0$; case III: $a_1 > 0$, $\tau_0 > 0$.

We have not plotted case IV because this case is ruled out by the fact that, when $\tau_0 < 0$, a_1 is necessarily positive (cf Figure 6.5.4 and Problem 10).

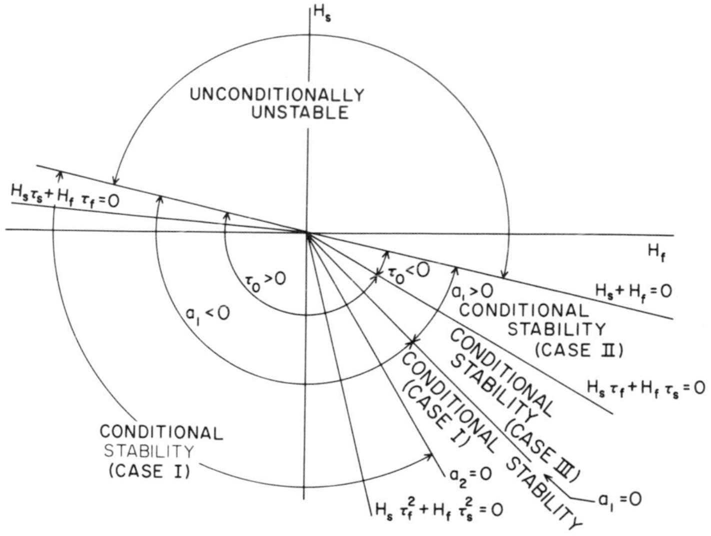

FIGURE 6.5.4. Regions of linear stability in the H_S, H_f plane assuming $\tau_s > \tau_f$.

We shall discuss these cases individually.

Case I: $a_1 < 0$, $\tau_0 > 0$. There are two possibilities in this case depending on the roots of $L_I(\omega) = 0$, which is a quadratic equation in ω^2, i.e.,

$$a_1\omega^4 + a_2\omega^2 - a_3 = 0 \tag{17}$$

Since $a_3 > 0$ and $a_1 < 0$, this equation has no roots if

$$\Delta \equiv a_2^2 - 4 \mid a_1 \mid a_3 < 0 \tag{18a}$$

or has two roots ω_1 and ω_2 if

$$\Delta > 0 \tag{18b}$$

In the former case, the locus of $L(i\omega)$ does not encircle the -1 point (curve A in Figure 6.5.3), and hence the system is stable for all values of the gain K (or the power level P_0). This type of stability is called "unconditional" stability. In the latter case, the locus intersects the real axis (curve B in Figure 6.5.3), and the critical point may or may not be encircled depending on the location of the intersection points M and M' on the real axis relative to the -1 point. One can show after some manipulations that $OM = L_R(\omega_1)$, which is given by

$$L_R(\omega_1) = Ka_3[(\tau_1 - \tau_2)/a_4](a_4^2 + \omega_1^2 a_S^2) \tag{19a}$$

where

$$a_4 = 1 + [\omega_1^2(H_f\tau_S^2 + H_S\tau_f^2)/(H_S + H_f)] \tag{19b}$$

$$a_5 = -\{\omega_1^2\tau_f\tau_S\tau_0 + [(H_S\tau_S + H_f\tau_f)/(H_f + H_S)]\} \tag{19c}$$

If both M and M' lie on one side of the -1 point, then the locus of $L(i\omega)$ does not encircle the critical point and the system is stable. If the initial point is between M and M', the system is unstable. Equation (19a) shows that the distances OM and OM' are proportional to K, and thus to P_0. Therefore, the reactor is stable at low and high power levels and unstable at the intermediate power levels. This type of stability is called "conditional" stability. If $L_R(\omega_1)$ and $L_R(\omega_2)$ are both positive, then the reactor is again unconditionally stable. A sufficient condition for $L_R(\omega)$ to be positive at all frequencies follows from (19a) and (19b) as

$$H_f\tau_S^2 + H_S\tau_f^2 < 0 \tag{20}$$

because $\tau_1 - \tau_2 > 0$, as can be seen from (10), and $H_S + H_f < 0$ by virtue of (12). However, $L_R(\omega_1)$ and $L_R(\omega_2)$ may both be positive even if (20) is not satisfied.

The critical power levels in the case of conditional stability are obtained from $|L_R(\omega_1)| = 1$ and $|L_R(\omega_2)| = 1$.

Case II: $a_1 > 0$, $\tau_0 < 0$. In this case, the quadratic equation (17) has one root, ω_0, and hence the locus of $L(i\omega)$ intersects the real axis at one point. The system is conditionally stable if $L_R(\omega_0) < 0$ with a critical power level determined by $|L_R(\omega_0)| = 1$. If $L_R(\omega_0) > 0$, then the system is unconditionally stable. The inequality (20) is again a sufficient condition for $L_R(\omega_0) > 0$, and thus for unconditional stability.

Case III: $a_1 > 0$, $\tau_0 > 0$. This case is similar to case II. However, in this case, $H_f \tau_S^2 + H_S\tau_f^2$ is necessarily positive because $\tau_0 > 0$ implies $H_f\tau_S + H_S\tau_f < 0$ and $a_1 > 0$ (cf Eq. 15b). The reactor is again conditionally stable if $L_R(\omega_0) < 0$ and unconditionally stable if $L_R(\omega_0) > 0$.

The conclusions of the above analysis are summarized in a parameter plane H_S vs. H_f in Figure 6.5.4. It is noted that there will be regions in the sectors labeled by conditional stability in this diagram where the system may be unconditionally stable, depending on the sign of $L_R(\omega)$ at the intersection points of the locus. In the region between $a_1 = 0$ and $a_2 = 0$, the sign of the discriminant also would define regions of unconditional stability.

An interesting physical conclusion is that the reactor is unconditionally stable when $H_S > 0$ (a slow, positive reactivity feedback) and $H_f < 0$ (a fast, negative reactivity feedback) so long as $H_S + H_f < 0$. Furthermore, there is a possibility of instability when $H_S < 0$ and $H_f > 0$ even if $H_s + H_f < 0$. The linear stability of a reactor with a two-temperature feedback path was also discussed by Bethe [18], Brittan [19], Ackroyd *et al.* [20], and Little and Schultz [21], with different approaches. (See Problem 10 for another application.)

6.6. Critical Power Level and the Effect of Delayed Neutrons on Linear Stability

In the previous sections, we have discussed the linear stability criteria which permit us to ascertain the stability of a reactor for small perturbations without actually solving the characteristic equation. Let us suppose that we find the reactor to be stable for a given set of parameters. It is now desirable to introduce quantities which will in a certain sense measure the degree of stability of the system.

These quantities are called the "stability" margins. One of the aims of this section is to define the "gain" and "phase" margins as two typical stability margins. Another objective of this section is to introduce the concept of critical power level.

We start our discussion by first showing that a reactor is always linearly stable at sufficiently low power levels. Indeed, at zero power, the characteristic equation (6.1, Eq. 10) is replaced by

$$Y(s) = 0 \tag{1}$$

where $Y(s)$ is the inverse of the zero-power transfer function. Equation (1) is just the inhour equation corresponding to a zero reactivity, and has $I + 1$ roots (if there are I delayed neutron groups), one at the origin and I on the negative real axis. In the presence of feedback, the characteristic equation can be written as

$$Q(s) = Y(s) - P_0 H(s) \tag{2}$$

The effect of the additional term $-P_0 H(s)$ on the root at the origin can be treated as a perturbation when P_0 is sufficiently small. Expanding (2) about $s = 0$, we get

$$sY'(0) + \tfrac{1}{2}s^2 Y''(0) + \cdots = P_0 H(0) + P_0 H'(0)\, s + \cdots \tag{3}$$

Since s vanishes with P_0, we are permitted to consider only the first terms on both sides to obtain s in the lowest order in P_0:

$$s = P_0(\beta/l^*)\,\gamma \tag{4}$$

where we have used $Y'(0) = (l^*/\beta)$ and $\gamma = H(0)$. Since we are considering reactors with negative power coefficients γ, $s < 0$ as asserted.

When the power level is increased gradually, there may be a critical value P_c at which the characteristic equation acquires, for the first time, roots with zero real parts. This is the threshold level at which the reactor ceases to be linearly stable. The critical power level is the solution of

$$P_c Z(i\omega_c)\, H(i\omega_c) = 1 \tag{5}$$

which also yields the critical frequency ω_c. This equation may have more than one set of values (P_c, ω_c). Conventionally, the critical power level refers to the lowest P_c that satisfies (5).

If Eq. (5), which can be rewritten as

$$Y_R(\omega_c) = P_c H_R(\omega_c) \tag{6a}$$

$$Y_I(\omega_c) = P_c H_I(\omega_c) \tag{6b}$$

by equating the real parts and imaginary parts, does not have a solution with real P_c and ω_c, then the reactor is linearly stable at all power levels. Since $Y_R(\omega_c)$ is always positive (cf 3.2, Eq. 10), a reactor whose feedback transfer function satisfies, for all ω,

$$H_R(\omega) \leqslant 0 \tag{7}$$

can then never be linearly unstable. The one-temperature feedback model described by

$$H(s) = \alpha\eta/(s + d) \tag{8a}$$

is the simplest reactor type which is linearly stable at all power levels because

$$H_R(\omega) = \alpha\eta d/(\omega^2 + d^2) < 0 \tag{8b}$$

[The temperature coefficient α is necessarily negative, for we are considering reactors with negative power coefficients, i.e., $\gamma = H(0) = (\alpha\eta/d) < 0$.]

The Effect of Delayed Neutrons on Linear Stability

In many reactor models, the presence of delayed neutrons has a very favorable influence on the stability properties of the reactor. This observation led to the conviction that delayed neutrons always provided additional damping to the natural oscillations of a reactor system [22]. Ergen [23] proved that, if a reactor has a feedback that satisfies

$$H_R(\omega) \leqslant 0$$

and has a periodic power oscillation in the absence of delayed neutrons, it can have only damped oscillations when the delayed neutrons are taken into account. However, as pointed out by Ruiz [24] in 1960, there are cases in which the presence of delayed neutrons has a destabilizing effect. For example, the response of a zero-power reactor to a pure sinusoidal reactivity variation is strictly periodic and hence bounded (cf 4.6, Eq. 41) when the delayed neutrons are neglected, whereas the power oscillations grow indefinitely when the delayed neutrons are taken into account. This destabilizing effect of delayed neutrons in the case of forced oscillations had long been recognized [18, 25]. However, their effect on the stability of the autonomous systems was not well understood until relatively recent times [26–28].

Smets [28] has reviewed the effect of delayed neutrons on the linear and nonlinear stability of reactor systems under various conditions. Here, we shall investigate their effect only on linear stability by considering the critical power level with and without delayed neutrons.

Let P_c and ω_c denote respectively the critical power level and the critical frequency in the absence of delayed neutrons. Since $Y(i\omega) = i\omega(\beta/l)$ in the absence of delayed neutrons (we keep β as a number which converts reactivity into dollars), P_c and ω_c satisfy (5),

$$i\omega_c(l/\beta) = P_c H(i\omega_c) \tag{9}$$

In the presence of delayed neutrons, the critical power and frequency P_c' and ω_c', respectively, satisfy

$$Y(i\omega_c') = P_c' H(i\omega_c') \tag{10}$$

Our purpose is to compare P_c and P_c' in order to draw some conclusions about the effect of the delayed neutrons on linear stability. The critical frequencies are obtained independently of the critical power levels from the phase angles:

$$\pi/2 = \arg H(i\omega_c) \qquad \text{without delayed neutrons} \tag{11a}$$

$$\arg Y(i\omega_c') = \arg H(i\omega_c') \qquad \text{with delayed neutrons} \tag{11b}$$

The phase angle of the inverse of the zero-power transfer function, $\arg Y(i\omega)$, is always between 0 and $\pi/2$. Hence,

$$\pi/2 \geqslant \arg H(i\omega_c') \tag{12}$$

Let us assume that the phase angle of the feedback transfer function, $\arg H(i\omega)$, is a nonincreasing function of the frequency. Thus, from (11a) and (12) we conclude that

$$\omega_c' \geqslant \omega_c \tag{13}$$

We can now compare P_c and P_c' from (9) and (10) as

$$P_c'/P_c = [|H(i\omega_c)|/|H(i\omega_c')|]\{|1 + (\beta/l) \sum [a_j/(\lambda_j + i\omega_c')]|\}(\omega_c'/\omega_c) \tag{14}$$

where we have substituted the explicit value of $|Y(i\omega)|$. The middle factor is always greater than unity because the real part of $\sum_j a_j/(\lambda_j + i\omega_c')$ is positive. Hence, if we assume that $|H(i\omega)|$ is also a nonincreasing function of frequency, then $\omega_c' \geqslant \omega_c$ implies $|H(i\omega_c)| \geqslant |H(i\omega_c')|$. We thus conclude the following:

If the amplitude and phase of the feedback transfer function $H(i\omega)$ are both nonincreasing when ω increases, then there is only one critical power level P_c in the absence of delayed neutrons, and the critical power level P_c' (or the smallest of all the critical power levels if there are more than one) when the delayed neutrons are taken into account is larger than P_c.

In order to relate this conclusion to the stability properties of the system, let us suppose that the reactor is operated at a power level $P_0 = P_c$ which is the threshold of instability in the absence of delayed neutrons. When delayed neutrons are taken into account, P_0 is smaller than the first critical power level, and hence the reactor is linearly stable.

Equation (14) indicates that, in general, P_c' can be less than P_c, depending on the variation of $|H(i\omega)|$ and $\varphi(\omega)$ with frequency, and the reactor may be linearly unstable with delayed neutrons, and stable with no delayed neutrons. For example, let us assume the Nyquist plot of the open-loop transfer function, $L(i\omega) = -P_0 Z(i\omega) H(i\omega)$, in the presence of delayed neutron intersects the negative real axis at two points, one above and one below the -1 point (Figure 6.6.1). We have

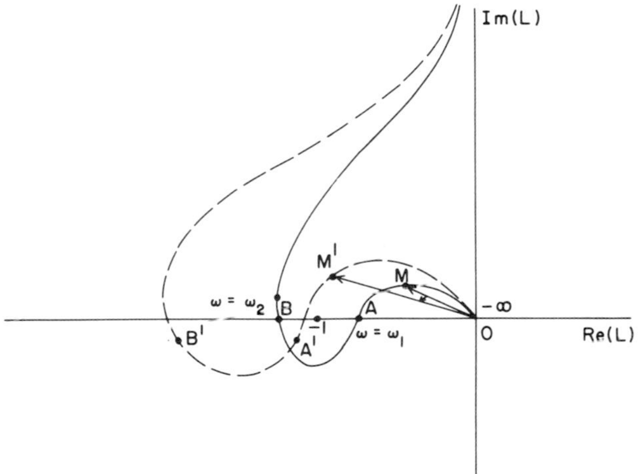

FIGURE 6.6.1. The effect of delayed neutrons on the open-loop transfer function; (——) with delayed neutrons, (— —) no delayed neutrons.

already encountered this situation in the two-temperature model (cf Figure 6.5.3 case I). In the absence of delayed neutrons, the open-loop transfer function is $-P_0H(i\omega)/(l/\beta)\, i\omega$. The Nyquist plot of the latter can be obtained from the plot with delayed neutrons as

$$L(i\omega)\left\{1 + (\beta/l)\sum_j [a_j/(\lambda_j + i\omega)]\right\} \tag{15}$$

Thus, a point M will be rotated in a counterclockwise sense for negative frequencies by an angle of $|\arg\{1 + (\beta/l)\sum [a_j/(\lambda_j + i\omega)]\}|$ and scaled up by a factor $|1 + \sum [a_j/(\lambda_j + i\omega)]|$ to obtain the point M'. The angle of rotation is small for large and small values of ω, approaching zero as $\omega \to 0$ and $\omega \to \mp\infty$. The scaling factor decreases from $1 + (\beta\bar{\tau}/l)$ to one when ω varies from 0 to $\mp\infty$. The new Nyquist plot may not encircle the -1 point and the reactor will be linearly stable in the absence of delayed neutrons.

We conclude from the above discussion that the effect of the delayed neutrons on linear stability is not always stabilizing.

6.7. Stability Margins

We see from (6.6, Eq. 5) that, at the linear instability threshold, $|L(i\omega_c)| = 1$ and $\arg L(i\omega_c) = 180°$ are satisfied simultaneously (Figure 6.7.1). The factor by which the open-loop gain must be increased to cause instability at a frequency at which the phase angle is 180° is

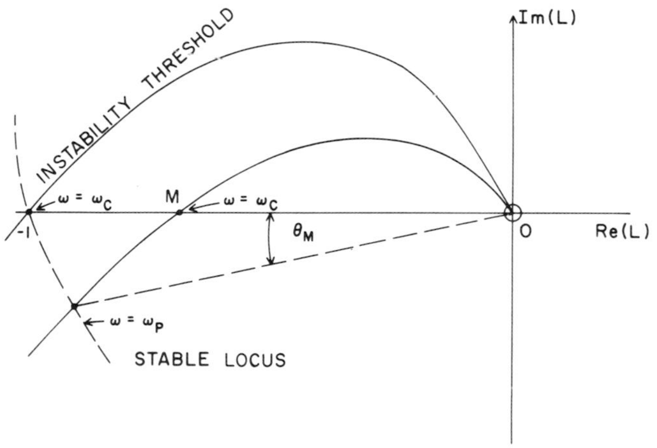

FIGURE 6.7.1. Stability margins. L-Plane; $\arg L(i\omega_c) = 0$; $|L(i\omega_P)| = 1$.

defined as the "gain" margin. It is usually expressed in decibels. We see from Figure 6.7.1 that the gain margin is

$$G_M = 20 \log(1/|\widehat{OM}|) \tag{1}$$

Since $|\widehat{OM}| = P_0 \, | H(i\omega_c) \, Z(i\omega_c)|$, we also express G_M as

$$G_M = 20 \log(P_c/P_0) \tag{2}$$

where P_c is the critical power level.

The phase margin is defined as the amount of additional phase lag required to cause instability at a frequency ω_p at which the gain is unity. In Figure 6.7.1, this is denoted by θ_M.

The concepts of gain and phase margin were used (see Sec. 5.4) by DeShong and Lipinski [29] to predict the critical power level of the EBWR (experimental boiling water reactor) during its initial power startup. They plotted the experimental values of G_M and θ_M, obtained from the measured frequency characteristic of the reactor at low power levels, as a function of P_0, and determined the critical power level as the intersection of the extrapolated curves with the horizontal straight lines corresponding to $G_M = 0$ and $\theta_M = 0$. We may point out that the curve G_M would be a linear function of $\log P_0$, according to (6.6, Eq. 10), if the feedback mechanism were truly linear because in this case H is independent of P_0. However, because of the nonlinearities, the feedback parameters, and thus the feedback transfer function $H(s)$, are dependent on the equilibrium power level. Consequently, the variation of G_M with $\log P_0$ was not available, and an empirical extrapolation procedure, as described above, had to be used.

6.8. Stability of Distributed-Parameter Systems

Our analysis has so far been restricted to determining the effect on stability of the various reactivity coefficients and the time constants associated with the feedback variables. We wish now to extend the analysis to determine some of the effects which are related to the fact that the power distribution and the feedback reactivity effects are not necessarily space-independent. In particular, we shall consider systems which contain a flowing coolant. In order to clarify the various phenomena, we shall consider systems with and without reflector feedback effects. In the first case, the feedback is due to some weighted value of the coolant temperature in the core, while in the second case, it will be due to some weighted average of the coolant temperature in the reflector (after it leaves the core). The origin of the weighting factors will be made clear in the development. All numerical results presented will be characteristic of ^{235}U delayed neutron data.

A. Characteristic Equation

The distributed-parameter description of the temperature feedback was discussed in Sec. 5.2C, and the temperature feedback transfer function $H(s)$ was obtained for a single-channel reactor (cf 5.2, Eq. 73). The linear stability of the reactor is determined by the roots of the characteristic equation $1 = P_0 H(s)\, Z(s)$. Before discussing the question of the stability of such a reactor, we shall summarize the general procedure of obtaining the characteristic equation in the case of a one-energy-group diffusion model in slab geometry. In this model, the flux $\phi(x, t)$ satisfies

$$(1/v_{\mathrm{T}})\, \partial\phi/\partial t = (\partial/\partial x)\, D(\partial\phi/\partial x) + (\nu\Sigma_{\mathrm{f}} - \Sigma_{\mathrm{a}})\,\phi + \sum_{i=1}^{I} \lambda_i c_i - \beta\nu\Sigma_{\mathrm{f}}\phi \qquad (1\mathrm{a})$$

$$\partial c_i/\partial t = \beta_i \nu\Sigma_{\mathrm{f}}\phi - \lambda_i c_i \qquad (1\mathrm{b})$$

We allow the cross sections to depend on the coolant temperature, which is assumed to obey

$$(\partial T/\partial t) + v(\partial T/\partial x) = (\Sigma_{\mathrm{f}}/H_{\mathrm{c}})\,\phi \qquad (2)$$

This equation is a special case of (5.2, Eq. 64), in which the heat content of the fuel (μ_{f}) is neglected. We also assume in this example that the dominant feedback mechanism is due to the coolant temperature. The temperature dependence of the cross section is approximated by

$$Q(x, t) = Q^0[1 + \alpha_Q T(x, t)] \qquad (3)$$

where Q stands for either Σ_{f}, Σ_{a}, or D.

The quantities Q^0 are the values of $\nu\Sigma_{\mathrm{f}}$, Σ_{a}, and D for the hot, zero-power reactor. If the overall temperature coefficient is negative, then a reactor described by Q^0 constants is supercritical at zero power. There is an "at power," just critical configuration which is characterized by the flux and temperature distributions $\phi_0(x)$, $T_0(x)$; where $T_0(x)$ satisfies

$$T_0(x) = (1/vH_{\mathrm{c}}) \int_0^x \Sigma_{\mathrm{f}_0}\phi_0(x')\, dx' \qquad (4)$$

that is to say, $T_0(x)$ is measured above the inlet coolant temperature, and the Σ_{f_0} are measured at inlet temperature. For simplicity, we assume the Q_0 to be space-independent (i.e., the reactor is homogeneous).

We proceed now as in Chap. 2; we multiply by $\phi_0(x)$ and integrate over $0 \leqslant x \leqslant h$ (h being the core length). Assuming separability of $\phi(x, t) = P(t)\,\phi_0(x)$, we obtain

$$l\dot{P} = \left[\delta_0 + \int_0^h dx\,\rho(x)\,T(x, t)\right] P - \beta P + \sum_i \lambda_i C_i \tag{5a}$$

$$\dot{C}_i = \beta_i P - \lambda_i C_i\,, \qquad i = 1,\dots, I \tag{5b}$$

where we have defined

$$l = \langle\phi_0 \mid \phi_0\rangle / v_T \langle\phi_0 \mid \nu\Sigma_f^0 \mid \phi_0\rangle = 1/v_T\nu\Sigma_f^0 \tag{6a}$$

$$\delta_0 = (\langle\phi_0 \mid \nu\Sigma_f^0 - \Sigma_a^0 \mid \phi_0\rangle - \langle D^0 \mid (d\phi_0/dx)^2\rangle)/\langle\phi_0 \mid \nu\Sigma_f^0 \mid \phi_0\rangle \tag{6b}$$

$$\rho(x) = [(\alpha_f\nu\Sigma_f^0 - \alpha_a\Sigma_a^0)\,\phi_0^2 - \alpha_D D^0 (d\phi_0/dx)^2]/\langle\phi_0 \mid \nu\Sigma_f^0 \mid \phi_0\rangle \tag{6c}$$

$$C_i(t) = \langle\phi_0 \mid c_i(x, t)\rangle/\langle\phi_0 \mid \nu\Sigma_f^0 \mid \phi_0\rangle \tag{7}$$

In obtaining (5), we have used the following approximation:

$$\bar{\beta}_i \equiv [\langle\phi_0 \mid \nu\Sigma_f \mid \phi_0\rangle/\langle\phi_0 \mid \nu\Sigma_f^0 \mid \phi_0\rangle]\,\beta_i \approx \beta_i \tag{8}$$

The excess reactivity at hot, zero power is given by Eq. (6b), and at equilibrium we find (Problem 11)

$$\delta_0 = -K_c \int_0^h dx\,\rho(x) \int_0^x dy\,\phi_0(y) \tag{9a}$$

where

$$K_c = \Sigma_{f_0}/vH_c \tag{9b}$$

The term δ_0 is the amount of supercriticality of the reactor at hot, zero power which is just compensated for at equilibrium by the negative reactivity due to the temperature feedback.

The time-dependent temperature in Eq. 5a satisfies

$$(\partial T/\partial t) + v(\partial T/\partial x) = \Sigma_{f_0}\phi_0(x)\,P(t)/H_c \tag{10a}$$

which can be solved with Laplace transforms as

$$\bar{T}(x, s) = \int_0^x dy\,[(1/v)T_0(y) + K_c\bar{P}(s)\phi_0(x - y)]\,e^{-sy/v} \tag{10b}$$

(cf 5.2, Eq. 68). In (10b), we assume that the inlet temperature does not change, i.e., $T(0, t) \equiv 0$. Since we consider small perturbations about equilibrium, we let $P(t) = 1 + p(t)$ in (5a) and linearize. After taking the Laplace transforms of (5a) and using (10b), we obtain

$$s\left[l + \sum_{i=1}^{6} \frac{\beta_i}{s + \lambda_i}\right] - K_c \int_0^h dx\, \rho(x) \int_0^x dy\, \phi_0(x - y)\, e^{-sy/v} = 0 \tag{11}$$

which is the characteristic equation.

We propose now to study the stability of the reactor model described above by considering the roots of (11). This model mocks up a core with feedback proportional to a weighted integral of the coolant temperature. The weighting function $\rho(x)$ is the local reactivity coefficient introduced in a more general way in (5.2, Eq. 9). Its exact spatial form depends on the solution of a multienergy-group model because $\alpha_D(x)$, $\alpha_f(x)$, and $\alpha_a(x)$ in (6b) contain spectral effects. We shall not solve this multigroup problem, but simply show the effects of various shapes for $\rho(x)$ on linear stability.

We note that the characteristic equation (11) depends also on the flux distribution $\phi_0(x)$ at equilibrium. Since $\rho(x)$ and $\phi_0(x)$ do not enter symmetrically, we consider their effects separately (Problem 12).

B. The Effect of Reactivity Shape

In this section, we shall assume that $\phi_0(x)$ can be represented by $\sin(\pi x/h)$.† This form states that the power production goes to zero on the boundaries of the core and is the type of solution one would expect for a bare homogeneous system at low power. We shall consider three functional forms for $\rho(x)$, each of which can occur under certain conditions.

Form 1: $\rho(x) = -\alpha_c$. This assumption implies that the reactivity effect of a temperature change is independent of position.

Form 2: $\rho(x) = -\alpha_c \phi^2(x)$. This approximation is essentially a one-group result arrived at by means of perturbation theory (statistical weight factor).

Form 3: $\rho(x) = -\alpha_c \delta(\frac{1}{2}h - x)$, Dirac delta. This result can be thought of as being due to an external control system; a thermocouple

† By this, $\phi(x)$ is taken to be dimensionless; the amplitude will be assumed to be absorbed by K_c.

senses the local temperature at the center of the core and moves a uniform bank of control elements in a direction transverse to the flowing coolant.

The characteristic equations corresponding to these three cases can be obtained by substituting $\phi_0(x) = \sin(\pi x/h)$ and $\rho(x)$ from above into (11).

We leave the details of the analysis to Problems 13a,b,c and proceed directly to the results. All three models show, in parameter space, multiple regions of instability separated by regions of linear stability. We consider first the case without delayed neutrons. If we choose the parameters to be† δ_0/l_e and t_c (the coolant transit time throught the core, h/v), then we find:

1. If $\rho(x) = -\alpha_c$, the reactor becomes unstable when $\delta_0 t_c/l_e > 3\pi^2/2$, but becomes linearly stable again when $\delta_0 t_c/l_e > 9\pi^2/2$ (higher boundaries also terminate this new stable region) (Problems 14a, 15a).

2. If $\rho(x) = -\alpha_c \phi_0^2(x)$, the first unstable parameter region is bounded by (Problems 14b, 15b)

$$45\pi^2/16 < \delta_0 t_c/l_e < 45\pi^2/8$$

3. If $\rho(x) = -\alpha_c \delta(\frac{1}{2}h - x)$ then the first unstable region is bounded by (Problems 14c, 15c)

$$12\pi < \delta_0 t_c/l_e < 60\pi$$

In Figure 6.8.1, we plot the lowest line only for the first and second cases [absorbing the α_c into $\rho(x)$ for convenience]; the third case lies even higher, yielding a larger stablc region. For the first case, we show several of the alternating stable–unstable regions in Figure 6.8.2.

The relationship between the location of the instability line and the shape of $\rho(x)$ is not a trivial one. Although one can assign an arbitrary multiplier [which is not related to the average value of $\rho(x)$] which will superimpose the first instability lines, the higher lines will not superimpose with the same multiplier (see Problem 15a,b,c). One concludes, then, that the detailed shape of $\rho(x)$ can alter the location of a stability line by a factor of two or more.

In the rest of this section, we shall study the effect of the flux shape on stability. For simplicity, we choose $\rho(x) \equiv -\alpha_c$, as this case gives the most conservative stability results.

† Here, l_e is simply the effective lifetime, i.e., $l_e = l + \Sigma_i(\beta_i/\lambda_i)$.

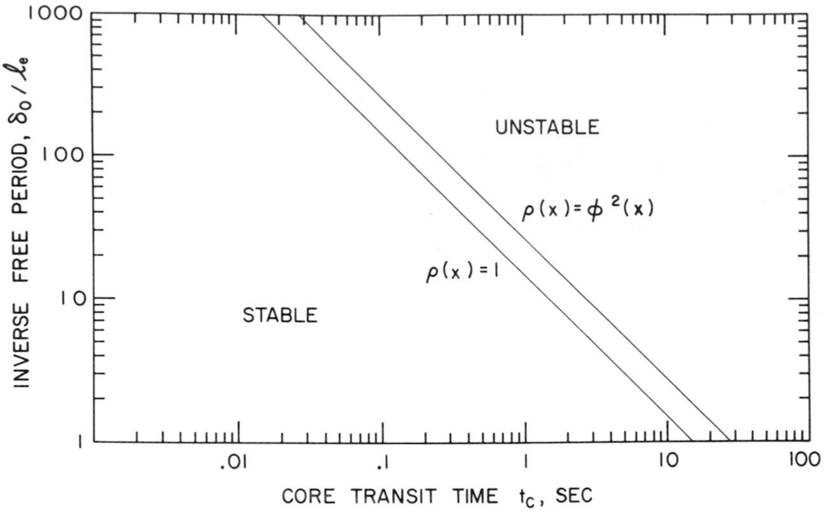

FIGURE 6.8.1. The effect of reactivity shape on the stability of a bare core.

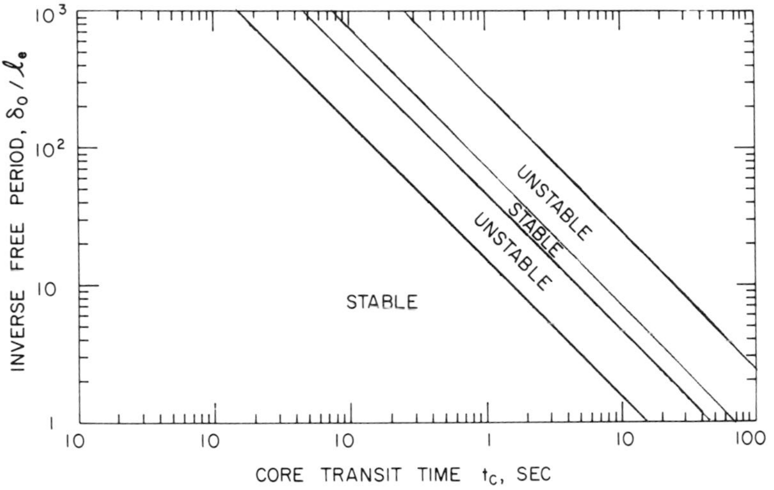

FIGURE 6.8.2. Alternating regions of stable–unstable response for $\rho(x) = -\alpha_c$.

C. The Effect of Power Distribution

We continue the analysis taking account of the delayed neutrons. We assume that the power distribution is given by

$$\phi(x) = \cos[2b(x - \tfrac{1}{2}h)/h] \tag{12}$$

As before, this is essentially a one-energy group result, but not necessarily for a bare core. The flux at the ends of the core is $\cos b$, which will not be zero unless $b = \pi/2$. If $b = 0$, the flux shape is flat, and $b^2 < 0$ mocks up an undermoderated shape (which is, however, not a one-group result). Using this form for $\phi(x)$, it is not difficult to show that the characteristic equation is

$$t_c s \left\{ l + \sum_{i=1}^{I} [\beta_i/(s+\lambda_i)] \right\} + \{\delta_0/2st_c[\tfrac{1}{4}(st_c)^2 + b^2]\}$$
$$\times [(st_c)^2 + b(1 - e^{-st_c})(2b - st_c \cot(b)] = 0 \tag{13}$$

This equation has a principal term (see Sect. 6.4) and exhibits instabilities for $b^2 \lessgtr 0$, but, interestingly enough, not for $b = 0$.[†] Within the validity of this model, one concludes that the presence of curvature in the power distribution is in itself destabilizing. In particular, on the instability boundary, if $b^2 > 0$ and $b \tan b > \frac{1}{2}\omega t_c \tan(\frac{1}{2}\omega t_c)$ (with $\pi < \frac{1}{2}\omega t_c$), a critical power level exists. A similar result holds for $b^2 = -c^2 < 0$ (see Problem 17). In other words, when $|\, b^2 \,|$ is sufficiently large, instabilities will appear at some power levels, but only if the transit time is sufficiently large or small. This result is shown in Figures 6.8.3 and 6.8.4. For core transit times in the physically meaningful range (0.01–1.0 sec), no instabilities appear.[‡] For $t_c \ll 1$, the straight portions of the boundaries of the unstable regions are characteristic of $l_e = l$ and also of $|\, \omega \,| \gg 1$, while for $t_c \gg 1$, we find $|\, \omega \,| \ll 1$ and $l_e = l + \sum_i (\beta_i/\lambda_i)$. This much is clear. Since the slopes are not the same, they must bend or pinch off.

D. Two-Region Reactor Models

We wish now to consider a two-region model composed of core and end reflector. While power production is assumed to occur in the core, the reflector temperature is the only one which effects the reactivity of the system. If $T_R(x, t)$ is the temperature in the reflector, then we assume that T_R satisfies

$$(\partial T_R/\partial t) + v_R(\partial T_R/\partial x) = 0, \qquad h \leqslant x \leqslant h + R \tag{14}$$

[†] This result is due solely to the fact that delayed neutrons are accounted for.

[‡] The numerical results are characteristic of $l = 10^{-5}$ sec, but the qualitative effects are (largely) independent of the magnitude of l.

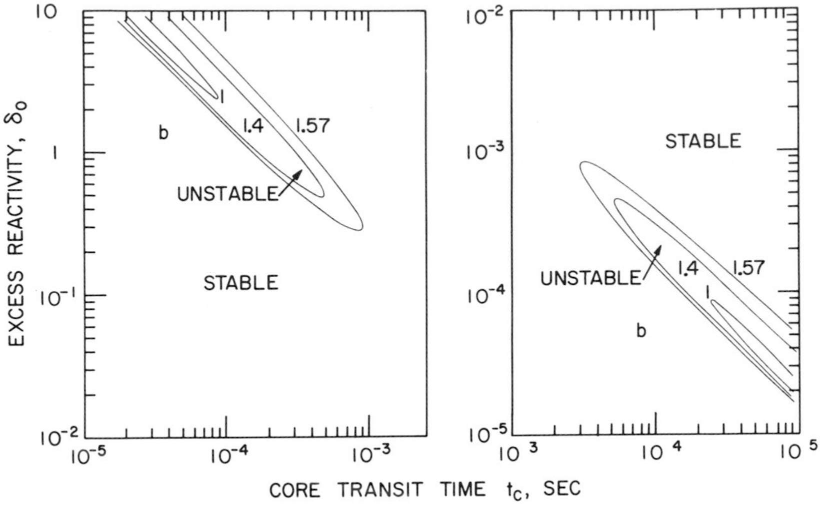

FIGURE 6.8.3. Effect of flux shape on stability; $\phi(x) = \cos 2bx$, $-\frac{1}{2} \leqslant x \leqslant \frac{1}{2}$.

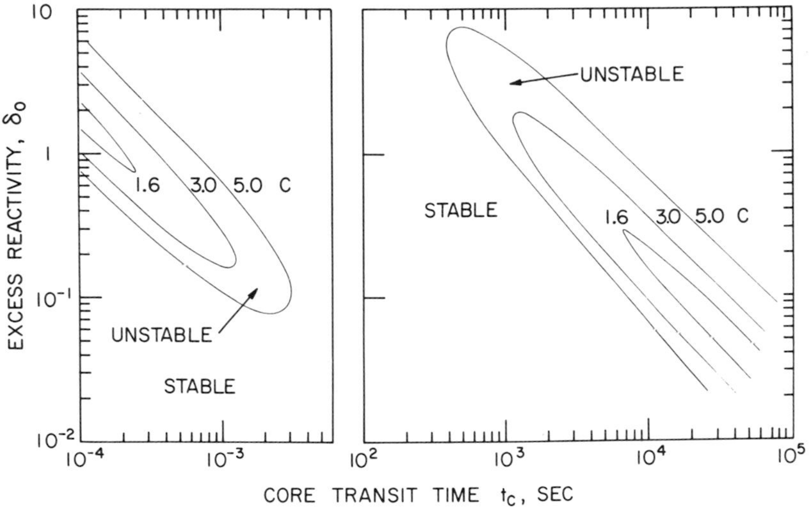

FIGURE 6.8.4. Effect of flux shape on stability; $\phi(x) = \cosh 2cx$, $-\frac{1}{2} \leqslant x \leqslant \frac{1}{2}$.

with the boundary condition

$$T_R(h, t) = T(h, t) \tag{15}$$

where T satisfies Eq. (2). Assuming that $\rho(x) = -\alpha_c$ as before, it is not difficult to show that the power amplitude satisfies the following equation:

$$l\dot{P} = \left[\delta_0 - \alpha_c \int_h^{h+R} T_R(x, t)\, dx\right] P + \sum_i \lambda_i C_i - \beta P \tag{16}$$

with the delayed emitters satisfying Eq. (5b).

It is also not difficult to show that the characteristic equation for this model is

$$s\left\{l + \sum_i [\beta_i/(s + \lambda_i)]\right\} + [\delta_0(1 - e^{-st_R})/\langle\phi\rangle\, st_R]\, e^{s-t_c} \int_0^1 e^{st_c x}\phi(x)\, dx = 0 \tag{17}$$

$$\langle\phi\rangle = \int_0^1 \phi(x)\, dx$$

Here, we have defined $t_R = R/v_R$ and $t_c = h/v_c$ as the reflector and core transit times, and we have normalized the core length to unity.

The solution of Eq. (17) on the instability boundary has several interesting facets. First, the critical relationship between ω (the critical frequency) and the two transit times is independent of $\phi(x)$ if $\phi(x)$ is symmetric about $x = h/2$. This relationship can be shown to be

$$\begin{aligned} Y(\omega) &= \cot[\tfrac{1}{2}\omega(t_c + t_R)] \\ &= -\omega \left\{\sum_i [\beta_i/(\omega^2 + \lambda^2)]\right\} \Big/ \left\{l + \sum_i [\beta_i\lambda_i/(\omega^2 + \lambda_i^2)]\right\} \end{aligned}$$

hence (since $\pm\omega$ are each solutions) we see that $Y(\omega)$ has a fixed sign $\leqslant 0$ for $\omega \geqslant 0$ and

$$(n - \tfrac{1}{2})\pi < |\tfrac{1}{2}\omega\,(t_c + t_R)| < n\pi, \qquad n \geqslant 1 \tag{18}$$

This result implies that, independently of $0 \leqslant t_R \leqslant \infty$, the frequency range is covered by $0 < |\omega t_c/2| < n\pi$, hence $|\omega| < n/t_c$ (in hertz) and, on the first boundary between stable and unstable regions, the frequency (for this model) is always less than $1/t_c$ Hz, which means that the critical frequencies are usually less than 10–15 Hz.

We wish now to consider how sensitive the location of the critical boundary is to the various physical parameters. At first, we take $\phi(x) = 1$ (a flat power profile). Figure 6.8.5 shows several curves of δ_0 vs. t_R at constant t_c. In the range of reasonable transit times through the reflector (0.05–10 sec), we observe the worst condition to occur when $t_c \approx t_R$. This worst condition occurs when $\delta_0 \approx 2.1$–2.4 \$. That the location of these lines is not sensitive to the value of the prompt lifetime can be seen in Figure 6.8.6. It is not until $l \geqslant 10^{-4}$ sec that significant depar-

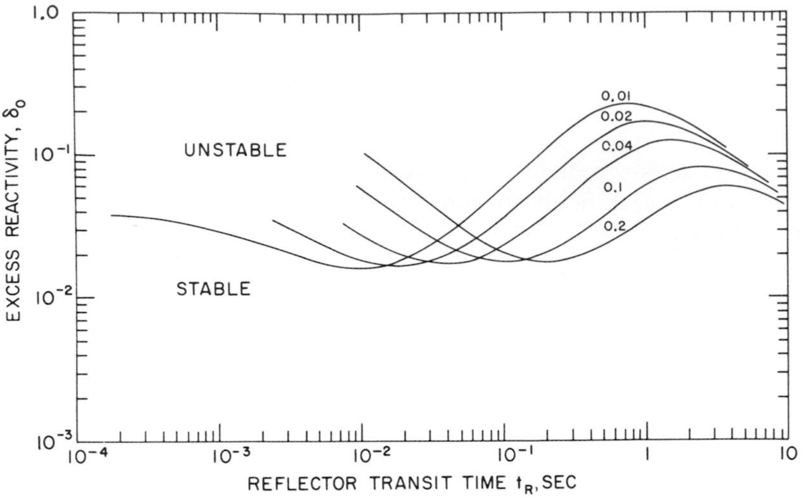

FIGURE 6.8.5. Effect of core transit time t_c on linear stability for a flat power profile; curves are for the respective values of t_c (in seconds); $l_p = 10^{-5}$ sec.

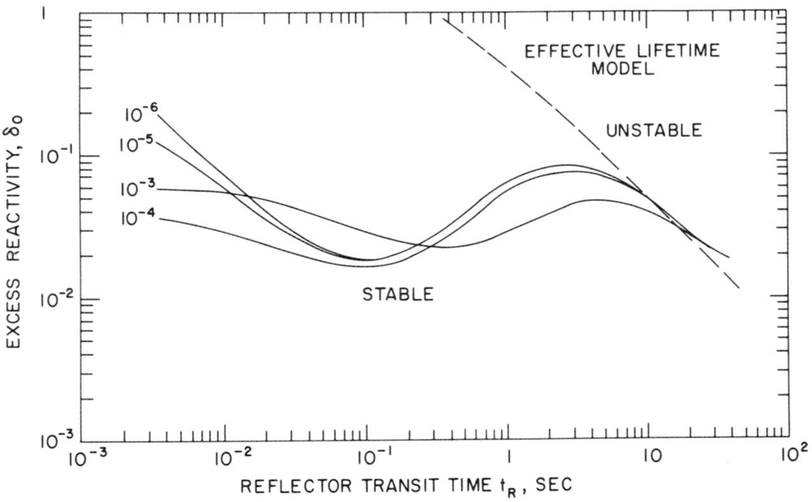

FIGURE 6.8.6. Effect of prompt neutron lifetime on linear stability; curves are for the respective values of prompt neutron lifetime (in seconds); $t_c = 0.1$ sec.

tures from the $l \leqslant 10^{-6}$ values occur. This does not imply, however, that an effective lifetime model can be used, since the dashed line in Figure 6.8.6 clearly is an unacceptable representation anywhere below $t_R = 20$ sec.

The choice $\phi(x) = 1$ is not a poor one; if we had chosen instead $\phi(x) = \cos[2b(-\frac{1}{2} + x)]$, $0 \leqslant x \leqslant 1$, we would find the results to be like those shown in Figure 6.8.7. As long as $t_R > 5t_c$, the reflector

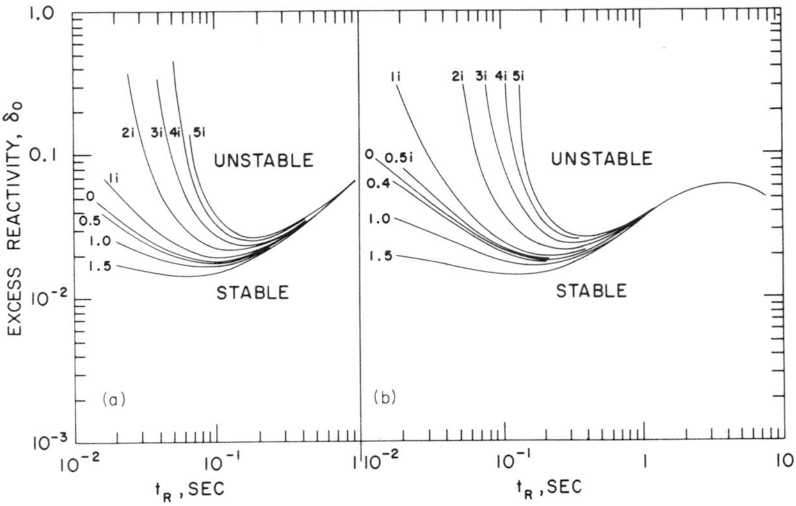

FIGURE 6.8.7. Some effects of power distribution on linear stability; $\phi(x) = \cos 2bx$, curves are for the respective values of b; (a) $t_c = 0.1$ sec, $l_p = 10^{-5}$ sec; (b) $t_c = 0.2$ sec, $l_p = 10^{-5}$ sec.

does not care whether the power shape has positive or negative curvature. Assuredly, if $t_R < t_c$, then significant effects occur for $|b| > 1$. Finally, we consider whether skewing of the power has any significant effects on stability, and we assume

$$\phi(x) = \cos[2b(x - \tfrac{1}{2})] + d \sin[2c(x - \tfrac{1}{2})] \tag{19}$$

The value of c is determined from b by assuming that the sine term is the first harmonic to the cosine term, as found by reflected reactor theory. The results of a single example are shown in Figure 6.8.8 and are typical of physically reasonable values of the parameters. Skewing does not have a strong effect on stability. We can summarize the results obtained so far. If $t_R > 5t_c$, then the exact (within reason) power distribution has a small effect on the location of the stability line, and similarly, the value of the prompt neutron lifetime is not of great importance as long as it is $<10^{-4}$ sec.

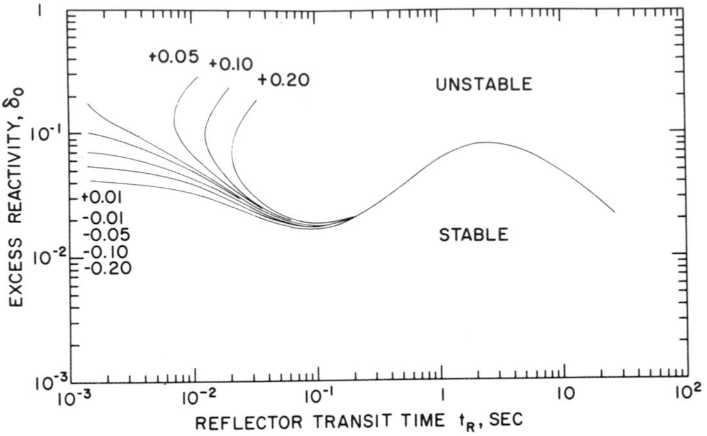

FIGURE 6.8.8. Further effects of power distribution on stability; $\phi(x) = \cos(2bx) + d \sin(2cx)$, $-\frac{1}{2} \leqslant x \leqslant \frac{1}{2}$; $t_c = 0.1$ sec, $b = 0.5$, $c = 1.702$, values of d shown on the respective curves.

E. Prediction of Stability

In the previous section, we used partial differential equations to describe the temperature equations, and subsequently derived linear stability criteria for the feedback system. The models so derived led to transcendental characteristic equations, in contrast to the point models, which led to algebraic characteristic equations. One can ask the following question: Under what circumstances can a continuous model be replaced by a point model and still yield a reasonably valid representation for the location of the stability lines in parameter space? This question is obviously tied in closely with the type of analysis performed in Sec. 5.7, where we considered conditions under which the time motion of the two models would yield answers within 1% of each other.

We have seen that the space-dependent models yield multiple regions of alternating stability and instability, and we do not expect to be able to reproduce such results; what we are primarily interested in is the location of the lowest stability line, and whether a single point model yields a conservative position for this line.

As before, we consider one-temperature, one- and two-region models.

SINGLE-TEMPERATURE, SINGLE-REGION MODEL

The usual approach taken in spatially averaging Eq. (2) yields a Newton's law of cooling model. Equally well, we know that if the feedback coefficient is stabilizing ($\alpha_c > 0$), no instability is possible for a Newton's law model. However, we saw in Figures 6.8.1–6.8.4 that the one-temperature, continuous model does show linearly unstable regions. The conclusion is simply that either one cannot produce a single-point model equivalent to the space-dependent one, or else some special type of averaging (unknown to us) is required.

Two-Region Model

We again consider the two-region model. We show now that the one lump per region model predicts total linear stability when the continuous model predicts the opposite. The transfer function implicit in (5.7, Eq. 25b) yields a characteristic equation

$$s\left\{l + \sum_i [\beta_i/(s + \lambda_i)]\right\} + \{4\delta_0/t_c t_R[s + (2/t_R)][s + (2/t_c)]\} = 0 \quad (20a)$$

For simplicity, we consider a one-delayed-group, prompt-jump approximations to Eq. (20a), which leads to

$$s^3 + [2(t_c + t_R)/t_c t_R]\, s^2 + (4/t_c t_R)[1 + (\delta_0/\beta)]\, s + (4\delta_0\lambda/\beta t_c t_R) = 0 \quad (20b)$$

Using the Routh–Hurwitz criterion $\Delta_3 > 0$, we find that Eq. (20b) exhibits no instabilities if

$$\lambda < 2[(1/t_c) + (1/t_R)] \quad (21)$$

since λ is approximately 0.1 sec^{-1}, this implies that the entire range $t_c < 20$ sec is linearly stable independently of t_R and vice versa. Equation (1b) also implies that, on the line of instability,

$$\omega^2 > 4/t_c t_R \qquad \text{(one lump)} \quad (22a)$$

Hence, in the notation of Sec. 5.7C,

$$x^2 > 4/\alpha \quad (22b)$$

But, independently of α, Eq. (22b) violates the validity criteria derived from (5.7, Eq. 28); hence, we can say that the one lump per region model is never accurate in predicting the *boundary* of instability.

Conclusion

The results obtained in above indicate quite clearly that a flowing coolant cannot be represented by a single point. The question of how many points are necessary to yield a valid representation in the neighborhood of the first instability line cannot of course be answered in general. Only the tools of Chap. 5 correctly applied can yield an answer in any particular case.

6.9. The Theory of Linearization

We now consider the connection between the stability of the linearized system and the stability of the original nonlinear system. Since linearization is effected by ignoring the second and higher powers of the incremental quantities, one intuitively believes that stability of the linearized systems always implies asymptotic stability of the nonlinear system for sufficiently small initial disturbances (asymptotic stability in the small). This intuitive justification disregards the possibility that a system may be linearly stable but become nonlinearly unstable even for infinitesimally small disturbances, if the linearization does not satisfy certain conditions. We shall give an example later to illustrate this point. However, when these conditions on the nonlinear terms in the equation are fulfilled, asymptotic stability in the small is implied by linear stability. Our next goal is to state these conditions in the form of a theorem.

A. Vector Form of the Point Kinetic Equations

In order to introduce the necessary mathematical concepts and use the theory of ordinary differential equations, we first express the point kinetic equations in vector notation.

In most of the feedback models, the feedback functional can be written in the following general form:

$$\delta k_{\mathrm{f}}[p] = \delta k_{\mathrm{f}}(X_1, X_2, \dots, X_n) \tag{1a}$$

where the X_j satisfy

$$\dot{X}_j = f_i(X_1, X_2, \dots, X_n, C_1, C_2, \dots, C_{\mathrm{I}}, p), \qquad j = 1, 2, \dots, n \tag{1b}$$

Here, $(X_1, X_2, \dots, X_n)$ denote a set of dynamical variables describing the feedback mechanism, e.g., the temperatures in the various regions in a heterogeneous reactor, the C_i are the delayed neutron precursor densities, and p is the reactor power. All the quantities are measured from their equilibrium values. The functions f_j and δk_f are in general nonlinear. Let us define

$$\mathbf{f} \equiv \mathrm{col}(f_1, f_2, ..., f_N) \tag{2a}$$

$$\mathbf{X} \equiv \mathrm{col}(X_1, ..., X_n ; X_{n+1}, ..., X_{n+I}, X_N) \tag{2b}$$

where

$$X_{n+i} \equiv C_i, \qquad i = 1, 2, ..., I \tag{3a}$$

and

$$X_N \equiv lp/\beta \tag{3b}$$

The functions f_j for $j \leqslant n$ are given by (1b). When $j > n$, we define them by

$$f_{n+i} \equiv a_i(\beta/l)\, X_N - \lambda_i X_{n+i}, \qquad i = 1, 2, ..., I \tag{4a}$$

and

$$f_N \equiv \delta k_f(X_1, X_2, ..., X_n)[P_0 + (\beta X_N/l)] - (\beta X_N/l) + \sum_{i=1}^{I} \lambda_i X_{n+i} \tag{4b}$$

one can easily verify that the kinetic equations reduce to the following matrix equation with the foregoing definitions:

$$\dot{\mathbf{X}} = \mathbf{f}(\mathbf{X}) \tag{5}$$

where $f_j(\mathbf{X}) = f_j(X_1, X_2, ..., X_N)$ with $N = n + I + 1$.

It must be noted, however, that the feedback functional cannot always be expressed in matrix form as in (1). We have already seen two examples in which this is not possible, namely temperature feedback with a time delay, and feedback in a circulating fuel reactor. In general, systems with time delays can be described by a set of first-order differential equations as in (5).

The advantage of expressing the reactor kinetic equations, whenever possible, in the form of a set of ordinary differential equations is that general theorems in the theory of differential equations can be applied directly to the stability analysis of reactor systems. We shall present the relevant theorems in the sequel without proof.

B. Existence and Uniqueness of the Solutions

First, we need the following definitions:

DEFINITION 1. The Euclidean "norm" of a vector $\mathbf{X} = (X_1, X_2, ..., X_N)$ is defined by $\| \mathbf{X} \| = (X_1^2 + X_2^2 + \cdots + X_N^2)^{1/2}$.

Definition 2. The absolute value norm [30] of vector is defined by

$$\| \mathbf{X} \| = \sum_{i=1}^{N} | X_i |$$

In general, the norm of a vector $\mathbf{X}$ is defined as a real number $\| \mathbf{X} \|$ such that (i) $\| \mathbf{X} \| > 0$, (ii) $\| \mathbf{X}_1 + \mathbf{X}_2 \| \leqslant \| \mathbf{X}_1 \| + \| \mathbf{X}_2 \|$ (triangular inequality), (iii) given a complex number α, $\| \alpha \mathbf{X} \| = | \alpha | \, \| \mathbf{X} \|$, and (iv) $\| \mathbf{X} \| = 0$ implies $\mathbf{X} = 0$. The explicit form of the norm is immaterial in questions of stability and convergence. We shall use one of the two explicit forms introduced above in calculations.

Definition 3. A vector function $\mathbf{f}(\mathbf{X}, t)$ is said to satisfy a Lipschitz condition in the region R defined by $\| \mathbf{X} - \mathbf{X}_0 \| \leq b$ and $t_0 \leqslant t \leqslant t_0 + a$ if there exists a constant m such that

$$\| f(\mathbf{X}_1 , t) - f(\mathbf{X}_2 , t) \| \leqslant m \| \mathbf{X}_1 - \mathbf{X}_2 \| \tag{6}$$

holds for all $\mathbf{X}$ and t in R. One observes that the Lipschitz condition implies continuity of $f(\mathbf{X}, t)$ in $\mathbf{X}$ because it implies the existence of bounded partial derivatives of $\mathbf{f}(\mathbf{X}, t)$, i.e., $\partial f_i / \partial X_j$ in R. It does not imply, however, continuity in t.

We now state the theorem concerning the existence and uniqueness of the solution of $\dot{\mathbf{X}} = \mathbf{f}(\mathbf{X})$. The notions involved in this theorem will be useful later in the statement of the question of stability. Although we are interested primarily in the stability of the autonomous systems, we shall state the existence and uniqueness theorem for nonautonomous systems, which are described by

$$\dot{\mathbf{X}} = \mathbf{f}(\mathbf{X}, t) \tag{7}$$

Theorem 1 (Existence and uniqueness). Let $\mathbf{f}(\mathbf{X}, t)$ be continuous in $\mathbf{X}$ and t and satisfy a Lipschitz condition in a region R. Let M be the maximum of the continuous function $\| \mathbf{f}(\mathbf{X}, t) \|$ in R. Then, $\dot{\mathbf{X}} = \mathbf{f}(\mathbf{X}, t)$ has a unique solution for $t_0 \leqslant t \leqslant t_0 + \alpha$ satisfying $\mathbf{X}(t_0) = \mathbf{X}_0$ in R, where $\alpha = \min(a, b/M)$.

It is noted that the theorem guarantees the existence and uniqueness of the solution only in a finite time interval. It does not imply that the solutions can be extended for arbitrarily large values of t. The following example will illustrate this point and enable us to introduce the concept of "finite escape time."

Consider the scalar equation $\dot{X} = X^2$, which has a solution

$$X(t) = 1/[(1/X_0) + t_0 - t] \tag{8}$$

which becomes infinite at $t_\infty = t_0 + (1/X_0)$. The requirements of the existence and uniqueness theorem are satisfied with the following constants:

$$a = \infty, \qquad |X - X_0| \leqslant b$$

$$|X_1^2 - X_2^2| = |X_1 + X_2|\,|X_1 - X_2| \leqslant 2(X_0 + b)\,|X_1 - X_2|$$
$$= k\,|X_1 - X_2|$$

$$M = (X_0 + b)^2, \qquad \alpha = \min[\infty, b/(X_0 + b)^2] = b/(X_0 + b)^2$$

Thus, the theorem guarantees the uniqueness and the existence of the solution in the interval $t_0 \leqslant t \leqslant t_0 + [b/(X_0 + b)^2]$. The time t_∞ at which $X(t_\infty) = \infty$ is not in this interval, because $(1/X_0) \geqslant b/(X_0 + b)^2$ for any value of X_0 and b.

Definition 4. A solution is said to have a "finite escape time" denoted by t_∞ if $X(t) \to \infty$ as $t \to t_\infty < \infty$.

The concept of finite escape time is important in reactor dynamics because, as will be seen later, the kinetic equations of some reactor models admit solutions with a finite escape time. Therefore, it is interesting to find sufficient conditions under which there is no finite escape time.

Theorem 2 (Nonexistence of finite escape time). If $\mathbf{f}(\mathbf{X}, t)$ satisfies a global Lipschitz condition, i.e., everywhere in $\mathbf{X}$ and uniformly in t for all $t \geqslant t_0$, and if $\mathbf{f}(0, t) \equiv 0$ for all $t \geqslant t_0$, then there can be no finite escape time.

Since the proof is short and enables us to introduce Gronwall's lemma (which provides a very useful inequality), we shall present it below.

By hypothesis,

$$\|\mathbf{f}(\mathbf{X}_1, t) - \mathbf{f}(\mathbf{X}_2, t)\| \leq m\,\|\mathbf{X}_1 - \mathbf{X}_2\|$$

holds for all $\mathbf{X}_1$, $\mathbf{X}_2$ and $t \geqslant t_0$. We integrate $\dot{\mathbf{X}} = \mathbf{f}(\mathbf{X}, t)$ and get

$$\mathbf{X}(t) = \mathbf{X}(t_0) + \int_{t_0}^{t} \mathbf{f}(\mathbf{X}(t'), t')\,dt'$$

whence

$$\|\mathbf{X}(t)\| \leq \|\mathbf{X}(t_0)\| + \int_{t_0}^{t} \|\mathbf{f}(\mathbf{X}(t'), t')\|\,dt'$$

Using the Lipschitz condition with $\mathbf{X}_2 = 0$ and employing $\mathbf{f}(0, t) = 0$, we get

$$\| \mathbf{X}(t) \| \leqslant \| \mathbf{X}(t_0) \| + m \int_{t_0}^{t} \| \mathbf{X}(t') \| \, dt' \tag{9}$$

Gronwall's lemma, which will be stated below, proves that, if (9) is true, so is the following inequality:

$$\| \mathbf{X}(t) \| \leqslant e^{m(t-t_0)} \| \mathbf{X}(t_0) \|, \qquad t > t_0 \tag{10}$$

This inequality implies that $\| \mathbf{X}(t) \|$ is finite for finite $(t - t_0)$, proving the assertion.

Equation (10) can also be interpreted to imply that the solutions of $\dot{\mathbf{X}} = \mathbf{f}(\mathbf{X}, t)$ are always of exponential order, and thus Laplace-transformable if $\mathbf{f}(\mathbf{X}, t)$ satisfies the Lipschitz condition everywhere in $\mathbf{X}$ and uniformly in t, and if $\mathbf{f}(0, t) = 0$, for all t.

Lemma 1 (Gronwall–Bellman lemma [31]). If $u(t) \geqslant 0$ and $v(t) \geqslant 0$ are given functions in $0 \leqslant t < +\infty$ such that u is continuous and v is L-integrable in every finite interval, and if, for some nonnegative constant c we have

$$u(t) \leqslant c + \int_0^t u(t') \, v(t') \, dt', \qquad t \geqslant 0 \tag{11a}$$

then we also have

$$u(t) \leqslant c \exp \int_0^t v(t') \, dt' \tag{11b}$$

The inequality (10) follows from (9) by taking $u(t) \equiv \| \mathbf{X}(t) \|$, $c = \| \mathbf{X}(t_0) \|$, $v(t) \equiv m$.

It is instructive to see clearly where the example $\dot{X} = X^2$ fails to satisfy the requirements of the Theorem 2, and thus admits solutions with a finite escape time. In this example, $f(X, t) \equiv X^2$, and there can be no constant $m > 0$ such that $| X_1^2 - X_2^2 | = | X_1 + X_2 | \, | X_1 - X_2 | \leqslant m \, | X_1 - X_2 |$ will hold for any X_1 and X_2. If, however, X_1, X_2 are restricted to a finite volume, then the Lipschitz condition is satisfied locally and existence and uniqueness is ensured in a finite interval as discussed previously.

We close this subsection by the following general remarks: As mentioned above, systems with time delays cannot be described by a finite number of first-order ordinary differential equations. A basic difference between systems with and without time delays is that, in the latter case, a unique solution is determined by specifying a finite number of initial values $X_1(0), X_2(0), \ldots, X_N(0)$, whereas a unique solution of systems with time delays requires, in general, a nondenumerably infinite number of values of the variables $X_1, X_2, \ldots, X_N$ assumed by them in the past. In other words, the "initial values" in the systems without delays must be replaced by the "initial curves" representing the variations of the dynamical variables in the past, when time delays are present. This distinction plays an important role in the definition of stability in these two types of systems.

C. Lyapunov's Theorem (The Indirect or First Method)

We are now ready to discuss the connection between the stability of the linear system and the asymptotic stability (in the small) of the original nonlinear system.

We start with $\dot{\mathbf{X}} = \mathbf{f}(\mathbf{X})$ and break $\mathbf{f}(\mathbf{X})$ into two parts as

$$\mathbf{f}(\mathbf{X}) = \mathsf{A}\mathbf{X} + \mathbf{g}(\mathbf{X}) \tag{12}$$

where A is any constant square matrix, and $\mathbf{g}(\mathbf{X})$ is the difference $\mathbf{f}(\mathbf{X}) - \mathsf{A}\mathbf{X}$. This separation may be achieved, for example, by expanding $\mathbf{f}(\mathbf{X})$ into Taylor series about the singular point $\mathbf{X} = 0$ and separating the linear terms. Thus, $\mathbf{g}(\mathbf{X})$ will represent the nonlinear terms.

Suppose that we can choose $\mathsf{A} \neq 0$ in such a way that

$$\lim_{\|\mathbf{X}\| \to 0} \|\mathbf{g}(\mathbf{X})\| / \|\mathbf{X}\| = 0 \tag{13}$$

where $\|\mathbf{g}(\mathbf{X})\| = \sum_i |g_i(\mathbf{X})|$, $\|\mathbf{X}\| = \sum_i |X_i|$, which will be used in this section to replace the Euclidean norm $(\sum_i |X_i|^2)^{1/2}$ for convenience.

When (13) is satisfied, the equation

$$\dot{\mathbf{X}} = \mathsf{A}\mathbf{X} \tag{14}$$

may be called the linearized equation because $\mathbf{g}(\mathbf{X})$ vanishes faster than $\mathbf{X}$. The condition (13) is referred to as the nonlinearity condition.

The stability of the linearized set in the vector form is determined by the real part of the roots of the characteristic equation

$$|s\mathsf{I} - \mathsf{A}| = 0 \tag{15}$$

where $|s\mathsf{I} - \mathsf{A}|$ is the determinant of the square matrix $s\mathsf{I} - \mathsf{A}$, and I is the identity matrix. If the roots have negative real parts, then the linearized system is asymptotically stable, as already discussed in the previous sections.

THEOREM 3 (Lyapunov). If the linearized system as defined above is asymptotically stable, then the equilibrium state $\mathbf{X}(t) \equiv 0$ of the nonlinear system $\dot{\mathbf{X}} = \mathbf{A}\mathbf{X} + \mathbf{g}(\mathbf{X})$ is also asymptotically stable, i.e., there is a region $\| \mathbf{X} \| < \delta$ such that all the solutions satisfying $\| \mathbf{X}(0) \| < \delta$ approach zero as $t \to \infty$.

If one of the roots of the characteristic equation is zero, or if two of them are purely imaginary, the stability of the nonlinear systems depends more on the specific nature of $\mathbf{g}(\mathbf{X})$ than on the nonlinearity condition (13).

The nonlinearity condition is satisfied in reactor kinetic equations when the feedback is linear, because the nonlinear term is quadratic. When the feedback is nonlinear, the nonlinearity condition must be verified by considering the specific form of the nonlinear terms. However, the nonlinearity condition is almost always met in the problems encountered in physical application.

As a simple example in which (13) is not satisfied, we consider a one-component system described by

$$\dot{X} = -X + | X |^{1/2} \tag{16}$$

This equation has two equilibrium states, $X = 0$ and $X = 1$. We investigate the stability of the origin. One may be tempted to linearize (16) by simply neglecting $| X |^{1/2}$. The linear portion $\dot{X} = -X$ of (16) is obviously asymptotically stable. The nonlinearity condition is not satisfied, however, because $| X |^{1/2}/| X | = 1/| X |^{1/2}$ diverges as $| X | \to 0$. It is easy to show that the equilibrium state $X \equiv 0$ of (16) is unstable because its solution is

$$X(t) = \{1 - [1 - X^{1/2}(0)] \, e^{-t/2}\}^2$$

and approaches one for all initial conditions.

Lyapunov's theorem also provides sufficient conditions to determine the size of the region of allowable initial disturbances for asymptotic stability. These conditions are different in various proofs of the theorem, and define different size regions. In all cases, the conditions are overly restrictive and difficult to apply to complicated models. We mention here the condition obtained by Coddington and Levinson [31]:

COROLLARY. Any solution $\mathbf{X}(t)$ of $\dot{\mathbf{X}} = \mathbf{A}\mathbf{X} + \mathbf{g}(\mathbf{X})$ asymptotically approaches zero if it satisfies

$$\| \mathbf{X}(0) \| < \delta/k \tag{17a}$$

where δ is determined by

$$\| g(\mathbf{X}) \| \leqslant \sigma \| \mathbf{X} \|/k \qquad \text{for} \quad \| \mathbf{X} \| < \delta(\sigma) \tag{17b}$$

and where σ is the magnitude of the least-negative, real part of the characteristic roots, and k is the number of dependent variables.

We shall now illustrate the application of this condition to reactors. We consider a one-temperature feedback model and ignore the delayed neutrons:

$$\dot{z}_1 = -\alpha(1 + z_1)\, z_2 \tag{18a}$$

$$\dot{z}_2 = d(z_1 - z_2) \tag{18b}$$

where

$$z_1 = (P - P_0/P_0), \qquad z_2 = (T - T_0)/T_0 \tag{18c}$$

clearly, we have

$$\mathsf{A} = \begin{bmatrix} 0 & -\alpha \\ d & -d \end{bmatrix}, \qquad \mathbf{g}(\mathbf{X}) = \begin{bmatrix} -\alpha z_1 z_2 \\ 0 \end{bmatrix}, \qquad \mathbf{X} = \begin{bmatrix} z_1 \\ z_2 \end{bmatrix} \tag{19}$$

The characteristic equation

$$\begin{vmatrix} s & \alpha \\ -d & s + d \end{vmatrix} = 0$$

yields

$$s_{1,2} = \tfrac{1}{2}[-d \pm (d^2 - 4\alpha d)^{1/2}] \tag{20a}$$

and

$$\sigma = \mathrm{Re}\{\tfrac{1}{2}[d - (d^2 - 4\alpha d)]^{1/2}\} \tag{20b}$$

Since $d > 0$, the linearized system is asymptotically stable.

The nonlinearity condition is satisfied at $\mathbf{X} = 0$. Indeed,

$$\| \mathbf{g}(\mathbf{X}) \| / \| \mathbf{X} \| = \alpha \mid z_1 \mid \mid z_2 \mid / (\mid z_1 \mid + \mid z_2 \mid)$$

which vanishes as $| z_1 | \to 0$ or $| z_2 | \to 0$.

To determine the size of the region of initial disturbance, we note that $k = 2$, and (17b) is

$$\alpha \mid z_1 z_2 \mid \leqslant \tfrac{1}{2}\sigma(\mid z_1 \mid + \mid z_2 \mid)$$

or

$$\mid z_1 z_2 \mid / (\mid z_1 \mid + \mid z_2 \mid) \leqslant \sigma/2\alpha \tag{21}$$

We want to find $\delta(\sigma)$ such that, when $| z_1 | + | z_2 | < \delta$, (21) will be true. Figure 6.9.1 shows the region corresponding to $| z_1 | + | z_2 | < \delta$: The left-hand side of (21) attains its maximum of $\delta/4$ in the indicated square region at $| z_1 | = | z_2 | = \delta/2$. Hence, if

$$\delta = 2\sigma/\alpha \tag{22}$$

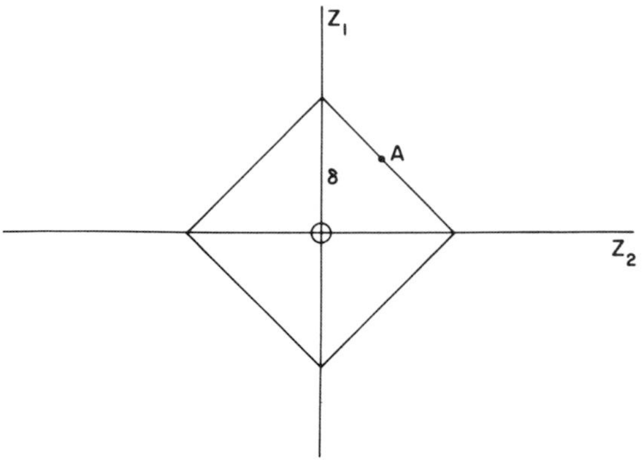

FIGURE 6.9.1. A region defined by $|z_1| + |z_2| < \delta$.

then $|z_1| + |z_2| < \delta$ will imply (21). Substituting (22) and (20) into (17a) yields the desired region

$$|z_1(0)| + |z_2(0)| \leqslant \tfrac{1}{2} \operatorname{Re}[y - (y^2 - 4y)^{1/2}] \tag{23a}$$

where

$$y = d/\alpha \tag{23b}$$

Recall that z_1 and z_2 are the deviations from the equilibrium power and temperature relative to their equilibrium values. Letting $z_2(0) = 0$, we find that the allowable disturbance in power, i.e., $|z_1(0)|$, is determined solely by the ratio d/α.

In order to point out that the region (23a) is overly restrictive, we mention that the reactor model considered here is asymptotically stable in the large, and thus the allowable region actually extends over the entire (z_1, z_2) plane. The condition (23a) yields reasonable estimates, however, in other reactor models where the operating power level is near the critical power level.

The foregoing analysis of linearization was based on the differential form. We can obtain similar results using the integrodifferential form with a linear feedback functional [32]. Let $G(t)$ and $H(s)$ denote the linear feedback kernel and its Laplace transform, and define $L(t)$ as the response of the linearized system to a unit reactivity impulse, i.e.,

$$L(t) \equiv \mathscr{L}^{-1}\{Z(s)/[1 - P_0 Z(s)\, H(s)]\} \tag{24}$$

Since the linearized system is assumed to be asymptotically stable, one can find a constant K such that

$$|L(t)| \leqslant K e^{-\sigma t}, \qquad t \geqslant 0 \tag{25}$$

where σ is the least-negative, real part of the characteristic equation $Y(s) - P_0H(s) = 0$. Since $L(0) = 1$ (when the unit of time is chosen to make $l/\beta = 1$), we have, from (25), $K \geqslant 1$.

Assume that the kernel $G(t)$ is absolute-integrable as required by the stability of the feedback (cf 5.1, Eq. 23b), i.e.,

$$N \equiv \int_0^\infty |G(t)|\,dt < \infty \tag{26}$$

and that the reactor is operated at the equilibrium flux P_0 until $t = 0$, at which time a power perturbation $z(0) = P(0)/P_0$ is introduced. The latter assumption implies that $p(t) = 0$ for $t < 0$ in (6.1, Eq. 1), which reduces with $l/\beta = 1$ to

$$\dot{z}(t) = (1 + z)\int_0^t G(t-u)\,z(u)\,du - z(t) + \int_0^t D(t-u)\,z(u)\,du \tag{27}$$

Then, we have the following statement by Akcasu [32]:

Corollary. Any solution $z(t)$ of (27) asymptotically approaches zero, as $t \to \infty$, as

$$|z(t)| < \sigma/NK \tag{28a}$$

if its initial value satisfies

$$|z(0)| < \sigma/NK^2 \tag{28b}$$

We shall apply this result to the one-temperature model described above. The feedback kernel and its Laplace transform are $G(t) = -\alpha d e^{-dt}$ and $H(s) = -\alpha d/(s + d)$. The characteristic equation is $s - H(s) = 0 = s + [\alpha d/(s + d)]$ (note that P_0 is normalized to unity). The linear response function $L(s)$ follows from (24) as

$$\begin{aligned} L(t) &= \mathscr{L}^{-1}\{(s + d)/(s^2 + sd + \alpha d)\} \\ &= [1/(s_1 - s_2)][(s_1 + d)\,e^{s_1 t} - (s_2 + d)\,e^{s_2 t}] \end{aligned} \tag{29}$$

where s_1, s_2 are given by (20a). Observe that $L(0) = 1$. One can easily verify, by finding the maximum of $L(t)$, that

$$|L(t)| \leqslant e^{-\sigma t}$$

which implies $K = 1$. Finally, one gets N, substituting $|G(t)| = \alpha d e^{-dt}$ into (26), as $N = \alpha$. Hence,

$$|z(0)| \leqslant \sigma/\alpha, \qquad |z(t)| \leqslant \sigma/\alpha \tag{30}$$

It is interesting to note that both methods yield the same region in this particular application. The second method is applicable to reactor models with time delays, whereas the first method applies only to reactor models without a memory.

In the derivation of (28), it was assumed that the reactor was operated at equilibrium for $t < 0$, and the perturbation was introduced in the power only. The method can be extended to include any arbitrary power variation (initial curve) in the past.

PROBLEMS

1. Show that all the roots of $a_0S^4 + a_1S^3 + a_2S^2 + a_3S + a_4 = 0$ lie to the left of the imaginary axis if and only if $a_0 > 0$, $a_1 > 0$, $a_3 > 0$, $a_4 > 0$ and $a_3(a_1a_2 - a_0a_3) > a_4a_1^2$ (cf Sect. 6.2A).

2. Find the additional conditions besides the positiveness of the coefficients of a fifth-order polynomial equation for all its roots to have negative real parts (cf Sect. 6.2A).

3. Investigate the stability of a fourth-order system discussed in Problem 1 using the Routh algorithm and establish the connection with Hurwitz determinants (cf Sect. 6.2B and Ref. [6–8]).

4. Show that K_p defined in (6.3, Eq. 8) is the net rate of prompt production of equilibrium xenon per fission.

5. Verify that in third- and fourth-order polynomials the conditions $a_1a_2 = a_0a_3$ and $a_3(a_1a_2 - a_0a_3) = a_4a_1^2$ respectively, yield the threshold for oscillatory instability (Hint: substitute $s = i\omega$ and eliminate ω^2) (cf Sect. (6.2, Eq. 3) and Ref. [7] for a general discussion).

6. Show that the equation $(z^2 + pz + q)e^z + rz^n = 0$ where p, q, and r are real and n is a positive integer cannot represent a stable system if $n \geqslant 3$ and $r \neq 0$. (See Ref. [14] for the necessary and sufficient conditions for all the roots to have negative real parts when $n \leq 2$ and $p > 0$, $q \geq 0$ and $r \neq 0$).

7. Using Thm. 4 of Sect. 6.4, show that $F(z) = p(\sin z) - z \cos z = 0$ cannot have all real roots if $p > 1$, and has all real roots if $p < 1$ (see Ref. [14]).

8. (a) Show that the necessary and sufficient condition for all the roots of $q + z\,e^z = 0$, where q is real and positive, to have negative real parts is $q < \pi/2$ (Hint: Use Thm. 2 of Sect. 6.4 for necessity and Thm. 3 for sufficiency).

(b) Using the conclusion of (a) show that $y^2 + q[\cos y - y \sin y] > 0$ holds for all real y if $0 < q < \pi/2$. (Hint: use 6.4A, Eq. 4.)

9. The necessary and sufficient conditions that all the roots of $z^2 e^z + pz + q = 0$ have negative real parts are (a) $0 < p < \pi/2$ and (b) $0 < q < a_p{}^2(\cos a_p)$ where a_p is the root of $\sin a = p/a$ in the open interval $(0, \pi/2)$.

10. Investigate the stability of the system described in (6.4, Eqs. 14) using the Nyquist stability criterion, and compare the conditions for stability obtained by this method to those derived in Sect. 6.4 using Pontryagin's technique. (Hint: First show that $H(s) = -[B + \epsilon A \exp(-s\tau)]/(s + \epsilon)]$ and $Z(s) = 1/s$, then plot $P_0 H(i\omega)\, Z(i\omega)$.)

11. Show that, if both $\rho(x)$ and $\phi(x)$ are symmetric about $x = h/2$, then

$$\delta_0/K_c = -\int_0^h \rho(x)\, dx \int_0^x \phi(y)\, dy = -\tfrac{1}{2}\int_0^h \rho(x)\, dx \int_0^h \phi(x)\, dx$$

Hint: Change the origin so that $-h/2 \leqslant x \leqslant h/2$.

12. Show that

$$\int_0^h \rho(x)\, dx \int_0^x \phi(x - y)\, e^{-sy/v_c}\, dy = \int_0^h dx\, e^{-sx/v_c} \int_x^h \rho(y)\, \phi(y - x)\, dy$$

13. Perform the integration for the three different reactivity shapes to show that:

(a) For $\rho(x) = -\alpha_c$,

$$s\left\{l + \sum_i [\beta_i/(s + \lambda_i)]\right\}$$
$$+ (K_c\alpha_c h^2 2/\pi)\{[s^2 t_c{}^2 + \pi^2(1 - e^{-st_c})/2]/st_c[(st_c)^2 + 4\pi^2]\} = 0$$
$$\delta_0 = K_c\alpha_c h^2/\pi$$

(b) For $\rho(x) = -\alpha_c\phi^2(x)$,

$$s\left\{l + \sum_i [\beta_i/(s + \lambda_i)]\right\} + \{\alpha_c K_c h^2/\pi[\pi^2 + (st_c)^2]$$
$$\times (\tfrac{4}{3}st_c + \{2\pi^4(1 - e^{-st_c})/st_c[(st_c)^2 + 4\pi^2]\}) = 0$$
$$\delta_0 = K_c\alpha_c h^2/2\pi$$

(c) For $\rho(x) = -\alpha_c \delta(\frac{1}{2}h - x)$,

$$s\left\{l + \sum_i [\beta_i/(s + \lambda_i)\right\} + K_c \alpha_c h^2 \{[st_c + \pi e^{-st_c/2}]/[\pi^2 + (st_c)^2] = 0$$

$$\delta_0 = K_c \alpha_c h^2/\pi$$

14. Show, in Problem 13, on the instability line where $s = i\omega$, $\omega^2 \geqslant 0$, that, if $\beta_i = 0$ (no delayed neutrons), then the following hold:

(a) For $\rho(x) = -\alpha_c$,

$$\omega t_c = \pm n\pi, \qquad n \geqslant 2$$

(b) For $\rho(x) = -\alpha_c \phi^2(x)$,

$$\omega t_c = \pm n\pi, \qquad n \geqslant 3$$

(c) For $\rho(x) = -\alpha_c \delta(\frac{1}{2}h - x)$,

$$\omega t_c = \pm(2n + 1)\,\pi, \qquad n \geqslant 1$$

15. Show that, if $Q_N = \delta_n t_c/l$, where δ_n is the critical value of δ on the stability line, then the following hold:

(a) For $\rho(x) = -\alpha_c$,

$$Q_n = 2n^2(n^2 - 1)\pi^2/\{4n^2 - 2[1 - (-1)^n]\}, \qquad n \geqslant 2$$

(b) For $\rho(x) = -\alpha_c \phi^2(x)$,

$$Q_n = 3n(n^2 - 1)\,\pi^2/(8n + \{12[1 - (-1)^n]/n(n^2 - 4)\}), \qquad n \geqslant 3$$

(c) For $\rho(x) = -\alpha_c \delta(\frac{1}{2}h - x)$,

$$Q_n = 2(2n + 1)[(2n + 1)^2 - 1]\,\pi/[2n + 1 - (-1)^n], \qquad n \geqslant 1$$

16. Show that, if $b^2 = -c^2 < 0$, if $c \tanh c > \frac{1}{2}\omega t_c \mid \tan(\frac{1}{2}\omega t_c)|$, and if $\frac{1}{2}(2n + 1)\pi < \frac{1}{2}\omega t_c < (n + 1)\pi$ with $n \geqslant 0$, then instabilities always exist at some power levels.

17. Show that (6.8, Eq. 22a) is valid independently of the number of delayed groups.

REFERENCES

1. J. C. Maxwell, *Proc. Roy. Soc.* **16**, 270 (1868).
2. P. H. Hammond, "Feedback Theory and Its Applications," p. 65. English Universities Press, London, 1958.

3. E. P. Popov, "The Dynamics of Automotive Control Systems." Addison-Wesley, Reading, Massachusetts, 1962.
4. E. J. Routh, "Dynamics of a System of Rigid Bodies," 6th ed., Part II, pp. 228–229. Dover, New York, 1955.
5. A. Hurwitz, Über die Bedingungen unter Welchen ein Gleichung nur Wurzeln mit negativen reeten Teilen beritz. *Math. Ann.* **46**, 273–284 (1895).
6. E. A. Guillemin, "The Mathematics of Circuit Analysis." Wiley, New York, 1949.
7. W. J. Duncan, "The Control and Stability of Aircraft," pp. 117–120. Cambridge Univ. Press, London and New York, 1952.
8. D. Graham and D. McRuer, "Analysis of Nonlinear Control Systems," pp. 457–460. Wiley, New York, 1961.
9. J. Chernick, *Nucl. Sci. Eng.* **8**, 233 (1960).
10. J. Chernick, G. Lellouche, and W. Wollman, The effect of temperature on xenon instability. *Nucl. Sci. Eng.* **10**, 120–131 (1961).
11. J. Chernick, The dependence of reactor kinetics on temperature. BNL-173. Brookhaven Nat. Lab., Upton, New York, 1951.
12. L. Shotkin and F. Abernathy, Linear stability of the thermal flux in a reflected core containing xenon and temperature reactivity feedback. *Nucl. Sci. Eng.* **15**, 197–212 (1963).
13. L. S. Pontriagin, "On the Zeros of Some Elementary Transcendental Functions" (Am. Math. Soc. Transl. Ser. 2), Vol. I, p. 95. 1955.
14. R. Bellman and K. L. Cooke, "Differential-Difference Equations." Academic Press, New York, 1963.
15. H. Jacobowitz, L. M. Shotkin, and Z. A. Akcasu, *Trans. Am. Nucl. Soc.* **9**, No. 1 (1966).
16. I. M. Horowitz, "Synthesis of Feedback Systems." Academic Press, New York, 1963.
17. J. Miida and M. Suda, *Nucl. Sci. Eng.* **11**, 55 (1961).
18. H. A. Bethe, APDA-117. 1956, Atomic Power Development Associates, Detroit, Michigan.
19. R. O. Brittan, ANL-5577. Argonne Nat. Lab., Argonne, Illinois, 1955.
20. R. T. Ackroyd, G. H. Kinchin, J. E. Mann, and J. D. McCullen, *Proc. U. N. Int. Conf. Peaceful Uses At. Energy, 2nd, Geneva,* 1958, P/1462. United Nations, New York, 1958.
21. D. Little and M. A. Schultz, *IRE Trans. Nucl. Sci.* **4**, 30 (1957).
22. E. P. Gyftopoulos and J. Devooght, *Nucl. Sci. Eng.* **8**, 244 (1960).
23. W. K. Ergen, *J. Appl. Phys.* **25**, 702 (1956).
24. A. Ruiz, HW-SA-2012. Hanford Atomic Products (1960).
25. A. Z. Akcasu, *Nucl. Sci. Eng.* **3**, 456 (1958).
26. T. A. Welton, System kinetics. *Nucl. Reactor Theory Proc. Symp. Appl. Math., 11th, New York,* 1959. Amer. Math. Soc., Providence, Rhode Island, 1961.
27. R. R. Smith, R. O. Haroldsen, and F. D. McGinnis, ANL-6863. Argonne Nat. Lab., Argonne, Illinois, May 1964.
28. H. B. Smets, *Nucl. Sci. Eng.* **25**, 236 (1966).
29. J. A. DeShong, Jr. and W. C. Lipinski, ANL-5850. Argonne Nat. Lab., Argonne, Illinois, July 1958.
30. A. R. Struble, "Nonlinear Differential Equations." McGraw-Hill, New York, 1962.
31. E. A. Coddington and N. Levinson, "Theory of Ordinary Differential Equations." McGraw-Hill, New York, 1955.
32. Z. Akcasu, *Nucl. Sci. Eng.* **24**, 88 (1966).

CHAPTER 7

Nonlinear Stability Analysis

In this chapter, we discuss the concepts of stability in nonlinear reactor systems and present criteria for asymptotic stability in the large as well as in a finite domain of initial disturbances. The nonlinearity of a reactor system is primarily due to the quadratic nature of the point kinetic equations in the presence of feedback. However, additional and more complex nonlinearities may be involved if the feedback mechanism is described by a nonlinear model. Both types of nonlinearities are considered in this chapter. The feedback models involving pure time delays, which require special treatment, are also included with examples.

The asymptotic stability of a reactor system in which the feedback is described as a lumped-parameter system is investigated using Lyapunov's second method. Systems involving distributed parameters are analyzed by a different approach in which the asymptotic behavior of the solution is discussed more directly using the kinetic and feedback equations. The effect of delayed neutrons on asymptotic stability in the large, as well as in a finite domain of initial values, is considered.

Finally, we discuss the questions of the boundedness of the solutions when they are not asymptotically stable, and of the existence of finite escape times.

A fair amount of mathematical detail is introduced in this chapter in order to be able to state precisely and to prove various stability theorems. We think that it is essential to know the proofs of these theorems in order to apply the resulting stability criteria correctly, as well as to appreciate why and how these criteria ensure asymptotic stability.

7.1. Concepts of Stability

In this section, we present the definitions of stability in nonlinear systems from the point of view of reactor applications. The intuitive idea of stability introduced in the previous chapter concerns the temporal behavior of a free dynamical system following perturbations in some of its dynamical variables. If all the dynamical variables eventually return to their equilibrium values before the occurrence of the disturbances, then the equilibrium state is stable. In linear systems, this simple concept of stability can easily be expressed as a precise mathematical statement in terms of the real parts of the roots of the characteristic equation, and several simple stability criteria can be derived to ascertain the stability of the system using precise definitions. In the case of nonlinear dynamical systems, however, the concept of stability turns out to be very complex, requiring distinctions among various kinds of stability. As a result, a large variety of precise definitions of stability have been proposed in the theory of nonlinear differential equations, often with very subtle differences among them. The purpose of this section is to present the principal definitions of stability which have been used in the analysis of reactor systems. In order to introduce the various concepts and definitions of stability, we shall first consider an autonomous nonlinear differential system of the form

$$\dot{\mathbf{x}} = \mathbf{f}(\mathbf{x}) \tag{1}$$

(the integrodifferential form of the kinetic equations will be considered later), and assume that $\mathbf{f}(0) = 0$, and that $\mathbf{f}(\mathbf{x})$ satisfies a Lipschitz condition in $\mathbf{x}$ for $\mathbf{x}$ in a region R containing the origin (cf 6.9, Eq. 6). Thus, $\mathbf{x}(t) \equiv 0$ is the unique solution satisfying the initial condition $\mathbf{x}(0) = 0$. We refer to this solution as the equilibrium state, and attempt to examine its stability.

Let the system (1) be perturbed slightly from its equilibrium state at the origin by forcing the state vector $\mathbf{x}(t)$ to be at $\mathbf{x}_0$ in R at $t = 0$. Then, there is a unique solution of (1) starting from $\mathbf{x}_0$. The simplest and perhaps the oldest concept of stability is that all the solutions of (1) starting sufficiently "close" to the origin remain in a correspondingly small neighborhood of the origin for all subsequent times. This idea of stability is expressed precisely in the following definition, due to Lyapunov [1]:

DEFINITION 1. The equilibrium state at the origin is "stable" if, for any given $\epsilon > 0$, there exists a $\delta(\epsilon) > 0$ such that $\| \mathbf{x}(0) \| < \delta(\epsilon)$ implies

$$\| \mathbf{x}(t) \| < \epsilon \qquad \text{for} \quad t \geqslant 0$$

We note that in this definition of stability, solutions are not required to return to equilibrium after any small perturbation. In most reactor applications, one is interested in the stronger kind of stability defined in the following way:

DEFINITION 2. The solution $\mathbf{x}(t) \equiv 0$ is "asymptotically stable" if it is stable, and in addition, if there exists a fixed $\delta > 0$ such that $\| \mathbf{x}(t) \| \to 0$ as $t \to \infty$ whenever $\| \mathbf{x}(0) \| < \delta$.

Loosely speaking, the stability of $\mathbf{x}(t) \equiv 0$ implies that a solution starting near the origin remains near the origin in the future ($t \geqslant 0$); and the asymptotic stability of $\mathbf{x}(t) \equiv 0$ implies, in addition, $\| \mathbf{x}(t) \| \to 0$ as $t \to \infty$.

The stability and the asymptotic stability of $\mathbf{x}(t) \equiv 0$ as defined above are called stability and asymptotic stability "in the small," respectively, because they involve small deviations from the equilibrium state, as characterized by $\delta > 0$. The size of the neighborhood of the origin, i.e., $\| \mathbf{x}(0) \| < \delta$, is unimportant insofar as there exists such a neighborhood.

It is natural to extend the definition of asymptotic stability as follows.

DEFINITION 3. The solution $\mathbf{x}(t) \equiv 0$ is asymptotically stable "in the large" (or "globally asymptotically stable") if it is stable, and if, in addition, every solution $\mathbf{x}(t)$, regardless of its initial magnitude, converges to zero as $t \to \infty$.

It is clear that a necessary condition for asymptotic stability in the large is that there is only one equilibrium state in the entire phase space, i.e., the algebraic set of equations $\mathbf{f}(\mathbf{x}) = 0$ has only one root ($\mathbf{x} = 0$). This is the reason for assuming a unique equilibrium state in Sec. 6.1.

It is noted that both stability and asymptotic stability have been defined with respect to the initial values $\mathbf{x}(0) = \mathbf{x}_0 = \{x_{0_1}, x_{0_2}, \ldots, x_{0_N}\}$. This is natural because the initial values completely and uniquely determine the solution for $t \geqslant 0$ when the system is described by a set of nonlinear first-order equations. However, as pointed out previously, a physical system containing time lags, or heredity, cannot be described by a nonlinear differential system. The future of such systems is determined by their entire past history. Thus, the definitions of stability and asymptotic stability must be modified in such systems.

For this purpose, we consider the standard form of the point kinetic equations (cf 3.1, Eq. 1), and assume, for the sake of clarity, that the feedback functional is linear. The free oscillations of such a reactor following arbitrary perturbations which terminate at $t = 0$ are governed by

$$(l/\beta)\dot{p} = (p + P_0)\left[\int_{-\infty}^{t} G(t-u)p(u)\,du\right] - p(t) + \sum_i \lambda_i C_i \tag{2a}$$

$$\dot{C}_i = a_i p - \lambda_i C_i\,, \qquad i = 1,\ldots, I; \qquad t > 0 \tag{2b}$$

In these equations, the rate of change of the incremental power $p(t)$ at a given time $t > 0$ depends on its temporal behavior at previous times. A unique solution of this set of integrodifferential equations requires specification of the initial values $p(0)$ and $C_i(0)$ as well as the variation of $p(t)$ for $t < 0$. Since the delayed neutron precursors are not produced or removed externally, we can in fact express $C_i(0)$ in terms of the past history of the power as

$$C_i(0) = a_i \int_{-\infty}^{0} e^{\lambda_i u} p(u)\,du \tag{3}$$

and thereby eliminate $C_i(0)$ as independent initial conditions. For the same reason, it is convenient to eliminate $C_i(t)$ between (2a) and (2b), and to work directly with the integrodifferential form of the kinetic equations (cf. 5.1, Eq. 25)

$$(l/\beta)\,\dot{p} = (p + P_0)\left[\int_{-\infty}^{t} G(t-u)\,p(u)\,du\right] - p(t) + \int_{-\infty}^{t} D(t-u)\,p(u)\,du, \qquad t > 0 \tag{4}$$

The temporal behavior of $p(t)$ for $t < 0$ is the response of the reactor to the external perturbations, and is determined by

$$(l/\beta)\,\dot{p} = (p + P_0)\left[\delta k(t) + \int_{-\infty}^{t} G(t-u)\,p(u)\,du\right] - p + \int_{-\infty}^{t} D(t-u)\,p(u)\,du + (l/\beta)S(t), \qquad t < 0 \tag{5}$$

where $\delta k(t)$ and $S(t)$ are the external reactivity change and the neutron source, respectively, and characterize the perturbation which terminates at $t = 0$. We shall refer to $p(t)$ for $t < 0$ as the "initial" curve and denote it by $\bar{p}(t)$, i.e.,

$$\bar{p}(t) \equiv p(t), \qquad t < 0 \tag{6}$$

An initial curve is the counterpart of the initial values in physical systems which are described by a set of ordinary differential equations.

In order to put the initial curve in evidence in (4), we break up the range of integration in (4) as $-\infty \leqslant u \leqslant 0$ and $0 \leqslant u \leqslant t$ and obtain

$$(l/\beta)\,\dot{p} = (p + P_0)\left[\int_0^t G(t-u)\,p(u)\,du + I_1(t)\right] + I_2(t) - p + \int_0^t D(t-u)\,p(u)\,du, \qquad t > 0 \tag{7}$$

where

$$I_1(t) \equiv \int_0^\infty G(t+u)\,\bar{p}(-u)\,du \tag{8a}$$

$$I_2(t) \equiv \int_0^\infty D(t+u)\,\bar{p}(-u)\,du \tag{8b}$$

The functions $I_1(t)$ and $I_2(t)$ are determined by the initial curves only, and represent the past history of the reactor. We can see from (7) that the initial slope of $p(t)$ at $t = 0$ depends on $I_1(0)$, $I_2(0)$, and $p(0)$. Therefore, a unique solution of (7) requires the specification of an initial curve $\bar{p}(t)$ as well as the initial value of $p(t)$. A jump in $p(t)$ at $t = 0$ will not affect $I_1(t)$ and $I_2(t)$ because it represents a point in the integration, but it will affect the evaluation of $p(t)$ for $t > 0$. Sufficient conditions for the existence and uniqueness of the solutions of (7) for a given initial curve $\bar{p}(t)$ and initial value $p(0)$ are discussed in general works on equations with time delays, for instance, by Krasovskii [2], and will not be presented here. We shall simply assume that these conditions are satisfied in the physical problems we shall be concerned with.

We now discuss the concept and the definition of stability in systems with a past history. In such a system, a statement of stability can only be made with respect to a certain class of initial curves because the temporal behavior of a solution of (7) as $t \to \infty$ is determined not only by the magnitude, but also by the shape of the power variations in the past. The class of initial curves must be sufficiently broad to include all the possible perturbations which may arise intentionally or accidentally during operation of the reactor. If we choose this class unrealistically broad from the physical point of view for the sake of mathematical generality, then the sufficient conditions for stability may turn out to be too restrictive to be of any practical interest in reactor applications. In general, the mathematical derivation of each stability criterion also dictates the associated class of initial curves.

Since an initial curve is the response of the reactor to an external disturbance either in the reactivity or in the source for $t < 0$, it satisfies of necessity the point kinetic equation as indicated by (5). We can therefore specify a physically realistic class of initial curves in terms of the possible intentional or accidental changes in the reactivity and the neutron source during operation of the reactor. First of all, the initial curves are bounded functions of time because the reactor power must be finite for $t \leqslant 0$, otherwise it would be too late to worry about the stability of the reactor in the future. Secondly, their first derivatives exist for all $t \leqslant 0$ provided $S(t)$ and $\delta k(t)$ are defined at t as indicated by (5). In fact, $\dot{p}(t)$ is bounded for all t if $S(t)$ and $\delta k(t)$ are bounded in $(-\infty, 0)$ where they are defined. Indeed, we obtain from (5) the following inequality:

$$(l/\beta) \mid \dot{p} \mid \leqslant \mid p_m + P_0 \mid \left[\mid \delta k(t) \mid + p_m \int_0^\infty \mid G(u) \mid du \right] + p_m + p_m \int_0^\infty \mid D(u) \mid du + (l/\beta) \mid S(t) \mid \tag{9}$$

where p_m is the maximum of $p(t)$ for $t \leqslant 0$. The integral of the delayed neutron kernel $D(u)$ is finite because it is a sum of decaying exponential terms. The absolute integrability of the feedback kernel $G(u)$, on the other hand, was a requirement for the stability of the feedback mechanism, as discussed previously. Thus, $\dot{p}(t)$ is bounded, and $p(t)$ satisfies

$$\mid p(t) - p(t') \mid \leqslant M \mid t - t' \mid, \qquad t, t' < 0 \tag{10}$$

if $\delta k(t)$ and $S(t)$ are bounded functions. In (10), M is a number which can be determined from (9) in terms of the bounds of $\delta k(t)$, $S(t)$, and p_m as well as the reactor parameters. We note that $\delta k(t)$ and $S(t)$ are not required to be continuous to obtain (10); they may have jump discontinuities corresponding to step reactivity insertions and sudden changes in the production rate of the external neutron source.

In light of the above discussion, it is reasonable to choose the class of initial functions as the set of bounded functions $\bar{p}(t) > -P_0$ that satisfy (10) (with a different M for each initial curve) and vanish for $t < -T$, where T is an arbitrarily large but finite time (different for each initial curve). The condition $\bar{p}(t) > -P_0$ follows from the positiveness of the power $P_0 + p(t)$ at all times, and hence is not a real restriction. The reason for choosing the initial curves to be zero outside a finite interval $(-T, 0)$ is a matter of expediency in the analysis. Physically, it implies that the perturbation is confined in a finite time interval, which is hardly a restriction in actual reactor applications. We shall refer to a function in the above class as a "physically admissible" initial curve. We shall attempt to establish criteria for asymptotic stability with respect to this class of initial functions. However, we shall point out in the course of the derivations how the restriction on the initial curves can be relaxed by imposing more stringent conditions on the feedback functional of the reactor.

Having specified the class of physically admissible initial curves, we can now describe the physical nature of the stability problem with which we shall be concerned. We consider a reactor which is operated at a constant power level P_0 prior to $t = -T$ [i.e., $\bar{p}(t) \equiv 0$ for $t \leqslant -T$]. We then perturb this reactor by an external reactivity insertion or by an external source in the interval $-T \leqslant t \leqslant 0$, and generate a physically admissible initial curve. The perturbation is terminated at $t = 0$ so that the reactor returns to its unperturbed conditions for $t > 0$. Finally, we observe the temporal behavior of the reactor power for $t > 0$ as the free motion of an autonomous system.

If the subsequent power variations always remain small whenever the initial perturbation is sufficiently "small," then we may consider the equilibrium $p(t) \equiv 0$ as stable. It is clear that we need a quantity to measure the magnitude of the initial perturbation, which is an initial curve defined in $[-T, 0]$, in order to be able to define stability in a precise manner. We therefore extend the concept of the norm of a vector to functions, as

$$\| \bar{p} \|_T \equiv \max(| \bar{p}(t) |) \qquad \text{for} \quad t \leqslant 0 \tag{11}$$

which is the maximum of $\bar{p}(t)$ in the interval $[-T, 0]$ [note that the value of $\bar{p}(t)$ at $t = 0$ is included in the norm]. Other choices of norm, such as the integral of $\bar{p}(t)$ in $[-T, 0]$, are also possible. The norm in (11) is more convenient in our analysis.

We are now in a position to define the stability of a reactor with a past history.

Definition 4. The solution $p(t) \equiv 0$ of (7) is stable if, for any given $\epsilon > 0$, there exists a $\delta(\epsilon) > 0$ such that $| p(t) | < \epsilon$ for all $t \geqslant 0$ whenever $\| \bar{p} \|_T \leqslant \delta(\epsilon)$.

In this definition, $p(t)$ is the solution of (7) that corresponds to the physically asmissible initial curve $\bar{p}(t)$, i.e., $p(t) \equiv \bar{p}(t)$ for $t \leqslant 0$.

Definition 5. The solution $p(t) = 0$ is asymptotically stable if it is stable, and in addition there exists a $\delta > 0$ such that $p(t) \to 0$ as $t \to \infty$ whenever $\| \bar{p} \|_T < \delta$.

Definition 6. The solution $p(t) \equiv 0$ is asymptotically stable in the large if it is stable, and in addition every solution $p(t) \to 0$ as $t \to \infty$ for all physically admissible initial curves.

7.2. General Criteria for Boundedness and Global Asymptotic Stability

In this section, we present two criteria for stability and global asymptotic stability of a reactor with an arbitrary feedback functional $\delta k_f[p]$. The criterion for global asymptotic stability reduces to the Welton criterion when the feedback functional is linear. However, the criterion is also applicable to reactor models with a nonlinear feedback functional, such as a xenon- and temperature-controlled reactor in which the feedback functional is quadratic.

The global asymptotic stability criterion discussed in this section was obtained by Akcasu and Dalfes [3] in 1960 in a heuristic manner. The results were justified mainly by energy considerations. Here, we present a rigorous derivation of the criterion in detail [3a, b]. The mathematical concepts and the auxiliary theorems involved in the following analysis will also be used in the subsequent sections.

A. Stability

We start with the standard form of the point kinetic equations (cf 3.1, Eq. 1), which we reproduce here, introducing

$$z(x) \equiv [P(t) - P_0]/P_0, \qquad z_i(x) \equiv [C_i(t) - C_0]/C_0 \tag{1a}$$

$$x = \beta t/l, \qquad h_i = l\lambda_i/\beta \tag{1b}$$

as follows:

$$\dot{z} = (1 + z)\, \delta k_f[z] - \sum_i a_i(z - z_i) \tag{2}$$

$$\dot{z}_i = h_i(z - z_i), \qquad i = 1,\ldots, I \tag{3}$$

We then consider a function $V(x)$ defined by

$$V(x) \equiv F(z) + \sum_{i=1}^{I} (a_i/h_i)F(z_i) - \int_{-x_0}^{x} z(x')\, \delta k_f[z(x')]\, dx', \qquad x > 0 \tag{4}$$

where

$$F(z) \equiv z(x) - \ln[1 + z(x)] \tag{5}$$

and where x_0 is an arbitrary positive number. We shall now show that $V(x)$ is a nonincreasing function of x if $z(x)$ and $z_i(x)$ satisfy (2) and (3). The derivative of $V(x)$ with respect to x is

$$\dot{V}(x) = [z\dot{z}/(1 + z)] + \sum_{i=1}^{I} [a_i z_i \dot{z}_i / h_i(1 + z_i)] - z\, \delta k_f[z] - \tag{6}$$

Substituting $\dot{z}$ and $\dot{z}_i$ from (2) and (3), we obtain

$$\dot{V}(x) = -\sum_i a_i\{[z(x) - z_i(x)]^2/[1 + z(x)][1 + z_i(x)]\} \tag{7a}$$

which is a continuous function of x whenever $z(x) \neq -1$ and $z_i(x) \neq -1$, inasmuch as $z(x)$ and $z_i(x)$ are continuous in x by virtue of (2) and (3). Since $(1 + z)$ and $(1 + z_i)$ are positive, we conclude that

$$\dot{V}(x) \leqslant 0 \tag{7b}$$

The equality occurs when $z = z_i$.

The function $F(z)$ defined by (5) is frequently encountered in nonlinear reactor dynamics. Its variation with z is depicted in Figure 7.2.1. It can

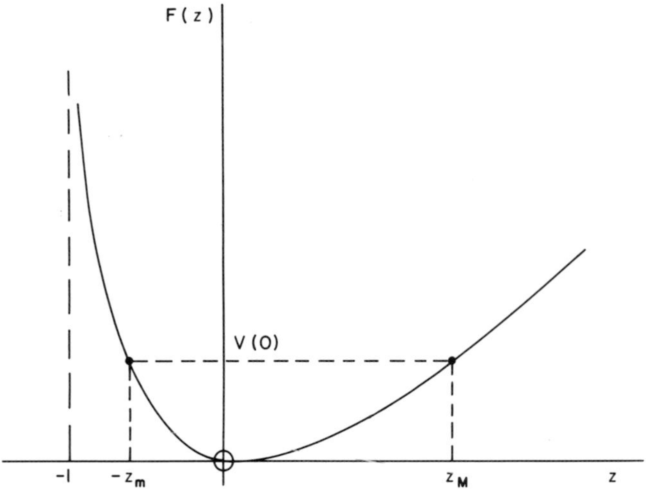

FIGURE 7.2.1. Variation of $F(z) = z - \ln(1 + z)$ with z.

be verified that: (i) $F(z)$ is positive and real in $(-1, \infty)$ except for $z = 0$, (ii) $F(0) = 0$, and (iii) $F(z)$ is continuous in $(-1, \infty)$.

A function with these properties is called "positive-definite." The positiveness of $F(z)$ for $z \neq 0$ and $z > -1$ can be seen from the following relations:

$$F(z) = \int_0^z [u/(1 + u)]\, du, \qquad z > 0$$

$$F(z) = \int_0^{|z|} [u/(1 - u)]\, du, \qquad z < 0; \quad |z| < 1$$

We now return to $V(x)$ in (4). Since $F(z)$ and $F(z_i)$ are nonnegative, $V(x)$ will also be nonnegative if we require

$$I[x, y] \equiv \int_{-x_0}^{x} y(x')\, \delta k_{\mathfrak{f}}[y]\, dx' \leqslant 0 \tag{8}$$

to hold for all $x \geqslant 0$, and for all functions $y(x) \geqslant -1$ defined in $(-\infty, +\infty)$ in such a way that it belongs to the class of physically admissible curves for $x \leqslant 0$, but is arbitrary for $x > 0$. For example, $y(x)$ may diverge as $x \to +\infty$. Clearly, the set of functions $y(x)$ will contain all the solutions $z(x)$ of (2) as a subset, and hence (8) will be necessarily true along a trajectory. Since the solutions of (2) are not known, we have to verify (8) for a wider class of functions in order to be able to use them in the study of the stability of $z(x)$. We shall call the functions $y(x)$ the "test" functions.

The lower limit x_0 of (8) is an arbitrary positive number whose value is not important in the proof. Although we always choose $x_0 = +\infty$ in the applications, we prefer to keep it, arbitrary in the general theory.

It is concluded from the above discussion that $V(x)$ is a nonnegative, monotonically decreasing function of time along a trajectory (Fig. 7.2.2), i.e.,

$$V(x) \leqslant V(0) \tag{9}$$

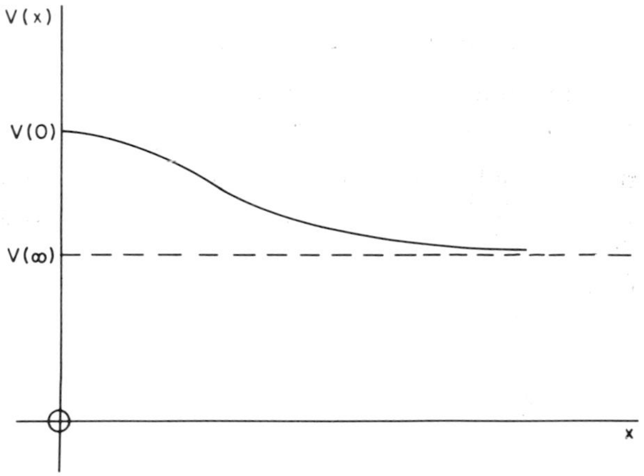

FIGURE 7.2.2. Variation of $V(x)$ along a solution of the kinetic equations.

if the condition (8) is satisfied for all the test functions $y(x)$. The initial value $V(0)$ is given by

$$V(0) = F(z_0) + \sum_{i=1}^{I} [a_i F(z_{i0})/h_i] - \int_{-x_0}^{0} \bar{z}(x') \, \delta k_f[\bar{z}] \, dx' \tag{10}$$

where z_0 and z_{i0} are the initial values of $z(x)$ and $z_i(x)$, respectively, and $\bar{z}(x)$ is an initial curve. Let us suppose that $V(0)$ is finite. Then, we find that $V(x)$ is also finite for all $x \geqslant 0$ provided the condition (8) holds. But $V(x)$ is the sum of three nonnegative terms (cf Eq. 4). Therefore, each term in (4) must be smaller than $V(0)$ for all $x \geqslant 0$, i.e.,

$$F(z) \leqslant V(0) \tag{11a}$$

$$a_i F(z_i)/h_i \leqslant V(0) \tag{11b}$$

$$\left| \int_{-x_0}^{x} z(x') \, \delta k_f[z(x')] \, dx' \right| \leqslant V(0) \tag{11c}$$

The inequalities (11a) and (11b) imply, with the help of Figure 7.2.1, that†

$$-1 < -z_{\mathrm{m}} \leqslant z(x) \leqslant z_{\mathrm{M}} \tag{12a}$$

$$-1 < -z_{i\mathrm{m}} \leqslant z_i(x) \leqslant z_{i\mathrm{M}} \tag{12b}$$

where z_{m} and z_{M} are positive numbers such that

$$F(z_{\mathrm{M}}) = F(-z_{\mathrm{m}}) = V(0), \qquad z_{\mathrm{M}} > z_{\mathrm{m}} \tag{13a}$$

$$F(z_{i\mathrm{M}}) = F(-z_{i\mathrm{m}}) = (h_i/a_i) V(0) \tag{13b}$$

We can now state the foregoing conclusions as a theorem.

THEOREM 1 (Boundedness). The response of a reactor is always bounded if (8) holds for all the test function $y(x)$ and if $V(0)$ is bounded.

We note that the only restriction we have to impose on the initial curves $\bar{z}(x)$ to ascertain boundedness is that they must lead to a finite $V(0)$ (cf Eq. 10).

As an application of this theorem, let us suppose that the reactor is operated at a constant power level for $x < 0$ and perturbed in power at $x = 0$ by an amount $z_0 = p(0)/P_0$. Clearly, $\bar{z}(x) \equiv 0$ for $x < 0$, and $z_{i0} = 0$. Hence, (10) reduces to

$$V(0) = z_0 - \ln(1 + z_0) \tag{14}$$

† In all real reactor systems, $a_i/h_i \equiv \beta_i/l\lambda_i > 1$, hence $z_{i\mathrm{M}} < z_{\mathrm{M}}$ and $-z_{i\mathrm{m}} > -z_{\mathrm{m}}$. Certainly, then, both $z(x)$ and $z_i(x)$ are bounded by z_{M} and z_{m}, i.e. the bounds of $z(x)$.

If $z_0 > 0$, then $z_M = z_0$, and from (12a) and (13b),

$$| z(x)| \leqslant z_0 \tag{15}$$

If $z_0 < 0$, then $z_m = | z_0 |$ and

$$-| z_0 | \leqslant z(x) \leqslant z_M \tag{16}$$

An interesting observation from (16) is that positive power excursions (dotted curve in Figure 7.2.3) may exceed the magnitude of the initial negative power disturbance because $z_M > z_0$, whereas positive power excursions are always smaller than the initial positive power disturbance (solid curve in Fig. 7.2.3).

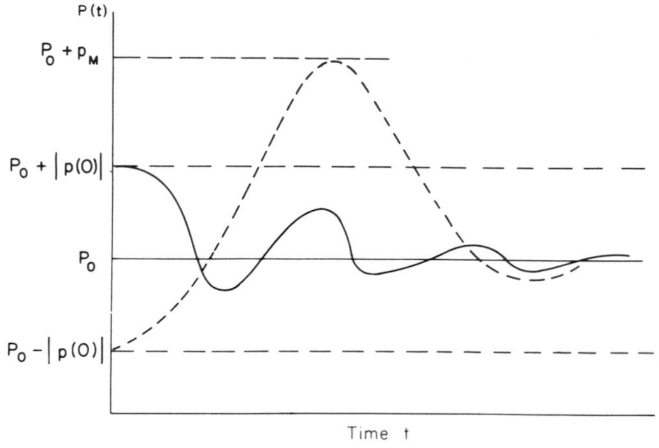

FIGURE 7.2.3. Bounds on power following positive and negative power disturbances. p_M is determined by $z_M - \ln(1 + z_M) = -| z(0)| - \ln[1 - | z(0)|]$ with $p_M = P_0 z_M$.

On the basis of Theorem 1, we can conclude that, if the conditions of the theorem are satisfied, we have, for all $x \geqslant 0$,

$$| z(x)| \leqslant z_M \tag{17a}$$

$$| z_i(x)| \leqslant z_M \tag{17b}$$

$$| \dot{z} | \leqslant (1 + z_M)W(M) + 2z_M \tag{18a}$$

[This follows from (2) and the stability of the feedback (cf Sec. 5.3), which states $| \delta k_f[z] | < W(M)$ for all x if $| z(x)| < M$ for all x. Here, M

is the greater of z_M and the upper bound of the physically admissible initial curve under consideration.]

$$|\dot{z}_i| \leqslant 2z_M h_i \tag{18b}$$

$$|\dot{V}(x)| \leqslant 4z^2_M/(1 - z_m)^2 \tag{19a}$$

$$|\ddot{V}(x)| \leqslant [4z_M/(1 - z_m)^3]\{(1 + z_M)W(M) + 2z_M[1 + (2l\lambda^*/\beta)]\} \tag{19b}$$

[This follows from (7a) (see Problem 1).]

Thus, the conditions of Theorem 1 are sufficient to prove that $\dot{V}(x)$ is continuous [because $z(x)$ and $z_i(x)$ are continuous and $(1 + z)$ and $(1 + z_i)$ are never zero, by virtue of (17a,b)] and bounded for all $x \geqslant 0$. Furthermore, it is uniformly continuous† because $\ddot{V}(x)$ is bounded.

The stability of the equilibrium state (cf 7.1, Def. 4) follows from the boundedness of the solutions (cf Eq. 17) and that $V(0)$, and hence the upper bound z_M of the solutions, can be made smaller than any desired number if the norm of the initial curve and the initial values of z_{i0} are sufficiently small.††

B. Asymptotic Stability

We now consider the question of global asymptotic stability of the equilibrium state $z(x) \equiv 0$. Our purpose is to determine sufficient conditions, in addition to those for boundedness of Theorem 1, such that $z(x) \to 0$ for any physically admissible initial curve.

† The continuity of $\dot{V}(x)$ at a point x implies that, given $\epsilon > 0$, we can find a $\delta(\epsilon, x)$ such that $|\dot{V}(x') - \dot{V}(x)| < \epsilon$ holds whenever $|x' - x| < \delta(\epsilon, x)$. The continuity is uniform for $x > a$ if $\delta(\epsilon, x)$ is independent of x for all $x > a$. When $\ddot{V}(x)$ is bounded for $x > a$, we have, from the mean value theorem,

$$|\dot{V}(x') - \dot{V}(x)| = |x' - x|\,|\ddot{V}(x + \theta)| \leqslant M\,|x' - x|$$

where $0 \leqslant \theta \leqslant |x - x'|$ and M is the bound of $|\ddot{V}(x)|$, i.e., $|\ddot{V}(x)| \leqslant M$ for all $x > a$. Thus, $\delta(\epsilon, x) = \epsilon/M$, proving uniform continuity.

†† It is clear from Figure 7.2.1 that we can find an $\eta(\epsilon) > 0$ for a given $\epsilon > 0$ such that $V(0) < \eta(\epsilon)$ implies $|z(x)| \leqslant z_M < \epsilon$. We can also determine $\delta_0(\eta) > 0$ and $\delta_i(\eta) > 0$ such that $F(z_0) \leqslant \eta/3$ and $a_i F(z_{i0}) \leqslant \eta h_i/3I$ if $|z_0| < \delta_0$ and $|z_{i0}| < \delta_i$. Thus, we have to show that the last term in (10) can be made smaller than $\eta/3$ by choosing $\|\bar{z}\|_T$ sufficiently small. Using the stability of the feedback (cf 5.3, Eq. 4b), we get

$$\left|\int_{+x_0}^{0} \bar{z}(x)\,\delta k_f[\bar{z}]\,dx\right| \leqslant W(\|\bar{z}\|_T)\int_{-x_0}^{0} |\bar{z}(x')|\,dx' \leqslant T\|\bar{z}\|_T W(\|\bar{z}\|_T)$$

Determine a $\delta > 0$ such that $\|\bar{z}\|_T < \delta$ implies $T\|\bar{z}\|_T\, W\|\bar{z}\|_T \leqslant \eta/3$. Then, if $\|\bar{z}\|_T < \delta(\epsilon)$, $|\bar{z}_0| < \delta_0(\epsilon)$, and $|z_{i0}| < \delta_i(\epsilon)$, then $|z(x)| < \epsilon$ for all $x \geqslant 0$.

We start with the observation that

$$\lim_{x\to\infty} V(x) = V_\infty \geqslant 0 \tag{20}$$

because $V(x)$ is monotonically decreasing and always nonnegative.[†] Furthermore, using (20) and the fact that $\dot{V}(x)$ is uniformly continuous [see the comments following (19)], we can also show that

$$\lim_{x\to\infty} \dot{V}(x) = 0 \tag{21}$$

along a trajectory.[††] Combining (7a) and (21), we conclude

$$\lim_{x\to\infty} [z(x) - z_i(x)] = 0, \qquad i = 1,..., I \tag{22}$$

which, by virtue of (3), leads to

$$\lim_{x\to\infty} \dot{z}_i(x) = 0 \tag{23}$$

It is our next task to show that, if $\dot{z}(x)$ is uniformly continuous for sufficiently large x, then (22) and (23) imply

$$\lim_{x\to\infty} \dot{z}(x) = 0 \tag{24}$$

This immediately follows from the lemma just proven in the footnote if we let $g_i(x) = z(x) - z_i(x)$. By virtue of (22), $g_i(x)$ has a limit, and $\dot{g}_i(x)$ is uniformly continuous if $\dot{z}(x)$ is uniformly continuous. The uniform continuity of $\dot{z}_i(x)$ follows from (3), i.e., $\dot{z}_i(x) = h_i(z(x) - z_i(x))$, because both $z(x)$ and $z_i(x)$ are uniformly continuous (they have bounded

[†] See, for example, Carslaw [4, p. 75].

[††] LEMMA. Let $g(x)$ be a real function of a real variable x which is defined for $x > a > 0$. If (i) $\lim_{x\to\infty} g(x) = g_\infty$ (g_∞ is finite) and (ii) $\dot{g}(x)$ is uniformly continuous for $x > a$, then $\lim_{x\to\infty} \dot{g}(x) = 0$.

PROOF. We start with $g(x) = g(a) + \int_a^x \dot{g}(x')\,dx'$. Since $g(x)$ has a limit, we have

$$\lim_{x\to\infty} \int_x^{x+\Delta} \dot{g}(x')\,dx' = \Delta \lim_{x\to\infty} \dot{g}[x + \theta(x)] \doteq 0$$

where Δ is an arbitrary positive number, and $0 \leqslant \theta(x) \leqslant \Delta$ (mean value theorem). Since $\dot{g}(x)$ is uniformly continuous for $x > a$, $\dot{g}(x)$ and $\dot{g}[x + \theta(x)]$ must have the same limit. Indeed, for a given $\epsilon > 0$, we have a $\delta(\epsilon)$ such that $|\dot{g}[x + \theta(x)] - \dot{g}(x)| < \epsilon$ for all x if $\theta(x) < \delta(\epsilon)$. Since $\theta(x) \leqslant \Delta$, we can choose $\Delta < \delta(\epsilon)$. Since the difference between $\dot{g}[x + \theta(x)]$ and $\dot{g}(x)$ can be made smaller than any desired $\epsilon > 0$, for any $x > a$, $\dot{g}(x)$ must also approach zero as $x \to \infty$. A different proof of this lemma was given by Barbălat [5].

derivatives). The conditions of the lemma are satisfied, and hence $\dot{g}_i(x) = \dot{z}_i(x) \to 0$ as $x \to \infty$. Using (23), we establish (24).

The uniform continuity of $\dot{z}(x)$ is ensured if we require that $\delta k_f[z]$ in (2) is, as a function of time, uniformly continuous along a trajectory, because the remaining terms in (2) are uniformly continuous. We can guarantee the uniform continuity of $\delta k_f[z]$, for example, by requiring $d\{\delta k_i[z]\}/dx$ to be bounded for sufficiently large times. Since the trajectories $z(x)$ are not known a priori, we must verify the uniform continuity of the feedback functional $\delta k_f[z]$ for a wider class of test functions. We have already established, by virtue of Theorem 1, that $z(x)$ is continuous, bounded, and has bounded, continuous first derivatives. Hence, it is sufficient to demonstrate the uniform continuity of a given feedback functional only for those test functions $v(x)$ that are bounded and have bounded first derivatives.

Let us suppose that the feedback functional satisfies the requirements of the Theorem 1 and of uniform continuity. Then, we can conclude that $\dot{z}(x) \to 0$ as $x \to \infty$. From this conclusion and from the boundedness of $z(x)$, one might expect that the limit of $z(x)$ for $x \to \infty$ exists. However, this is not the case, as can be seen considering $\sin \sqrt{x}$, which does not have a limit even though its derivative $(\cos \sqrt{x})/(2\sqrt{x})$ vanishes for $x \to \infty$. We must therefore impose more restrictions on $\delta k_f[z]$ to ensure asymptotic stability.

Using (22) and (24) in (2), we obtain

$$\lim_{x\to\infty} \delta k_f(z) = 0 \tag{25}$$

along a trajectory. Under mild conditions (continuity of functionals for a certain class of functions), (25) will also imply

$$\lim_{x\to\infty} z(x) = 0 \tag{26}$$

along a trajectory. In order to ensure that (26) follows from (25), it is sufficient to demonstrate that $\delta k_f[w] \to 0$ implies $w(x) \to 0$ as $x \to \infty$ for all test functions $w(x)$ which are bounded and whose first derivatives are uniformly continuous and vanish as $x \to \infty$.

We are now in a position to state the following theorem:

THEOREM 2 (Asymptotic stability). The finite and nonzero equilibrium state of a reactor is asymptotically stable in the large if its feedback functional satisfies, in addition to the conditions of Theorem 1, the following conditions:

(i) $\delta k_f[v]$ is uniformly continuous for sufficiently large x, and for all test functions $v(x)$ which are bounded and have bounded first derivatives for $x > 0$.

(ii) The limit

$$\lim_{x \to \infty} \delta k_f[w] = 0 \tag{27a}$$

implies

$$\lim_{x \to \infty} w(x) = 0 \tag{27b}$$

for all test functions $w(x)$ which are bounded and have uniformly continuous first derivative which vanish as $x \to \infty$.

Remarks

We observe from (7a) that $\dot{V}(x) \equiv 0$ along a trajectory in the absence of delayed neutrons ($a_i = 0$). Hence, when the delayed neutrons are not present, we can ascertain the boundedness of the solutions of the kinetic equations (Theorem 1), but we cannot claim asymptotic stability even if the additional conditions of Theorem 2 are satisfied, because the proof of the latter makes explicit use of the delayed neutrons. An interesting consequence of this remark is that, if a reactor satisfies the conditions of Theorem 2, and displays periodic power oscillations in the absence of delayed neutrons after some disturbance, then it must exhibit damped oscillations when delayed neutrons are taken into account. This conclusion was reached before by Ergen [6], Smets [7], and Popov [8, 9] by different methods in the case of a linear feedback functional.

7.3. Global Asymptotic Stability of Reactors with Linear Feedback

In this section, we investigate asymptotic stability in the large of a reactor with a linear feedback using the general stability criterion derived in the previous section (7.2, Theorem 2).

A. Welton's Criterion

A linear feedback is described by the following linear functional (cf 5.1, Eq. 21)

$$\delta k_f[z] = \int_{-\infty}^{x} du\, G(x - u) z(u) \tag{1}$$

The stability of the feedback mechanism requires the feedback kernel to be absolute-integrable:

$$\int_0^\infty |G(u)| \, du < \infty \tag{2}$$

and the negativeness of the power coefficient of reactivity γ requires

$$\gamma \equiv \int_0^\infty G(u) \, du < 0 \tag{3}$$

We shall now determine a condition on the Laplace transform $H(s)$ of $G(u)$ which will ensure that $\delta k_f[z]$ defined by (1) fullfils the requirement (7.2, Eq. 8) for asymptotic stability in the large. Substituting (1) into (7.2, Eq. 8), and choosing the arbitrary number x_0 to be ∞, we obtain

$$I[x, y] = \int_{-\infty}^{x} y(x') \, dx' \int_{-\infty}^{x'} G(x' - u) y(u) \, du \leqslant 0 \tag{4}$$

Since $G(u)$ is a causal function, $G(u) = 0$ for $u < 0$. Hence, we can replace the upper limit of the second integral, x', by x:

$$I[x, y] = \int_{-\infty}^{x} y(x') \, dx' \int_{-\infty}^{x} G(x' - u) y(u) \, du \leqslant 0$$

Using the Fourier representation of $G(u)$, i.e.,

$$G(u) = (1/2\pi) \int_{-\infty}^{\infty} e^{i\omega u} H(i\omega) \, d\omega \tag{5}$$

we find

$$I[x, y] = (1/2\pi) \int_{-\infty}^{\infty} d\omega \, H(i\omega) \left| \int_{-\infty}^{x} dx' \, y(x') \exp i\omega x' \right|^2 \leqslant 0$$

or

$$I[x, y] = (1/\pi) \int_{0}^{\infty} \mathrm{Re}[H(i\omega)] \, d\omega \left| \int_{-\infty}^{x} dx' \, y(x') \exp i\omega x' \right|^2 \leqslant 0 \tag{6}$$

It immediately follows from (6) that $I(x, y) \leqslant 0$ holds for all $x > 0$ and for all test functions if

$$\mathrm{Re}[H(i\omega)] \leqslant 0 \tag{7}$$

holds for all ω in $[0, \infty]$. According to Sect. 7.2, Theorem 1, (7) is a sufficient condition for the boundedness of power oscillations in a reactor.

In order to establish the additional conditions for asymptotic stability, we consider Theorem 2, Sect. 7.2. We have to show that

$$\delta k_{\mathrm{f}}(x) = \int_0^\infty G(u)v(x-u)\,du \tag{8}$$

is uniformly continuous for all bounded test functions that have bounded first derivatives for $x > 0$. Since $v(x)$ is a physically admissible initial curve, it also has bounded first derivatives for $x < 0$. Hence,

$$|\,v(x) - v(x')| \leqslant M|\,x - x'\,| \tag{9}$$

(cf 7.1, Eq. 10) holds for all x and x'. From (8) and (9), we obtain

$$\begin{aligned} |\,\delta k_{\mathrm{f}}(x) - \delta k_{\mathrm{f}}(x')| &\leqslant \int_0^\infty |\,G(u)|\,|\,v(x-u) - v(x'-u)|\,du \\ &\leqslant M|\,x - x'\,|\int_0^\infty |\,G(u)|\,du \end{aligned} \tag{10}$$

which proves the uniform continuity of $\delta k_{\mathrm{f}}[x]$.

Finally, we have to show that

$$\lim_{x\to\infty}\int_0^\infty G(u)w(x-u)\,du = 0 \tag{11}$$

implies $w(x) \to 0$ for all bounded test functions with uniformly continuous first derivatives which vanish as $x \to 0$. This immediately follows from Wiener's fundamental Tauberian theorem in a form due to Pitt [11]. For the sake of completeness, we reproduce the statement of the theorem.

Pitt's Form of Wiener's Tauberian Theorem. If $G(x)$ is absolute-integrable in $(-\infty, +\infty)$ and if its Fourier transform $H(i\omega)$ does not vanish there, and if $w(x)$ is bounded and has a first derivative which remains greater than a negative constant, then

$$\lim_{x\to\infty}\int_{-\infty}^\infty G(x-u)w(u)\,du = A\int_{-\infty}^\infty G(x)\,dx \tag{12a}$$

implies

$$\lim_{x\to\infty} w(x) = A \tag{12b}$$

It is clear that (11) can be written in the form of (12a) because $G(u) = 0$ for $u \leqslant 0$. Furthermore, $w(u)$ satisfies the requirements of the theorem because it has bounded first derivative everywhere. Thus, we obtain the following criterion from (11) and (12a):

If, in addition to (7), we can demonstrate that

$$H(i\omega) \neq 0, \qquad -\infty < \omega < \infty \tag{13}$$

then the reactor is asymptotically stable for all physically admissible initial curves.

It is clear that if we require $\mathrm{Re}[H(i\omega)] < 0$ without the equality sign, then $H(i\omega)$ cannot vanish at any frequency, and the condition

$$\mathrm{Re}\, H(i\omega) < 0 \tag{14}$$

alone is sufficient for asymptotic stability. This condition is known as Welton's criterion and was first obtained in 1952 by Welton [10].

We note that Wiener's theorem does not make explicit use of the fact that, in our problem, the first derivative of $w(x)$ approaches zero as $x \to \infty$. By exploiting this property, we can replace the condition $H(i\omega) \neq 0$ for all ω by[†]

$$\int_0^\infty |\, G(u)| u \, du < \infty \tag{15}$$

[†] LEMMA. First, we observe that, since $w(u)$ is bounded,

$$\lim_{x\to\infty} \int_{-\infty}^{T} G(x-u)[w(u) - w(x)]\, du = 0$$

This is shown by a change of variable as

$$\left| \int_{-\infty}^{T} G(x-u)[w(u) - w(x)]\, du \right| \leqslant 2M \int_{x-T}^{\infty} |\, G(u)|\, du$$

where M is the maximum of $w(x)$ in $(-\infty, T)$. The right-hand side vanishes because $G(u)$ is absolute-integrable.

Secondly, we show that, if (15) holds, then

$$\lim_{x\to\infty} \int_{T}^{x} G(x-u)[w(u) - w(x)]\, du = 0$$

Expanding $w(x)$ as $w(x) = w(u) + (x-u)\, \dot{\omega}(u + \theta(u))$, where $0 \leqslant \theta(u) \leqslant x - u$, and choosing $T(\epsilon)$ for a given $\epsilon > 0$ such that $\dot{w}(u) < \epsilon$ for $u > T(\epsilon)$, we have

$$\left| \int_{T}^{x} G(x-u)[w(u) - w(x)]\, du \right| \leqslant \epsilon \int_{T}^{x} |\, G(x-u)|\, |\, x-u\, | \leqslant \epsilon \int_0^\infty u\, |\, G(u)|\, du$$

which proves the second assertion. From the first and second observations, we find

$$\lim_{x\to\infty} \left[\int_0^\infty G(u)\, w(x-u)\, du - w(x) \int_0^\infty G(u)\, du \right] = 0$$

Hence, if (15) holds, and if $w(x)$ is bounded and $\dot{\omega}(x) \to 0$ as $x \to \infty$, then

$$\int_0^\infty G(u)\, w(x-u)\, du$$

behaves as $w(x)$ for large values of x. Combining this result with (11), we find $w(x) \to 0$ as $x \to \infty$.

Hence, (7) and (15) constitute another set of sufficient conditions for asymptotic stability when the equality in $\mathrm{Re}[H(i\omega)] \leqslant 0$ is allowed. The condition (15) is very relaxed. For example, it is always satisfied if $G(t)$ can be expressed as a sum of exponential terms with negative exponents, such as the case when the feedback can be described by a set of coupled linear differential equations.

It is interesting to note some of the implications of condition (7) in the time domain. The real part of $H(i\omega)$, which is denoted by $H_{\mathrm{R}}(\omega)$, is, by definition,

$$H_{\mathrm{R}}(\omega) \equiv \int_0^\infty G(t)\cos(\omega t)\,dt \tag{16a}$$

which is the cosine transform of $G(t)$. The inverse transform is (Problem 2)

$$G(t) \equiv (2/\pi)\int_0^\infty H_{\mathrm{R}}(\omega)\cos(\omega t)\,d\omega \tag{16b}$$

Taking the absolute values of (16b) we obtain

$$|\,G(t)| \leqslant (2/\pi)\int_0^\infty |\,H_{\mathrm{R}}(\omega)|\,d\omega = -G(0) \tag{17}$$

where we have used $|\,H_{\mathrm{R}}(\omega)\,| = -H_{\mathrm{R}}(\omega)$ by virtue of (7). Thus, if $H_{\mathrm{R}}(\omega) \leqslant 0$, the feedback kernel is bounded by its value at zero. It also follows from (17) that $G(0)$ is necessarily negative if (7) holds. We shall see later that $G(0) < 0$ is a sufficient condition for the nonexistence of solutions with a finite escape time. Thus, the condition (7) is also sufficient to rule out solutions with finite escape time (cf. 7.13).

Positive (Negative) Real Functions

Definition 1. A function $F(s)$ is called positive (negative) real if (i) it is the Laplace transform of a real causal function $f(t)$, (ii) $\mathrm{Re}\,F(i\omega) \geqslant 0$ ($\leqslant 0$) for all ω. Since the feedback kernel $G(t)$ is a real causal function, then $H(s)$ becomes a negative real function if the condition (7) is satisfied.†

Negative (positive) real functions possess several interesting properties, some of which are as follows [11a]: (1) $\mathrm{Re}\,H(s) \leqslant 0$ for $\mathrm{Re}\,s \geqslant 0$.

† Since we have defined the feedback kernel $G(t)$ in such a way that a positive $G(t)$ implies positive (destabilizing) feedback, $H(s)$ is negative real.

This property is a consequence of the following relation for real, causal functions [Problem 3]:

$$H(s) = (1/\pi) \int_{-\infty}^{\infty} d\omega \, [H_{\mathrm{R}}(\omega)/(s - i\omega)], \qquad \operatorname{Re} s > 0 \tag{18}$$

The real parts of this equality yields

$$\operatorname{Re} H(s) = (1/\pi) \int_{-\infty}^{\infty} \{xH_{\mathrm{R}}(\omega)/[x^2 + (y-\omega)^2]\} \, d\omega, \quad x > 0, \; s = x + iy \tag{19}$$

which leads to $\operatorname{Re} H(s) \leqslant 0$ for $x > 0$ if $H_{\mathrm{R}}(\omega) \leqslant 0$ for all ω. In particular, we find that $H(s) \leqslant 0$ on the positive real axis, because, being the Laplace transform of a real function, $H(s)$ is real when s is real.

(2) The function $|H(s)/s|$ decreases and the function $|sH(s)|$ increases monotonically as s increases from zero to infinity along the real axis.

The proof follows from (18), which can also be written as

$$H(s) = (2s/\pi) \int_{0}^{\infty} [H_{\mathrm{R}}(\omega)/(s^2 + \omega^2)] \, d\omega, \qquad \operatorname{Re} s > 0 \tag{20}$$

Since $|H_{\mathrm{R}}(\omega)| = -H_{\mathrm{R}}(\omega)$, we have for real s

$$|H(s)/s| = (2/\pi) \int_{0}^{\infty} [|H_{\mathrm{R}}(\omega)|/(s^2 + \omega^2)] \, d\omega$$

Thus, the integrand decreases monotonically at each ω. A moment's reflection will show that this property is additive, and hence true also for the integral.

The proof of the second assertion pertaining to $|sH(s)|$ is now trivial.

(3) The argument of $-H(s)$ satisfies the following inequality:

$$|\arg(s)| \geqslant |\arg(-H(s))|, \qquad \operatorname{Re} s > 0 \tag{21}$$

The proof follows from (20) in two steps:

$$-H(s) = (2s/\pi) \int_{0}^{\infty} d\omega \, [(s^{*2} + \omega^2) |H_{\mathrm{R}}(\omega)| / |s^2 + \omega^2|^2]$$

from which we obtain $-H(s) = As + Bs^*$, where A and B are positive numbers depending on s and $|H_{\mathrm{R}}(\omega)|$. The desired inequality is established with the help of the diagram in Figure 7.3.1.

The criterion (7) was first derived by Welton [10] in 1952 using an analogy between the equations satisfied by the logarithm of the reactor

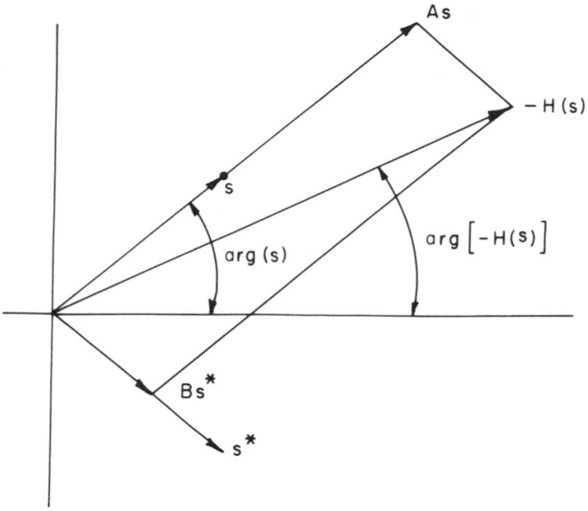

FIGURE 7.3.1. Proof of a inequality (21).

power and the equation of motion of a material particle in a certain force field. Several different derivations have been proposed in the literature [3, 7, 12] since that time, some of which will be presented in later sections. The present derivation allows $G(t)$ to contain delta functions, and $H(s)$ to be a transcendental function of s, and thereby permits one to apply Welton's criterion to systems with distributed, as well as lumped parameters. The following two examples illustrate the application of this criterion to some reactor models.

B. Asymptotic Stability of a Reactor with a Delayed Temperature Feedback

This reactor model is described by the following feedback kernel:

$$G(t) = \gamma\,\delta(t) + \alpha\eta U(t-\tau)\,e^{-d(t-\tau)} \tag{22a}$$

whose Laplace transform is

$$H(s) = \gamma + [\alpha\eta\,e^{-s\tau}/(s+d)] \tag{22b}$$

The real part of $H(i\omega)$ readily follows from (22b) as

$$\text{Re}[H(i\omega)] = \gamma + \alpha\eta[d(\cos\omega\tau) - \omega(\sin\omega\tau)]/(\omega^2 + d^2) \leqslant 0 \tag{23}$$

The necessary and sufficient conditions for (23) to hold for all frequencies are†

$$\gamma < 0 \tag{24}$$

$$|\gamma| \geqslant |\alpha\eta|/\{d[1 + (\omega_0^2/d^2)]^{1/2}\} \tag{25}$$

† The condition (15) is satisfied by (22a). Hence, (23) is sufficient for asymptotic stability.

where ω_0 is the lowest root of $\omega_0 = d \tan(\omega_0 \tau)$. The condition (24) follows from (23) with $\omega = \infty$. The condition (25) is obtained by first casting (23) into the form

$$-|\gamma|(\omega^2 + d^2)^{1/2} + \alpha\eta \cos[\omega\tau + \phi(\omega)] \leqslant 0 \tag{26}$$

where $\tan \phi(\omega) = \omega/d$. We distinguish between two cases. When $\alpha > 0$, (26) is satisfied if $|\gamma| d \geqslant \alpha\eta$. The latter inequality is contained in $H(0) = \gamma d + \alpha\eta < 0$, which ensures that the power coefficient of reactivity is negative. When $\alpha < 0$, (26) holds for all ω if $|\gamma|(\omega_0{}^2 + d^2)^{1/2} \geqslant |\alpha|\eta$, where ω_0 is the root of $\omega_0\tau + \phi(\omega_0) = \pi$, which is identical to (25). The region of asymptotic stability in the large is displayed in the parameter plane characterized by $A = -\gamma d$ and $B = -\alpha\eta$ in Figure 7.3.2.

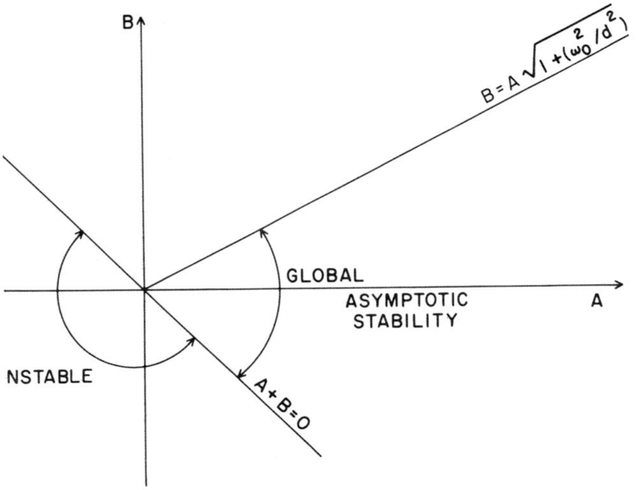

FIGURE 7.3.2. Asymptotic stability in a reactor with a delayed temperature feedback.

C. Circulating Fuel Reactor

The feedback functional of an idealized circulating fuel reactor was obtained before as [5.2, Eq. 90]

$$G(t) = \alpha\eta U(\theta - t)[1 - (t/\theta)], \qquad \theta \geqslant 0 \tag{27}$$

The associated feedback transfer function is

$$H(s) = \alpha\eta\{(1/s) - [(1 - e^{-s\theta})/s^2\theta]\} \tag{28}$$

The real part of $H(i\omega)$ readily follows from (28) as

$$H_R(\omega) = \alpha\eta(1 - \cos\omega\theta)/\omega^2\theta \leqslant 0 \tag{29}$$

We can easily show that $H(0) = \alpha\eta\theta/2$, hence the negativeness of the power coefficient of reactivity, $H(0) < 0$, requires α to be negative. With $\alpha < 0$, we see that criterion (29) is always satisfied in an idealized circulating fuel reactor. However, we cannot conclude from this observation that any equilibrium state is globally asymptotically stable,† because the delayed neutrons are neglected in an idealized circulating fuel reactor. As pointed out in Chap. 1, the point kinetic equations (7.2, Eqs. 2 and 3) are not applicable to the moving fuel reactors, and the conclusions concerning the effect of delayed neutrons on stability are not valid in the case of nonsolid-fuel reactors. In view of these remarks, we can only conclude that an equilibrium state in an idealized circulating fuel reactor is stable, and the power oscillations are always bounded for any initial curve. We shall show later that the circulating fuel reactors may have periodic power oscillations (limit cycles) when the equilibrium power level exceeds a certain value (cf Sec. 7.13B).

7.4. Asymptotic Stability of Reactors with a Nonlinear Feedback

This section illustrates the application of the general stability criterion for asymptotic stability in the large to reactors with a nonlinear feedback.

A. Xenon and Temperature Feedback

It was shown in Chap. 5 that the combined effect of xenon and temperature feedback is described by the following nonlinear functional:

$$\delta k_f[p] = \int_{-\infty}^{t} K(t-u)p(u)\,du - \alpha_p p(t) + \gamma\sigma_{xe}\int_{-\infty}^{t} du\,p^2(u)\,e^{-\lambda_{xe}(t-u)} \tag{1}$$

where the various symbols were defined previously.

Equilibrium states of this reactor are the roots of

$$\delta k_f(z_0) = z_0[\bar{K}(0) - \alpha_p + (\gamma\sigma_{xe}z_0/\lambda_{xe})] = 0 \tag{2}$$

This equation has only one root, $z_0 = 0$, if $\gamma < 0$, i.e., if the temperature coefficient of reacitivity is negative. (Recall that the temperature feedback

† The condition (15) is trivially satisfied in this case. We also note that $H(i\omega) \neq 0$ (cf 7.3, Eq. 13) because $H_I(\omega) = -\alpha\eta/\omega$ when $H_R(\omega) = 0$.

effects are treated quasistatically.) When $\gamma > 0$, there are two equilibrium states, as shown by Chernick *et al.* [13]. Since we are interested in global asymptotic stability (which requires a unique equilibrium state as a necessary condition), we shall consider only the case of $\gamma < 0$.

Substituting (1) into (7.2, Eq. 8), and choosing x_0 to be infinity, we obtain

$$I[t, y] = \int_{-\infty}^{t} du \int_{-\infty}^{u} y(u)y(v)K(u-v)\,dv - \alpha_{\mathrm{p}} \int_{-\infty}^{t} y^2(u)\,du$$
$$+ \gamma\sigma_{\mathrm{xe}} \int_{-\infty}^{t} y(u)\,du \int_{-\infty}^{u} y^2(v)\,\mathrm{e}^{-\lambda_{\mathrm{xe}}(u-v)}\,dv \leqslant 0 \tag{3}$$

where we have used the real time t rather than $x = \beta t/l$. Our task is now to determine [14, 15] sufficient conditions under which (3) will be nonpositive for all $t \geqslant 0$ and for all physically admissible functions $y(t)$.

We recall that $y(u) \geqslant -P_0$ at all times. Multiplying both sides of this inequality by

$$\gamma\sigma_{\mathrm{xe}} \int_{-\infty}^{u} y^2(v)\,\mathrm{e}^{-(u-v)\lambda_{\mathrm{xe}}}\,dv$$

integrating over u in $(-\infty, t)$, and making use of $\gamma < 0$, we obtain an upper bound for the third term in (3):

$$\gamma\sigma_{\mathrm{xe}} \int_{-\infty}^{t} du\,y(u) \int_{-\infty}^{u} dv\,y^2(v)\,\mathrm{e}^{-\lambda_{\mathrm{xe}}(u-v)}$$
$$\leqslant P_0|\gamma|\sigma_{\mathrm{xe}} \int_{-\infty}^{t} du \int_{-\infty}^{u} dv\,y^2(v)\,\mathrm{e}^{-\lambda_{\mathrm{xe}}(u-v)}$$
$$= P_0|\gamma|\sigma_{\mathrm{xe}} \int_{-\infty}^{t} du \int_{0}^{\infty} dv\,\mathrm{e}^{-\lambda_{\mathrm{xe}}v} y^2(u-v)$$
$$= P_0|\gamma|\sigma_{\mathrm{xe}} \int_{0}^{\infty} dv\,\mathrm{e}^{-\lambda_{\mathrm{xe}}v} \int_{-\infty}^{t-v} du\,y^2(u)$$
$$\leqslant P_0|\gamma|\sigma_{\mathrm{xe}} \int_{0}^{\infty} dv\,\mathrm{e}^{-\lambda_{\mathrm{xe}}v} \int_{-\infty}^{t} du\,y^2(u) = (P_0|\gamma|\sigma_{\mathrm{xe}}/\lambda_{\mathrm{xe}}) \int_{-\infty}^{t} du\,y^2(v) \tag{4}$$

Using (4) to replace the third term in (3), we find

$$I[t, y] \leqslant \int_{-\infty}^{t} du\,y(u) \int_{-\infty}^{t} dv\,y(v)\{K(u-v) + [(P_0|\gamma|\sigma_{\mathrm{xe}}/\lambda_{\mathrm{xe}}) - \alpha_{\mathrm{p}}]\,\delta(u-v)\} \leqslant 0 \tag{5}$$

The right-hand side of (5) is equivalent to (7.3, Eq. 4), with

$$G(u) = K(u) + [(P_0|\gamma|\sigma_{\mathrm{xe}}/\lambda_{\mathrm{xe}}) - \alpha_{\mathrm{p}}]\,\delta(u) \tag{6}$$

and can be made nonpositive by requiring Re $H(i\omega) \leqslant 0$. Hence, if

$$-\mathrm{Re}\ \overline{K}(i\omega) \leqslant (P \mid \gamma \mid \sigma_{\mathrm{xe}}/\lambda_{\mathrm{xe}}) - \alpha_{\mathrm{p}} \tag{7}$$

holds, then the equilibrium state P_0 of a xenon- and temperature-controlled reactor is asymptotically stable in the large.† Substituting $K(t)$ from (5.3, Eq. 29), we can express (7) more explicitly as

$$-\left\{\frac{\sigma_{\mathrm{xe}}\lambda_{\mathrm{xe}}}{\omega^2+\lambda_{\mathrm{xe}}^2}\left(\alpha_2+\sum_{j=1}^{I}\frac{\lambda_j a_j}{\lambda_j-\lambda_{\mathrm{xe}}}\right)-\sigma_{\mathrm{xe}}\sum_{j=1}^{I}\frac{a_j}{\lambda_j-\lambda_{\mathrm{xe}}}\frac{\lambda_j^2}{\omega^2+\lambda_j^2}+\frac{\alpha_I y_I\lambda_I\sigma_{\mathrm{f}}}{\omega^2+\lambda_I^2}\right\}$$

$$\leqslant \frac{P_0 \mid \gamma \mid \sigma_{\mathrm{xe}}}{\lambda_{\mathrm{xe}}} - \alpha_{\mathrm{p}} \tag{8}$$

where α_2, α_{I}, and α_p are defined by (5.3, Eqs. 25). The region of global asymptotic stability is depicted in Figure 7.4.1. The double-humped curve in this figure separates the region of linear stability and instability. It was shown by Chernick *et al.* [13] that, in portions of the region of linear stability, the equilibrium state was asymptotically stable for small perturbations but unstable for large perturbations. An example of such a situation is shown in Figure 7.4.2, which represent the response of the reactor to a power perturbation when $P_0 = 1.6 \times 10^{11}$ neutrons cm^{-2} sec^{-1} and $\gamma = -3 \times 10^{-16}$ (the point Q in Figure 7.4.1). When the initial power perturbation is 10^4 times the equilibrium value, the subsequent power oscillations are damped (curve A). But when the initial disturbance is increased to 10^5 times the equilibrium value, the power response does not return to its equilibrium value, but exhibits a periodic motion (curve B). Hence, the equilibrium state is only asymptotically stable when the reactor is operated at the point Q in the parameter plane, but not asymptotically stable in the large. The condition (8) separates a portion of the region of linear stability in which the equilibrium states are globally asymptotically stable.

† The conditions of Sec. 7.2, Theorem 2 for asymptotic stability are satisfied without any additional restrictions on the feedback functional. The first condition, can be readily demonstrated for each term in (1) in a similar manner to that used in (7.3, Eqs. 8–10). The second condition, (7.2, Eq. 27), can be proved using the lemma in the footnote on p. 336. According to this lemma, the first two terms in (1) behave as $w(t)$, whereas the last terms behave as $w^2(t)$ as $t \to \infty$. Hence,

$$\delta k_{\mathrm{f}}(w) \to w(t)[\overline{K}(0) - \alpha_{\mathrm{p}} + (\gamma\sigma_{\mathrm{xe}}/\lambda_{\mathrm{xe}})\, w(t)] \to 0$$

which implies either $w(t) \to 0$ or $w(t) \to [\alpha_{\mathrm{p}} - \overline{K}(0)]\, \lambda_{\mathrm{xe}}/\gamma\sigma_{\mathrm{xe}}$. The latter possibility is ruled out because the constant that $w(t)$ approaches is always less than —1 for $\gamma < 0$ (cf Eq. 2). The details of the proof are left to the reader.

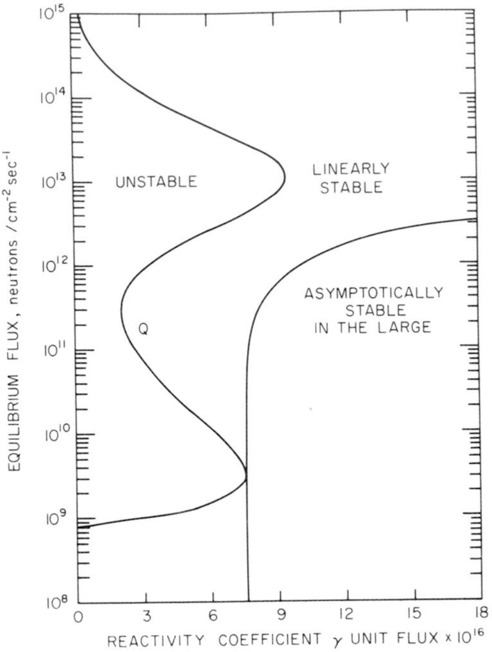

FIGURE 7.4.1. Regions of stability for a xenon–temperature-controlled reactor. At Q, $P_0 = 1.6 \times 10^{11}$ neutrons cm^{-2} sec^{-1}, $\gamma = -3 \times 10^{-16}$.

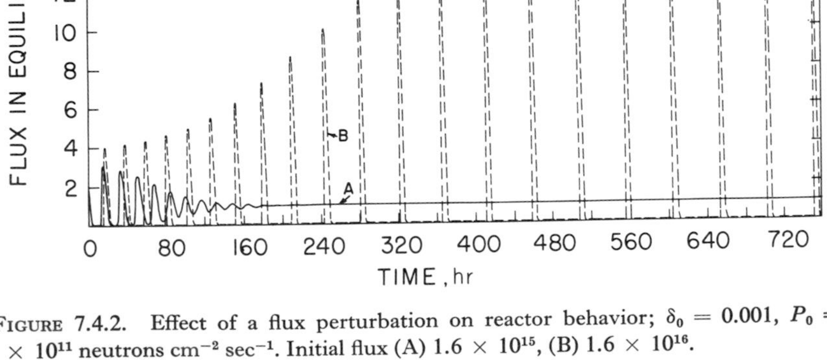

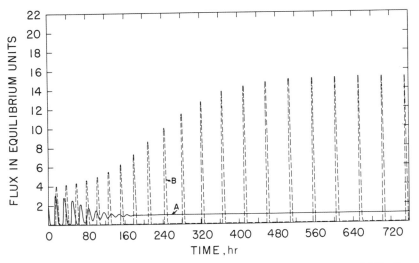

FIGURE 7.4.2. Effect of a flux perturbation on reactor behavior; $\delta_0 = 0.001$, $P_0 = 1.6 \times 10^{11}$ neutrons cm^{-2} sec^{-1}. Initial flux (A) 1.6×10^{15}, (B) 1.6×10^{16}.

B. Nonlinear One-Temperature Feedback

The nonlinear feedback that we consider here is described by

$$\delta k_f[p] = F(T) \tag{9a}$$

$$\dot{T} = ap - H(T) \tag{9b}$$

where $a > 0$ is a constant, $T(t)$ is the incremental temperature measured from equilibrium, and $F(T)$ and $H(T)$ are arbitrary continuous functions. Our task is to determine sufficient conditions on these functions such that the equilibrium state $p = 0$, $T = 0$ is asymptotically stable in the large.

The uniqueness of the equilibrium state requires the equations $H(T) = 0$ and $F(T) = 0$ to be satisfied simultaneously only at $T = 0$.

Let us consider the general stability criterion (7.2, Eq. 8):

$$I[t, y] = \int_{-\infty}^{t} ay(u)F[T(u)]\, du \leqslant 0 \tag{10}$$

where $y(u)$ is an admissible function which is an initial curve for $t \leqslant 0$, and arbitrary for $t > 0$. The corresponding function $T(t)$ is to be obtained from (9b) with $p(t)$ replaced by $y(t)$, i.e., $\dot{T} = ay(t) - H(T)$. Expressing $y(t)$ from the latter as

$$y(t) = [\dot{T} + H(T)]/a$$

and substituting into (10), we obtain

$$I[t, y] = \int_{-\infty}^{t} \dot{T}F(T)\, du + \int_{-\infty}^{t} H(T)F(T)\, du \tag{11}$$

where $T = T(u)$. We perform a change of variable in the first integral from u to T. Since $T(-\infty) = 0$ (inital curve) we get

$$I[t, y] = \int_{0}^{T} F(T')\, dT' + \int_{-\infty}^{t} H(T)F(T)\, du \leqslant 0 \tag{12}$$

Note that T may be positive or negative. However, the first integral is always negative if $TF(T) < 0$. The second integral is nonpositive if we require $H(T)F(T) < 0$ for all T. We can therefore conclude that the equilibrium state $p = T = 0$ of (9) is asymptotically stable in the large if

$$TF(T) < 0 \tag{13a}$$

$$TH(T) > 0 \tag{13b}$$

holds for all $T \neq 0$.

7.5. Asymptotic Stability of an Integral Equation

Corduneanu [16] investigated the asymptotic stability of the following integral equation:

$$\sigma(t) = f(t) + \int_0^t l(t-\tau)\phi[\sigma(\tau)]\, d\tau \tag{1}$$

where $f(t)$, $l(t)$, and $\phi(\sigma)$ are real functions which satisfy the following conditions: (i) $f(t)$ is defined for $t \geqslant 0$, and $f'(t)$ and $f''(t)$ are absolute-integrable in $(0, \infty)$; (ii) $l(t) = j(t) - \rho$, where $\rho > 0$, and $j(t)$ is defined for $t \geqslant 0$, and $j(t)$ and $j'(t)$ are absolute- as well as square-integrable; (iii) $\phi(\sigma)$ is continuous for all real σ and satisfies

$$\sigma\phi[\sigma] > 0, \qquad \sigma \neq 0, \qquad \phi(0) = 0 \tag{2}$$

and (iv) if

$$J(i\omega) \equiv \int_0^\infty j(t)\, e^{-i\omega t}\, dt \tag{3a}$$

and

$$L(i\omega) \equiv J(i\omega) - (\rho/i\omega) \tag{3b}$$

then there exists a $q \geqslant 0$ such that

$$\mathrm{Re}[(1 + i\omega q)L(i\omega)] \leqslant 0, \qquad \omega \neq 0 \tag{4}$$

We shall show later that the point kinetic equations in the absence of delayed neutrons can be cast into the form of (1). Corduneanu proves the following theorem.

THEOREM 1. If the conditions (i)–(iv) are satisfied, then (1) has at least one solution $\sigma(t)$ defined for $t \geqslant 0$, and such a solution satisfies

$$\lim_{t\to\infty} [\sigma(t)] = 0 \tag{5}$$

We shall skip the proof of the existence of a solution of (1), and consider the question of asymptotic stability. The proof follows a method due to Popov [9] which we shall discuss further in a later section.

Let us define

$$\begin{aligned} \phi_t(\tau) &\equiv \phi[\sigma(\tau)], & 0 \leqslant \tau \leqslant t \\ &\equiv 0 & \tau > t \end{aligned} \tag{6}$$

and consider the function

$$\lambda_t(\tau) = \int_0^\tau [j(\tau - \xi) + qj'(\tau - \xi)]\phi_t(\xi)\, d\xi + ql(0)\phi_t(\tau), \qquad \tau \geqslant 0 \tag{7}$$

Differentiation of (1) yields

$$\sigma'(t) = f'(t) + l(0)\phi[\sigma(t)] + \int_0^t j'(t - \tau)\phi[\sigma(\tau)]\, d\tau \tag{8}$$

Combining (1) and (8) with (7), we obtain $\lambda_t(\tau)$ as

$$\lambda_t(\tau) = \sigma(\tau) + q\sigma'(\tau) - [f(\tau) + qf'(\tau)] + \rho \int_0^\tau \phi[\sigma(x)]\, dx, \qquad 0 \leqslant \tau \leqslant t \tag{9}$$

$$\lambda_t(\tau) = \int_0^t [j(\tau - x) + qj'(\tau - x)]\phi[\sigma(x)]\, dx, \qquad \tau > t \tag{10}$$

It is clear that $\lambda_t(\tau)$ is a function of τ for a fixed value of t, which is defined for $\tau \geqslant 0$. It can be easily verified, using condition (ii) concerning the absolute- and square-integrability of $j(\tau)$ and $j'(\tau)$ that $\lambda_t(\tau)$ is also absolute- and square-integrable in $(0, \infty)$ for each t. We can therefore define its one-sided Fourier transform as

$$\Lambda_t(i\omega) = \int_0^\infty \lambda_t(\tau)\, e^{-i\omega\tau}\, d\tau = [(1 + i\omega q)J(i\omega) - \rho q]F_t(i\omega) \tag{11a}$$

where

$$F_t(i\omega) = \int_0^t \phi[\sigma(x)]\, e^{-i\omega x}\, dx \tag{11b}$$

With the aid of the Parseval relation, we obtain from (11)

$$\begin{aligned} \eta(t) &= \int_0^t \lambda_t(\tau)\phi[\sigma(\tau)]\, d\tau = \int_0^\infty \lambda_t(\tau)\phi_t(\tau)\, d\tau \\ &= (1/2\pi) \int_{-\infty}^\infty \Lambda_t(i\omega)F_t^*(i\omega)\, d\omega \end{aligned} \tag{12}$$

where $F_t^*(i\omega)$ is the complex conjugate of $F_t(i\omega)$. Since $\eta(t)$ is real, we obtain from (12)

$$\eta(t) = (1/2\pi) \int_{-\infty}^\infty \mathrm{Re}[(1 + i\omega q)L(i\omega)]\, |F_t(i\omega)|^2 \leqslant 0 \tag{13}$$

The inequality follows from condition (iv). Substituting $\lambda_t(\tau)$ from (9) into (12), and using $\eta(t) \leqslant 0$, we find

$$\int_0^t \phi[\sigma(\tau)]\sigma(\tau)\,d\tau + q\int_0^t \phi[\sigma(\tau)]\sigma'(\tau)\,d\tau$$
$$+ \rho\int_0^t d\tau \left\{\int_0^\tau \phi[\sigma(x)]\,dx\right\}\phi[\sigma(\tau)] - \int_0^t [f(\tau) + qf'(\tau)]\phi[\sigma(\tau)]\,d\tau \leqslant 0 \quad (14)$$

Defining

$$F(\sigma) \equiv \int_0^\sigma \phi(u)\,du \quad (15a)$$

$$\Phi(t) \equiv \int_0^t \phi[\sigma(x)]\,dx \quad (15b)$$

we can express (14) as

$$\int_0^t \phi[\sigma(\tau)]\sigma(\tau)\,d\tau + qF[\sigma(t)] + \tfrac{1}{2}\rho\Phi^2(t)$$
$$- \int_0^t [f(\tau) + qf'(\tau)]\phi[\sigma(\tau)]\,d\tau - qF[f(0)] \leqslant 0 \quad (16)$$

where we have used $f(0) = \sigma(0)$ and

$$\int_0^t \phi[\sigma(\tau)]\sigma'(\tau)\,d\tau = \int_{\sigma(0)}^{\sigma(t)} \phi(u)\,du = F[\sigma(t)] - F[\sigma(0)]$$

On the other hand, we can verify by integration by parts and by (15b) that

$$\int_0^t [f(\tau) + qf'(\tau)]\phi[\sigma(\tau)]\,d\tau = [f(t) + qf'(t)]\Phi(t)$$
$$- \int_0^t [f'(\tau) + qf''(\tau)]\Phi(\tau)\,d\tau \quad (17)$$

Taking the absolute values of both sides in (17), we get

$$\left|\int_0^t [f(\tau) + qf'(\tau)]\phi[\sigma(\tau)]\,d\tau\right| \leqslant K\sup[|\Phi(\tau)|, 0 \leqslant \tau \leqslant t] \quad (18)$$

where K is a positive number which depends only on the function f. Use of (18) in (16) yields

$$\int_0^t \phi[\sigma(\tau)]\sigma(\tau)\,d\tau + qF[\sigma(t)] + \tfrac{1}{2}\rho\Phi^2(t) - K\sup_{0\leqslant\tau\leqslant t}|\Phi(\tau)| - qF[f(0)] \leqslant 0 \quad (19)$$

Since the first two terms of (19) are nonnegative by virtue of (2), we are led to

$$\Phi^2(t) - (2K/\rho) \sup_{0 \leqslant \tau \leqslant t} |\Phi(\tau)| - (2q/\rho)F[f(0)] \leqslant 0 \tag{20}$$

Let $T(t)$ be the time at which $\Phi(t)$ attains its maximum in $(0, t)$, i.e.,

$$\Phi[T(t)] \equiv \sup\{|\Phi(\tau)|, 0 \leqslant \tau \leqslant t\} \tag{21}$$

Clearly, $0 \leqslant T(t) \leqslant t$. Since Eq. (20) holds for all t, it also holds at the instant Φ attains its maximum. Hence,

$$\Phi^2[T(t)] - (2K/\rho)\Phi[T(t)] - 2(q/\rho)F[f(0)] \leqslant 0 \tag{22}$$

from which we conclude that

$$\Phi(t) \leqslant \Phi[T(t)] \leqslant (K/\rho) + \{(K^2/\rho^2) + 2qF[f(0)/\rho\}^{1/2} \equiv \alpha[F[f(0)]] \tag{23}$$

where $\alpha(r)$ is a continuous function of its argument. We conclude from (23) that $\Phi(t)$ is bounded for all $t \geqslant 0$. From (1), we also find

$$\sigma(t) = f(t) + l(0)\Phi(t) - \int_0^t j'(t-\tau)\Phi(\tau)\, d\tau \tag{24}$$

and

$$|\sigma(t)| \leqslant M + \left[|l(0)| + \int_0^t |j'(u)|\, du\right] \alpha[F[f(0)]] \equiv \beta[F[f(0)]] \tag{25}$$

where M is the upper bound of $|f(t)|$ for $t \geqslant 0$, i.e., $|f(t)| \leqslant M$ for $t \geqslant 0$. Thus, $\sigma(t)$ is bounded for $t \geqslant 0$. On the other hand, (8) indicates that $\sigma'(t)$ is also bounded for $t \geqslant 0$, and hence $\sigma(t)$ is uniformly continuous (cf the footnote on p. 330) for $t \geqslant 0$.

Using (19) and (23), we can show that

$$\int_0^t \phi[\sigma(\tau)]\sigma(\tau)\, d\tau \leqslant qF[f(0)] + K\alpha[F[f(0)]] \tag{26}$$

If we define

$$\Psi(t) \equiv \int_0^t \phi[\sigma(\tau)]\sigma(\tau)\, d\tau$$

we find that $\Psi(t)$ is nonnegative (cf Eq. 2) and bounded. Thus, $\Psi(t)$ admits a finite limit as $t \to \infty$. Furthermore, $\Psi'(t) = \phi(\sigma(t))\sigma(t)$ is uniformly continuous. The lemma presented in the footnote of p. 331 shows that

$$\lim_{t\to\infty} \phi[\sigma(t)]\sigma(t) = 0 \tag{27}$$

which implies $\sigma(t) \to 0$ as $t \to \infty$, proving the asymptotic stability of the solutions of (1).

Application to Reactors

Point reactor kinetic equations can be reduced to the form in (8) in the absence of delayed neutrons and when the feedback functional is linear. Defining

$$\sigma(x) = \ln[1 + z(x)] \tag{28}$$

and ignoring the delayed neutrons ($a_i = h_i = 0$) in (7.2, Eq. 2), we obtain

$$\sigma'(x) = \int_{-\infty}^{x} G(x - u)[e^{\sigma(u)} - 1]\, du \tag{29}$$

In order to allow a power coefficient in the feedback mechanism, we break up the feedback kernel $G(x)$ into two parts:

$$G(x) = -\gamma\delta(x) + K(x) \tag{30}$$

where $-\gamma$ is the power coefficient, and $K(x)$ is the bounded part of the feedback kernel accounting for the reactivity effects due to nuclear energy produced in the past. Substituting (30) into (29), we obtain

$$\sigma'(x) = -\gamma\phi[\sigma(x)] + \int_{-\infty}^{x} K(x - u)\phi[\sigma(u)]\, du \tag{31a}$$

where we have defined

$$\phi(\sigma) \equiv e^{\sigma} - 1 \tag{31b}$$

By introducing

$$f'(x) \equiv \int_{-\infty}^{0} K(x - u)\bar{z}(u)\, du \tag{32a}$$

$$j'(x) \equiv K(x) \tag{32b}$$

$$-\gamma \equiv l(0) = j(0) - \rho \tag{32c}$$

we can write (31a) as

$$\sigma'(x) = f'(x) + l(0)\phi(\sigma) + \int_{0}^{x} j'(x - u)\phi[\sigma(u)]\, du \tag{33}$$

which is identical to (8).

We observe from (32a) that $f'(x)$ depends on the past history of the reactor. Let us assume that the physically admissible initial curves

$\bar{z}(x)$ [$\bar{z}(x) = e^{\sigma(x)} - 1$, for $x \leqslant 0$] are so chosen that $f'(x)$ and $f''(x)$ are absolute-integrable in $(0, \infty)$, i.e.,

$$\int_0^\infty \left| \int_{-\infty}^0 du\, K(x-u)\bar{z}(u) \right| dx < \infty \tag{34a}$$

$$\int_0^\infty \left| \int_{-\infty}^0 du\, \dot{K}'(x-u)\bar{z}(u) \right| dx < \infty \tag{34b}$$

Then, condition (i) is satisfied. In order to verify condition (ii), we must first define $j(x)$ using (32b), which specifies only the derivative of $j(x)$. Integrating (32b) on (x, ∞), we obtain

$$j(x) = -\int_x^\infty K(u)\, du \tag{35}$$

Condition (ii) requires $j(x)$ and $j'(x)$ to be both square- and absolute-integrable. Hence, we impose the following restrictions on $K(x)$ to guarantee the above requirements:

$$\int_0^\infty |K(x)|\, dx < \infty \tag{36a}$$

$$\int_0^\infty |K(x)|^2\, dx < \infty \tag{36b}$$

$$\int_0^\infty \left| \int_x^\infty K(u)\, du \right| dx < \infty \tag{36c}$$

$$\int_0^\infty \left| \int_x^\infty K(u)\, du \right|^2 dx < \infty \tag{36d}$$

Condition (iii) is automatically satisfied because

$$\sigma(e^\sigma - 1) > 0, \qquad \sigma \neq 0 \tag{37}$$

In order to verify the last condition, i.e., Eq. (4), we need the Fourier transform of $j(x)$:

$$\begin{aligned} J(i\omega) &= -\int_0^\infty dx\, e^{-i\omega x} \int_x^\infty K(u)\, du \\ &= [j(0)/i\omega] + [\kappa(i\omega)/i\omega] \end{aligned} \tag{38}$$

where $\kappa(i\omega)$ is the Fourier transform of $K(x)$. Substituting (38) into (36) and using (32c), we obtain

$$L(i\omega) = [\kappa(i\omega)/i\omega] - (\gamma/i\omega) \tag{39}$$

We are now in a position to state the following criterion.

THEOREM 2. If the linear feedback functional of a reactor described by $-\gamma\delta(t) + K(t)$ or by its transform $-\gamma + \kappa(s)$ satisfies the conditions (36), if

$$-\gamma + \kappa(0) < 0 \tag{40a}$$

and if there exists a $q \geqslant 0$ such that

$$q[-\gamma + \operatorname{Re} \kappa(i\omega)] + (1/\omega) \operatorname{Im} \kappa(i\omega) \leqslant 0, \qquad \omega \neq 0 \tag{40b}$$

holds for all ω, then the equilibrium of the reactor, whose existence is guaranteed by (40a), is asymptotically stable in the absence of delayed neutrons for all initial curves for which (34) holds.

It is noted that the nonnegative q in (40b) is a number which must be found, if it exists, for a given reactor such that the inequality (40b) holds for all ω. It is not an arbitrary number which one picks up a priori. However, one may specify certain limiting values of q, such as 0, 1, etc., and then determine the condition to be imposed on the reactor parameters entering $\kappa(i\omega)$, and γ for which (40b) is true for all ω.

It is very important to note that the condition (40a) is not implied by (40b), and therefore appears explicitly in the statement of the theorem. It guarantees the existence of a unique, finite equilibrium value, which is a necessary condition for asymptotic stability in the large. It is used implicitly in the proof of the theorem because (32c) and (35) lead to

$$-\gamma + \kappa(0) = -\rho \tag{40c}$$

where ρ is assumed to be strictly positive in the proof of the Corduneanu theorem (condition ii).

Since $q = 0$ is allowed in the theorem, it is of interest to find the conditions on γ and $\kappa(i\omega)$ such that (40b) will hold for all ω with $q = 0$. Clearly, this condition is

$$[\operatorname{Im} \kappa(i\omega)]/\omega \leqslant 0 \tag{40d}$$

The latter alone, however, is not sufficient to secure asymptotic stability; we must also satisfy (40a). It is interesting, and indeed very useful, to introduce the Hilbert transform relation (cf 4.4, Eqs. 23 and 24) between the real and imaginary parts of $\kappa(i\omega)$:

$$\operatorname{Re} \kappa(i\omega) = (1/\pi) \int_{-\infty}^{\infty} \{[\operatorname{Im} \kappa(i\omega')]/(\omega - \omega')\} \, d\omega'$$

where we assume that $\operatorname{Re} \kappa(i\infty) = 0$. Evaluating this at $\omega = 0$, we find [using (40d)]

$$\operatorname{Re} \kappa(0) \equiv \kappa(0) = -(1/\pi) \int_{-\infty}^{\infty} \{[\operatorname{Im} \kappa(i\omega')]/\omega'\} \, d\omega' \geqslant 0 \qquad (40e)$$

Hence, in the absence of a power coefficient, i.e., $\gamma = 0$, (40a) cannot be satisfied because it requires $\kappa(0) < 0$, in contrast to (40e). When $\gamma \neq 0$, then (40d) guarantees asymptotic stability if

$$-\gamma - (1/\pi) \int_{-\infty}^{\infty} \{[\operatorname{Im} \kappa(i\omega')]/\omega'\} \, d\omega' < 0 \qquad (40f)$$

also holds. The foregoing analysis illustrates the fact that (40b) does not contain (40a) in general (note that, if $[\operatorname{Im} \kappa(i\omega)]/\omega \to 0$ as $\omega \to 0$, then (40a) is implied by (40b)).

We may also point out in passing that (40d) is sufficient to guarantee linear stability at all power levels P_0 in the absence of delayed neutrons if (40f) is also satisfied [the characteristic equation is $s + \gamma - \kappa(s) = 0$, where γ and $\kappa(s)$ are proportional to P_0] (Problem 4).

It may be tempting to divide (40b) by q and express the condition with $p = (1/q)$:

$$-\gamma + [\operatorname{Re} \kappa(i\omega)] + (p/\omega) \operatorname{Im} \kappa(i\omega) \leqslant 0 \qquad (40g)$$

This condition is of course equivalent to (40b) provided we do not allow p to be zero. This restriction is due to the fact that Corduneanu's theorem is proven for a finite nonnegative q. For example, if $-\gamma + \operatorname{Re} \kappa(i\omega) \leqslant 0$ for a certain reactor for all ω, then $p = 0$ serves to make (40g) true for all $\omega \neq 0$. This condition is identical to (7.3, Eq. 7), and guarantees only the boundedness of the solutions according to Theorem 1, Sec. 7.2. If the restriction $p \neq 0$ is not taken into account, then (40g) would seemingly imply asymptotic stability in the large. The following example will show that this is not true in certain reactor types. We shall, however, rederive condition (40g) using Kalman's theory in Sec. 7.11, in which p will be allowed to be zero with some additional restriction on the feedback model (Problem 5).

Example. We shall illustrate the application of Theorem 2 to the circulating fuel reactor which was discussed in (7.3c). It can easily be verified using (7.3, Eq. 27) that the conditions (36) are satisfied for a circulating fuel reactor. The condition (40b) yields [with the substitution of $H(i\omega)$ from (7.3, Eq. 28)]

$$\operatorname{Re}\{[q + (1/i\omega)]\alpha\eta[(1/i\omega) + (1 - \cos\theta\omega + i \sin\theta\omega)/\theta\omega^2]\} \leqslant 0$$

which reduces to

$$q[1 - \cos x] + [(\sin x)/x] \geqslant 1, \qquad x \neq 0 \tag{41}$$

where we have used the fact that $\alpha < 0$ and substituted $q\theta \Rightarrow q$ and $\theta\omega \Rightarrow x$. Clearly, (41) cannot be satisfied for all $x > 0$ for any choice of q, and thus Theorem 2 does not apply to the circulating fuel reactor. It is interesting to note that Theorem 2 is noncommittal, and does not shed any light on the behavior of the circulating fuel reactor, whereas the condition (7.3, Eq. 29) ascertains boundedness of the solutions by virtue of Theorem 1 of Sec. 7.2.

7.6. Lyapunov's "Second Method"

A. Basic Ideas

The purpose of this section is to give a concise exposition of Lyapunov's second method, which is a powerful mathematical tool for tackling linear and nonlinear stability problems, with particular attention to applications in reactor dynamics. Lyapunov's original work was published in a Russian journal in 1892, was translated into French in 1907, and was reprinted in the U.S.A. in 1949 [1]. Since then, several papers and books in English have appeared dealing with the exposition and application of Lyapunov second's method [17, 18a], and new theorems extending Lyapunov's original theorems to a wider class of problems have been proven.

The objective of Lyapunov's second method is to examine the stability of a set of ordinary differential equations of the form†

$$\dot{\mathbf{x}} = \mathbf{f}(\mathbf{x}) \tag{1}$$

utilizing the specific form of $\mathbf{f}(\mathbf{x})$ but without an explicit knowledge of the solutions. In this respect, the second method is a stability criterion. The name "second method" is given to differentiate this "direct" approach to stability problems from the "indirect" approach, called the "first method" of Lyapunov's, in which the stability question is examined with an explicit representation of the solutions of (1).

The basic idea in the second method can be explained in terms of the energy of an isolated physical system [19]. Let $E(\mathbf{x})$ denote the total

† While we consider only autonomous systems the "second method" can be extended to include nonautonomous systems as well.

energy of such a system (when it is in a state $\mathbf{x}$) whose evaluation is governed by (1). If the rate of change $dE(\mathbf{x})/dt$ is negative for every possible state $\mathbf{x}$, except for a single equilibrium state $\mathbf{x}_e$, then the energy of the system will gradually decrease until it finally assumes its minimum value $E(\mathbf{x}_e)$. In more physical terms, a "dissipative" system perturbed from its equilibrium state will always return to it. In mechanical systems, one can construct the energy function $E(\mathbf{x})$ in terms of the state variables $\mathbf{x}$ without much difficulty. However, there is no natural way of defining energy when the equations (1) are given in a purely mathematical form. In certain simple systems, one can cast the original equations into a form which might be interpreted as equations of motion of a mechanical system, and an energy function can then be constructed by analogy. In order to illustrate this approach, let us consider a reactor which is described by the following equations:

$$dP(t)/dt = -(\alpha/l)T(t)P(t) \tag{2}$$

$$\mu\, dT(t)/dt = P(t) - P_0 - P_e(T) \tag{3}$$

in which $T(t)$ denotes the departures from the equilibrium temperature T_0, and $P_e(T)$ characterizes the law of heat removel from the reactor. Equations (2) and (3) represent the temporal behavior of a reactor in the absence of delayed neutrons, with a simple temperature feedback. Let us introduce the following change of dependent variables:

$$x(t) \equiv -\ln[P(t)/P_0], \qquad v(t) = \alpha T/l \tag{4}$$

and define

$$M = l\mu/\alpha \tag{5}$$

Then, Eqs. (2) and (3) are replaced by

$$v(t) = dx/dt \tag{6a}$$

$$M\, dv/dt = F(x, v) \tag{6b}$$

where

$$F(x, v) \equiv P_0\, e^{-x} - P_0 - P_e(v) \tag{6c}$$

and where P_0 is the equilibrium power level. Equations (6) may be interpreted as the equation of motion of a material particle of mass M under the influence of a force $F(x, v)$ which consists of two parts: $P_0(e^{-x} - 1)$ and $P_e(v)$. The first part is derived from a potential $P_0(e^{-x} + x)$ and represents a conservative force. The second part is a frictional force and causes dissipation of energy. The potential and kinetic energies of the particle are

$$E_{\mathrm{KE}} = \tfrac{1}{2}Mv^2 = \tfrac{1}{2}\mu\alpha T^2/l \tag{7a}$$

$$E_{\mathrm{PE}} = -\int_0^x P_0[(\exp x') - 1]\, dx' = P - P_0 - P_0 \ln(P/P_0) \tag{7b}$$

The total energy of the system is $V(x, v) = E_{\mathrm{KE}} + E_{\mathrm{PE}}$, or

$$V(P, T) = \tfrac{1}{2}(\mu\alpha/l)T^2 + P - P_0 - P_0 \ln(P/P_0) \tag{7c}$$

Clearly, the total energy vanishes at the equilibrium $P = P_0$ and $T = 0$, i.e., $V(P_0, 0) = 0$. Furthermore, $V(P, T)$ is always positive because $P - P_0 - P_0 \ln(P/P_0)$ is a positive-definite function. We can now determine conditions under which the rate of change of energy is always nonpositive. By differentiation of (7c), we obtain

$$\dot{V}(P, T) = (\partial V/\partial T)(dP/dt) + (\partial V/\partial T)(dT/dt)$$

Substituting dP/dt and dT/dt from (2) and (3), we get

$$\dot{V}(P, T) = -(\alpha/l)TP_e(T)$$

If the cooling law is such that $TP_e(T) > 0$ for all $T \neq 0$, then $\dot{V}(P, T) \leqslant 0$, and the system described by (2) and (3) is dissipative. Any perturbation from the equilibrium state will eventually decay to zero (Problem 6b).

B. Stability Theorems

In more complicated dynamical systems, which are described by equations in a purely mathematical form, the construction of an energy function by a mechanical analogy is usually not feasible. However, one may still expect that the equilibrium state $\mathbf{x}_e$ of a set of ordinary differential equations will be asymptotically stable if one can find a scalar function $V(\mathbf{x})$ (irrespective of any physical interpretation) which possesses the following two properties: (a) $V(\mathbf{x}) > 0$ and $\dot{V}(\mathbf{x}) < 0$ for $\mathbf{x} \neq \mathbf{x}_e$, (b) $V(\mathbf{x}) = \dot{V}(\mathbf{x}) = 0$ when $\mathbf{x} = \mathbf{x}_e$. This expectation, based on purely physical intuition, is the principal idea of the second method of Lyapunov's, and was proven to be true by Lyapunov [1] (Theorems 1 and 2).

Thus, the question of stability is reduced to the construction of a special scalar function $V(\mathbf{x})$, called a Lyapunov function, from the given form of Eq. (1) without an explicit knowledge of solutions. Unfortunately, there is no systematic way at present to find a Lyapunov function for a given dynamical system. Therefore, the success of the second method depends on the ingenuity and the experience of the user.

Before presenting the Lyapunov stability theorems, we have to introduce the precise definition of a Lyapunov function.

DEFINITION 1. A scalar function $V(\mathbf{x})$ is said to be "positive-definite" in a region R about the origin if $V(\mathbf{x}) = 0$ when $\mathbf{x} = 0$, and $V(\mathbf{x}) > 0$ in R whenever $\mathbf{x} \neq 0$.

DEFINITION 2. The derivative of $V(\mathbf{x})$ along a solution of (1) is

$$\dot{V} = \mathbf{f} \cdot \nabla V \equiv \sum_i f_i(x)(\partial V/\partial x_i) \tag{8}$$

provided the partial derivatives exist.

DEFINITION 3. A positive-definite function in R is called a Lyapunov function if it is continuous together with its first partial derivatives and $\dot{V} \leqslant 0$ in R.

We are now in a position to state the Lyapunov stability theorems for an autonomous system described by $\dot{\mathbf{x}} = \mathbf{f}(\mathbf{x})$. We assume without loss of generality that $\mathbf{f}(0) = 0$, and hence $\mathbf{x} \equiv 0$ in an equilibrium state.

THEOREM 1. If there exists a Lyapunov function $V(\mathbf{x})$ in an open region R containing the origin, then the solution $\mathbf{x} \equiv 0$ of (1) is stable.

PROOF. Let $\epsilon > 0$ be a positive number such that (Figure 7.6.1)

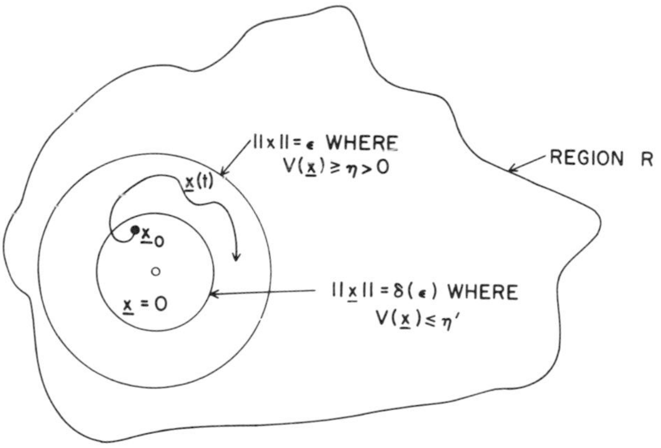

FIGURE 7.6.1. Proof of stability.

the set $\| \mathbf{x} \| \leqslant \epsilon$ is in the region R in which the Lyapunov function $V(\mathbf{x})$ exists, and $\mathbf{f}(\mathbf{x})$ is defined. Since $V(\mathbf{x})$ is continuous with its first partial derivatives in the entire region R, it is also continuous with its first partial derivatives (as a function of $n - 1$ independent variables) on the surface $\| \mathbf{x} \| = \epsilon$. Clearly, the set of values of these $n - 1$ variables is a spherical region about the origin in the $(n - 1)$-dimensional space, and hence is closed and bounded. Therefore, $V(\mathbf{x})$ has a minimum on the surface $\| \mathbf{x} \| = \epsilon$, i.e., $V(\mathbf{x}) \geqslant \eta > 0$ when $\| \mathbf{x} \| = \epsilon$. That this minimum η is different from zero is a consequence of $V(\mathbf{x}) > 0$ when $\| \mathbf{x} \| \neq 0$.

The values of $V(\mathbf{x})$ will be smaller than η in certain portions of the spherical region $\| \mathbf{x} \| < \epsilon$ because $V(0) = 0$. Hence, there exists a $0 < \delta(\epsilon) < \epsilon$ such that $V(\mathbf{x}) \leqslant \eta' < \eta$ (e.g., $\eta' = \eta/2$) for all $\| \mathbf{x} \| \leqslant \delta(\epsilon)$. Clearly, there may exist some points in the spherical shell $\delta(\epsilon) < \| \mathbf{x} \| < \epsilon$ at which $V(\mathbf{x}) \leqslant \eta'$ still holds.

Consider a trajectory of (1) which starts from a point $\mathbf{x}_0$ inside $\| \mathbf{x} \| < \delta(\epsilon)$, i.e., $\mathbf{x}(0) = \mathbf{x}_0$ and $\| \mathbf{x}_0 \| < \delta(\epsilon)$. Since V is nonincreasing $(\dot{V} \leqslant 0)$ along the trajectory $\mathbf{x}(t)$, we have

$$V(\mathbf{x}(t)) \leqslant V(\mathbf{x}_0) \leqslant \eta' < \eta$$

so long as $\| \mathbf{x}(t) \| \leqslant \epsilon$. But the trajectory $\mathbf{x}(t)$ cannot cross the surface $\| \mathbf{x} \| = \epsilon$ on which $V(\mathbf{x}) \geqslant \eta$. Hence, $\| \mathbf{x}(t) \| \leqslant \epsilon$ for all $t \geqslant 0$ when $\| \mathbf{x}_0 \| \leqslant \delta(\epsilon)$, which proves the theorem.

REMARKS

Since $V(\mathbf{x}(t))$ is a positive, monotonically decreasing function of t along a solution which initiates at a point $\| \mathbf{x}_0 \| < \delta(\epsilon)$, the following limit exists:

$$\lim_{t \to \infty} V(\mathbf{x}(t)) = V_\infty \geqslant 0 \tag{9}$$

where V_∞ is a constant. If $\mathbf{x}(t) \to 0$ as $t \to \infty$, then V_∞ is necessarily zero. However, the above theorem simply states that $\| \mathbf{x}(t) \| < \epsilon$ for all $t \geqslant 0$, and hence $\mathbf{x}(t)$ does not have to return to the origin for large times. Suppose that $\mathbf{x}(t)$ approaches a nontrivial periodic solution. Then, the constant V_∞ is greater than zero, and the asymptotic trajectory is on a closed surface determined by $V(\mathbf{x}) = V_\infty$, which is entirely encompassed by the spherical surface $\| \mathbf{x} \| = \epsilon$.

We can attach a geometric meaning to the derivative along a trajectory (Definition 2) by considering the surfaces $V(\mathbf{x}) = C (C > 0)$ and a point $\mathbf{x}_0$ on one of these surfaces (Fig. 7.6.2), i.e., $V(\mathbf{x}_0) = C$. Let $\mathbf{x}_1$ be a point on an adjacent surface corresponding to $C + \delta C$ (Fig. 7.6.2).

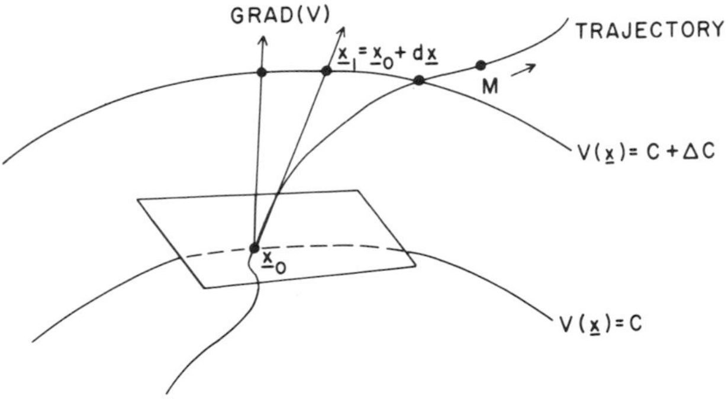

FIGURE 7.6.2. Directional derivative.

Hence, $V(\mathbf{x}_1) = C + \delta C$. Expanding $V(\mathbf{x}_1) = V(\mathbf{x}_0 + \delta\mathbf{x})$ into a Taylor series, we obtain

$$\delta\mathbf{x} \cdot \nabla V \,|_{\mathbf{x}=\mathbf{x}_0} \cong \delta C \tag{10}$$

which indicates that ∇V is a vector normal to the surface $V(\mathbf{x}) = C$, in the direction of increasing C (outward direction). Since $d\mathbf{x}/dt$ is the velocity of the representative point on a trajectory, we conclude that, if $\dot{V} \leqslant 0$, a representative point $\mathbf{x}(t)$ either intersects the family of surfaces inward ($\dot{V} < 0$), or remains during the entire time on some surface $V = V_\infty$ ($\dot{V} = 0$). Since $V(\mathbf{x}) = 0$ at $\mathbf{x} = 0$, the family of surfaces $V = C$ are concentric about the origin, and are necessarily closed for sufficiently small values of C (we prove this statement later; these surfaces may not be closed for large values of C, as we shall illustrate by an example). Thus, if a trajectory initiates within some closed surface and if $\dot{V} < 0$ for all $\mathbf{x} \neq 0$, then it must intersect all the surfaces $V = C > 0$ it encounters inward. This result is expressed as a theorem.

THEOREM 2. If there exists a Lyapunov function $V(\mathbf{x})$ in some neighborhood of the origin, and if

$$\dot{V} = \mathbf{f} \cdot \nabla V < 0 \tag{11}$$

for all $\mathbf{x} \neq 0$ in this neighborhood, then the solution $\mathbf{x}(t) \equiv 0$ of (1) is asymptotically stable.

We shall illustrate the applications of these theorems later. Because of their importance in stability analysis, we shall first investigate some properties of the positive-definite functions, some of which were already introduced above.

C. Positive-Definite Functions

Let $V(\mathbf{x})$ be a scalar function of $x_1, x_2, \ldots, x_n$ in some neighborhood of the origin. Assume that $V(\mathbf{x})$ has a power series expansion about the origin in x_i. Then, by rearranging the terms in this expansion, we have [17]

$$V(\mathbf{x}) = V_p(\mathbf{x}) + V_{p+1}(\mathbf{x}) + \cdots \tag{12}$$

where $V_p(\mathbf{x})$ is a homogeneous polynomial of degree p, i.e., $V_p(\lambda\mathbf{x}) = \lambda^p V(\mathbf{x})$, where λ is an arbitrary number. For sufficiently small values of $\mathbf{x}$, the terms $V_{p+1}(\mathbf{x})$, $V_{p+2}(\mathbf{x})$,... are dominated by $V_p(\mathbf{x})$, and hence the sign of V will be determined by the sign of $V_p(\mathbf{x})$. We can now show that, if $V(\mathbf{x})$ is positive-definite, p must be even. Following La Salle and Lefschetz [17], we substitute

$$x_1 = u_1 x_n, \qquad x_2 = u_2 x_n, \qquad \ldots, \qquad x_{n-1} = u_{n-1} x_n$$

in V_p and obtain

$$V_p(\mathbf{x}) = x_n{}^p V_p(u_1, u_2, \ldots, u_{n-1}, 1) \tag{13}$$

Suppose we replace $\mathbf{x}$ by $-\mathbf{x}$ in (13). This transformation does not change the signs of $u_1, u_2, \ldots, u_{n-1}$; hence,

$$V_p(-\mathbf{x}) = (-1)^p V_p(\mathbf{x})$$

is established, provided $V_p(u_1, u_2, \ldots, u_{n-1}, 1) \neq 0$. However, since $V_p(\mathbf{x})$ is not identically zero, we can always choose $u_1, u_2, \ldots, u_{n-1}$ such that $V_p(u_1, u_2, \ldots, u_{n-1}, 1) \neq 0$. If $V(\mathbf{x})$ is positive-definite, then both $V_p(-\mathbf{x})$ and $V_p(\mathbf{x})$ must be positive, hence p must be even. This is of course only a necessary condition, as we can see from $V_2 = x_1{}^2 - x_2{}^2$, which obviously is not positive-definite.

The lowest value of p is two because, when $p = 0$, V_p is a constant which must be zero if $V(\mathbf{x}) = 0$ at $x = 0$. In many applications we shall encounter, the expansion of $V(\mathbf{x})$ will start with $V_2(\mathbf{x})$, which is a quadratic form

$$V_2(\mathbf{x}) = \sum_{i,j} a_{ij} x_i x_j \tag{14}$$

In matrix notation,

$$V_2(\mathbf{x}) = \tilde{\mathbf{x}} \mathsf{A} \mathbf{x} \tag{15}$$

where $\tilde{\mathbf{x}}$ is the transposed $\mathbf{x}$ and is a row matrix. The elements of the square matrix A are a_{ij}. The necessary and sufficient conditions for $V_2(\mathbf{x})$ to be positive-definite are (Sylvester's theorem) that the principal minors of A are positive, i.e.,

$$a_{11} > 0, \qquad \begin{vmatrix} a_{11} & a_{12} \\ a_{21} & a_{22} \end{vmatrix} > 0, \quad \ldots, \quad \begin{vmatrix} a_{11} & a_{12} & \cdots & a_{1n} \\ a_{21} & a_{22} & \cdots & a_{2n} \\ \vdots & & & \\ a_{n1} & a_{n2} & \cdots & a_{nn} \end{vmatrix} > 0 \tag{16}$$

D. Properties of $V(\mathbf{x}) = C$ Surfaces

We shall now show that the surface $V(\mathbf{x}) = C$ are necessarily closed for sufficiently small values of C. Let us consider a continuous curve Γ which starts from the origin and extends to some point $\mathbf{x}_0$ on the spherical surface defined by $\| \mathbf{x} \| = \epsilon$ (Figure 7.6.3). If we can establish

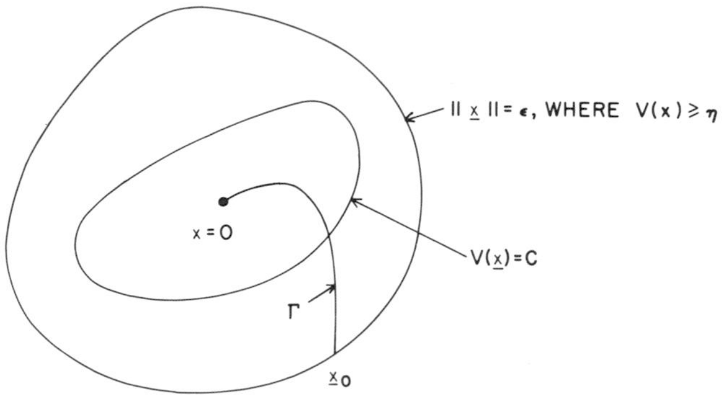

FIGURE 7.6.3. Closure of $V(x) = C$ for C sufficiently small.

that this arbitrary curve must inevitably intersect the surface $V(\mathbf{x}) = C$, as long as C is sufficiently small, then the surface $V(\mathbf{x}) = C$ is closed. We have already shown that $V(\mathbf{x})$ has a positive minimum η on $\| \mathbf{x} \| = \epsilon$, so that $V(\mathbf{x}_0) \geqslant \eta$. If we follow the variation of $V(\mathbf{x})$ along Γ, we find that it starts from zero at the origin and increases gradually to a value not smaller than η; hence, if $C < \eta$, it necessarily acquires the value C in between. In other words, the curve Γ intersects all of the family of surfaces $V(\mathbf{x}) = C$ with $C < \eta$, proving that they are all closed and enclose the origin. Clearly, if $C_2 > C_1$, then the surface $V(\mathbf{x}) = C_1$ is contained within the surface $V(\mathbf{x}) = C_2$. The following example will clarify the proof of Theorem 1 and illustrate a case in which the $V = C$ surfaces are not always closed.

EXAMPLE 1. Let us suppose that the function

$$V(\mathbf{x}) = x_1^2 + [x_2^2/(1 + x_2^2)] \tag{17}$$

is a Lyapunov function for a certain dynamical system. We first prove that $V(\mathbf{x}) > 0$ for all (x_1, x_2) and $V(0) = 0$. If we assume that $\dot{V} \leqslant 0$ also holds everywhere, then the region R in the statement of Theorem 1 is the entire plane. We can therefore choose the $\epsilon > 0$ as large as we please. The circle corresponding to $\| \mathbf{x} \| = \epsilon$ is indicated in Figure 7.6.4.

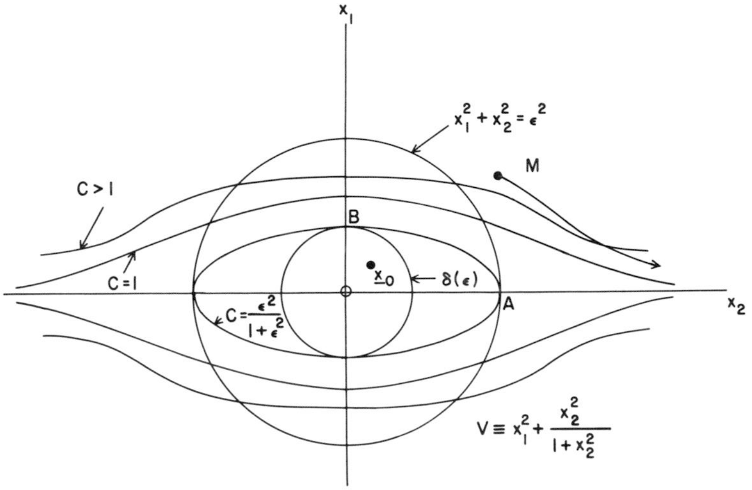

FIGURE 7.6.4. A Family of curves $V = C$.

The values of $V(\mathbf{x})$ on the circle are given by

$$V(x_1, (\epsilon^2 - x_1^2)^{1/2}) = x_1^2 + [(\epsilon^2 - x_1^2)/(1 + \epsilon^2 - x_1^2)] \tag{18}$$

which is a continuous function in the closed and bounded interval $-\epsilon \leqslant x_1 \leqslant \epsilon$ (Figure 7.6.5). The minimum η of V on the circle $\| \mathbf{x} \| = \epsilon$ is

$$\eta = \epsilon^2/(1 + \epsilon^2) > 0$$

We shall now determine $\delta(\epsilon)$ such that $V \leqslant \eta' < \eta$ for all $x \leqslant \delta(\epsilon)$. For this, we consider the closed curve $V = \eta$ which is tangent to the circle $\| \mathbf{x} \| = \epsilon$ at the point A. The radius of the circle which is wholly contained by this curve is $\overline{OB}^2 = \epsilon^2/(1 + \epsilon^2)$. Hence, $\delta(\epsilon)$ can be chosen as any number satisfying

$$\delta(\epsilon) < \epsilon^2/(1 + \epsilon^2) \tag{19}$$

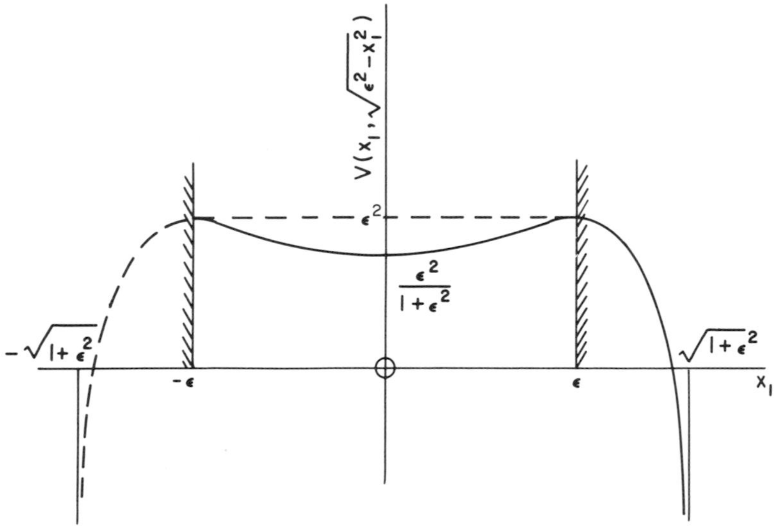

FIGURE 7.6.5. Variation of $V(\mathbf{x})$ on $\| \mathbf{x} \| = \epsilon$.

Thus, any trajectory starting within the circle $\| \mathbf{x} \| = \delta(\epsilon)$ will remain, according to Theorem 1, within the circle $\| \mathbf{x} \| = \epsilon$ if $\dot{V} \leqslant 0$ (stability), and will converge to the origin if $\dot{V} < 0$ for $\mathbf{x} \neq 0$ (asymptotic stability).

Although the region of stability (or asymptotic stability) is expressed as a circular disk in the proof of the stability Theorem 1 and 2, we know from the geometric interpretation described above that any solution which starts within a closed surface $V = C$ remains within that surface, or converges to the origin. It is important to note that a trajectory starting from a point which is not contained by any closed curve (such as the point M in Figure 7.6.5) may tend to infinity even though $V(\mathbf{x}) > 0$ and $\dot{V}(\mathbf{x}) < 0$ hold in the entire space. Thus, the boundaries of the region of stability (or asymptotic stability) is determined by that closed surface $V(\mathbf{x}) = C$ that has the largest C ($C = 1$ in the foregoing example).

If $\delta(\epsilon) \to \infty$ as $\epsilon \to \infty$, and $\dot{V} < 0$ everywhere, then the equilibrium state at the origin is asymptotically stable in the large. It is clear from (19) that $\delta(\epsilon) \to 1$ as $\epsilon \to \infty$, and hence the asymptotic stability in the large of a dynamical system cannot be assured with a Lyapunov function given by (17) even if $\dot{V} < 0$ holds in the entire space. However, one may find other Lyapunov functions for the same dynamical system which might ensure asymptotic stability in the large.

The closure of the surfaces $V = C$ is assured if the Lyapunov function $V(\mathbf{x})$ approaches infinity as $\| \mathbf{x} \| \to \infty$. Then, for any arbitrarily large number C, we can always find an $R(C)$ such that $V(\mathbf{x}) > C$ for all

$\| \mathbf{x} \| \geqslant R(C)$, and thus none of the surfaces $V \leqslant C$ can extend beyond $\| \mathbf{x} \| = R(\epsilon)$. In other words, all surfaces $V = C$ are closed independently of the magnitude of C. Thus, we obtain the following theorem.

THEOREM 3. If there exists a Lyapunov function in the entire space, and if $\dot{V} < 0$ everywhere when $\mathbf{x} \neq 0$, and if, in addition, $V(\mathbf{x}) \to \infty$ as $\| \mathbf{x} \| \to \infty$, then the solution $\mathbf{x} \equiv 0$ of (1) is asymptotically stable in the large.

E. Extension of the Asymptotic Stability Criterion

In applications of Lyapunov's second method, we sometimes can construct a positive-definite function $V(\mathbf{x})$ in a neighborhood of the origin which has only a semidefinite derivative, i.e., $\dot{V}(\mathbf{x}) \leqslant 0$, in the region in question. In such cases, we can assert only stability of the origin with the help of Theorem 1, but we cannot ascertain asymptotic stability. The theorem 4 which we are going to prove below (first proven by Barbasin and Krasovskii [2a] enables us to investigate both asymptotic stability and stability with a Lyapunov function of this type.

Before stating and proving this theorem, we must introduce the concept of "positive limiting set" Γ^* of a solution $\mathbf{x}(t)$ of $\dot{\mathbf{x}} = f(\mathbf{x})$. When there exists a Lyapunov function with a semidefinite derivative, we can only assert that $\mathbf{x}(t)$ is bounded at all times, i.e., $\| \mathbf{x}(t) \| < \epsilon$ if it starts from a point $\mathbf{x}_0$ sufficiently close to the origin. The trajectory may approach the origin or may approach, loosely speaking, a curve. For example, if the system admits a periodic solution, the trajectory will approach a closed curve in the phase space which does not contain the origin. Then, the set of points on this curve is the positive limiting set associated with the solution $\mathbf{x}(t)$. If $\mathbf{x}(t)$ tends to a point, then its positive limiting set consists of only this point.

DEFINITION. The positive limiting set associated with a bounded solution $\mathbf{x}(\mathbf{x}_0, t_0, t)$, specified by the initial condition $\mathbf{x}(t_0) = \mathbf{x}_0$, is a set of points Γ^* such that, if a point $\mathbf{x}_0{}^*$ belongs to Γ^*, then there exists a sequence of times $t_0 + \theta_k$, with $\theta_k \to \infty$ as $k \to \infty$, such that the limit of points $\mathbf{x}(\mathbf{x}_0, t_0, t_0 + \theta_k)$ is $\mathbf{x}_0{}^*$, i.e.,

$$\lim_{k\to\infty} \mathbf{x}(\mathbf{x}_0, t_0, t_0 + \theta_k) = \mathbf{x}_0{}^* \tag{20}$$

The indicated limit exists because $\mathbf{x}(t)$ is bounded for $t \geqslant 0$. We shall now prove an important property enjoyed by the positive limiting set.

LEMMA 1. Let $\mathbf{x}_0{}^*$ be a point in the positive limiting set of $\mathbf{x}(\mathbf{x}_0, t_0, t)$.

Then, a half-trajectory initiating at $\mathbf{x}_0{}^*$ at t_0 lies entirely in Γ^* for all $t \geqslant t_0$.

PROOF. Let $\mathbf{x}(\mathbf{x}_0{}^*, t_0, t)$ denote such a half-trajectory. Since the solutions of the differential equations $\dot{\mathbf{x}}(t) = \mathbf{f}(\mathbf{x})$ depends continuously on the initial conditions, we have

$$\mathbf{x}(\mathbf{x}_0{}^*, t_0, t) = \lim_{k\to\infty} \mathbf{x}(\mathbf{x}^k, t_0, t) \tag{21}$$

where $\mathbf{x}^k$ is the sequence $\mathbf{x}(\mathbf{x}_0, t_0, t_0 + \theta_k)$ that defines the point $\mathbf{x}_0{}^*$. Clearly, the points $\mathbf{x}_0$ and $\mathbf{x}^k$ lie on the same trajectory (see Figure 7.6.6a).

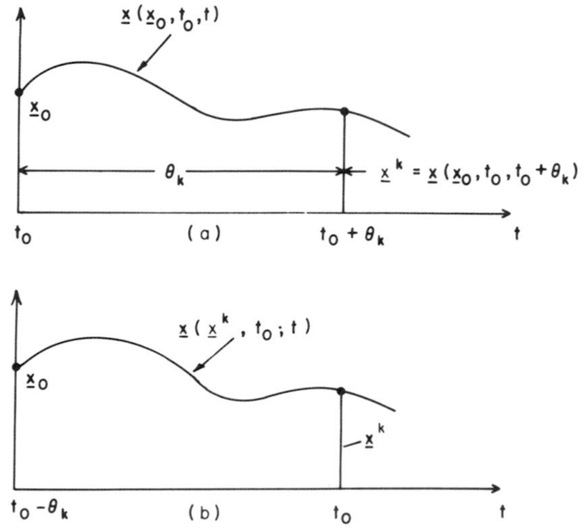

FIGURE 7.6.6. Characteristics of the functions (a) $\mathbf{x}(\mathbf{x}_0, t_0, t)$ and (b) $\mathbf{x}(\mathbf{x}^k, t_0, t)$.

Hence, the trajectory $\mathbf{x}(\mathbf{x}^k, t_0, t)$ also passes through the point $\mathbf{x}_0$ at $t = t_0 - \theta_k$ (Figure 7.6.6b):

$$\mathbf{x}(\mathbf{x}^k, t_0, t) \equiv \mathbf{x}(\mathbf{x}_0, t_0 - \theta_k, t) \tag{22}$$

On the other hand, if $\mathbf{x}(t)$ is a solution of $\dot{\mathbf{x}} = \mathbf{f}(\mathbf{x})$, then so is $\mathbf{x}(t + T)$, where T is an arbitrary time displacement, because $\mathbf{f}(\mathbf{x})$ does not depend on time explicitly (autonomous system). Therefore, we can shift the time origin on the right-hand side of (22) by an amount θ_k, and obtain the crucial relationship

$$\mathbf{x}(\mathbf{x}^k, t_0, t) = \mathbf{x}(\mathbf{x}_0, t_0, t + \theta_k) \tag{23}$$

Substitution of (23) into (21) yields

$$\mathbf{x}(\mathbf{x}_0{}^*, t_0, t) = \lim_{k\to\infty} \mathbf{x}(\mathbf{x}_0, t_0, t + \theta_k) \tag{24}$$

which proves that $\mathbf{x}(\mathbf{x}_0{}^*, t_0, t)$ belongs to the positive limiting set Γ^* associated with the solution $\mathbf{x}(\mathbf{x}_0, t_0, t)$.

Using the foregoing lemma, we can establish the following theorem for asymptotic stability.

THEOREM 4. Let $V(\mathbf{x})$ be a scalar function with first partial derivatives. Let R_l denote the region for a given $l > 0$ such that $V(\mathbf{x}) < l$: (i) R_l is bounded; (ii) $V(\mathbf{x}) > 0$ ($\mathbf{x} \in R_l$ and $\mathbf{x} \neq 0$); (iii) $\dot{V}(\mathbf{x}) \leqslant 0$ ($\mathbf{x} \in R_l$); (iv) the set of points Γ at which $\dot{V}(\mathbf{x}) = 0$ contains no nontrivial half-trajectory that lies entirely in its own positive limiting set; then every solution starting in R_l tends to the origin as $t \to \infty$, and hence the equilibrium $\mathbf{x}(t) \equiv 0$ is asymptotically stable.

PROOF. Since $V(t) \equiv V(\mathbf{x}(t))$ is monotonically decreasing $[\dot{V}(t) \leqslant 0]$, then $V(t) \leqslant V(0) = V(\mathbf{x}_0)$, which implies that any solution $\mathbf{x}(\mathbf{x}_0, t_0, t)$ starting in $R_l[V(\mathbf{x}_0) < l]$ must remain in R_l $[V(\mathbf{x}(t)) < l]$. Since R_l is bounded, we find that $\mathbf{x}(t)$ is also bounded. The limit of $V(t)$ as $t \to \infty$, denoted by V_∞, exists because $V(t) \geqslant 0$ and $V(t)$ is monotonically decreasing. Clearly, $V_\infty < l$. By continuity of $V(\mathbf{x})$, we have

$$\begin{aligned} V(\mathbf{x}_0{}^*) &= V(\lim_{k\to\infty} \mathbf{x}(\mathbf{x}_0, t_0, t + \theta_k)) \\ &= \lim_{k\to\infty} V(\mathbf{x}(\mathbf{x}_0, t_0, t + \theta_k)) \\ &= \lim_{t\to\infty} V(\mathbf{x}(\mathbf{x}_0, t_0, t)) = V_\infty \end{aligned} \tag{25}$$

which implies that $V(\mathbf{x}) = V_\infty$ on the positive limiting set Γ^* of $\mathbf{x}(\mathbf{x}_0, t_0, t)$, and Γ^* is in R_l. Consider the trajectory $\mathbf{x}(\mathbf{x}_0{}^*, t_0, t)$ starting at $\mathbf{x}_0{}^*$ at t_0. Since all the points of this trajectory for $t \geqslant t_0$ are in Γ^* by virtue of Lemma 1, $V(\mathbf{x}(\mathbf{x}_0{}^*, t_0, t)) \equiv V_\infty$ and $\dot{V}(\mathbf{x}(\mathbf{x}_0{}^*, t_0, t) \equiv 0$, from which we conclude that $\dot{V}(\mathbf{x}) = 0$ on Γ^*, and that Γ^* belongs to the set Γ. Thus, the entire half-trajectory $\mathbf{x}(\mathbf{x}_0, t_0, t)$ is in Γ. But, according to condition (iv) of the theorem, there can be no such nontrivial half-trajectory in Γ. Hence, $\mathbf{x}(\mathbf{x}_0{}^*, t_0, t) \equiv 0$, which implies that

$$\mathbf{x}_0{}^* = \lim_{k\to\infty} \mathbf{x}(\mathbf{x}_0, t_0, t_0 + \theta_k) = 0 \tag{26}$$

for all sequences $t_k = t_0 + \theta_k$. This result establishes the asymptotic stability.

REMARK

(a) If the set of points on which $\dot{V}(\mathbf{x}) = 0$ happens to be a surface described by

$$F(\mathbf{x}) = 0 \tag{27}$$

then condition (iv) is satisfied if

$$\mathbf{f}(\mathbf{x}) \cdot \nabla F \neq 0 \tag{28}$$

holds on the surface $F(\mathbf{x}) = 0$. Indeed, if a half-trajectory $\mathbf{x}(\mathbf{x}_0, t_0, t)$ lies entirely on this surface for all $t \geqslant t_0$, then $F(\mathbf{x}(\mathbf{x}_0, t_0, t)) = 0$ must hold for all $t \geqslant t_0$. By differentiation, we obtain $\dot{\mathbf{x}} \cdot \nabla F = 0$, which implies $\mathbf{f}(\mathbf{x}) \cdot \nabla F(\mathbf{x}) = 0$. Thus, if (28) holds, the half-trajectory cannot remain on the surface $F(\mathbf{x}) = 0$.

(b) It is observed in the proof of the theorem that the boundedness of the region R_l was needed only to ensure the boundedness of the solutions starting in R_l. In some applications, the boundedness of the solutions can be established as a separate problem. In such cases, the asymptotic stability can be guaranteed by a weaker theorem [20].

THEOREM 5. Let $V(\mathbf{x})$ be a scalar function with continuous first partial derivatives for all $\mathbf{x}$, and assume that $V(\mathbf{x})$ is positive-definite, and $\dot{V}(\mathbf{x})$ is nonpositive for all $\mathbf{x}$. If the set of points at which $\dot{V}(\mathbf{x}) = 0$ contains no nontrivial, bounded half-trajectory which lies entirely in its own positive limiting set, then every bounded solution tends to the origin as $t \to \infty$.

F. Concluding Remarks

The foregoing discussions indicate that Lyapunov's second method reduces the question of the stability of dynamical systems to the construction of a family of closed surfaces which surround the origin and each other, and possess the property that the trajectories of the dynamical system intersect each of these surfaces inwardly. Once the existence of such surfaces has been established, one can determine the maximum region in the phase space for stability and asymptotic stability, and hence the maximum permissible perturbation on the dynamical variables describing a state of the system.

The size of the region of stability depends on the choice of the Lyapunov function. One may obtain more relaxed conditions on the system parameters to guarantee asymptotic stability by constructing different Lyapunov functions. Since there is no systematic method to construct Lyapunov functions for a given dynamical system, the success of Lyapunov's second method depends on the experience and ingenuity of the user. In the following sections, we shall present Lyapunov functions that can be used to investigate the question of stability of reactor systems.

7.7. The Lurie–Letov Function

In this and the subsequent sections we shall study some particular Lyapunov functions which have been used in reactor dynamics to investigate asymptotic stability of reactor systems whose temporal behavior can be adequately described by a set of equations of the following form:

$$(l/\beta)\dot{P} = \left(k_0 - \sum_{j=1}^{N} \alpha_j X_j - \gamma P\right) P + \sum_{i=1}^{I} \lambda_i C_i - P \tag{1a}$$

$$\dot{C}_i = a_i P - \lambda_i C_i , \qquad i = 1,..., I \tag{1b}$$

$$\dot{X}_j = b_j P - \sum_{i=1}^{N} h_{ji} X_i , \qquad j = 1,..., N \tag{1c}$$

In this set of $I + N + 1$ equations, k_0 is the externally introduced positive reactivity; the $-\alpha_j$ are the reactivity coefficients associated with the respective feedback variables X_j; $-\gamma$ is the power coefficient of reactivity; a_i , λ_i , and β are the usual parameters describing the delayed neutrons; l is the mean generation time; and b_j and h_{ji} are parameters characterizing the feedback mechanism in the reactor. The functions $P(t)$, $C_i(t)$, and $X_j(t)$ are the dynamical variables describing a state of the system.

It is noted that Eqs. (1) describe a reactor with a linear feedback model. The associated linear feedback functional is given by

$$\delta k_{\mathrm{f}}[p] = -\sum_{j=1}^{N} \alpha_j \, \delta X_j - \gamma p \tag{2a}$$

$$\delta \dot{X}_j = b_j p - \sum_{i=1}^{N} h_{ji} \, \delta X_i \tag{2b}$$

where δX_j and p are the incremental changes of the feedback variables and the power from their equilibrium values. A comparison of (2) with (5.2, Eqs. 43 and 46) reveals that the multiregion model for temperature feedback with a separate reactivity coefficient in each region is a special case of the above feedback model with $\gamma = 0$. Indeed, the set of differential equations describing the temperatures in various regions can be written as [(5.2, Eq. 43)]

$$\mu_j \, \delta \dot{T}_j = (H_{0j}/P_0)p + \sum_{i=1}^{N} X_{ji}(\delta T_i - \delta T_j) - X_{j0} \, \delta T_j \tag{3}$$

where X_{ji} is the heat transfer coefficient between the jth and ith regions. If we substitute $H_{0j}/P_0\mu_j = b_j$, $h_{ji} = -X_{ji}/\mu_j$, and $h_{jj} = \sum_{i=0}^{N} X_{ji}/\mu_j$ in (3), we reproduce (2b). In reducing (3) into a diagonal form in Sec. 5.2, we explicitly used the symmetry of X_{ij}, i.e., $X_{ij} = X_{ji}$, which is a natural consequence of the concept of heat transfer coefficient. However, when the feedback variables δX_j in (2) represent feedback effects other than temperatures, such as voids (cf Sec. 5.6) and external automatic controls, then the coefficients h_{ij} in (2b) will in general lack the symmetry property $h_{ij}\mu_i = h_{ji}\mu_j$ enjoyed by temperatures. Hence, we shall use a slightly different transformation to reduce Eqs. (2) to a standard canonical form as used in (5.2, Eqs. 53).

We may also point out the feedback model (2) also includes the xenon feedback described in (5.3, Eqs. 24) if the nonlinear term γp^2 in (5.3, Eq. 24b) is neglected, and some of the δX_j in (2) are interpreted as the iodine and xenon concentrations [or their linear combination as in (5.3, Eq. 22)]. Hence, by leaving the number and physical nature of the generalized feedback variables δX_j in (2) unspecified, we can account in general for a large class of feedback phenomena with the same set of equations. However, feedback effects that are described by a transcendental transfer function, such as those involving pure time lags (cf Sec. 5.2, Eq. 79 and 6.4, Eq. 14), and the circulating fuel reactor (cf. Secs. 5.2e and 7.3c), cannot be investigated by the set of equations (2).

A. Reduction to Canonical Form

We observe that Eqs. (1) contain $N^2 + 2N + 2I + 3$ parameters: (l/β), α_j, γ, λ_i, a_i, b_j, k_0, and h_{ij}. Some of these parameters are mathematically redundant insofar as stability analysis is concerned. It is therefore desirable to transform these equations to a standard canonical form with fewer parameters.

For this purpose, we shall always assume that the matrix[†] $H \equiv [h_{ij}]$

[†] Throughout Section 7.7 matrices will be set in italic rather than the bold type used previously.

is nonsingular ($| h_{ij} | \neq 0$). From the physical point of view, this assumption is not a restriction because, if H were singular, then the steady-state equations, obtained from (1c) with $\dot{X}_j = 0$, and $P(t) \equiv P_0$ as

$$\sum_{i=1}^{N} h_{ji} X_{i0} = b_j P_0 \tag{4}$$

would be inconsistent, that is, would have no solution unless $P_0 = 0$. Since we are interested in the stability of the reactor at a finite power level, the above possibility is ruled out in any realistic reactor application.

We shall further assume that the characteristic roots of H are all distinct, real, and positive. Thus, H can be diagonalized by a similarity transformation [21] as

$$S^{-1}HS = D = \operatorname{diag}(\eta_1, \eta_2, \ldots, \eta_N) \tag{5}$$

where S is a real, nonsingular $N \times N$ matrix, and where $\eta_1, \eta_2, \ldots, \eta_N$ are the characteristic roots of H, i.e., the roots of the polynomial $| \eta E - H | = 0$. The assumption concerning the positiveness of the characteristic roots physically implies that the feedback mechanism described by (2), which determines the input–output relationship [$p(t)$ is input and $\delta X_j(t)$ are the outputs] of a linear system, is stable. The linear feedback kernel associated with (2) in fact can be obtained easily by eliminating δX_j as follows. We first define the following matrices and vectors:

$$\alpha = \operatorname{col}(\alpha_i) \tag{6a}$$

$$b = \operatorname{col}(b_i) \tag{6b}$$

$$\delta X = \operatorname{col}(\delta X_i) \tag{6c}$$

$$\delta\theta = S^{-1}\,\delta X \tag{6d}$$

Multiplying (2b) from the left by S^{-1} and using (6), we obtain

$$\delta\dot{\theta} + D\,\delta\theta = S^{-1}bp(t)$$

which yields

$$\delta\theta_j(t) = \int_0^\infty (S^{-1}b)_j\, e^{-\eta_j u} p(t-u)\, du \tag{7}$$

Substituting (7) into (2a), we find

$$\delta k_{\mathrm{f}}[p] = \int_0^\infty G(u)\, p(t-u)\, du \tag{8a}$$

where

$$G(u) = -\gamma\,\delta(u) - \sum_{i=1}^{N} K_i \epsilon_i\, e^{-\eta_i u} \tag{8b}$$

and where

$$K_i = \sum_{j=1}^{N} \alpha_j S_{ji} = (\alpha' S)_i\,, \qquad \alpha' = \text{row}(\alpha_i) \tag{8c}$$

$$\epsilon_i = \sum_{j=1}^{N} S_{ij}^{-1} b_j = (S^{-1}b)_i \tag{8d}$$

In order for $G(u)$ to be absolute-integrable, which is necessary and sufficient for the stability of the linear feedback mechanism, $\eta_j > 0$ must hold for all j.

We now assume that $K_i\epsilon_i \neq 0$ for all i and discuss its implications. Let us suppose that $K_i\epsilon_i = 0$ for some $i = m$. Then, the feedback kernel (8b) will not contain the mth term. Physically, this implies that the variation of $\delta\theta_m(t)$ does not affect the feedback reactivity. We can therefore investigate the stability of the system in terms of the remaining $N - 1$ variables (cf Eqs. 12 below), and then obtain the behavior of $\delta\theta_m(t)$ using (7) for $j = m$.†

A consequence of the assumption that $(S^{-1}b)_i \neq 0$ for all i is that none of the elements of the canonical equilibrium vector

$$\theta_0 \equiv S^{-1}X_0 = S^{-1}H^{-1}bP_0 = D^{-1}S^{-1}bP_0 \tag{9a}$$

is zero, where $X_0 = \text{col}\,(X_{01}\,, X_{02}\,,\ldots, X_{0N})$. Indeed, from (9a), we obtain

$$\theta_{0j} = (S^{-1}b)_j P_0/\eta_j = (\epsilon_j/\eta_j)P_0 \tag{9b}$$

which is nonzero for all j. In fact, we can further assume without loss of generality that $\theta_{0j} > 0$ for all j, because if they are not so for some $j = j^*$, then a simple change of sign of the elements of the j^*th row in S^{-1} makes them so. Clearly, such a sign change does not affect the similarity transformation in (5).

With the foregoing remarks in mind, we are now ready to transform the

† Since the stability of $p(t)$ is independent of $\delta\theta_m(t)$, we may choose $\delta\theta_m(t) \equiv 0$. From (6d), we obtain $(S^{-1}\delta X)_m = 0$, or $\sum_{j=1}^{N} (S^{-1})_{mj}\delta X_j = 0$, from which we can eliminate one of the feedback variables $X_j(t)$.

original set of equations (1) into a standard canonical form. Substituting in (1)

$$x_0 = [P(t) - P_0]/P_0$$

$$x_i = [C_i(t) - C_{i0}]/C_{i0}, \qquad i = 1, \ldots, I \tag{10a}$$

$$y_j = \delta\theta_j(t)/\theta_{0j}, \qquad j = 1, \ldots, N \tag{10b}$$

where the equilibrium values C_{i0} and P_0 are given by

$$C_{i0} = a_i P_0/\lambda_i \tag{11a}$$

$$P_0 = k_0 \Big/ \Big[\gamma + \sum_{j=1}^{N} \alpha_j (H^{-1}b)_j\Big] \tag{11b}$$

we obtain

$$(l/\beta)\dot{x}_0 = -(\kappa_0 x_0 + \kappa' y)(1 + x_0) + \sum_{i=1}^{I} a_i(x_i - x_0) \tag{12a}$$

$$\dot{x}_i = \lambda_i(x_0 - x_i), \qquad i = 1, \ldots, I \tag{12b}$$

$$\dot{y}_j = \eta_j(x_0 - y_j), \qquad j = 1, \ldots, N \tag{12c}$$

where

$$\kappa_0 = \gamma P_0, \qquad \kappa = QK, \qquad Q = \mathrm{diag}(\theta_{0i}) \tag{13}$$

Since $P(t)$ and $C_i(t)$ are always positive, we have

$$x_{\mathrm{m}}(t) > -1, \qquad m = 0, 1, \ldots, I \tag{14}$$

In general, we cannot write a similar inequality for the feedback variables purely on a physical basis because the canonical variables $\theta_j(t) = \theta_{j0} + \delta\theta_j$ are not necessarily positive.

As we have mentioned earlier on several occasions (e.g., Sect. 6.1), Eqs. (12) have two equilibria, $(x_i = 0, y_i = 0)$ and $(x_i = -1, y_i = -1)$ so long as the equilibrium condition

$$k_0 = \kappa_0 + \sum_{j=1}^{N} \kappa_j = \kappa_0 + \kappa' I > 0 \tag{15}$$

where $I = \mathrm{col}(1, 1, \ldots, 1)$, is satisfied. The equilibrium state $x_i = y_i = -1$ is totally unstable because it implies $P = C_i = X_j = 0$, i.e., a zero-power reactor without an operating feedback but with a positive excess of reactivity. Such a system is exponentially unstable to infinitesimal perturbations.

We now proceed to investigate the stability of the null solution using Lyapunov's second method. Equations (12) are in the desired canonical form, which contain only $2(N + I + 1)$ parameters, as opposed to the noncanonical form in (1) involving $N^2 + 2N + 2I + 3$ parameters.

B. The Lurie–Letov Function

Consider the following scalar function

$$V(\mathbf{x}, \mathbf{y}) \equiv (l/\beta)[x_0 - \ln(1 + x_0)] + \sum_{i=1}^{I} (a_i/\lambda_i)[x_i - \ln(1 + x_i)]$$

$$+ y'Fy + \sum_{j=1}^{N} A_j y_j^2 \tag{16a}$$

where the $N \times N$ matrix is defined by

$$F_{ij} = q_i q_j/(\eta_i + \eta_j) \tag{16b}$$

and where the q_j are real and $A_j > 0$. This function is slightly different from a class of functions studied by Lurie [22] and Letov [23]. The terms $x_i - \ln(1 + x_i)$, $i = 0, 1,..., I$ are each always positive for $x_i > -1$ and zero only if $x_i = 0$ (cf Sec. 7.2). The last sum is positive since $A_j > 0$ and zero only if each $y_i = 0$. The term $y'Fy$, however, is only nonnegative, as can be verified as

$$y'Fy = \sum_{i,j=1}^{N} [q_i q_j/(\eta_i + \eta_j)] y_i y_j$$

$$= \int_0^\infty du \sum_{i,j=1}^{N} q_i q_j y_i y_j \, e^{-(\eta_i+\eta_j)u}$$

from which we have

$$y'Fy = \int_0^\infty du \left[\sum_{j=1}^{N} q_j y_j \, e^{-\eta_j u}\right]^2 \tag{17}$$

Although $y'Fy$ is only nonnegative,† the function $V(\mathbf{x}, \mathbf{y})$ is positive-

† Note that $y'Fy$ would be positive-definite if $q_j \neq 0$ for all j and all $\eta_j > 0$ are distinct. The first requirement is obvious because, if one of the q_j is zero for some $j = m$, then (17) does not contain y_m, so that $y'Fy$ can vanish if all $y_j = 0$ but $y_m \neq 0$. The requirement for the distinctness of η_j can be seen as follows: $y'Fy$ can vanish only if the integrand is identically zero in (17), i.e., $\Sigma_j\, q_j y_j e^{-\eta_j u} \equiv 0$. Let η_1 be the smallest of the η_j assuming that the η_j are all distinct. Dividing by $e^{-\eta_j u}$, we get $q_1 y_1 + \Sigma_{j\neq 1}\, q_j y_j e^{-(\eta_j - \eta_1)u} \equiv 0$. Since $\eta_j - \eta_1 > 0$, we obtain $y_1 = 0$ by taking the limit $u \to \infty$. Repeating this procedure, we prove that $y'Fy = 0$ implies $y_j = 0$ for all j if the η_j are distinct. However, if the latter is not true, we must have at least a pair $\eta_1 = \eta_2$. Then the integrand would still be identically zero even if $y_1, y_2 \neq 0$, provided $q_1 y_1 + q_2 y_2 = 0$.

definite by virtue of the remaining terms in a region of the $(N+I+1)$-dimensional phase space defined by $x_i > -1$ for $i = 0, 1, \ldots, I$. This region, which clearly contains the origin, is the region R used in the stability theorems in the previous section. We shall prove presently that a trajectory initiating in this region always remains there.

We now consider the derivative of $V(\mathbf{x}, \mathbf{y})$ along a trajectory. Differentiating (16) with respect to time and substituting $\dot{x}_i$ and $\dot{y}_i$ from (12) (cf 7.6, Eq. 8), we obtain, after some manipulation,

$$\dot{V}(t) = -\Big(\sum_{j=1}^{N} q_j y_j + x_0 \sqrt{\kappa_0}\Big)^2 - \sum_{j=1}^{N} a_i[(x_i - x_0)^2/(1+x_i)(1+x_0)]$$

$$- 2\sum_{j=1}^{N} A_j \eta_j y_j^2 - x_0 \sum_{j=1}^{N} y_i W_j \tag{18a}$$

where

$$W_j \equiv \kappa_j - 2q_j \Big(\sqrt{\kappa_0} + \sum_{k=1}^{N} [q_k \eta_k/(\eta_j + \eta_k)]\Big) - 2A_j \eta_j \tag{18b}$$

Let us suppose that we can choose $A_j > 0$ and real q_j such that $W_j = 0$ for all $j = 1, 2, \ldots, N$. Then, $\dot{V}(t) \leqslant 0$ in the region R defined above, for all the remaining terms (18a) are nonpositive.

It immediately follows from the above conclusions that (cf Sec. 7.2A): (i) $V(t) \leqslant V(0)$ provided $x_i(0)$ and $y_j(0)$ are in R; (ii) $x_i(t)$ and $y_j(t)$ are bounded and hence remain in R, and in particular, $x_i(t) > -1$ for $i = 0, 1, \ldots, I$ (cf 7.2, Eqs. 11, 12, and 17); (iii) all the derivatives of $x_i(t)$ and $y_j(t)$ are bounded [from (ii) and Eq. (12)] [we note for future use that the boundedness of $\ddot{y}_j$ (t) implies the uniform continuity of $\dot{y}_j(t)$]; (iv) all the derivatives of $V(t)$ are bounded [from (18) and (ii)], and in particular, $\dot{V}(t)$ is uniformly continuous because $\ddot{V}(t)$ is bounded (cf footnote on p. 330);

(v) $$\lim_{t\to\infty} V(t) = V_\infty \geqslant 0 \tag{19}$$

(vi) $$\lim_{t\to\infty} \dot{V}(t) = 0 \tag{20}$$

(this follows from the lemma in the foornote on p. 331).

At this stage we must distinguish between two cases.

Case 1. $\kappa_0 > 0$. In this case, $\dot{V}(t) = 0$ implies $x_i = y_j = 0$, i.e.,

$$\lim_{t\to\infty} x_i(t) = 0, \qquad i = 0, 1,\ldots, I \tag{21a}$$

$$\lim_{t\to\infty} y_j(t) = 0, \qquad j = 1,\ldots, N \tag{21b}$$

proving global asymptotic stability. This conclusion is of course also a direct consequence of the second Lyapunov theorem (Sec. 7.6, Theorem 2) because $\dot{V}(t) < 0$ for x_i, $y_j \neq 0$ and zero only if $x_i = y_j = 0$, and $V(\mathbf{x}, \mathbf{y}) \to \infty$ as $\| \mathbf{x} \| + \| \mathbf{y} \| \to \infty$.

Case 2. $\kappa_0 = 0$. In this case, Eq. (20) implies, in addition to (21b),

$$\lim_{t\to\infty} (x_i(t) - x_0(t)) = 0, \qquad i = 1,\ldots, I \tag{22}$$

Clearly, the last conclusion does not ensure asymptotic stability. However, we can still establish asymptotic stability in the large using the lemma in the footnote on p. 331. Since $y_j(t) \to 0$, and $\dot{y}_j(t)$ is uniformly continuous [cf (iii)], this lemma guarantees $\dot{y}_j(t) \to 0$. Substituting the latter into $\dot{y}_j = \eta_j(x_0 - y_j)$ [i.e., Eq. (12c)], we immediately obtain $x_0(t) \to 0$ by virtue of (21b) (note that we use $\eta_j \neq 0$ here), and from (22), $x_i(t) \to 0$.

It is interesting to note that asymptotic stability in the case of $\kappa_0 = 0$ can also be ascertained with the help of Theorem 4 of Sec. 7.6. We can easily varify from (16) that the region R in which $V(\mathbf{x}, \mathbf{y}) < M$ is finite for any finite $M > 0$ because each term in (16) is bounded separately, and except for $y'Fy$, they are all positive-definite. Furthermore, $V(\mathbf{x}, \mathbf{y}) > 0$ for $\mathbf{x} \neq 0$, $\mathbf{y} \neq 0$, and $\dot{V}(\mathbf{x}) \leqslant 0$ in R. The set of points Γ at which $\dot{V}(\mathbf{x}, \mathbf{y}) = 0$ follows from (18a) as

$$x_0(t) \equiv x_i(t), \qquad y_j(t) = 0 \tag{23}$$

Substituting these into (12), we obtain the equations of the half-trajectories in Γ as $\dot{x}_i(t) \equiv 0$, $i = 0, 1,\ldots, I$, and $\dot{y}_j(t) \equiv \eta_j x_0(t) \equiv 0$, which imply $x_i(t) \equiv y_j(t) \equiv 0$. Hence, there are no nontrivial half-trajectories in Γ, and, according to Theorem 4, Sec. 7.6, any trajectory starting in R tends to the origin. Since $V(\mathbf{x}, \mathbf{y}) \to \infty$ as $\| \mathbf{x} \| + \| \mathbf{y} \| \to \infty$, the region R can be made as large as desired, and asymptotic stability in the large is established.

The results of the foregoing discussions can be summarized as the following theorem.

THEOREM 1. The equilibrium $x_i(t) \equiv y_j(t) \equiv 0$ of Eq. (12) is asymptotically stable in the large if we can choose a set of positive constants A_j and a set of real numbers q_j such that

$$\kappa_j - 2A_j\eta_j = 2q_j \left\{ \sqrt{\kappa_0} + \sum_{k=1}^{N} [q_k\eta_k/(\eta_k + \eta_j)] \right\}, \qquad j = 1, 2, \ldots, N \quad (24)$$

hold.

Equation (24) represent N relations between the system parameters κ_j, η_j, and κ_0 and the $A_j > 0$ and q_j real numbers. If we specify somehow the values of A_j as a set of positive numbers, then (24) can be considered as equations for q_j (i.e., N quadratic equation in the N unknowns q_j). Looked at in this way, we are interested in the conditions that must be placed on the κ_j, η_j, κ_0, and A_j so as to guarantee the existence of real solutions for the q_j, and not in the actual values of the q_j themselves. These conditions will depend on κ_j and A_j only through $(\kappa_j - 2A_j\eta_j)$ as indicated in (24). Hence, if we can determine them with $A_j = 0$ for all j, by considering

$$\kappa_j = 2q_j \left\{ \sqrt{\kappa_0} + \sum_{k=1}^{N} [q_k\eta_k/(\eta_j + \eta_k)] \right\}, \qquad j = 1, 2, \ldots, N \qquad (25)$$

then the conditions for $A_j \neq 0$ can be obtained simply by replacing κ_j by $\kappa_j - 2A_j\eta_j$.

Let us suppose that we know the necessary and sufficient conditions for the existence of real q_j satisfying (25), and that they are expressible in the form of a set of inequalities

$$f_k(\kappa_j, \eta_j, \kappa_0) > 0, \qquad k = 1, 2, \ldots, N \qquad (26a)$$

then the conditions

$$f_k(\kappa_j - 2A_j\eta_j, \eta_j, \kappa_0) > 0, \qquad k = 1, \ldots, N \qquad (26b)$$

guarantee the reality of q_j satisfying (24). Let us also assume that we can determine the boundaries of the regions in the $(2N + 1)$-dimensional parameter space with κ_j, η_j, and κ_0 as coordinates such that the inequalities (26a) are all satisfied at points in these regions. The regions corresponding to (26b) with a given A_j can be obtained by shifting the boundary points by decreasing their κ_j coordinates by an amount $2A_j\eta_j$. This is actually equivalent to a translation of the original regions in the direction given by $(-2A_j\eta_j, 0, 0)$.

Since the proof of Theorem 1 makes use of only the positiveness of A_j, but does not involve its magnitude, to guarantee asymptotic stability in the large, we can choose $2A_j\eta_j$ infinitesimally small, i.e., $2A_j\eta_j = \epsilon > 0$; then, except for the boundaries, any point in a region defined by (26a) will remain in the same region after an infinitesimal translation $\kappa_j \to \kappa_j - \epsilon$.

Since (26b) are satisfied by the new coordinates after this translation, we may conclude that the conditions (26a) will determine the regions with the parameter space for asymptotic stability in the large, by virtue of Theorem 1, provided the points on those portions of the boundaries affected by $\kappa_j \to \kappa_j - \epsilon$ are exluded. These points can not be investigated by Theorem 1, because at these points, $A_j = 0$ for all j, and the derivative of $V(\mathbf{x}, \mathbf{y})$ along a trajectory is no longer negative-definite, even though $V(\mathbf{x}, \mathbf{y})$ remains >0 with $A_j = 0$ (cf Eq. 18a).

We can, however, modify the proof of Theorem 1 to include the boundaries as well. Before presenting this extension, it is instructive to illustrate the above discussion by the case of a one-temperature model characterized by κ_0, κ_1, η_1. In this case, Eq. (25) reduces to

$$q_1^2 + 2\sqrt{\kappa_0}q_1 - \kappa_1 = 0 \tag{27a}$$

which has a real root if $\kappa_0 > 0$ and

$$f_1(\kappa_0, \kappa_1) \equiv \kappa_0 + \kappa_1 > 0 \tag{27b}$$

We observe that f_1 does not contain η_1, hence the appropriate phase space is the (κ_0, κ_1) plane shown in Figure 7.7.1. The boundaries of the region for asymptotic stability the κ_1 axis and the line $\kappa_0 + \kappa_1 = 0$.

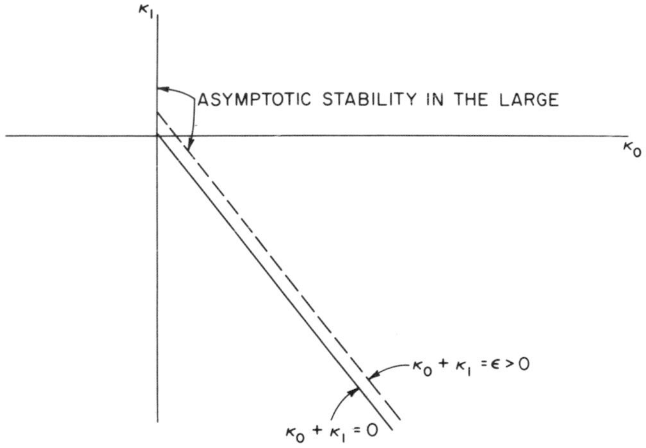

FIGURE 7.7.1. Regions of asymptotic stability in the large.

The points on the κ_1 axis are included, whereas those on the oblique line are excluded from this region. The dashed line corresponds to the shifted boundary. We may note in passing that (27b) is identical to (15a), i.e., the condition for finite equilibrium.

C. Extension of Theorem 1 to Include Boundaries

Let us consider the following V-function and its derivative:

$$V(\mathbf{x}, \mathbf{y}) = (l/\beta)[x_0 - \ln(1 + x_0)] + \sum_{i=1}^{I} (a_i/\lambda_i)[x_i - \ln(1 + x_i)] + y'Fy \tag{28a}$$

$$\dot{V}(\mathbf{x}, \mathbf{y}) = -[q'y + x_0 \sqrt{\kappa_0}]^2 - \sum_{i=1}^{I} a_i[(x_i - x_0)^2/(1 + x_i)(1 + x_0)] \tag{28b}$$

which are obtained from (16a) and (18a) with $A_j = 0$ for all j. The matrix F is defined in (16b), and q' is the transpose of the real N-vector $q = (q_1, q_2, \ldots, q_n)$, which is assumed to satisfy

$$\kappa_j = 2q_j \left\{\sqrt{\kappa_0} + \sum_{k=1}^{N} [q_k \eta_k/(\eta_k + \eta_j)]\right\} \tag{28c}$$

in obtaining (28b) (cf Eq. 18a).

If $\kappa_j \neq 0$ for all j (as we always assume), none of the q_j can vanish, according to (28c). Using this conclusion and the fact that all $\eta_j > 0$ are distinct [also part of our assumption in Eqs. (12)], $y'Fy$ is positive-definite (cf the footnote on p. 373). Hence, $V(\mathbf{x}, \mathbf{y})$ is again positive-definite. However, $\dot{V}(\mathbf{x}, \mathbf{y})$ in (28b) is only nonpositive and can vanish in the set Γ defined by

$$x_i(t) \equiv x_0(t), \qquad i = 1, 2, \ldots, I \tag{29a}$$

and

$$-q'y(t) \equiv x_0(t)\sqrt{\kappa_0} \tag{29b}$$

In order to ascertain asymptotic stability in the large using Theorem 4, Sec. 7.6, we must show that there is no nontrivial half-trajectory in Γ. The other conditions of the theorem are satisfied as in the case of $A_j \neq 0$.

Using (29a) in (12b), we get $\dot{x}_i(t) = 0$ for $i = 1, 2, \ldots, I$. From (29a), we have $\dot{x}_0(t) \equiv 0$. Differentiating (29b) and using the latter, we find $q'\dot{y}(t) \equiv 0$. On the other hand, (12c) yields $\ddot{y}_j = -\eta_j \dot{y}_j$, or $\dot{y}_j(t) =$

$\dot{y}_j(0)e^{-\eta_j t}$. Hence, $qe^{-D(t)}\,\dot{y}(0) \equiv 0$ implies $\dot{\mathbf{y}}(0) = 0$ (recall that $q_j \neq 0$, $\eta_j > 0$ are distinct), or $\dot{\mathbf{y}}(t) \equiv 0$. From (12c), we get $y_j(t) \equiv x_0(t)$. Thus, we have established $x_0(t) \equiv x_i(t) \equiv y_j(t) = \text{const}$. From (12a), we have $x_0(\kappa_0 + \kappa' I) = 0$. But $\kappa_0 + \kappa' I > 0$ for equilibrium to exist. Hence, $x_0(t) \equiv y_j(t) \equiv x_i(t) \equiv 0$ is the only half-trajectory in Γ, and asymptotic stability in the large is established.

THEOREM 2. The equilibrium ($x_0 \equiv y_j \equiv 0$) of (12) is asymptotically stable in the large if we can choose a set of real constants q_j satisfying (28c), and if $\kappa_0 + \kappa' I > 0$.

It is important to note that the condition $\kappa_0 + \kappa' I > 0$ now appears explicitly in the statement of the Theorem 2 because it was used in its proof. (The condition of Theorem 1 contains $\kappa_0 + \kappa' I > 0$ automatically).

It appears from Theorem 2 that there is no need to distinguish between the conditions for $f_k(\kappa_j\,, \eta_j\,, \kappa_0) > 0$ and for $f_k(\kappa_j - 2A_j\eta_j\,, \eta_j\,, \kappa_0) > 0$ with infinitesimal A_j, because the former includes the boundary points.

REMARKS ON THE EXISTENCE OF REAL q_j

A direct algebraic solution of (24) for the q_j in order to determine the inequalities for $f_k(\kappa_j\,, \eta_j\,, \kappa_0) > 0$ is not in general an easy task when $N \geqslant 3$. However, a few necessary conditions for the existence of real q_j can still be established without actually solving (24). These conditions are imposed on certain sums of κ_j and η_j as will be shown below.

The first condition follows from (15) by substituting κ_j from (24):

$$k_0 = \kappa_0 + 2\sqrt{\kappa_0}\sum_{j=1}^{N} q_j + \sum_{i,j=1}^{N} 2q_iq_j[\eta_i/(\eta_i + \eta_j)] + 2\sum_{j=1}^{N} A_j\eta_j > 0 \tag{30}$$

The double summation in (30) can be written as

$$2\sum_{i,j=1}^{N} q_iq_j[\eta_i/(\eta_i + \eta_j)] = \Big[\sum_{j=1}^{N} q_j\Big]^2 \tag{31}$$

Defining

$$q \equiv \operatorname{col}(q_i) \tag{32a}$$

$$A \equiv \operatorname{col}(A_i) \tag{32b}$$

We obtain from (30)

$$\kappa_0 + \kappa' I = (\sqrt{\kappa_0} + q' I)^2 + 2A' D I > 0 \tag{33}$$

where $D = \operatorname{diag}(\eta_1, \eta_2, \ldots, \eta_n)$ (cf Eq. 5), $I = \operatorname{col}(1, 1, \ldots, 1)$ (cf Eq. 15a), and the primes denote transposed matrices.

In order to obtain other conditions, we multiply both sides of (25c) by η_j and η_j^2, and sum on j:

$$\kappa' DI = 2\sqrt{\kappa_0}\, q' DI + 2\int_0^\infty \left(\sum_{j=1}^{N} q_j \eta_j\, e^{-\eta_j s}\right)^2 ds + 2A' D^2 I \tag{34}$$

$$\kappa' D^2 I = 2\sqrt{\kappa_0}\, q' D^2 I + (q' DI)^2 + 2A' D^3 I \tag{35}$$

We now consider the case of $\kappa_0 = 0$. We find, then, that the right-hand sides of (33), (34), and (35) are all positive. This implies that, for any order N, we must have the following hold true (if $\kappa_0 = 0$)

$$\kappa' I = \sum_{j=1}^{N} \kappa_j > 0 \tag{36a}$$

$$\kappa' DI = \sum_{j=1}^{N} \kappa_j \eta_j > 0 \tag{36b}$$

$$\kappa' D^2 I = \sum_{j=1}^{N} \kappa_j \eta_j^2 > 0 \tag{36c}$$

It may be noted that we cannot obtain any more inequalities of the type $\kappa' D^n I > 0$ with $n \geqslant 3$ by multiplying (24) by η_j^n and summing over j, because the quadratic form

$$2\sum_{i,j=1}^{N} q_i q_j [\eta_i \eta_j^{\,n}/(\eta_i + \eta_j)] = \sum_{i,j=1}^{N} q_i q_j [\eta_i \eta_j/(\eta_i + \eta_j)](\eta_j^{n-1} + \eta_i^{n-1}) \tag{37}$$

which is nonnegative for $n = 0$, $n = 1$, and $n = 2$ [cf (31), (34), and (35)], but ceases to be so for $n \geqslant 3$, where η_j and q_j are arbitrary real numbers.† The condition (36a) is a necessary condition for asymptotic

† Indeed, the first two Sylvester determinants for this quadratic are $\Delta_1 = \eta_1^{\,n} > 0$ and

$$\Delta_2 = \begin{vmatrix} \eta_1^{n-2} & (\eta_1^{n-1} + \eta_2^{n-1})/(\eta_1 + \eta_2) \\ (\eta_1^{n-1} + \eta_2^{n-1})/(\eta_1 + \eta_2) & \eta_2^{n-2} \end{vmatrix} (\eta_1 \eta_2)^2 > 0$$

Our assumption has been $\eta_j > 0$ for all j. Hence, the first condition is satisfied. But the second determinant is nonnegative only for $n = 0, 1, 2$, and nonpositive for all other positive or negative values of n. Indeed, letting $\alpha = (\eta_2/\eta_1)$, we can show that $\Delta_2 \sim [\eta_1^{n-1}/(1 + \alpha)^2](\alpha^n - 1)(1 - \alpha^{n-2})$, which is zero for $n = 0$ and $n = 2$, positive for $n = 1$, and negative for all $n \neq 0, 1, 2$.

stability in the large when $\kappa_0 = 0$, because it guarantees the existence of a finite equilibrium power. The conditions (35b) and (36c) are clearly necessary conditions for the roots q_j of (24) to be real in the case of $\kappa_0 = 0$. Thus, if one of them is not satisfied, the roots of (24) cannot be all real, and Theorem 2 is not applicable. However, this does not imply that the equilibrium state is not asymptotically stable, because Theorem 2 provides only a sufficient condition for asymptotic stability. Let us now suppose that (36b) and (36c) are both satisfied. We can still not conclude that the q_j are all real, guaranteeing asymptotic stability in the large, because (36b) and (36c) are only necessary for the realness of q_j. These considerations indicate that the inequalities (36b) and (36c) have little practical utility in general when N exceeds 2.

Application to Two-Temperature Feedback Model

As an application of the algebraic asymptotic stability criterion (Theorem 2) derived from the Lurie–Letov function, and in order to illustrate the construction of inequalities (26a), we consider now the two-temperature feedback model with $\kappa_0 > 0$.

Substituting

$$\beta_j = \kappa_j/\kappa_0\,, \qquad j = 1, 2 \tag{38a}$$

$$x_j = q_j/\sqrt{\kappa_0}\,, \qquad j = 1, 2 \tag{38b}$$

in (28c), we obtain explicitly

$$\beta_1 = 2x_1 + x_1^2 + [2x_1x_2\eta_2/(\eta_1 + \eta_2)] \tag{39a}$$

$$\beta_2 = 2x_2 + x_2^2 + [2x_1x_2\eta_1/(\eta_1 + \eta_2)] \tag{39b}$$

In order to find the condition for the realness of x_1 and x_2 in (39), we put, following Letov,

$$x_j \equiv (y_j - 1), \qquad j = 1, 2$$

and rewrite (39) in the new variables y_j:

$$\beta_1 = (y_1^2 - 1) + (y_1 - 1)(y_2 - 1)[2\eta_2/(\eta_1 + \eta_2)] \tag{40a}$$

$$\beta_2 = (y_2^2 - 1) + (y_1 - 1)(y_2 - 1)[2\eta_1/(\eta_1 + \eta_2)] \tag{40b}$$

Adding these two equations, we find

$$(y_1 + y_2 - 1)^2 = 1 + \beta_1 + \beta_2 = (1/\kappa_0)[\kappa_0 + \kappa' I] = k_0/\kappa_0 \tag{41}$$

where the last equality follows from (15) and represents the condition for the existance of a finite equilibrium, k_0 being the external reactivity.

This result shows that the conditions (33) and (36a) which follow from it are included in the set of inequalities (26).

Next, multiply (40a) and (40b) by η_1 and η_2, respectively, and subtract:

$$\eta_1 y_1^2 - \eta_2 y_2^2 = \eta_1(\beta_1 + 1) + \eta_2(\beta_2 + 1) \tag{42}$$

Clearly, the two equations (41) and (42) are identical to the original set (39). The realness condition for y_1 and y_2 can be obtained by eliminating y_2 between (41) and (42):

$$(\eta_1 - \eta_2)y_1^2 + 2\eta_2[1 \pm (k_0/\kappa_0)^{1/2}]y_1 - \eta_2[1 \pm (k_0/\kappa_0)^{1/2}]^2 + \eta_1(\beta_1 + 1) + \eta_2(\beta_2 + 1) = 0 \tag{43}$$

The condition for the realness of y_1 is easily established by considering the discriminant of (43) and taking into account (41) as

$$\eta_1^2 + \eta_2^2 + \eta_1^2\beta_1 + \eta_2^2\beta_2 \pm 2\eta_1\eta_2(k_0/\kappa_0)^{1/2} > 0 \tag{44}$$

The inequalities in (26a) can now be written explicitly from (41) and (44) as

$$f_1(\kappa_0, \kappa_1, \kappa_2, 0) \equiv \kappa_0 + \kappa' I > 0 \tag{45a}$$

$$f_2(\kappa_0, \kappa_1, \kappa_2, 0) \equiv \kappa_0 \operatorname{Tr}(D^2) + \kappa' D^2 I \pm 2 \mid D \mid [\kappa_0(\kappa_0 + \kappa' I)]^{1/2} > 0 \tag{45b}$$

These conditions are necessary and sufficient for the realness of the roots of (39). It is interesting to note that (45a) and (45b) reduce respectively to (36a) and (36c) when $\kappa_0 = 0$, indicating that (36b) cannot be independent of (36a) and (36c) when $N = 2$. It is of course easily verified that (36b) can be obtained from the latter immediately if $N = 2$.

As it stands, (45) contains five parameters, and a representation in two dimensions would seem out of question. However, (45b) can be written in the form

$$\operatorname{Tr}(D)[\kappa_0 \operatorname{Tr}(D) + \kappa' DI] > \mid D \mid [\pm(\kappa_0 + \kappa' I)^{1/2} - (\kappa_0)^{1/2}]^2 \tag{46}$$

Hence, defining [24]

$$X^2 = \operatorname{Tr}(D)[\kappa' DI + \kappa_0 \operatorname{Tr}(D)]/\kappa_0 \mid D \mid \tag{47a}$$

$$Y^2 = 1 + (\kappa' I/\kappa_0) \tag{47b}$$

We can easily reduce (46) with positive sign to

$$(X + 1 - Y)(X - 1 + Y) > 0 \tag{48}$$

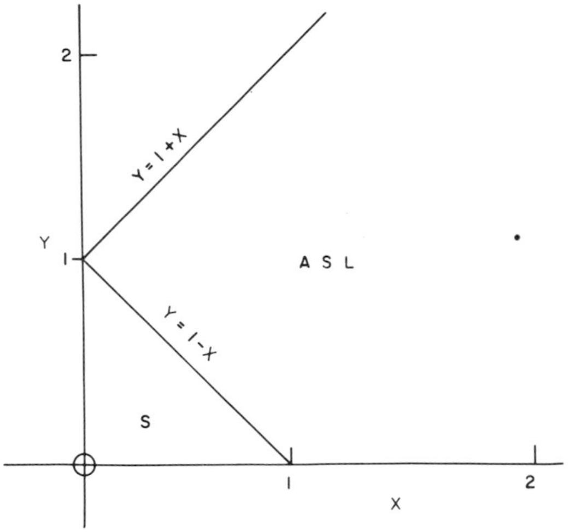

FIGURE 7.7.2. Stability phase space.

Since (47) involves only X^2 and Y^2, it suffices to consider only $X > 0$ and $Y > 0$. Then (48) can be mapped in the X, Y plane as in Figure 7.7.2. Note that the region corresponding to the inequality

$$(X - Y - 1)(X + Y + 1) > 0,$$

which is obtained from (46) with the negative sign, is already included in this region. The region marked A.S.L., to the right of the oblique lines, defines a parameter region; the equilibrium solution of (12) is asymptotically stable in the large for any values of κ_0, κ_1, κ_2, η_1, and η_2 which make X and Y lie inside this region.

The usefulness of the results obtained above depends on the simplicity with which the $f_k > 0$ can be applied to a real problem. If the user must first transform to the canonical framework, the labor (especially for $N > 2$) increases significantly. It is characteristic that under linear, nonsingular transformations, inner products do not change their value; hence, if the f_k are written only as inner products, they may be inverted immediately. We have already written (36) and (45) in this form. For example, transforming back to the parameter space of Eq. (1), we have

$$\begin{aligned}
\operatorname{Tr}(D) &= \operatorname{Tr}(H) \qquad \text{(cf Eq. 5)}\\
\operatorname{Tr}(D^2) &= \operatorname{Tr}(H^2)\\
\kappa' I &= \alpha' H^{-1} b P_0 \qquad \text{(cf Eqs. 13, 8c, 9a)}\\
\kappa' D^m I &= \alpha' H^{m-1} b P_0
\end{aligned}$$

This result permits algebraic analysis to be performed in the simpler canonical frame, and all numerical analysis to be performed in the physical frame. In simple cases, where the matrix H is small enough ($N \leqslant 3$), analysis can usefully be performed in the physical framework [25]; this point will be taken up in Problem 7.

Obviously, the conditions $f_k > 0$ become more difficult to find for $N > 2$. This leads us to the natural question as to whether there is some other way of choosing the q_j or A_j so as to guarantee asymptotic stabily in the large. This leads us to the approach taken by Lefschetz [20, 26].

7.8. The Lefschetz Criterion

A. Global Asymptotic Stability

Let us introduce the transformation

$$\sigma_i = \ln(1 + x_i), \qquad i = 0, 1, \ldots, I \tag{1}$$

and define

$$\phi(\sigma_i) \equiv e^{\sigma_i} - 1 \tag{2}$$

in the canonical set (7.7, Eq. 12). (The noncanonical form will be discussed later.) The equations then read

$$(l/\beta)\dot{\sigma}_0 = -\left[\kappa_0\phi(\sigma_0) + \sum_{j=1}^{N} \kappa_j y_j\right] + \sum_{i=1}^{I} a_i\{[\phi(\sigma_i) - \phi(\sigma_0)]/[1 + \phi(\sigma_0)]\} \tag{3a}$$

$$\dot{\sigma}_i = \lambda_i[\phi(\sigma_0) - \phi(\sigma_i)]/[1 + \phi(\sigma_i)], \qquad i = 1, 2, \ldots, N \tag{3b}$$

$$\dot{y}_j = \eta_j[\phi(\sigma_0) - y_j], \qquad j = 1, 2, \ldots, N \tag{3c}$$

We notice that the function $\phi(\sigma_i)$ satisfies

$$\sigma_i\phi(\sigma_i) > 0 \qquad \text{for} \quad \sigma_i \neq 0 \tag{4a}$$

$$\phi(0) = 0 \tag{4b}$$

and

$$\int_0^{\sigma_i} \phi(\sigma)\, d\sigma = x_i - \ln(1 + x_i) \tag{4c}$$

Let us now consider the following scalar function of **y** and **σ**:

$$V = y'Fy + \sum_{i=0}^{I} \epsilon_i \int_0^{\sigma_i} \phi(\sigma)\, d\sigma \tag{5a}$$

where

$$y = \text{col}(y_j) \tag{5b}$$

$$\epsilon_0 = l/\beta \tag{5c}$$

$$\epsilon_i = a_i/\lambda_i\,, \qquad i = 1,\ldots, I \tag{5d}$$

and where F is an arbitrary symmetric, positive-definite matrix,† i.e., $y'Fy > 0$ for any $\| y \| \neq 0$. For simplicity, if F is positive-definite, we write $F > 0$.

Since each integral term in (5a) is positive-definite, the function $V(\mathbf{y}, \boldsymbol{\sigma})$ is positive-definite also.

The derivative of V along a trajectory of (3) is

$$-\dot{V} = y'Gy + \kappa_0\phi^2(\sigma_0) + \sum_{i=1}^{I} a_i\{[\phi(\sigma_0) - \phi(\sigma_i)]^2/[1 + \phi(\sigma_0)][1 + \phi(\sigma_i)]\}$$
$$+ y'(\kappa - 2FDI)\phi(\sigma_0) \tag{6}$$

where

$$G \equiv DF + FD \tag{7}$$

and $D = \text{diag}(\eta_1, \eta_2, \ldots, \eta_n)$ as before. We now define

$$d \equiv \tfrac{1}{2}\kappa - FDI \tag{8}$$

and rewrite (6) as

$$-\dot{V} = y'Gy + \kappa_0\phi^2(\sigma_0) + 2\phi(\sigma_0)\, d'y$$
$$+ \sum_{i=1}^{I} a_i\{[\phi(\sigma_0) - \phi(\sigma_i)]^2/[1 + \phi(\sigma_0)][1 + \phi(\sigma_i)]\} \tag{9}$$

Equation (9) indicates that $-\dot{V}$ is not a priori positive-definite. However, it can be made positive-definite by placing further restrictions on F through G. To make the first term positive-definite, we must require first $G > 0$. But this is not sufficient for $-\dot{V} > 0$, because of the third

† Note that any arbitrary quadratic form $y'Ay$ in which A is not symmetric can be written as $y'Fy$ where F is symmetric, i.e., $y'Ay \equiv y'Fy$. It is easy to show that $F = \frac{1}{2}(A + A')$.

term, which is still not sign definite. In order to handle this term, we cast (9) in the following form:

$$-\dot{V} = z'Qz + \sum_{i=1}^{I} a_i\{[\phi(\sigma_0) - \phi(\sigma_i)]^2/[1 + \phi(\sigma_0)][1 + \phi(\sigma_i)]\} \tag{10}$$

where

$$z \equiv \text{col}(y_1, \ldots, y_N, \phi(\sigma_0)) \tag{11a}$$

$$Q \equiv \begin{bmatrix} G & d \\ d' & \kappa_0 \end{bmatrix} \tag{11b}$$

We now require Q to be positive-definite. Let us consider the Sylvester determinates of Q. Since $G > 0$ is already assumed, all the diagonal minors of G including $|G|$ are positive. These are the first $N - 1$ Sylvester determinants of Q. Thus, the only additional requirement for Q, and hence $-\dot{V}$, to be positive is $|Q| > 0$. It is not difficult to show† that $|Q|/|G| = \kappa_0 - d'G^{-1}d$, so that the above condition becomes

$$\kappa_0 > d'G^{-1}d \tag{12}$$

Since $G > 0$, so is G^{-1}. Hence, $d'G^{-1}d$ is a positive-definite quadratic form, and vanishes only if $d = 0$. Thus, according (12), κ_0 must be

† We have the following determinantal relation [20, 26]:

$$|Q| = |G|(\kappa_0 - d'G^{-1}d)$$

which can be proved by multipling Q from the left by $\text{diag}(G^{-1}, 1)$,

$$\begin{pmatrix} G^{-1} & 0 \\ 0 & 1 \end{pmatrix} Q = \begin{pmatrix} E & G^{-1}d \\ d' & \kappa_0 \end{pmatrix}$$

The determinant of this matrix relation gives

$$|G^{-1}||Q| = \begin{vmatrix} E & G^{-1}d \\ d' & \kappa_0 \end{vmatrix} \equiv \Delta_{N+1}$$

By expanding the determinant on the right-hand side, we can show that $\Delta_{N+1} = \Delta_N - d_1(G^{-1}d)_1$. Repeating this procedure, we get $\Delta_{N+1} = \kappa_0 - d'G^{-1}d$. Using $|G^{-1}| = |G|^{-1}$, we obtain the desired relation. A generalization of this result is as follows [20, 26]: If

$$Q = \begin{pmatrix} G & P \\ S' & R \end{pmatrix}$$

where G is $N \times N$, R is $M \times M$, and P and S are $N \times M$ matrices, and if G is nonsingular, then we have

$$|Q| = |G||R - S'G^{-1}P|$$

positive† in order for us to be able to construct a Lyapunov function $V(\mathbf{y}, \boldsymbol{\sigma})$ of the form of (5a) with a negative $\dot{V}$.

Let us suppose that $\kappa_0 > 0$, and summarize the procedure for constructioning this Lyapunov function. We first note that (7) can be solved for F in terms of G as

$$F = \int_0^\infty dt\, e^{-Dt} G\, e^{-Dt} \tag{13a}$$

because the matrix $-D$ is stable (Lemma 1 in Sec. 7.8B below). In the case of a diagonal D as we have, F can be expressed as

$$F_{ij} = G_{ij}/(\eta_i + \eta_j) \tag{13b}$$

which can also be obtained directly from (7). Since we assumed F to be symmetric, G_{ij} is also symmetric. This is true in general whenever D is symmetric in (13a). The Lemma 2 which will be proved below shows that $F > 0$ if $G > 0$. Hence, any arbitrary choice of a symmetric $G > 0$ will automatically guarantee a symmetric $F > 0$.

We now substitute (13b) into (5) to express d in terms of G as

$$d_j \equiv \tfrac{1}{2}\kappa_j - \sum_{i=1}^{N} [G_{ij}\eta_i/(\eta_j + \eta_i)] \tag{14}$$

Using (14) in (12), we obtain the condition $G > 0$ must satisfy as:

$$\kappa_0 > \sum_{j,j'=1}^{N} \left\{ \left[\tfrac{1}{2}\kappa_j - \sum_{i=1}^{N} G_{ij}\eta_i/(\eta_i + \eta_j) \right] \times G_{jj'}^{-1} \left[\tfrac{1}{2}\kappa_{j'} - \sum_{i=1}^{N} G_{j'i}\eta_i/(\eta_i + \eta_{j'}) \right] \right\} \tag{15}$$

Hence, if a symmetric $G > 0$ can be found such that (15) holds, then we have a positive-definite V-function given by (5a), in which F is to be obtained from (13b), such that $-\dot{V}$ is negative-definite. Using Theorem 2, Sect. 7.6, we establish asymptotic stability in the large, because $V(\mathbf{y}, \boldsymbol{\sigma}) \to \infty$ with $\mathbf{y}$ and $\boldsymbol{\sigma}$. We can state this result as the following theorem.

† In the case of $\kappa_0 = 0$, the present method does not offer any advantages. When $\kappa_0 = 0$, (12) yields $2d = \kappa - 2FDI = 0$, and the problem becomes a special case of that solved in Sec. 7.7C.

THEOREM 1. If the reactor parameters κ_0, η_j, and κ_j are such that one can find a symmetric $G > 0$ satisfying (15), then the equilibrium solution of (12) is asymptotically stable in the large.

There are certain advantages to the Lefschetz method of matrices. If κ_0 is at our disposal, then the choice of any symmetric $G > 0$ will yield the value of $\kappa_0 > 0$, by (15), which is sufficient to produce asymptotic stability in the large. Clearly, given κ and D, we wish to know the minimum value of κ_0 sufficient for asymptotic stability in the large. Then, we must minimize the right-hand side of (15) with respect to G_{ji}. As an illustration, let us restrict ourselves to diagonal $G > 0$, so that (15) reduces to

$$\kappa_0 > \sum_{j=1}^{N} \tfrac{1}{4}(G_{jj} - \kappa_j)^2/G_{jj} \tag{16}$$

where $G_{jj} > 0$. Since (16) involves the sum of positive terms, its minimum is the sum of the minima of the various terms. Hence, terms with $\kappa_j > 0$ disappear because their minimum vanishes, i.e., $G_{jj} = \kappa_j$. The minimum of the terms with $\kappa_j < 0$ occurs for $G_{jj} = |\kappa_j|$, and is equal to $|\kappa_j|$. Thus, the minimum of κ_0 follows as

$$\kappa_0 > \kappa_{0,\mathrm{MIN}} = \tfrac{1}{2} \sum_{j=1}^{N} (|\kappa_j| - \kappa_j) \tag{17}$$

This result is certainly crude because it is obtained by minimizing (15) with respect to diagonal G-matrices with positive elements, which constitutes only a very restricted subset of the class of symmetric, positive-definite G-matrices. As a consequence of this constraint on G, (17) involves only $\kappa_j < 0$ without taking into account the stabilizing effect of the $\kappa_j > 0$, and hence it is overly sufficient. Nevertheless, it yields an interesting result which implies that, if the power coefficient exceeds the sum of the negative canonical temperature coefficients, then the reactor is asymptotically stable in the large. The condition for a finite equilibrium to exist is, as we recall, $\kappa_0 + \sum_i \kappa_i > 0$, which implies $\kappa_0 + \sum_j (\text{positive } \kappa_j) > \sum_j |\text{negative } \kappa_j|$. It follows that (17) would be identical to $\kappa_0 + \kappa' I > 0$ if all κ_j were negative. Hence, (17) implies that all such systems, which possess a finite equilibrium state, are also asymptotically stable in the large.

Finally, we may note that, if (17) holds, the condition $\kappa_0 + \kappa' I > 0$ is certainly satisfied, i.e., (17) guarantees the existence of a finite equilibrium level as well, as it must.

B. Proof of the Lemmas

LEMMA 1. The solution of the matrix equation

$$RF + FS = -G \tag{18a}$$

when G, F, R, and S are $N \times N$ matrices, and R and S have negative characteristic values (i.e., R and S are stable) is

$$F = \int_0^\infty e^{Rt} G \, e^{St} \, dt \tag{18b}$$

PROOF. Multiply (18a) by e^{Rt} from the left and by e^{St} from the right, and verify

$$\begin{aligned} -e^{Rt} G \, e^{St} &= [(d/dt) \, e^{Rt}] F \, e^{St} + e^{Rt} F [(d/dt) \, e^{St}] \\ &= (d/dt)(e^{Rt} F \, e^{St}) \end{aligned} \tag{18c}$$

Integrating both sides from $t = 0$ to ∞ and noticing that

$$\lim_{t \to \infty} e^{Rt} F \, e^{St} = 0$$

because R and S are stable, one obtains (18b). The uniqueness of F is obvious from the above proof.

LEMMA 2. If

$$F = \int_0^\infty dt \, (\exp R't) G \exp Rt \tag{19a}$$

and if G is positive-definite, then F is also positive-definite.

PROOF. Let X be an arbitrary column vector. Consider

$$X'FX = \int_0^\infty dt \, X'(\exp R't) G (\exp Rt) X$$

If we let $Y = (\exp Rt)X$, then $Y' = X' \exp(R't)$, and

$$X'FX = \int_0^\infty dt \, Y'(t) G Y(t) \tag{19b}$$

But $Y'(t)GY(t) > 0$ by hypothesis for all Y. Hence, $X'FX > 0$ for $X \neq 0$ (because Y vanishes if and only if $X = 0$).

Clearly, the above proof indicates that, if G is semidefinite, so is F (this result will be useful in the proof of Kalman's lemma in Sec. 7.10).

Equation (13a) follows from (19a) with $R = -D$, $R' = -D$. We note that the characteristic values of R (and thus of D) need not be distinct for (19a) [and (13a)] to be true.

C. Nondiagonal Form of Lefschetz's Method

Let us briefly resketch the matrix method for a nondiagonal system such as the original set (7.7, Eq. 1). For simplicity, we ignore delayed neutrons by setting $a_i = 0$, and rewrite (7.7, Eq. 1) as

$$(l/\beta)\dot{\sigma}_0 = -\alpha' y - P_0\gamma\phi(\sigma_0) \tag{20a}$$

$$\dot{y} = P_0 b\phi(\sigma_0) - Hy \tag{20b}$$

where we have defined as before

$$\sigma = \sigma_0 = \ln(P/P_0)$$

$$\phi(\sigma) = e^{\sigma} - 1$$

$$y = X - X_0$$

with P_0 and X_0 the equilibrium values of $P(t)$ and $X(t)$. Equations (3) with $a_i = 0$ are the canonical form of (20), and are obtained from the latter by a similarity transformation as described in Sec. 7.7.

The Lyapunov function in the nondiagonal form is similar to (5a):

$$V = y'Fy + (l/\beta)\int_0^{\sigma} \phi(\sigma')\, d\sigma' \tag{21a}$$

whose derivative is (cf Eq. 6)

$$-\dot{V} = yGy + P_0\gamma\phi^2(\sigma) + 2\phi(\sigma)\, d'y \tag{21b}$$

where

$$d = \tfrac{1}{2}\alpha - P_0 Fb \tag{21c}$$

$$G = H'F + FH \tag{21d}$$

In this case, we find, in place of (12),

$$\gamma P_0 > d'G^{-1}d \tag{22}$$

Obviously, we have made no comment on the eigenvalues of H except that $-H$ must be stable and nonsingular. Hence, we need not demand the eigenvalues be distinct (zero multiplicity) to make use of (22).

7.9. Asymptotic Stability in a Finite Region

In the foregoing analysis, the function $V \to \infty$ with $\| y \|$, $\| \sigma \| \to \infty$, and hence yields conditions for asymptotic stability in the large. We can obtain more relaxed conditions if we require asymptotic stability for initial perturbations that are restricted to a finite region in the phase space. We shall now present a worked example to illustrate the method of approach to such a problem, by considering a reactor model containing only one feedback variable and a power coefficient, in the absence of delayed neutrons.

Following Smets [7], we start with the following function:

$$V = (l/\beta) \int_0^{\sigma_0} \phi(\sigma)\, d\sigma + [\eta_1/2(\kappa_0 + \kappa_1)][(\kappa_1 y_1/\eta_1) - (l\sigma_0/\beta)]^2 \tag{1}$$

which is always positive for all y_1, $\sigma_0 \neq 0$, and vanishes only for $y_1 = \sigma_0 = 0$. The time derivative of V can be obtained using (7.8, Eq. 3) with $A_i = 0$, $j = 1$ as

$$-\dot{V} = \phi(\sigma_0)[(\eta_1 l \sigma_0/\beta) + \kappa_0 \phi(\sigma_0)] \tag{2}$$

Since $\dot{V}$ does not contain y_1, it is not sign-definite and thus we cannot apply the stability theorem (7.6, Theorem 2) directly. However, if $\kappa_0 \geqslant 0$, then $\dot{V} < 0$ for all $\sigma_0 \neq 0$, hence $\dot{V} \leqslant 0$ for all σ_0, y_1, and $\dot{V} = 0$ for $\sigma_0 = 0$ regardless of the values of y_1. We can easily verify that the equations

$$(l/\beta)\dot{\sigma}_0 = -[\kappa_0 \phi(\sigma_0) + \kappa_1 y_1] \tag{3a}$$

$$\dot{y}_1 = \eta_1(\phi(\sigma_0) - y_1) \tag{3b}$$

do not have any nontrivial half-trajectory in the set of points $\{\sigma_0 = 0, y_1 = \text{arbitrary}\}$, where $\dot{V} = 0$. (Indeed, if $\sigma_0 \equiv 0$, then (3a) yields $y_1 \equiv 0$.) Hence, the stability theorem (7.6, Theorem 4) applies, and the equilibrium point $\sigma_0 = y_1 = 0$ is asymptotically stable in the large.

We now consider the case of $\kappa_0 < 0$, in which $\dot{V}$ is negative only for the values of σ_0 determined (when $\eta_1 l > \beta \mid \kappa_0 \mid$) by

$$\phi(\sigma_0) < \phi(\sigma^*) = \eta_1 l \sigma^*/\beta | \kappa_0 | \tag{4}$$

Let us now consider the family of closed curves $V(y_1, \sigma_0) = C$. Since $\dot{V} \leq 0$ everywhere inside such a curve, then the stability theorem (7.6, Theorem 4) applies in this region, and the equilibrium $\sigma_0 = y_1 = 0$ is asymptotically stable for all initial conditions $\sigma_0(0)$ and $y_1(0)$ inside such a surface.

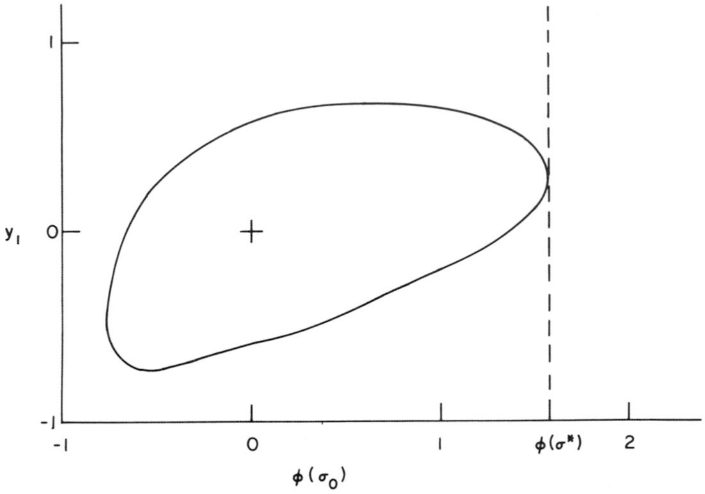

FIGURE 7.9.1. Local asymptotic stability; $\beta V(y_1, \sigma_0)/l = C^*$, $\kappa_0 < 0$.

The maximal region of asymptotic stability is determined by the curve $V(y_1, \sigma_0) = C^*$, which is tangent to the straight line $\sigma_0 = \sigma^*$ (see Figure 7.9.1) across which $\dot{V}(y_1, \sigma_0)$ changes sign. At the point of contact, $d\sigma_0/dy_1 = 0$ on the curve. Hence, using $V(y_1, \sigma_0) = C^*$ and

$$(\partial V/\partial y_1) + (\partial V/\partial \sigma_0)(d\sigma_0/dy_1) = 0$$

we obtain

$$[\partial V(y_1, \sigma_0)/\partial y_1]|_{y_1=y_1^*, \sigma_0=\sigma^*} = 0 \tag{5}$$

Substitution of (1) into (5) yields

$$y_1^* = l\eta_1\sigma^*/\beta\kappa_1 \tag{6}$$

where σ^* is determined from the equality in (4). Evaluating (1) at $\sigma_0 = \sigma^*$, $y_1 = y_1^*$, we obtain the value of C^* as

$$C^* = (l/\beta) \int_0^{\sigma^*} d\sigma\, \phi(\sigma) \tag{7}$$

As a specific example, we choose $\beta\kappa_1/l\eta_1 = 3.0$ and $|\kappa_0|/\kappa_1 = 0.2$ and calculate σ^* from (4) as $\sigma^* = 0.948$ and $\beta C^*/l$ from (7) as $\beta C^*/l = 1.632$. The resulting curve for $V(y_1, \sigma_0) = C^*$ is plotted in Figure 7.9.1. Of some interest is the fact that far larger perturbations are permitted in $P(0)$ than are permitted in $T(0)$.

We now ask whether this is the widest possible region for asymptotic stability that can be achieved with this type of V-function. If we replace the coefficient of the second term of (1) by $\alpha/2(\kappa_1 + \kappa_0)$, where $\alpha \geqslant 0$ and arbitrary, we would expect (again for $\kappa_0 < 0$ and $\kappa_1 + \kappa_0 > 0$) that a choice of α greater than η_1 would yield a larger region for asymptotic stability. However, it turns out, as we shall show presently, that the largest region is obtained when $\alpha = \eta_1$.

The derivative of V when $\alpha \neq \eta_1$ can be shown to be

$$-\dot{V} = \phi(\sigma_0)\{(l/\beta)\alpha\sigma_0 + \kappa_1[1 - (\alpha/\eta_1)]y_1 - |\kappa_0|\phi(\sigma_0)\} \tag{8}$$

which reduces to (2) if $\alpha = \eta_1$. Our problem is to determine the largest value $C^*(\alpha)$ for given α such that $\dot{V} < 0$ will hold everywhere inside the closed curve $V(\sigma_0, y_1) = C^*(\alpha)$. For this value of $C^*(\alpha)$, the latter curve becomes tangent to the curve $\dot{V} = 0$ whose equation is given by (8). The coordinates (σ^*, y_1^*) of the point of contact and $C^*(\alpha)$ are obtained by solving the following set of equations:

$$V(\sigma^*, y_1^*) = C^*(\alpha) \tag{9a}$$

$$\dot{V}(\sigma^*, y_1^*) = 0 \tag{9b}$$

$$[(\partial V/\partial\sigma_0)(\partial\dot{V}/\partial y_1) - (\partial V/\partial y_1)(\partial\dot{V}/\partial\sigma_0)]|_{\sigma_0=\sigma^*, y_1=y_1^*} = 0 \tag{9c}$$

However, we are not interested directly in the value of $C^*(\alpha)$, but rather in the value of α that maximizes $C^*(\alpha)$. Hence, differentiating (9a) with respect to α, and using (1), we obtain

$$\begin{aligned} dC^*/d\alpha = {} & (l/\beta)(d\sigma^*/d\alpha)\phi(\sigma^*) + (1/2k_0)[(\kappa_1 y_1^*/\eta_1) - (l\sigma^*/\beta)]^2 \\ & + (\alpha/k_0)[(\kappa_1 y_1^*/\eta_1) - (l\sigma^*/\beta)][(\kappa_1/\eta_1)(dy_1^*/d\alpha) - (l/\beta)(d\sigma^*/d\alpha)] \end{aligned} \tag{10}$$

On the other hand, differentiation of (9b) with respect to α gives

$$\begin{aligned} (d\sigma^*/d\alpha)\exp\sigma^* = {} & (l/\beta|\kappa_0|)[\sigma^* + \alpha(d\sigma^*/d\alpha)] \\ & - (\kappa_1 y_1^*/\eta_1|\kappa_0|) + (\kappa_1/|\kappa_0|)[1 - (\alpha/\eta_1)](dy_1^*/d\alpha) \end{aligned} \tag{11}$$

Now, at $\alpha = \eta_1$, Eqs. (11) and (9b) yield $d\sigma^*/d\alpha = 0$ if $dy_1^*/d\alpha$ is finite [this can be verified by differentiating (9c)]. Hence, we find from (10) with $d\sigma^*/d\alpha = 0$ that $dC^*(\alpha)/d\alpha = 0$ at $\alpha = \eta_1$.

It is also possible to extend the analysis to a multivariable feedback model when the delayed neutrons are ignored. In this case, (1) is replaced by

$$V = (l/\beta)\int_0^{\sigma_0}\phi(\sigma)\,d\sigma + [\eta_j/2(\kappa' I - |\kappa_0|)][\kappa' D^{-1}y - (l\sigma_0/\beta)]^2 \tag{12a}$$

where we have chosen $\kappa_0 < 0$, and $\alpha = \eta_j$ (the reason for choosing for α one of the η_i will be explained later). The time derivative of (12a) is

$$-\dot{V} = \phi(\sigma_0)[\eta_j(l\sigma_0/\beta) - |\kappa_0|\phi(\sigma_0)] + \sum_{i=1}^{N} \kappa_i[1 - (\eta_j/\eta_i)]\, y_i \tag{12b}$$

The problem is to determine the point of contact of the closed surface $V(\sigma_0, y_1, ..., y_n) = C^*$ and the surface $\dot{V}(\sigma_0, y_1, ..., y_n) = 0$ by solving

$$\dot{V}(\sigma^*, y_i^*) = 0 \tag{13a}$$

$$\partial V/\partial\sigma^* = \lambda\, \partial\dot{V}/\partial\sigma^* \tag{13b}$$

$$\partial V/\partial y_i^* = \lambda\, \partial\dot{V}/\partial y_i^*, \qquad i = 1, ..., N \tag{13c}$$

There are $N + 2$ equations in this set and $N + 2$ unknowns, i.e., σ^*, $y_1^*, ..., y_n^*$, and λ. The maximum region for asymptotic stability is determined by the closed surface $V(\sigma_0, y_1, ..., y_n) = C^*$, where $C^* = V(\sigma^*, y_1^*, ..., y_n^*)$, i.e., the value of V at the point of contact. An inspection of (12b) reveals that $\dot{V}$ does not contain y_j; hence, $\partial\dot{V}/\partial y_j^* = 0$. (This conclusion is a direct consequence of choosing $\alpha = \eta_j$.) From (13), we obtain $\partial V/\partial y_j^* = 0$ which leads to

$$\kappa' D^{-1} y^* \equiv \sum_{i=1}^{N} (\kappa_i/\eta_i)\, y_i^* = l\sigma^*/\beta$$

This in turn results in $\partial V/\partial y_i^* = 0$ for all i. Using this information in (13), we find that all $y_i^* = 0$ except y_j^*, and hence

$$C^*(\eta_j) = (l/\beta) \int_0^{\sigma^*} \phi(\sigma)\, d\sigma \tag{14a}$$

$$|\kappa_0|\phi(\sigma^*) = l\eta_j\sigma^*/\beta \tag{14b}$$

$$y_j^* = (l\sigma^*/\beta)\eta_j/\kappa_j \tag{14c}$$

The question of maximizing $C^*(\eta_j)$ over η_i by choosing for α a different one of the η_i can be handled easily by means of (14b).

The particular V-function used in Eq. (12a) is interesting in that it contains terms that do not appear in the Lurie–Letov function, in particular, the terms σ^2 and $\mathbf{y}\sigma$. This observation leads us naturally to ask whether a more general V-function exists which is a quadratic form plus an integral of a nonlinearity. We consider this question in the next section.

7.10. Popov Function

In order to secure a weaker criterion for asymptotic stability in the large, we shall have to consider our basic set of equations in the absence of delayed neutrons. We shall return to the effects of delayed neutrons in the next section.

We reproduce, for convenience, the basic equations as

$$\dot{\sigma} = -\kappa' y - \kappa_0 \phi(\sigma) \tag{1a}$$

$$\dot{y} = D(I\phi(\sigma) - y) \tag{1b}$$

which are identical to (7.8, Eq. 3) with $a_i = 0$.† The factor l/β is absorbed in κ' and κ_0 for convenience, and the subscript of σ_0, which becomes redundant in the absence of delayed neutrons, is dropped in (1a) and (1b). We recall that

$$\sigma\phi(\sigma) > 0, \qquad \sigma \neq 0 \tag{2a}$$

and

$$k_0 \equiv \kappa_0 + \kappa' I > 0 \tag{2b}$$

the latter being the criterion for the existence of a unique equilibrium state.

With Popov [8] we now prove the following lemma.

LEMMA 1. If the equilibrium solution ($\sigma \equiv 0$, $\| y \| \equiv 0$) of the system

$$\dot{\sigma} = -\kappa'(E - D^{-1}p) y - \kappa_0 \phi(\sigma) \tag{3a}$$

$$\dot{y} = D(I\phi(\sigma) - y) \tag{3b}$$

where p is some positive constant, is asymptotically stable in the large, and if this fact can be established by means of a Lyapunov function of the form

$$V_{\text{II}} = \Phi(y) + \int_0^\sigma \phi(\sigma')\, d\sigma' \tag{4}$$

where $\Phi(y)$ is a positive-definite function, then the equilibrium solution of (1) is also asymptotically stable.

† There seems to be no possible extension of the subsequent development to include the delayed neutrons unless $p = 0$.

Proof. Let the first set of equations be subscribed I and the second, II. Then, the time derivative of (4) with respect to (3), i.e.,

$$\dot{V}_{II} = \dot{\Phi}(y) - \phi(\sigma)[\kappa' y + \kappa_0 \phi(\sigma)] + p\phi(\sigma)\kappa' D^{-1} y \tag{5}$$

is negative-definite for $\sigma \neq 0$, $\| y \| \neq 0$ by the hypothesis of Lemma 1.

We now prove that, if $k_0 > 0$ and $p \geqslant 0$, then

$$V_I = V_{II} + (p/2k_0)(\kappa' D^{-1} y - \sigma)^2 \tag{6}$$

(which is also positive-definite, since $V_{II} > 0$) is a Lyapunov function for the first set of equations. Indeed, the total derivative of V_I along the trajectory of (1) is just

$$\dot{V}_I = \dot{\Phi} - \phi(\sigma)[\kappa' y + \kappa_0 \phi(\sigma)] + p(\kappa' D^{-1} y - \sigma)\phi(\sigma)$$

or

$$\dot{V}_I = \dot{V}_{II} - p\sigma\phi(\sigma) \tag{7}$$

Using (2a), we see that $p\sigma\phi(\sigma) > 0$ for $\sigma \neq 0$; hence, $\dot{V}_I$ is negative-definite, and the first system is also asymptotically stable.

The foregoing lemma indicates that asymptotic stability in the original system will be established by the same criteria that guarantee asymptotic stability in the second system. In order to investigate the asymptotic stability of the second system, we may choose the positive-definite function $\Phi(y)$ as

$$\Phi(y) \equiv \sum_{i,j=1}^{N} [q_i q_j/(\eta_i + \eta_j)] y_i y_j + \sum_{j=1}^{N} A_j y_j^2 \tag{8}$$

where the q_j are real and $A_j > 0$. Then, the function V_{II} becomes a Lurie–Letov function, which was discussed in Sec. 7.7 (cf 7.7, Eq. 16). The total derivative of V_{II} can be obtained from (5) by substituting $\Phi(y)$ from (8) and following the same manipulative steps as those involved in getting (7.7, Eq. 18):

$$-\dot{V}_{II} = [q' y + \sqrt{\kappa_0}\phi(\sigma)]^2 + 2\sum_{j=1}^{N} A_j \eta_j y_j^2 + \phi(\sigma)\sum_{j=1}^{N} y_j W_j + p\sigma\phi(\phi) \tag{9a}$$

where

$$W_j = \kappa_j(I - p/\eta_j) - 2q_j\left\{\sqrt{\kappa_0} + \sum_{i=1}^{N} [q_i \eta_i/(\eta_i + \eta_j)]\right\} - 2A_j\eta_j \tag{9b}$$

Clearly, Eqs. (9) reduce to (7.7, Eq. 18) when $p = 0$ provided delayed neutrons are ignored. Here also we find that, if $W_j = 0$ for all j, then

$\dot{V}_{\text{II}}$ is negative-definite, provided the q_j are real and $A_j > 0$, and the stability of the second system is established. On the basis of Lemma 1, we can state the following theorem, which is a generalization of Theorem 1 of Sec. 7.7.

THEOREM 1. If a value of $p \geqslant 0$ exists such that

$$\kappa_j[1 - (p/\eta_j)] = 2q_j \left\{\sqrt{\kappa_0} + \sum_{i=1}^{N} [q_i\eta_i/(\eta_i + \eta_j)]\right\} + 2A_j\eta_j, \quad j = 1,..., N \tag{10}$$

can be satisfied for some $A_j > 0$ and q_j real, and if $\kappa_0 + \kappa' I > 0$, then the equilibrium $\sigma \equiv \| y \| \equiv 0$ of (1) is asymptotically stable in the large.

REMARKS

(a) It is important to notice that $k_0 = \kappa_0 + \kappa' I$ appears explicitly in the Lyapunov function V_{I} defined by (6), in contrast with that defined by (7.7, Eq. 16) in the Lurie–Letov theory. Consequently, the condition given by (7.7, Eq. 24) in the statement of Theorem 1, Sec. 7.7 automatically includes the condition $\kappa_0 + \kappa' I > 0$ for the existence of a finite equilibrium, whereas in the Popov theorem, it must be stated as an additional condition to (10).

In order to illustrate this point, which is often overlooked, let us again consider the simplest case of a one-temperature reactor model characterized by κ_0, κ_1, and η_1. The condition (10) reduces in this case to

$$\kappa_1[1 - (p/\eta_1)] - 2A_1\eta_1 = 2q_1 \sqrt{\kappa_0} + q_1^2 \tag{11}$$

which is satisfied by $A_1 > 0$, $p \geqslant 0$, and a real q_1 if $\kappa_0 > 0$ and

$$\kappa_0 + \kappa_1[1 - (p/\eta_1)] - 2A_1\eta_1 \geqslant 0 \tag{12}$$

Suppose we choose $p = \eta_1$. Then, (12) would be satisfied whenever $\kappa_0 > 0$ because $A_1 > 0$ is allowed to be infinitesimal. This result would seem to imply that the reactor with $\kappa_0 > 0$ would be asymptotically stable in the large regardless of the value of κ_1, even if $\kappa_0 + \kappa_1 < 0$. But we know that the reactor is monotonically unstable if the latter inequality holds, seemingly contradicting the theorem in the absence of $\kappa_0 + \kappa' I > 0$.

(b) It is observed that the necessary and sufficient conditions for (10) to be satisfied can be written as

$$f_k(\kappa_j[1 - (p/\eta_j)] - 2A_j\eta_j, \kappa_0, \eta_j) \geqslant 0 \tag{13}$$

where $f_k(\kappa_j, \kappa_0, \eta_j) \geqslant 0$ are the necessary and sufficient conditions for the reality of q_j (cf 7.7, Eq. 26a) in

$$\kappa_j = 2q_j \left\{ \sqrt{\kappa_0} + \sum_{i=1}^{N} [q_i \eta_i / (\eta_i + \eta_j)] \right\}$$

Therefore, the parameter region described by (13) can again be obtained by modifying the boundaries of the region determined by $f_k(\kappa_j, \kappa_0, \eta_j)$ as discussed in Sect. 7.7B. The region for asymptotic stability in the large is of course determined by the intersection of this region and that defined by $\kappa_0 + \kappa' I > 0$.

We now ask whether the condition (13), in which p is allowed to be any nonnegative number, yields weaker conditions on the parameters κ_j, κ_0 for fixed η_j (a larger parameter region) than the Lurie–Letov condition, in which $p = 0$. The answer is of course obvious, because (13) contains the Lurie–Letov condition as a special case, and therefore must cover at least the same, if not larger, parameter regions. This question has been discussed in detail for the two- and three-temperature models [8, 27].

Since the conditions for asymptotic stability in the large (ASL) with $p \geq 0$ are independent of the equilibrium power level, the region for ASL in parameter space can at most be equal to the parameter region for linear stability at all power levels. Therefore, if the conditions with $p = 0$ already cover this maximal parameter region, then the extention to $p \geqslant 0$ certainly cannot yield any weaker condition for ASL. It has been shown that this indeed happens in the case of the two-temperature reactor model ($N = 2$) with $\kappa_0 = 0$. When $\kappa_0 = 0$ and $N \geqslant 3$, the maximal parameter region is not covered by $p = 0$, and hence one may extend the parameter region for ASL by allowing $p \geqslant 0$ as in (13). It has indeed been shown [8] that, if $\kappa_0 > 0$, $N \leqslant 2$, the entire region for linear stability at all power levels is also asymptotically stable. The same conclusion has also been shown [27] to be true for $\kappa_0 = 0$, $N = 3$ (Problem 9).

The value of the Lurie–Letov, Lefschetz, or Popov criteria rests ultimately on the existence of two positive-definite matrices (the G and F matrices of Lefschetz) we now ask whether there is some way of avoiding the necessity of choosing any such matrices at all. The answer is yes, and the method yields, instead of the N inequalities $f_k > 0$, a single function $P(\omega)$ of order $2N$ in the frequency variable ω. The positiveness of the function $P(\omega)$ for all ω will yield precisely the N inequalities $f_k > 0$ and in a much simpler way.

7.11. Criteria for Asymptotic Stability in the Frequency Domain

The purpose of this section is to derive criteria for global asymptotic stability of a reactor without delayed neutrons, described by

$$\dot{\sigma} = -\kappa' y - \kappa_0 \phi(\sigma) \tag{1a}$$

$$\dot{y} = D(I\phi(\sigma) - y) \tag{1b}$$

where $\kappa_j \neq 0$ for all j, $\kappa_0 \geqslant 0$, $\kappa_0 + \kappa' I > 0$, and $D = \operatorname{diag}(\eta_1, \eta_2, ..., \eta_n)$, with $\eta_j > 0$ and distinct for all j. We have already encountered this system in Sec. 7.10. The connection between the frequency criteria to be obtained in this section and the algebraic conditions derived in Sec. 7.7 and expressed as a set of inequalities $f_k(\kappa_0, \eta_i, \kappa_i) > 0$ (cf 7.7, Eq. 26) will also be established.

We have already derived a frequency criterion for global asymptotic stability of reactors without delayed neutrons and with an arbitrary linear feedback in Sec. 7.5, which was expressed (cf 7.5, Eqs. 40a and 40b) as

$$q \operatorname{Re}\{H(i\omega)\} + (1/\omega) \operatorname{Im}\{H(i\omega)\} \leqslant 0, \qquad \omega \neq 0 \tag{2a}$$

with $q \geqslant 0$, or (cf 7.5, Eq. 40g) as

$$\operatorname{Re}\{H(i\omega)\} + (p/\omega) \operatorname{Im}\{H(i\omega)\} \leqslant 0, \qquad \omega \neq 0 \tag{2b}$$

with $p > 0$.

In these equations, $H(i\omega)$ is the feedback transfer function. The feedback mechanism described by (1b) is clearly linear, and is characterized by the following feedback kernel:

$$G(t) = -[\kappa_0\, \delta(t) + \kappa'\, e^{-Dt}\eta], \qquad \eta \equiv DI \tag{3}$$

(cf 7.7, Eq. 8b). The feedback transfer function associated with (3) is

$$H(i\omega) = -[\kappa_0 + \kappa'(i\omega E + D)^{-1} DI] \tag{4}$$

Substituting (4) into (2), we have explicitly

$$q[\kappa_0 + \kappa' D(E\omega^2 + D^2)^{-1} D\eta] - p\kappa'(E\omega^2 + D^2)^{-1} DI \geqslant 0 \tag{5}$$

where either $q \geqslant 0$ and $p = 1$, or $q = 1$ and $p > 0$.

It is important to remember that the condition (2a) or (2b) must be supplemented (cf 7.5, Eq. 40a) by

$$H(0) < 0 \tag{6}$$

which is required for the existence of a unique finite power level. This condition reduces to

$$\kappa_0 + \kappa' I > 0 \tag{7}$$

in the present model, and already assumed in (1).

Making use of the particular form of the feedback mechnism in the system described by (1), we shall be able, in this section, to extend the frequency condition (2b) by allowing p to be zero. Such an extension is possible because the feedback mechanism in (1) is a lumped-parameter system, whose description involves only a finite number of linear differential equations, whereas Corduneanu's theorem is applicable to distributed-parameter systems as well as lumped-parameter systems.

The derivation of this extended frequency condition for global asymptotic stability of lumped-parameter systems is based on a lemma by Kalman [28]. We shall first present the proof of this lemma as a mathematical preliminary.

Lemma 1 (Kalman). Given a real number $\kappa_0 \geqslant 0$, two real N-vectors η and κ, and a real unstable $N \times N$ matrix D ($-D$ stable, i.e., all the eigenvalues of D are positive) such that† $\det[\eta, D\eta, D^2\eta, \ldots, D^{N-1}\eta] \neq 0$, then a real N-vector q and an $N \times N$ matrix F satisfying

$$D'F + FD = qq' \tag{8a}$$

$$2\sqrt{\kappa_0}\, q = \kappa - 2FDI - p(D')^{-1}\kappa \tag{8b}$$

exist if and only if

$$P(i\omega) \equiv \kappa_0 + \mathrm{Re}\{\kappa'(E - pD^{-1})(D + i\omega E)^{-1}\eta\} \geqslant 0 \tag{9}$$

holds for all ω.

We observe that F in (8a) must be nonnegative-definite and symmetric by virtue of Lemmas 1 and 2 of Sec. 7.8 because $-D$ is stable by hypothesis and qq' is nonnegative-definite [use $x'qq'x = (\sum_i q_i x_i)^2$ with q_i real].

Although D is diagonal in our problem, the proof of the lemma does hold for nondiagonal D and arbitrary η and κ.

Proof of the Lemma. (a) *Necessity.* The transposed form of (8b) is

$$2\sqrt{\kappa_0}\, q' = \kappa' - 2\eta' F - p\kappa' D^{-1} = \kappa'(E - pD^{-1}) - 2\eta' F \tag{10}$$

† Kalman [29, 30], refers to this condition as the "complete controllability" of the pair (D, η). It implies that the Vectors $\eta, D\eta, \ldots, D^{N-1}\eta$ are linearly independent.

where we have used $F = F'$. Multiplying (10) by $(D + i\omega E)^{-1}\eta$ from the right, taking the real parts of the resulting equation, and rearranging, we obtain

$$\operatorname{Re}\{\kappa'(E - pD^{-1})(D + i\omega E)^{-1}\eta\}$$
$$= 2\sqrt{\kappa_0}\operatorname{Re}\{q'(D + i\omega E)^{-1}\eta\} + 2\operatorname{Re}\{\eta' F(D + i\omega E)^{-1}\eta\} \tag{11}$$

The last term in (11) can be expressed in terms of q' by using (8a). Multiplying the latter by $\eta'(D' - i\omega E)^{-1}$ from the left and by $(D + i\omega E)^{-1}\eta$ from the right, we obtain

$$|q'(D + i\omega E)^{-1}\eta|^2 = \eta'(D' - i\omega E)^{-1}(D'F + FD)(D + i\omega E)^{-1}\eta \tag{12}$$

The following operator identity†

$$(D' - i\omega E)^{-1}D' \equiv E + (D' - i\omega E)^{-1}i\omega \tag{13a}$$

and its transposed and conjugated from

$$D(D + i\omega E)^{-1} \equiv E - i\omega(D + i\omega E)^{-1} \tag{13b}$$

lead to

$$(D' - i\omega E)^{-1}(D'F + FD)(D + i\omega E)^{-1} \equiv F(D + i\omega E)^{-1} + (D' - i\omega E)F \tag{14}$$

from which we conclude that (multiply both sides by η' from the right, by η from the left, and recall that η and F are real, and F is symmetric)

$$|q'(D + i\omega E)^{-1}\eta|^2 = 2\operatorname{Re}\{\eta' F(D + i\omega E)^{-1}\eta\} \tag{15}$$

Substituting (15) into (11), adding κ_0 to both sides, and expressing the right-hand side as a complete square, we obtain

$$P(i\omega) = |q'(D + i\omega E)^{-1}\eta + \sqrt{\kappa_0}|^2 \geqslant 0 \tag{16}$$

which proves that $P(i\omega) \geqslant 0$ for all real ω provided a real n-vector q exists. This result establishes the necessity of the lemma.

(b) *Sufficiency.* In order to prove the sufficiency of (9), we must demonstrate the existence of a real N-vector q when $P(i\omega) \geqslant 0$ for all real ω. Following Kalman [28], we shall prove the existence of q by exhibiting a procedure for its construction.

† Consider $D' = D' - i\omega E + i\omega E$ and multiply both sides by $(D' - i\omega E)^{-1}$ from the left to obtain (13a). This identity is frequently used in fluctuation theory in connection with Langevin's technique (see, for example, Lax [31]).

Let a_k be the coefficient of s^k in the polynomial $\det(sE + D) \equiv \varphi(s)$, i.e.,

$$\varphi(s) = |\, sE + D \,| = \sum_{k=0}^{N} a_k s^k, \qquad a_N = 1; \quad a_0 = +|\, D \,| \tag{17}$$

and introduce the following N vectors:

$$\begin{aligned} e_1 &= \eta \\ e_2 &= (-D + a_{N-1})\eta \\ &\vdots \\ e_{N-1} &= [(-D)^{N-2} + a_{N-1}(-D)^{N-3} + \cdots + a_3(-D) + a_2]\eta \\ e_N &= [(-D)^{N-1} + a_{N-1}(-D)^{N-2} + \cdots + a_2(-D) + a_1]\eta \end{aligned} \tag{18a}$$

these vectors are linearly independent because they can be written as

$$\begin{aligned} e_2 &= -D\eta + a_{N-1}e_1 \\ e_3 &= -De_2 + a_{N-2}e_1 \\ &\vdots \\ e_N &= -De_{N-1} + a_1 e_1 \end{aligned} \tag{18b}$$

Since each e_j starts with $D^{j-1}\eta$, linear dependence of $e_1, e_2, \ldots, e_n$ would violate the linear independence of the vectors $\eta, D\eta, \ldots, D^{n-1}\eta$.

The vectors $\{e_j\}$ form a basis. The components of η and the elements of D relative to this basis can be obtained from (18b) by casting the latter into the following form:

$$\begin{aligned} De_1 &= -e_2 + a_{N-1}e_1 \\ De_2 &= -e_3 + a_{N-2}e_1 \\ &\vdots \\ De_{N-1} &= -e_N + a_1 e_1 \\ De_N &= a_0 e_1 \end{aligned} \tag{18c}$$

The last line follows from the fact that every matrix satisfies its own characteristic equation, and hence

$$\varphi(-D) \equiv (-D)^N + a_{N-1}(-D)^{N-1} + \cdots + a_1(-D) + a_0 E = 0 \tag{19}$$

Using $e_{ij} = \delta_{ij}$ in this basis, we obtain the elements of D as

$$D_{ij} = -\delta_{i,j+1} + a_{N-j}\,\delta_{i,1}\,, \qquad j = 1, 2,\ldots, N-1$$
$$D_{iN} = a_0\,\delta_{i1}$$

Hence, we have

$$D = \begin{bmatrix} a_{N-1} & a_{N-2} & \cdots & & a_1 & a_0 \\ -1 & 0 & & & 0 & 0 \\ 0 & -1 & & & 0 & 0 \\ \vdots & \vdots & & & \vdots & \vdots \\ 0 & 0 & & -1 & 0 & 0 \\ 0 & 0 & \cdots & 0 & -1 & 0 \end{bmatrix} \tag{20a}$$

and

$$\eta' = [1, 0,\ldots, 0] \tag{20b}$$

Let $q_1, q_2,\ldots, q_n$ denote the components of the N-vector q relative to the basis $e_1, e_2,\ldots, e_n$. Then, the polynomial $q'(sE + D)^{-1}\eta$ can easily be evaluated using the forms of D and η in (20a) and (20b), respectively:

$$q'(sE + D)^{-1}\eta = \sum_{j=1}^{N} q_j(sE + D)^{-1}_{j1} \tag{21a}$$

$$= \sum_{j=1}^{N} q_j[\Delta_{j1}/\varphi(s)] \tag{21b}$$

where the Δ_{j1} are the cofactors of the first row of the matrix $(sE + D)$, and are related to the minors of the respective elements by $(-1)^{j+1}M_{1j}$. Using (20a), we can easily verify that $\Delta_{11} = s^{N-1}$, $\Delta_{12} = s^{N-2},\ldots, \Delta_{1N-1} = s$, $\Delta_{1N} = 1$, and hence

$$q'(sE + D)^{-1}\eta = (q_1 s^{N-1} + q_2 s^{N-2} + \cdots + q_{N-1}s + q_N)/\varphi(s) \tag{22}$$

It is clear from (22) that the components of q relative to $e_1, e_2,\ldots, e_n$ can be identified as the coefficients of the numerator of the rational function $q'(sE + D)^{-1}\eta$. Thus, we can prove the existence of q if we can solve (16) for $q'(D + i\omega E)^{-1}\eta$ in terms of $P(i\omega)$, and express the solution as a rational function.

For this purpose, we shall show that $P(i\omega)$ defined by (9) can be written as

$$P(i\omega) \equiv \kappa_0 + \mathrm{Re}[\kappa'(E - pD^{-1})(D + i\omega E)^{-1}\eta]$$
$$= |\,\Theta(i\omega)|^2/|\,\varphi(i\omega)|^2 \tag{23}$$

where $\Theta(i\omega)$ is a polynomial in $i\omega$ of degree N with real coefficients and $\varphi(i\omega)$, as we recall, is the determinant of $(D + i\omega E)$.

We first note that

$$\text{Re}(D + i\omega E)^{-1} = (D^2 + \omega^2 E)^{-1} D \tag{24a}$$

and

$$|\varphi(i\omega)|^2 = |D^2 + \omega^2 E| \tag{24b}$$

Thus, the numerator of the left-hand side of (23) is the following polynomial:

$$\Gamma(\omega^2) = P(i\omega)|D^2 + \omega^2 E| \tag{25}$$

Since $\Gamma(\omega^2)$ has real coefficients and is nonnegative by hypothesis $[P(i\omega) \geqslant 0]$ for all ω, its zeros λ_k are complex conjugate pairs, or of even multiplicity if they are real and positive, or negative with arbitrary multiplicity. Thus, $\Gamma(\omega^2)$ can be written as the product of factors of the following form:

$$|\omega^2 + \delta^2|^2, \qquad (\omega^2 - \alpha)^{2n}, \qquad (\omega^2 + \beta^2)$$

where δ is complex, $\alpha > 0$, and β is real. Each of these factors can be decomposed as

$$(\delta + i\omega)(\delta^* + i\omega)(\delta - i\omega)(\delta^* - i\omega),$$

$$[\alpha + (i\omega)^2]^n[\alpha + (i\omega)^2]^n, \qquad (\beta + i\omega)(\beta - i\omega)$$

If we consider the product of factors like

$$(\delta + i\omega)(\delta^* + i\omega), \qquad [\alpha + (i\omega)^2]^n, \qquad (i\omega + \beta)$$

we obtain a polynomial $\Theta(i\omega)$ in $i\omega$ such that $|\Theta(i\omega)^2| \equiv \Gamma(\omega^2)$. Since

$$(\delta + i\omega)(\delta^* + i\omega) = [|\delta|^2 + 2i\omega \,\text{Re}\,\delta + (i\omega)^2]$$

$\Theta(i\omega)$ has only real coefficients.

The forgoing discussion shows that there always exists a $\Theta(i\omega)$, a polynomial in $i\omega$ with real coefficients, such that (23) is true. The choice of $\Theta(i\omega)$ is not unique because one can arrange the factors of $\Gamma(\omega^2)$ in different ways.† The important point is that we can always construct $\Theta(i\omega)$ by the procedure described above.

Substituting (23) into (16) and setting $s = i\omega$, we find

$$\varphi(s)[q'(D + sE)^{-1}\eta] = \Theta(s) - \varphi(s)\sqrt{\kappa_0} \tag{26}$$

† Indeed, $(D + i\omega E)^{-1} \equiv (D + i\omega E)^{-1}(D - i\omega E)^{-1}(D - i\omega E) = (D^2 + \omega^2 E)^{-1}(D - i\omega E)$.

It can be seen from (23) that the leading coefficient of $\Theta(s)$ (the coefficient of s^N) is $\sqrt{\kappa_0}$, whereas the leading coefficient of $\varphi(s)$ is unity (cf Eq. 17). Hence, the right-hand side of (26) is a polynomial in s of degree $N - 1$. If we compare (26) to (22), we can actually determine the components of q as the coefficients of the descending powers of s, with respect to the basis $e_1, e_2, \ldots, e_N$. This construction procedure completes the proof of the existence of a real N-vector q when $P(i\omega) \geqslant 0$ for all real ω, and hence establishes the sufficiency of the lemma.

Kalman's main lemma can also be stated in a generalized form as follows.

LEMMA 2. Given a real number $\kappa_0 \geqslant 0$, two real N-vectors η and κ, a real unstable matrix D, and a symmetric, positive-definite matrix R such that (D, η) is completely controllable, then a real N-vector q and an $N \times N$ matrix F satisfying

$$D'F + FD = qq' + \epsilon R \tag{27a}$$

$$2\sqrt{\kappa_0}q = \kappa - 2F\eta - p(D')^{-1}\kappa \tag{27b}$$

exist if and only if $\epsilon > 0$ is sufficiently small and

$$P(i\omega) \equiv \kappa_0 + \text{Re}[\kappa'(E - pD^{-1})(D + i\omega E)^{-1}\eta] > 0 \tag{27c}$$

is satisfied for all ω, and for some $p \geqslant 0$.

Clearly, (27a) implies $F = F'$ and $F > 0$ because of $R > 0$. The proof of this form of the lemma is similar to that of the previous form. The only difference is that (15) in the proof of necessity is now replaced by

$$P(i\omega) \equiv |\, q'(D + i\omega E)^{-1}\eta + \sqrt{\kappa_0}\,|^2 + \epsilon A^2 \tag{28a}$$

where

$$A^2 \equiv \eta'(D' - i\omega E)^{-1}R(D + i\omega E)^{-1}\eta \tag{28b}$$

and (23) by

$$P(i\omega) - \epsilon A^2 = |\,\Theta(i\omega)|^2/|\,\varphi(i\omega)|^2 \tag{29}$$

Since $P(i\omega) > 0$ (strictly positive) this time, $P(i\omega) - \epsilon A^2$ can be made nonnegative so that a polynomial $\Theta(i\omega)$ can be constructed to satisfy (29). The procedure of finding $\Theta(i\omega)$, which is the basis for the sufficiency proof of the lemma, is identical to that used in the previous case (cf Eq. 25).

We shall use the original form of the lemma, which allows the equality in the frequency condition $P(i\omega) \geqslant 0$ because it permits us to establish a weaker condition for global asymptotic stability, and illustrates the fine points of the Lyapunov stability theorems. The price we have to pay, however, is a more complicated analysis, though a more instructive one.

We are now ready to derive a frequency condition for global asymptotic stability of a reactor described by (1) using the original form of Kalman's lemma. We note that D and η in (1) are of a special form, i.e.,

$$D = \mathrm{diag}(\eta_i) \qquad \text{and} \qquad \eta = DI = \mathrm{col}(\eta_i)$$

Since the η_j are all assumed to be positive, $-D$ is stable, as required by the hypothesis of the lemma. Furthermore, the determinant $\det[\eta, D\eta, \ldots, D^{N-1}\eta]$ reduces in this case to

$$|D| \begin{vmatrix} 1 & \eta_1 & \eta_1^2 & \cdots & \eta_1^{N-1} \\ 1 & \eta_2 & \eta_2^2 & & \eta_2^{N-1} \\ \vdots & & & & \\ 1 & \eta_N & \eta_N^2 & & \eta_N^{N-1} \end{vmatrix} \tag{30}$$

as one can see from $D^{n-1}\eta = \mathrm{col}(\eta_1^n, \ldots, \eta_N^n)$. If all $\eta_1, \eta_2, \ldots, \eta_N$ are distinct and are strictly positive, this determinant is nonzero, and the requirement of Kalman's lemma is fulfilled.

We now construct a function $V(y, \sigma)$ which is a quadratic form in (y, σ) plus the integral of $\phi(\sigma)$:

$$V = y'Fy + [p/2(\kappa_0 + \kappa' I)](\kappa' D^{-1} y - \sigma)^2 + \int_0^\sigma \phi(\sigma')\, d\sigma' \tag{31}$$

where $F \geqslant 0$ (nonnegative-definite, i.e., the quadratic form $y'Fy$ is positive-semidefinite) and $F' = F$ (symmetric). Furthermore, $\kappa_0 + \kappa' I > 0$ (required for the existence of a finite power level) and $p \geqslant 0$. The derivative of V along a trajectory of (1) is

$$-\dot{V} = y'Gy + \kappa_0 \phi^2(\sigma) + y'(\kappa - pD^{-1}\kappa - 2F\eta)\phi(\sigma) + p\sigma\phi(\sigma) \tag{32a}$$

where

$$G \equiv DF + FD \tag{32b}$$

Since $\kappa_0 \geqslant 0$ (stabilizing power coefficient), we introduce a real q-vector as

$$2\sqrt{\kappa_0}\, q \equiv \kappa - 2F\eta - pD^{-1}\kappa \tag{33}$$

In terms of q, (32a) becomes

$$-\dot{V} = y'(G - q'q)y + [\sqrt{\kappa_0}\,\phi(\sigma) + q'y]^2 + p\sigma\phi(\sigma) \tag{34}$$

By its construction, V is nonnegative, i.e., $V \geqslant 0$. On the other hand, for $\kappa_0 > 0$, $-\dot{V} \geqslant 0$ if and only if $G - qq' \geqslant 0$ (nonnegative-definite). For $\kappa_0 = 0$, $-\dot{V} \geqslant 0$ if and only if $G \geqslant 0$ and $\kappa - 2F\eta = pD^{-1}\kappa$, as can be seen from (32a).

Suppose that the system described by (1) satisfies the frequency condition (9), which can also be written as

$$-P(i\omega) \equiv \mathrm{Re}[H(i\omega)] + (p/\omega)\,\mathrm{Im}[H(i\omega)] \leqslant 0 \tag{35a}$$

for some $p \geqslant 0$ and for all ω. Then, Kalman's lemma guarantees the existence of a real q and a symmetric, nonnegative matrix F satisfying

$$DF + FD = qq' \tag{35b}$$

$$2\sqrt{\kappa_0}\,q = \kappa - 2F\eta - pD^{-1}\kappa \tag{35c}$$

With this choice of q, (32b) implies $G - q'q = 0$ in (34):

$$-\dot{V} = [\sqrt{\kappa_0}\,\phi(\sigma) + q'y]^2 + p\sigma\phi(\sigma) \tag{36}$$

Hence, the condition $P(i\omega) \geqslant 0$ is sufficient to construct a positive-semidefinite function V with a negative-semidefinite first derivative. This information is sufficient to prove the boundedness of the solution even though V is not positive-definite.

BOUNDEDNESS Since $\dot{V} \leqslant 0$, $V(t)$ is nonincreasing, so that

$$V(t) \leqslant V(0) \tag{37}$$

holds along a trajectory. Using (31) and the fact that each term in it is nonnegative, we find

$$\int_0^{\sigma} \phi(\sigma')\,d\sigma' \leqslant V(0) \tag{38}$$

which proves the boundedness of $\sigma(t)$. However, we cannot draw the same conclusion, at least easily, for $\| y(t)\|$ though we have $y'Fy \leqslant V(0)$, because F is only nonnegative-definite and $\| y(t)\|$ can become large with $y'Fy$ remaining bounded. Using the boundedness of $\sigma(t)$ and $\phi[\sigma(t)]$ and (1b), we can still prove that $\| y(t)\|$ is bounded. Indeed, Eq. (1b) shows that

$$y_j(t) = y_j(0)\,e^{-\eta_j t} + \int_0^t du\, e^{-\eta_j u}\eta_j\phi[\sigma(t-u)] \tag{39a}$$

from which we have

$$| y_j(t)| \leqslant | y_j(0) e^{-\eta_j t} | + \phi_M \tag{39b}$$

where ϕ_M is the upper bound of $\phi[\sigma(t)]$. Hence, we can conclude that the Kalman condition $P(i\omega) \geqslant 0$ is sufficient for the boundedness of $\| y \| + | \sigma |$ regardless of whether $p = 0$ or $p > 0$, and $\kappa_0 = 0$ or $\kappa_0 > 0$.

It is interesting to point out in passing that the extended form of the Kalman lemma with the modified condition $P(i\omega) > 0$ would be sufficient to construct a positive-definite V-function because $F > 0$ in this case. Then, $V(t) \leqslant V(0)$ would also imply a bounded $\| y(t)\|$ without recourse to the equation of motion.

GLOBAL ASYMPTOTIC STABILITY. In order to prove global asymptotic stability, we must first seek a Lyapunov function V which is positive-definite. For this purpose, we prove the following lemma.

LEMMA 3. The matrix $F + \alpha D^{-1}\kappa\kappa' D^{-1}$ is positive-definite if $\alpha > 0$; $\kappa_j \neq 0$; and $\eta_j > 0$ are distinct.

PROOF. From (35b) (cf Lemma 1, Sec. 7.8) we have

$$F = \int_0^\infty e^{-Dt} qq' e^{-Dt} \, dt \tag{40}$$

from which we obtain

$$y'Fy = \int_0^\infty y' e^{-Dt} qq' e^{-Dt} y \, dt = \int_0^\infty [q' e^{-Dt} y]^2 \, dt \tag{41}$$

Hence, if $y'Fy = 0$, then

$$q' e^{-Dt} y = 0 \tag{42}$$

holds for all t. Substitution of q' from (35c) into (42) yields

$$\kappa'(E - pD^{-1}) e^{-Dt} y - 2\eta' F e^{-Dt} y \equiv 0 \tag{43}$$

Using (40), we can show that the last term in (43) is identically zero. Indeed, multiplying (40) from the right by $e^{-Dt}y$, we get

$$F e^{-Dt} y = \int_0^\infty du \, e^{-Du} qq' e^{-D(t+u)} y(t) \tag{44}$$

But $q' e^{-D(t+u)} y \equiv 0$ by virtue of (42). Hence, (43) reduces to

$$\kappa' e^{-Dt} y \equiv p\kappa' D^{-1} e^{-Dt} y \tag{45}$$

If we can show that $\kappa' D^{-1}y$ cannot be zero unless $y = 0$, then the positive-definiteness of $F + \alpha D^{-1}\kappa\kappa' D^{-1}$ will be established [recall that y in (45) satisfies $y'Fy = 0$]. If $\kappa' D^{-1}y$ is zero, then (45) yields $\kappa' y = 0$, $\kappa' Dy = 0, \ldots, \kappa' D^n y = 0, \ldots$, by differentiation at $t = 0$. Since $\kappa' D^n y = 0$ for any $n = 0, 1, 2, \ldots$, we have

$$\kappa' \, e^{-Dt} y \equiv 0 \tag{46}$$

But this identity can be satisfied only if $y = 0$ because (D, κ) is completely controllable, i.e.,

$$\det(\kappa, D\kappa, \ldots, D^{N-1}\kappa) = \prod_i \kappa_i \det(I, DI, \ldots, D^{N-1}I)$$

Since none of the κ_i are zero, and all η_i are distinct,

$$\det(\kappa, D\kappa, \ldots, D^{N-1}\kappa) \neq 0$$

Hence, $y'Fy$ and $(y'D^{-1}\kappa)(\kappa' D^{-1}y) = (\kappa' D^{-1}y)^2$ cannot vanish simultaneously unless $y = 0$.

The foregoing conclusion is still valid even if $\kappa_0 = 0$. In this case, we have

$$DF + FD = qq' \tag{47a}$$

$$\kappa - 2F\eta - pD^{-1}\kappa = 0 \tag{47b}$$

Substituting F from (40) into (47b), we get

$$\kappa'(E - pD^{-1}) = 2\eta' \int_0^\infty e^{-Du} qq' \, e^{-Du} \, du \tag{48}$$

Multiplying (48) by $e^{-Dt}y$ from the right and using (42), we obtain

$$\kappa'(E - pD^{-1}) \, e^{-Dt} y = 0 \tag{49}$$

which is identical to (45). The rest of the proof is unchanged.

We now return to the question of global asymptotic stability. We shall distinguish two cases, as follows.

(a) $p \neq 0$. In this case, V is positive-definite because it can vanish only when $\sigma = 0$ and $y = 0$ by virtue of the second lemma. Since we have already established the positive-definiteness of V, and the boundedness of the solutions, we can use Theorem 5 of Sec. 7.6 to secure global asymptotic stability if we can also show that the set of points defined by $\dot{V} = 0$ does not contain any nontrivial half-trajectory which can be entirely in its positive limiting set.

By (36), $\dot{V} = 0$ only if $\sigma(t) \equiv 0$ and $q'y(t) \equiv 0$. Substituting $\sigma(t) = \phi(\sigma(t)) \equiv 0$ into (1b), we obtain

$$y(t) = e^{-Dt}y(0) \tag{50}$$

Equation (1a) yields

$$\kappa' e^{-Dt}y(0) \equiv 0 \tag{51}$$

which can only be satisfied if $y(0) = 0$. Hence, the only half-trajectory in the set of $\dot{V} = 0$ is $\sigma(t) \equiv 0$, $y(t) \equiv 0$.

(b) $p = 0$. In this case, V is again positive-definite because $V = 0$ only if $\sigma = 0$ and $y'Fy = 0$. But the latter implies $q'e^{-Dt}y(t) \equiv 0$ and thus $\kappa'e^{-Dt}y \equiv 0$ as shown in the third lemma (cf Eq. 45, $p = 0$). Complete controllability of (D, κ) implies $y = 0$.

In order to use Theorem 5 of Sec. 7.6, we must again show that there is no half-trajectory other than the origin in the set $\dot{V} = 0$.

By (36), $\dot{V} = 0$ implies

$$\sqrt{\kappa_0}\phi(\sigma(t)) \equiv -q'y(t) \tag{52}$$

Taking the Laplace transform of (39a) and (52), and eliminating $\tilde{y}(s)$, we have

$$\tilde{\phi}(s) = q'(sE + D)^{-1}y(0)/[\sqrt{\kappa_0} + q'(sE + D)^{-1}\eta] \tag{53}$$

where $\tilde{\phi}(s)$ is the Laplace transform of $\phi(\sigma(t))$. Since we already established the boundedness of $\sigma(t)$, $\phi(\sigma(t))$ can lie in its own positive limit set only if it is almost periodic. Hence, at least one pair of roots of the denominator must be pure imaginary, i.e.,

$$\sqrt{\kappa_0} + q'(i\omega_k E + D)^{-1}\eta = 0 \tag{54}$$

must hold for some $\mp\omega_k \neq 0$. Substitution of (54) into (16) shows that $P(i\omega_k) = 0$. This implies that the frequency condition (35a), which is given in the present case ($p = 0$) by

$$\operatorname{Re} H(i\omega) \leqslant 0 \tag{55}$$

holds with the equality sign at $\omega = \omega_k$.

Consequently, if $\operatorname{Re} H(i\omega) < 0$ holds for all ω, then global asymptotic stability is established without any further discussion.

Suppose that the equality occurs at $\omega = \omega_k$. Then, (53) implies that $\phi(\sigma(t))$ is a sinusoidal function of time which can be put into the following form:

$$\phi[\sigma(t)] = \alpha \cos(\omega_k t) \tag{56}$$

If (54) has more than one pair of roots, then (56) will contain as many terms as the number of ω_k. The important point is that there can exist only a finite number of ω_k. For simplicity, we assume only one ω_k.

We now substitute (56) into the integrodifferential form of the system (1), i.e.,

$$\dot{\sigma}(t) = \int_0^\infty du\, G(u)\phi[\sigma(t-u)] \tag{57}$$

and obtain the periodic solution

$$\sigma(t) = \sigma_{\text{ave}} + \{\alpha[\text{Im}\, H(i\omega_k)](\cos \omega_k t)/\omega_k\} \tag{58}$$

where σ_{ave} is a constant depending on the initial values. In (58), we have used $\text{Re}\, H(i\omega_k) = 0$. Since $\phi(\sigma) = e^\sigma - 1$, we are led to the following identity:

$$\exp(\sigma_{\text{ave}} + \{\alpha[\text{Im}\, H(i\omega_k)]/\omega_k\} \cos \omega_k t) - 1 \equiv \alpha \cos \omega_k t \tag{59}$$

We already know that $\exp(\alpha \cos \omega_k t)$ contains infinitely many harmonics. Hence, the above identity can only be satisfied with $\sigma_{\text{ave}} = \alpha = 0$. The latter implies $\sigma(t) = 0$, $y(t) = 0$, proving that the set of points for which $\dot{V} = 0$ does not contain any half-trajectory but the equilibrium solution.

The foregoing results can be stated as the following theorem.

THEOREM 1. If $\kappa_j \neq 0$, $\eta_j > 0$ and distinct, if

$$-H(0) \equiv \kappa_0 + \kappa' I > 0 \tag{60a}$$

and if

$$[\text{Re}\, H(i\omega)] + (p/\omega)\, \text{Im}\, H(i\omega) \leqslant 0 \tag{60b}$$

for some $p \geqslant 0$ and for all ω, then the equilibrium $\sigma \equiv 0$, $y(t) \equiv 0$ of (1) is asymptotically stable in the large.

DISCUSSION

(a) This theorem is applicable to lumped-parameter systems which are described by a set of equations of the form of (1). This fact was explicitly used in the proof of the theorem in order to eliminate limit cycles (cf Eq. 59) which may arise if Re $H(i\omega)$ can vanish. In a lumped-parameter system, Re $H(i\omega) = 0$ is a polynomial equation and can have only a finite number of zeros, and therefore the identity in (59) cannot be satisfied with finite amplitudes.

The above remark explains the seemingly contradictory result that Ergen's circulating fuel reactor model, discussed in Sec. 7.5, admits semistable limit cycles even though Re $H(i\omega) \leqslant 0$ and $H(0) < 0$, whereas a reactor satisfying the latter would be asymptotically stable according to the above theorem. The explanation lies in the fact that the circulating fuel reactor is an distributed-parameter system, and thus outside the range of applicability of the theorem. Interestingly enough, Re $H(i\omega) = 0$ has infinitely many zeros, at $\omega = \omega_0$, $2\omega_0$,... for this reactor model, implying that reasoning such as used in (59) would not apply.

(b) The condition (2a) due to Corduneanu, which is applicable to both distributed- and lumped-parameter systems, is obtained by dividing (60b) by p and letting $q = (1/p)$. Since $q = 0$ is allowed in his derivation (cf Sec. 7.5), one concludes that (60b) can be replaced in a lumped-parameter system by

$$q[\text{Re } H(i\omega)] + (p/\omega) \text{ Im } H(i\omega) \leqslant 0$$

in which $p + q > 0$, $p \geqslant 0$, and $q \geqslant 0$. This conclusion could also be obtained in the framework of the Kalman approach presented above. The interested reader is referred to Kalman's paper [28], or to Lefschetz's book [20].

As we recall, in the case of distributed-parameter systems, we cannot allow p to be zero.

(c) In control theory, one is interested in global asymptotic stability of systems described by (1) in which the function $\phi(\sigma)$ belongs a wide class of functions [i.e., $\phi(0) = 0$, $\phi(\sigma) \neq 0$ for $\sigma \neq 0$; $\sigma\phi(\sigma) > 0$]. Since $\phi(\sigma)$ is allowed to be completely linear, or to have linear portions in symmetric intervals such as $(-\sigma_0, \sigma_0)$, the identity (59) can be satisfied with finite values of α. Indeed, when $\phi(\sigma) \equiv m\sigma$ $(m > 0)$, (59) can be satisfied with $\sigma_{\text{ave}} = 0$ and arbitrary α if

$$m[\text{Im } H(i\omega_k)]/\omega_k = 1 \tag{61}$$

This and Re $H(i\omega_k) = 0$ together are precisely the conditions for linear instability. Since (61) can always be satisfied for some value of m (m corresponds to the equilibrium power level in our case, which was absorbed in H) if Im $H(i\omega_k)]/\omega_k > 0$, Kalman eliminates this possibility by requiring

$$[\text{Im } H(i\omega)]/\omega \leqslant 0 \tag{62}$$

to hold for the values of ω at which Re $H(i\omega) = 0$.

Since the form of $\phi(\sigma)$ is known in reactor applications, one can dispense with the condition (62).

(d) The proof of Theorem 1 would be simpler if we did not allow the equality in the frequency condition $P(i\omega) > 0$. Then, we could use the extended form of Kalman's lemma, (lemma 2) to construct a positive definite V-function with a negative-definite $\dot{V}$. This approach, which was taken by Lefschetz [26], somewhat simplifies the proof of the theorem by avoiding the difficulties arising from working with semidefinite V-functions, even though it leads to a slightly more restrictive condition for asymptotic stability in the large.

(e) It is interesting to compare the frequency condition $P(i\omega) > 0$ to the Popov condition given by (7.10, Eq. 10) using Lemma 2. We recall from (27) that the choice of the matrix R is arbitrary so long as it is positive-definite. Furthermore, the condition $P(i\omega) > 0$ guarantees the existence of a real q and a symmetric $F > 0$ satisfying (27). Since R is arbitrary, we can choose it as

$$\epsilon R_{ij} = (\eta_i + \eta_j) A_i \, \delta_{ij} \tag{63}$$

where $A_i > 0$ but otherwise arbitrary. Since R does not appear in the frequency condition $P(i\omega) > 0$, this choice does not introduce any restriction on the lemma. We can now solve (27a) for F as

$$F_{ij} = [q_i q_j/(\eta_i + \eta_j)] + A_i \, \delta_{ij} \tag{64}$$

and substitute it into (27b). Then, the latter becomes

$$2\sqrt{\kappa_0} q_j = \kappa_j[1 - (p/\eta_j)] - 2q_j \sum_{i=1}^{N} [q_i \eta_i/(\eta_j + \eta_i)] - \eta_j A_j \tag{65}$$

According to Lemma 2, the condition $P(i\omega)$ is necessary and sufficient for the existence of real q_j for sufficiently small $A_j > 0$. But (65) is identical to (7.10, Eq. 10), which is a generalization of the Lurie–Letov condition (cf Sec. 7.7, Theorem 1). On the other hand, the sufficient and necessary conditions for the existence of real q_j in (7.10, Eq. 10) for some

$p \geqslant 0$ and for sufficiently small $A_j > 0$ were expressed in the form of a set of inequalities

$$f_i\{\kappa_j[1 - (p/\eta_j)] - 2A_j\eta_j\,,\, \eta_j\,,\, \kappa_0\} > 0$$

Hence, the frequency condition $P(i\omega)$ and the above algebraic conditions are equivalent in the sense that they both cover the same parameter region in the parameter space.

(f) We may demonstrate, as a consistency check, that the frequency condition (60b) automatically implies linear stability at all equilibrium power levels, which is a necessary condition for asymptotic stability in the large. Since neither (60a) nor (60b) of Theorem 1 depend on P_0 (recall that κ_0, κ_j, and hence H, are proportional to P_0), they guarantee asymptotic stability in the large at all power levels.

From the linear theory, we have the characteristic equation in the absence of delayed neutrons as

$$s - P_0 H(s) = 0 \tag{66}$$

where we have put the proportionality of H to P_0 in evidence. Let $s = i\omega + \alpha$ and $H(i\omega + \alpha) = H_R(\omega, \alpha) + iH_I(\omega, \alpha)$. Then, (66) implies

$$\alpha - P_0 H_R(\omega, \alpha) = 0 \tag{67a}$$

$$\omega - P_0 H_I(\omega, \alpha) = 0 \tag{67b}$$

Since α and ω vanish as P_0 when $P_0 \to 0$, we have $\alpha \approx P_0 H_R(0, 0) < 0$ and $\omega \approx 0$ for small P_0 [note $H_I(0, 0) = 0$]; $H_R(0, 0) < 0$ follows from $H(0) < 0$. Hence, at P_0, the characteristic equation has a zero at the origin, but for sufficiently small $P_0 > 0$, the roots have negative real parts.

Since $H(s)$ is analytic for Re $s \geqslant 0$, (66) defines a continuous function $s(P_0)$ where Re $s \geqslant 0$. Thus, the root must cross the imaginary axis as a continuous function of P_0 for linear instability to occur. If we require that

$$[\operatorname{Im} H(i\omega)]/\omega \leqslant 0 \qquad \text{when} \quad \operatorname{Re} H(i\omega) = 0 \tag{68}$$

then, there can be no value of P_0 at which the root locus crosses the imaginary axis. Therefore, (68) is a necessary and sufficient condition for linear stability at all power levels.

We now return to the frequency condition (60b), i.e.,

$$\operatorname{Re} H(i\omega) + (p/\omega) \operatorname{Im} H(i\omega) \leqslant 0$$

If this is satisfied for some $p > 0$, then (68) must necessarily be true.

If $p = 0$, then we have $\operatorname{Re} H(i\omega) \leqslant 0$. Since $H(s)$ has no poles for $\operatorname{Re} s \geqslant 0$, this condition implies that $H(i\omega)$ is a negative real function, and hence $\operatorname{Re} H(s) = H_R(\omega, \alpha) \leqslant 0$ for all ω and $\alpha \geqslant 0$ (cf 7.3, Eq. 19). Thus, if $\operatorname{Re} H(i\omega) \leqslant 0$, α in (67a) can never be positive, and the reactor can never be linearly unstable. However, $\alpha = 0$ (neutral stability) is allowed. In this case, the nonlinear nature of the reactor eliminates the possibility of periodic oscillations.

We have already mentioned that $\operatorname{Re} H(i\omega)$ can be positive in a given reactor model; then, the frequency condition (60b) can be satisfied for some $p > 0$ only if $\operatorname{Im} H(i\omega) \leqslant 0$ for all ω for which $\operatorname{Re} H(i\omega) \geqslant 0$. Otherwise, one cannot ascertain the asymptotic stability of the given reactor model from the above frequency condition.

We shall discuss as an illustration the variation of $\operatorname{Re} H(i\omega)$ and $\operatorname{Im} H(i\omega)$ in the case of a two-temperature ($N = 2$) reactor model with $\kappa_0 > 0$. In this reactor model, we have

$$-[\operatorname{Re} H(i\omega)] = (\omega^2 + \eta_1{}^2)^{-1}(\omega^2 + \eta_2{}^2)^{-1} \times \{\kappa_0\omega^4 + [\kappa_0(\operatorname{Tr} D^2) + \kappa' D^2 I]\omega^2 + |D|^2(\kappa_0 + \kappa' I)\} \tag{69}$$

$$[\operatorname{Im} H(i\omega)]/\omega = (\omega^2 + \eta_1{}^2)^{-1}(\omega^2 + \eta_2{}^2)^{-1}[\kappa' D I \omega^2 + |D^2| \kappa' D^{-1} I] \tag{70}$$

where $\kappa_0 + \kappa' I > 0$ is assumed to satisfy (60a). We note that $-\operatorname{Re} H(i\omega) \to \kappa_0 > 0$ as $\omega \to \infty$, and $\operatorname{Re} H(0) < 0$. Furthermore, the numerator of (69) is quadratic in ω^2, and hence $\operatorname{Re} H(i\omega) = 0$ can have at most only two zeros for $\omega^2 \geqslant 0$. Similarly, (70) can have only one zero for $\omega^2 \geqslant 0$. In Figure 7.11.1, we consider the case in which $\operatorname{Re} H = 0$ has two zeros, because otherwise $\operatorname{Re} H(i\omega) \leqslant 0$, and $p = 0$ guarantees asymptotic stability in the large.

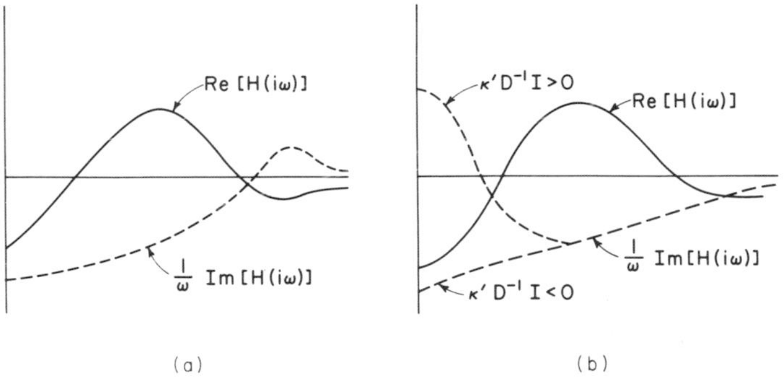

FIGURE 7.11.1. Possible graphs of $\operatorname{Re} H(i\omega)$ and $[\operatorname{Im} H(i\omega)]/\omega$ for $\kappa_0 > 0$ and $N = 2$. (a) $\kappa' D I > 0$, $\kappa' D^{-1} I < 0$; (b) $\kappa' D I < 0$.

One concludes from these figures that one can hope to find a $p > 0$ to satisfy the frequency condition only in the cases (a) and (b) indicated in this figure. The possible values of p in these cases can easily be restricted as

$$0 \leqslant p \leqslant p_{\max} = \min(\infty, (\kappa_0 + \kappa' I)/\kappa' D^{-1} I) \tag{71}$$

which is obtained from the frequency condition considering the limits as $\omega \to 0$ and $\omega \to \infty$. The fact that the signs of $\operatorname{Re} H(i\omega)$ and $[\operatorname{Im} H(i\omega)]/\omega$ are correct does not necessarily mean that a value of p satisfying the frequency condition exists. Further analysis is necessary to prove this fact. It has been shown, however, that for $\kappa_0 \geqslant 0$, $N \leqslant 2$, a p always exists that guarantees ASL whenever the linear system is stable (asymptotically) at all power levels [8], and similarly if $\kappa_0 = 0$, $N = 3$ [27]. It has also been shown that, for $N \geqslant 4$, this is no longer true [27].

7.12. The Effects of Delayed Neutrons

The effect of delayed neutrons has been discussed on various occasions in the preceding chapters. In Chap. 6, we have seen that delayed neutrons may have a destabilizing effect on the linear stability of certain reactor types. In Sec. 7.2, we have shown that, if a reactor whose feedback kernel satisfies $\operatorname{Re} H(i\omega) \leqslant 0$ for all ω displays periodic power oscillations in the absence of delayed neutrons, it can have only damped oscillations when the delayed neutrons are taken into account. Similarly, Popov showed that, if a reactor admits a certain Lyapunov function $V > 0$ such that $\dot{V} = 0$ in the absence of delayed neutrons, then the reactor is asymptotically stable in the large in the presence of delayed neutrons.

In the analysis of the Lurie–Letov function in Sec. 7.7B, we explicitly included the delayed neutron equations. However, the results obtained either in the form of a set of algebraic equations $f_k(\kappa_j, \kappa_0, \eta_j) > 0$ or the equivalent frequency criterion with $p = 0$, are independent of the effects of delayed neutrons. The reactor is asymptotically stable in the large if these conditions are satisfied regardless of whether the delayed neutrons are present or not.

The shortcoming of the above analyses is that they do not indicate any way to use the parameters of the delayed neutron equations for determining the allowable regions in parameter space. In order to secure a criterion for asymptotic stability which does include the effects of delayed neutrons explicitly, we introduce a theorem due to Popov [12]. This theorem also provides bounds of the permissible perturbations in dynamical variables for asymptotic stability in reactors which are not necessarily stable in the small at all power levels.

We present a slightly extended proof of this difficult but very significant theorem in some detail because it does not seem to be available in the English-language literature. Furthermore, it is perhaps the only way to really appreciate the power and intrinsic value of the theorem. The less mathematically oriented reader may skip the following proof, and study the simple example which illustrates the application of the theorem to the one-temperature reactor model.

A. Details of the Proof of Popov's Theorem

It is convenient to introduce the basic method in general and then specify without detailed proof the application to the reactor equation (7.7, Eq. 12).

We consider the following system of ordinary differential equations:

$$\dot{\mathbf{x}} = \mathbf{f}(\mathbf{x}) \tag{1a}$$

where $\mathbf{x}$ is an N-vector, and $\mathbf{f}(\mathbf{x})$ is a continuous vector function with N components. Existence and uniqueness of the solutions are assumed (cf Sec. 6.8).

Let

$$\mathbf{f}(0) = 0 \tag{1b}$$

The system then admits the solution

$$\mathbf{x}(t) \equiv 0 \tag{1c}$$

We will study the stability of this solution. In order to do this, it is convenient to define certain scalar functions of the vector $\mathbf{x}$.

Definition of Certain Classes of Functions

(i) Functions of Class K_D. A scalar function $\alpha(\mathbf{x})$ belongs to the class K_D (where D is some domain of the vector space $\mathbf{x}$ which includes $\mathbf{x} = 0$) if it is defined and continuous for any $\mathbf{x}$ in D, and satisfies

$$\alpha(0) = 0 \tag{2a}$$

$$\alpha(\mathbf{x}) > 0 \quad \text{for} \quad \mathbf{x} \neq 0 \quad \text{in} \quad D \tag{2b}$$

Let μ be any positive number. We associate with μ a subdomain of D, and denote it by $\Delta(\mu)$, such that is contains the origin $\mathbf{x} = 0$, and has the property that, for every vector $\mathbf{x}$ in $\Delta(\mu)$, the inequality

$$\alpha(\mathbf{x}) < \mu \tag{3a}$$

is satisfied, and for every vector on the boundary of $\Delta(\mu)$,

$$\alpha(\mathbf{x}) = \mu \tag{3b}$$

holds.

Since $\alpha(\mathbf{x})$ belongs to K_D [and thus $\alpha(0) = 0$ and $\alpha(\mathbf{x})$ is continuous at $\mathbf{x} = 0$], it is clear that such a domain $\Delta(\mu)$ exists at least for sufficiently small values of μ. For large values of μ, (3b) may not be satisfied [e.g., if μ exceeds the maximum of $\alpha(\mathbf{x})$ in D], and a $\Delta(\mu)$ with the desired properties may not exist.

Let $\mu_{\max}$ be the largest number for which a $\Delta(\mu)$ exists. It follows that, for every two numbers μ_1, μ_2 satisfying the inequality

$$\mu_1 < \mu_2 < \mu_{\max} \tag{4a}$$

we also have the relation

$$\Delta(\mu_1) \subset \Delta(\mu_2) \subset \Delta(\mu_{\max}) \subset D \tag{4b}$$

i.e., the domains $\Delta(\mu)$ for various μ are concentric.

(ii) Functions of Class K_{DM}. A scalar function $\beta(\mathbf{x})$ of the vector $\mathbf{x}$ belongs to the class K_{DM} (D is the same domain as above, M is a set of points contained in D, and itself containing the origin) if it is defined and continuous for all $\mathbf{x}$ in D, satisfies the equation

$$\beta(\mathbf{x}) = 0 \tag{5a}$$

for every point in M, and the inequality

$$\beta(\mathbf{x}) > 0 \tag{5b}$$

for every point in D but not contained in M.

With these definitions, it is possible to introduce two lemmas which will make the subsequent analysis easier.

Lemma 1. If there exists a function $\alpha(\mathbf{x})$ in K_D, a function $\beta(\mathbf{x})$ in K_{DM}, and a third scalar function $\gamma(\mathbf{x})$ which is continuous and zero for

$\mathbf{x} = 0$, such that all the solutions of the system (1a) satisfy the inequallty

$$\alpha(\mathbf{x}(T)) + \int_0^T \beta(\mathbf{x}(t))\, dt \leqslant \gamma(\mathbf{x}(0)) \tag{6}$$

for all $T \geqslant 0$ and if the set M corresponding to $\beta(\mathbf{x})$ does not contain any positive half-trajectory of (1a) other than $\mathbf{x}(t) \equiv 0$, which lies in its positive limiting set, then all the solutions of (1a) that satisfy for $t = 0$ the conditions

$$\gamma(\mathbf{x}(0)) < \mu < \mu_{\max} \tag{7a}$$

$$\mathbf{x}(0) \subset \Delta(\mu) \tag{7b}$$

remain in $\Delta(\mu)$ for every $t > 0$, and

$$\lim_{t\to\infty} \mathbf{x}(t) = 0 \tag{8}$$

holds. Hence, the origin is asymptotically stable for all $\mathbf{x}(0)$ in $\Delta(\mu)$.

PROOF. We first show that $\mathbf{x}(t)$ always remains in $\Delta(\mu)$ for all $t > 0$. Indeed, if it did not for some $t = t_1$, then $\mathbf{x}(t_1)$ would be on the boundary of $\Delta(\mu)$ so that

$$\alpha(\mathbf{x}(t_1)) = \mu \tag{9}$$

would hold.

Since, for $t < t_1$, $\mathbf{x}(t)$ is wholly contained in $\Delta(\mu)$, which is contained in D, we also have $\beta(\mathbf{x}(t)) \geqslant 0$ for all $0 \leqslant t \leqslant t_1$, and hence

$$\int_0^{t_1} \beta(\mathbf{x}(t))\, dt \geqslant 0$$

But, from (7) and (9), we would have

$$\alpha(\mathbf{x}(t_1)) + \int_0^{t_1} \beta(\mathbf{x}(t))\, dt \geqslant \mu > \gamma(\mathbf{x}(0))$$

which contradicts the hypothesis (6). Thus, the trajectory $\mathbf{x}(t)$ is wholly contained in $\Delta(\mu)$.

We next prove asymptotic stability. Since $\alpha(\mathbf{x}(t)) \geqslant 0$, the inequality (6) holds without the first term, i.e.,

$$I(T) \equiv \int_0^T \beta(\mathbf{x}(t))\, dt \leqslant \gamma(\mathbf{x}(0)) \tag{10}$$

Since $I(T)$ is a bounded, monotone increasing function of T, $I(T) \to$ const as $T \to \infty$. Furthermore, its derivative $\beta(\mathbf{x}(t))$ is uniformly continuous because $\beta(\mathbf{x})$ is a continuous function of $\mathbf{x}(t)$, and $\mathbf{x}(t)$ itself is uniformly continuous by virtue of the fact that $\dot{\mathbf{x}} = \mathbf{f}(\mathbf{x})$ is bounded when $\mathbf{x}(t)$ is contained in $\Delta(\mu)$. Hence, using Barbalat's lemma, we deduce from (10) that

$$\lim_{t\to\infty} \beta(\mathbf{x}(t)) = 0 \tag{11}$$

Therefore, the positive limiting set of each trajectory is contained both in M and $\Delta(\mu)$ (cf Sect. 7.6E). We already know that this limiting set is invariant with respect to the system considered, i.e., any half-trajectory initiating at a point of this set remains in it for all subsequent times. However, by virtue of the hypothesis of Lemma 1, the only such set contained in M is $\mathbf{x} = 0$, hence, $\mathbf{x}(t) \to 0$ as $t \to \infty$ is established.

A second lemma is about an inequality involving Fourier transforms.

LEMMA 2. Consider four functions $f_1(t)$, $f_2(t)$, $f_3(t)$, and $f_4(t)$ which are zero for $t < 0$ (causal functions), and their Fourier transforms $F_1(i\omega)$, $F_2(i\omega)$, $F_3(i\omega)$, and $F_4(i\omega)$, which are assumed to exist.

Let us suppose there exist six functions $U_1(i\omega)$, $U_2(i\omega)$, $U_3(i\omega)$ and $V_1(i\omega)$, $V_2(i\omega)$, $V_3(i\omega)$ such that the following relation holds:

$$F_j(i\omega) = U_j(i\omega)F_4(i\omega) + V_j(i\omega), \qquad j = 1, 2, 3 \tag{12}$$

If the following inequality holds for all real, finite ω

$$\operatorname{Re} \kappa(i\omega) > 0 \tag{13}$$

where

$$\kappa(s) \equiv -U_1(s)U_2(-s) - U_3(s) \tag{14}$$

and if the integral

$$\int_{-\infty}^{\infty} [W(i\omega)W(-i\omega)/\operatorname{Re} \kappa(i\omega)]\, d\omega \tag{15}$$

converges, where

$$W(s) \equiv V_1(s)U_2(-s) + V_2(s)U_1(-s) + V_3(s) \tag{16}$$

then the inequality

$$\int_0^{\infty} f_1(t)f_2(t)\, dt + \int_0^{\infty} f_3(t)f_4(t)\, dt$$
$$\leqslant (1/8\pi) \int_{-\infty}^{\infty} [|W(i\omega)|^2/\operatorname{Re} \kappa(i\omega)]\, d\omega$$
$$+ 1/4\pi) \int_{-\infty}^{\infty} [V_1(i\omega)V_2(-i\omega) + V_1(-i\omega)V_2(i\omega)]\, d\omega \tag{17}$$

is always satisfied.

PROOF. Noting that the Parseval theorem states

$$\int_{-\infty}^{\infty} f_j(t)f_k(t)\,dt \equiv \int_0^{\infty} f_j(t)f_k(t)\,dt = (1/2\pi)\int_{-\infty}^{\infty} F_j(i\omega)F_k(-i\omega)\,d\omega$$
$$= (1/2\pi)\int_{-\infty}^{\infty} F_j(-i\omega)F_k(i\omega)\,d\omega \qquad (18)$$

where we used the causality of $f_j(t)$.

Using Eq. (18) and the definition of $F_j(i\omega)$ in (12), and combining terms so as to introduce $\kappa(i\omega)$ and $W(i\omega)$ [using their definitions in (14) and (16), respectively] we at once obtain

$$\int_0^{\infty} (f_1f_2 + f_3f_4)\,dt = -(1/4\pi)\int_{-\infty}^{\infty} \Big\{|F_4(i\omega)|^2[\kappa(i\omega) + \kappa(-i\omega)] - \Big[W(i\omega)F_4(-i\omega) + W(-i\omega)F_4(i\omega)\Big]\Big\}\,d\omega + R \qquad (19)$$

where R denotes the second term on the right-hand side of (17). Adding and subtracting the integral in (15), we have the righthand-side of (19) as

$$\text{RHS} = -(1/2\pi)\int_{-\infty}^{\infty} d\omega\,|\,[\text{Re}\,\kappa(i\omega)]^{1/2}F_4(i\omega) - \tfrac{1}{2}[W(i\omega)/[\text{Re}\,\kappa(i\omega)]^{1/2}]\,|^2 + R + (1/8\pi)\int_{-\infty}^{\infty} [|W(i\omega)|^2/\text{Re}\,\kappa(i\omega)]\,d\omega$$

Since the first term is always nonpositive, we obtain the desired inequality in (17).

We now consider the system described by (7.7, Eqs. 12), which we reproduce here for convenience:

$$(l/\beta)\dot{x}_0 = -(\kappa_0 x_0 + \kappa' y)(1 + x_0) + \sum_{i=1}^{I} a_i(x_i - x_0) \qquad (20a)$$

$$\dot{x}_i = \lambda_i(x_0 - x_i), \qquad i = 1, 2, \ldots, I \qquad (20b)$$

$$\dot{y} = D(Ix_0 - y) \qquad (20c)$$

We shall introduce one more lemma concerning the existence of half-trajectories of Eqs. (20) in a certain set, which will appear in the proof.

LEMMA 3. The set of points for which

$$x_0 - x_i = 0, \qquad i = 1, \ldots, I \tag{21a}$$

$$\kappa_0 x_0 + \kappa' y = 0 \tag{21b}$$

does not contain any nontrivial, positive half-trajectory of the system (20).

The proof of this lemma has already been given in Sec. 7.7C in a slightly different form. We can, however, sketch it here briefly as follows: From (20) and (21) we have $\dot{x}_j(t) = 0$ for $j = 0, 1, \ldots, I$. From (20c) we find $\ddot{y} + D\dot{y} = 0$ whose solution is $\dot{\mathbf{y}}(t) = \exp(-Dt)\,\dot{y}(0)$. Differentiating (21b) we get $\kappa'\dot{y}(t) = 0$, which leads to $\kappa' \exp(-Dt)\,\dot{y}(0) = 0$. Since $\eta_j > 0$ and distinct and $\kappa_j \neq 0$, the latter implies $\dot{y}(0) = 0$, and thus $\dot{y}(t) \equiv 0$ follows from $e^{-Dt}\dot{y}(0) \equiv 0$. Hence, using (20c), we have $x_0(t) \equiv y_j(t)$, and $x_0(t) = x_i(t) = y_j(t) \equiv$ const. But (21b) yields $\kappa_0 x_0 + \sum_j \kappa_j y_j = 0$ or $(\kappa_0 + \kappa' I)x_0 = 0$, which implies $x_0 = 0$ provided $\kappa_0 + \kappa' I \neq 0$, establishing the lemma.†

Besides the original set of equations (20), we shall also need the following set of linear equations:

$$(l/\beta)\dot{\bar{x}}_0 = -h(\kappa_0 \bar{x}_0 + \kappa' \bar{y}) + \sum_{i=1}^{I} a_i(\bar{x}_i - \bar{x}_0) \tag{22a}$$

$$\dot{\bar{x}}_i = \lambda_i(\bar{x}_0 - \bar{x}_i), \qquad i = 1, 2, \ldots, I \tag{22b}$$

$$\dot{\bar{y}} = D(I\bar{x}_0 - \bar{y}) \tag{22c}$$

where we have introduced h, which is a positive number. Clearly, this new set with $h = 1$ is the linearized form of (20), and enables one to investigate linear stability at the equilibrium power level P_0. The latter is a factor common to κ_0 and κ_j as their definition in (7.7, Eqs. 13) indicate. The meaning of h in (22a) now becomes quite straightforward; an $h_1 < 1$ implies another power level $P_{01} = h_1 P_0 < P_0$ (similarly, $h_2 > 1$ implies $P_{02} = h_2 P_0 > P_0$). Hence, we can investigate linear stability of the reactor system in (20) at different power levels just by adjusting the values of h.

We assume that the null solution ($\bar{x}_i \equiv \bar{y}_j \equiv 0$) of the linear system (22) is asymptotically stable for any value of h satisfying $h_1 \leqslant h \leqslant h_2$, where $h_1 < 1$ and $h_2 > 1$. Clearly, this assumption implies that the reactor is linearly stable at all power levels in the range $P_{01} \leqslant P_0 \leqslant P_{02}$.

† One can also show in a similar way that the set (21a) does not contain any nontrivial half-trajectory.

After the foregoing preliminaries, we now consider a solution of (20) whose initial values satisfy

$$x_i(0) > -1, \qquad i = 0, 1, \ldots, I \tag{23}$$

We know already that (23) implies $x_i(t) > -1$ also for all $t > 0$. Furthermore, we define the functions $\tilde{x}_i(t)$ and $\tilde{y}_j(t)$ by

$$\tilde{x}_i(t) = \begin{cases} x_i(t) & \text{for} \quad 0 \leqslant t < T \\ \bar{x}_i(t) & \text{for} \quad t \geqslant T \end{cases} \tag{24a}$$

$$\tilde{y}(t) = \begin{cases} y(t) & \text{for} \quad 0 \leqslant t \leqslant T \\ \bar{y}(t) & \text{for} \quad t > T \end{cases} \tag{24b}$$

in which T is arbitrary, $x_i(t)$, $y(t)$ represent solutions of (20), and $\bar{x}_i(t)$, $\bar{y}(t)$ represent those solutions to (22) for $h = h_2$ that satisfy

$$\bar{y}(T) = y(T), \qquad \bar{x}_i(T) = x_i(T) \tag{25}$$

Since the solutions of (22) are asymptotically stable for $h = h_2$, the functions $\tilde{x}_i$, $\tilde{y}$ tend exponentially to zero as $t \to \infty$; they are absolute and square integrable. Furthermore, they satisfy

$$(l/\beta)\dot{\tilde{x}}_0 = -f_T(t) + \sum_{i=1}^{I} a_i(\tilde{x}_i - \tilde{x}_0) - h_1(\kappa_0\tilde{x}_0 + \kappa'\tilde{y}) \tag{26a}$$

$$\dot{\tilde{x}}_i = \lambda_i(\tilde{x}_0 - \tilde{x}_i) \tag{26b}$$

$$\dot{\tilde{y}} = D(I\tilde{x}_0 - \tilde{y}) \tag{26c}$$

where

$$f_T(t) = (\kappa_0 x_0 + \kappa' y)(1 + x_0 - h_1), \qquad 0 \leqslant t \leqslant T \tag{27a}$$

$$f_T(t) = (h_2 - h_1)(\kappa_0\bar{x}_0 + \kappa'\bar{y}), \qquad t > T \tag{27b}$$

Indeed, for $t \leqslant T$, Eqs. (26) coincide with (20), and for $t > T$, with (22) with $h = h_2$.

Consider now the function

$$\xi_1(T) \equiv -\int_0^\infty \tilde{x}_0(t)[\kappa_0\tilde{x}_0 + \kappa'\tilde{y}]\, dt$$

$$= -\int_0^T x_0[\kappa_0 x_0 + \kappa' y]\, dt - \int_T^\infty \bar{x}_0[\kappa_0\bar{x}_0 + \kappa'\bar{y}]\, dt$$

$$= -I_1 - I_2 \tag{28}$$

Solving (20a) for $\kappa_0 x_0 + \kappa' y$, we find

$$-I_1 = (l/\beta) \int_0^T [x_0 \dot{x}_0/(1 + x_0)]\, dt + \sum_{i=1}^{I} a_i \int_0^T [x_0(x_0 - x_i)/(1 + x_0)]\, dt \qquad (29a)$$

Using (20b), we find a further reduction of the right-hand side:

$$-I_1 = (l/\beta) \int_0^T [x_0 \dot{x}_0/(1 + x_0)]\, dt + \sum_{i=1}^{I} (a_i/\lambda_i) \int_0^T [x_i \dot{x}_i/(1 + x_i)]\, dt$$

$$+ \sum_{i=1}^{I} a_i \int_0^T [(x_0 - x_i)^2/(1 + x_0)(1 + x_i)]\, dt \qquad (29b)$$

Similarly, from (22a) above with $h = h_2$, we get

$$-I_2 = (l/\beta h_2) \int_T^\infty \bar{x}_0 \dot{\bar{x}}_0\, dt + (1/h_2) \sum_{i=1}^{I} a_i \int_T^\infty (\bar{x}_0 - \bar{x}_i)\bar{x}_0\, dt \qquad (30a)$$

Using (22b) and noting that, by assumption,

$$\lim_{t\to\infty} \bar{x}_i(t) = 0, \qquad i = 0, 1, 2, \ldots, I$$

we evaluate the integral on the right-hand side of (30a) as

$$-I_2 = -(l/2\beta h_2)[\bar{x}_0(T)]^2 - (1/2h_2) \sum_{i=1}^{I} (a_i/\lambda_i)[\bar{x}_i(T)]^2$$

$$+ (1/h_2) \sum_{i=1}^{I} a_i \int_T^\infty (x_0 - x_i)^2\, dt \qquad (30b)$$

We define now another function

$$\xi_2(T) \equiv \int_0^\infty dt\, \{\kappa_0 \tilde{x}_0 + \kappa' \tilde{y} - [1/(h_2 - h_1)] f_T(t)\} f_T(t)\, dt \qquad (31a)$$

Using the definition of $f_T(t)$ in (27), we find

$$\xi_2(T) = \int_0^T [(1 + x_0 - h_1)(h_2 - 1 - x_0)/(h_2 - h_1)](\kappa_0 x_0 + \kappa' y)^2\, dt \qquad (31b)$$

Again, we define

$$\xi(T) \equiv \xi_1(T) + \delta \xi_2(T) \qquad (31c)$$

where δ is some nonnegative number.

We now choose the functions $f_i(t)$ in Lemma 2 as

$$f_1 = \tilde{x}_0 \tag{32a}$$

$$f_2 = -(\kappa_0\tilde{x}_0 + \kappa'\tilde{y}) \tag{32b}$$

$$f_3 = -\delta\{\kappa_0\tilde{x}_0 + \kappa'\tilde{y} - [f_T/(h_2 - h_1)]\} \tag{32c}$$

$$f_4 = -f_T \tag{32d}$$

and note that $\xi(T)$ in (31c) corresponds to the left-hand side of the inequality in (17) which occurs in Lemma 2.

We proceed now more rapidly, and take into account the definitions (32) and (26) to determine the Laplace transforms of $f_i(t)$. Taking the Laplace transform of (26a), eliminating $\tilde{x}_i(s)$ and $\tilde{y}(s)$, and replacing s by $i\omega$, we get

$$\{[1/Z(i\omega)] - h_1H(i\omega)\}F_1(i\omega) = F_4(i\omega) + K(i\omega) \tag{33a}$$

where

$$-H(i\omega) \equiv \kappa_0 + \sum_{j=1}^{N} [\kappa_j\eta_j/(\eta_j + i\omega)] \tag{33b}$$

$$Z^{-1}(i\omega) \equiv i\omega \left\{(l/\beta) + \sum_{i=1}^{I} [a_i/(\lambda_i + i\omega)]\right\} \tag{33c}$$

$$K(i\omega) \equiv (l/\beta)x_0(0) + \sum_{i=1}^{I} [a_i/(\lambda_i+i\omega)]x_i(0) - h_1 \sum_{j=1}^{N} [\kappa_j y_j(0)/(\eta_j+i\omega)] \tag{33d}$$

clearly, $Z(i\omega)$ and $H(i\omega)$ are the zero-power and feedback transfer functions, respectively. Since the reactor is linearly stable at h_1, the equation

$$Z^{-1}(s) - h_1H(s) = 0$$

has no roots with nonnegative real parts. Comparing (33a) to (12), we obtain the functions $U_1(i\omega)$ and $V_1(i\omega)$.

Similarly, from the definition of $f_2(t)$ in (32b), we obtain

$$F_2(i\omega) = -\kappa'(D + i\omega E)^{-1}y(0) + H(i\omega)F_1(i\omega)$$

Eliminating $F_1(i\omega)$ from (33a), we express $F_2(i\omega)$ in terms of $F_4(i\omega)$, and comparing the resulting equation to (12), we extract the definitions of

$U_2(i\omega)$ and $V_2(i\omega)$. Repeating a similar procedure for $f_3(t)$ defined by (32c), we determine $U_3(i\omega)$ and $V_3(i\omega)$. The results can be written as follows:

$$U_1(i\omega) \equiv Z(i\omega)[1 - h_1 Z(i\omega)H(i\omega)]^{-1},$$

$$V_1(i\omega) \equiv U_1(i\omega)K(i\omega)$$

$$U_2(i\omega) \equiv H(i\omega)U_1(i\omega),$$

$$V_2(i\omega) \equiv H(i\omega)V_1(i\omega) - \sum_{j=1}^{N} [\kappa_j y_j(0)/(\eta_j + i\omega)]$$

$$U_3(i\omega) \equiv \delta\{U_2(i\omega) - [1/(h_2 - h_1)]\},$$

$$V_3(i\omega) = \delta V_2(i\omega)$$

Using (15) and (16), we also obtain the function Re $\kappa(i\omega)$ and $W(i\omega)$ as

$$-\text{Re}\,\kappa(i\omega) = |U_1(i\omega)|^2[\text{Re}\,H(i\omega)] - \delta\{[1/(h_2 - h_1)] - \text{Re}[U_1(i\omega)H(i\omega)]\} \tag{34a}$$

$$W(i\omega) = K(i\omega)[2|U_1|^2(\text{Re}\,H) + \delta H(i\omega)U_1(-i\omega)] - [\delta + U_1(-i\omega)]\kappa'(D + i\omega E)^{-1}y(0) \tag{34b}$$

We note for future use that $K(i\omega)$ in (33d) and, hence, $W(i\omega)$ in (34b) are linear homogeneous functions of the initial values $x_j(0)$ and $y_j(0)$.

The convergence of the integral in (15) can now be discussed in terms of (34). Considering large values of ω^2, we find from (34b) that $|W(i\omega)|^2$ behaves as $[\kappa' y(0)]^2 \delta^2/\omega^2$ if $\delta > 0$ and $y(0) \neq 0$; as $|K|^2\kappa_0^2/\omega^4$ if $y(0) = 0$, $\kappa_0 \neq 0$, $\delta \geqslant 0$; as $|\kappa|^2 \delta^2/\omega^4$ if $y(0) = 0$, $\kappa_0 = 0$, $\delta > 0$; and as $(\kappa' I)^2/\omega^8$ if $\kappa_0 = 0$, $\delta = 0$, $y(0) \neq 0$. On the other hand, Re $\kappa(i\omega) \sim \delta/(h_2 - h_1)$ if $\delta > 0$; Re $\kappa(i\omega) \sim \kappa_0/\omega^2$ if $\delta = 0$, $\kappa_0 \neq 0$; and Re $\kappa(i\omega) \sim (\kappa' I)^2/\omega^4$ if $\delta = 0$, $\kappa_0 = 0$.

Hence, $|W|^2/\text{Re}\,\kappa(i\omega)$ vanishes as $1/\omega^2$ or faster in all cases except for $\delta = 0$, $y(0) \neq 0$, $\kappa_0 = 0$. In the latter case, the ratio under consideration approaches a constant, and the integral (15) does not converge. Therefore, when $y(0) \neq 0$ and $\kappa_0 = 0$, we must require the inequality (13) to be satisfied with a $\delta > 0$.

The inequality in (17), which is guaranteed by Lemma 2, can now be written as

$$\xi(T) \leqslant B[x(0), y(0)] \tag{35}$$

where[†]

$$B[x(0), y(0)] \equiv (1/8\pi) \int_{-\infty}^{\infty} [| W |^2/\text{Re}\, \kappa(i\omega)]\, d\omega$$
$$+ (1/4\pi) \int_{-\infty}^{\infty} [V_1(i\omega)\, V_2(-i\omega) + V_1(-i\omega) V_2(i\omega)]\, d\omega \quad (36)$$

Now, if

$$S_j(T) \equiv x_j(T) - \ln[1 + x_j(T)] - x_j(0) + \ln[1 + x_j(0)] - (1/2h_2)x_j^2(T),$$
$$j = 0, 1,..., I$$

then we may express $\xi(T)$ combining (31c), (31b), and (28); we find

$$\xi(T) = (l/\beta)S_0(T) + \sum_{i=1}^{I} (a_i/\lambda_i)\, S_i(T) + (1/h_2) \sum_{i=1}^{I} a_j \int_T^{\infty} (\bar{x}_0 - \bar{x}_i)^2 dt$$
$$+ \int_0^T \beta[Z(t)]\, dt \quad (37)$$

where Z denotes (x_i, y_j) and

$$\beta(Z) \equiv \beta(x_i, y_i) = \sum_{i=1}^{I} [a_i(x_0 - x_i)^2/(1 + x_0)(1 + x_i)]$$
$$+ \delta[(1 - h_1 + x_0)(h_2 - 1 - x_0)/(h_2 - h_1)](\kappa_0 x_0 + \kappa' y)^2 \quad (38)$$

Since $\bar{x}_i(t)$ in (37) are solutions of (22) with $h = h_2$ whose initial values satisfy (25), we may express the corresponding term in (37) as

$$(1/h_2) \sum_{j=1}^{I} a_j \int_T^{\infty} [\bar{x}_0 - \bar{x}_j]^2\, dt \equiv P(x(T), y(T)) \quad (39)$$

where $P(x, y)$ is a positive-definite quadratic function[††] of its arguments,

[†] Since V_1, V_2, and W are linear, homogeneous functions of $x_j(0)$ and $y_j(0)$, $B[x(0), y(0)]$ is a quadratic form of its arguments, and hence $B(0, 0) = 0$. Although the first term in (36) is always positive, the second may be negative. Hence, $B(x, y)$ is not necessarily positive.

[††] The positiveness of $P(x, y)$ is obvious from its definition because the integral on the left-hand side of (39) is positive or zero. The latter is the case only when $\bar{x}_0(t) \equiv \bar{x}_j(t)$, $j = 1, 2,..., I$, for $t \geqslant T$. But the functions $\bar{x}_0(t)$, $\bar{x}_j(t)$ satisfy (22) for $h = h_2$. From (22b), we conclude $\bar{x}_j(t) = \text{const}$, $j = 0, 1,..., I$. But the linear system (22) is asymptotically stable for $h = h_2$, so that $\bar{x}_j(t) \equiv 0$, $j = 0, 1,..., I$, and in particular $x_j(T) = 0$ for $j = 0, 1,..., I$. Using (22a) and $\bar{x}_j(t) \equiv 0$, we get $\kappa'\bar{y}(t) \equiv 0$, $t \geqslant T$. Solving (22c) with the initial condition $\bar{y}(T) = y(T)$, we get $\kappa' e^{-D(t-T)}\, \bar{y}(T) \equiv 0$ for all $t \geqslant T$. We conclude from the latter that $\bar{y}(T) = 0$. We have thus established that $P(x(T), y(T)] = 0$ only if $x(T) = y(T) = 0$. The explicit form of $P(x, y)$ can be obtained explicitly, if desired, by actually solving (22) and performing the indicated integrals in (39).

whose coefficients do not depend on T, but depend only on the system parameters.

Taking into account the relations (37) and (39), we see that the inequality (35) can be written as

$$\alpha(Z(T)) + \int_0^T \beta(Z(t))\, dt \leqslant \gamma(Z(0)) \tag{40}$$

if we define

$$\alpha(Z) \equiv (l/\beta)F(x_0) + \sum_{i=1}^{I} (a_i/\lambda_i)F(x_i) + P(x, y) \tag{41a}$$

$$\gamma(Z) \equiv (l/\beta)[x_0 - \ln(1 + x_0)] + \sum_{i=1}^{I} (a_i/\lambda_i)[x_i - \ln(1 + x_i)] + B(x, y) \tag{41b}$$

where

$$F(x_j) \equiv x_j - \ln(1 + x_j) - (1/2h_2)x_j^2, \qquad j = 0, 1,..., I \tag{41c}$$

We can summarize the above results as follows: Every solution $Z(t)$ of the system (20) that starts at a point $Z(0)$ in the region $x_j > -1$, $j = 0, 1,..., I$ [no restriction on $y_j(0)$], remains in the same region for all subsequent times, and, moreover, satisfies the inequality (40) for all $T \geqslant 0$, provided the conditions of Lemma 2 are satisfied.

The inequality (40) is precisely of the type of (6) of Lemma 1. However, the functions $\alpha(Z)$ and $\beta(Z)$ do not belong to the class of functions K_D and K_{DM}, respectively, if the domain D is taken as $x_j > -1$, $j = 0, 1,..., I$. However, we can find a domain D in which $\alpha(Z)$ and $\beta(Z)$ possess the desired properti to be in K_D and K_{DM}.

We first observe that $F(x_j)$ in (41c) has a positive maximum at $x_j = h_2 - 1$, and hence is positive at least in the region

$$-1 < x_j < h_2 - 1, \qquad j = 0, 1,..., I \tag{42}$$

Since $P(x, y)$ is a positive-definite quadratic function, we find that $\alpha(Z) > 0$ for $Z \neq 0$ in the above region, and $\alpha(0) = 0$.

Secondly, the function $\beta(Z)$ defined by (38) is continuous, and is nonnegative in the region

$$h_1 - 1 < x_0 < h_2 - 1 \tag{43a}$$

$$-1 < x_j, \qquad j = 1, 2,..., I \tag{43b}$$

which includes the origin (since $h_1 < 1 < h_2$). We now define the domain D as the intersection of (42) and (43), i.e.,

$$h_1 - 1 < x_0 < h_2 - 1 \tag{44a}$$

$$-1 < x_j < h_2 - 1, \qquad j = 1, 2, \ldots, I \tag{44b}$$

The fact that $\alpha(Z)$ belongs to the class K_D in D is obvious. In order to show that $\beta(Z)$ belongs to K_{DM}, we must investigate the set M at which $\beta(Z) = 0$. From (38), we find that $x_0 - x_j = 0$, $j = 1, 2, \ldots, I$, and $\kappa_0 x_0 + \kappa' y = 0$ defines the set M, and that the origin is contained in M. According to Lemma 3, this set does not contain any nontrivial, positive half-trajectory. In the case of $\delta = 0$ in (38), M reduces to $x_j - x_0 = 0$, $j = 1, 2, \ldots, I$, which also does not contain any nontrivial, positive half-trajectory, as one can easily show using the procedure described in Lemma 3. We conclude from the above discussions that $\beta(Z)$ belongs to the class K_{DM} and the set M associated with $\beta(Z)$ satisfies the conditions of Lemma 1.

The function $\gamma(Z)$ defined by (41) is continuous, and zero when $Z = 0$, since $B(x, y)$ is a quadratic function in x and y.

Since all the conditions of Lemma 1 are satisfied in D, inequality (40) implies $Z(t) \to 0$ as $t \to \infty$ if the initial value $Z(0)$ satisfies (7).

The foregoing results can be summarized in Popov's theorem.

B. Statement of Popov's Theorem

Consider the reactor model described by Eqs. (20), and require that (a) all $\eta_j > 0$ and distinct, $j = 1, 2, \ldots, N$; (b) all $\kappa_j \neq 0$, $j = 1, 2, \ldots, N$; (c) $\kappa_0 + \kappa' I > 0$; (d) there exist two positive numbers $h_1 < 1$ and $h_2 > 1$ such that the linear system described by (22) is asymptotically stable for any h satisfying $h_1 \leqslant h \leqslant h_2$; (e) there exists a nonnegative number δ such that

$$\operatorname{Re} \kappa(i\omega) \equiv | U_1 |^2 \operatorname{Re}[-H(i\omega)]\, \delta\{\operatorname{Re}[-U_1(i\omega)H(i\omega)] + [l/(h_2 - h_1)] > 0 \tag{45a}$$

holds for all finite real ω; (f) there exists a finite $K_0 > 0$ such that

$$| W(i\omega)|^2/\operatorname{Re} \kappa(i\omega) < K_0/\omega^2 \tag{45b}$$

holds for sufficiently large ω^2.

With these conditions, the equilibrium state of (20) is asymptotically stable.

The range of the permissible initial values for asymptotic stability can be found using the functions $\alpha(Z)$ and $\gamma(Z)$ defined by (41a) and (41b) as follows. First, we choose h_1 and h_2, and find a $\delta \geqslant 0$ such that (45a) is satisfied. Secondly, we determine the subdomain $\Delta(\mu)$ and $\mu_{\max}$ in the domain D defined by (44) using the expression of $\alpha(Z)$ in (41a). Then, according to the theorem, every solution of (20) that satisfies at $t = 0$

$$\gamma[Z(0)] < \mu < \mu_{\max} \tag{46a}$$

$$Z(0) \subset \Delta(\mu) \tag{46b}$$

also satisfies for all $t > 0$

$$Z(t) \subset \Delta(\mu) \tag{46c}$$

$$\lim_{t\to\infty} Z(t) = 0 \tag{46d}$$

While the choice of h_1 and h_2, which decisively affects the region of the permissible initial values, is straightforward, the implicit method of determining the actual bounds on the initial values in a given problem is extremely difficult except for the choices of $h_1 \to 0$ and $h_2 \to \infty$.

As an illustration of Popov's theorem, we consider a particularly simple problem: a one-temperature reactor model in the absence of delayed neutrons with a destabilizing power coefficient κ_0, i.e., $\kappa_0 < 0$. This model is described by (with $l/\beta = 1$)

$$\dot{x}_0 = -(\kappa_0 x_0 + \kappa_1 y_1)(1 + x_0) \tag{47a}$$

$$\dot{y}_1 = \eta_1(x_0 - y_1) \tag{47b}$$

Conditions (a) and (b) require $\eta_1 > 0$ and $\kappa_1 \neq 0$. Condition (c) requires $\kappa_0 + \kappa_1 > 0$. The functions $Z(s)$ and $H(s)$ are given by $Z(s) = (1/s)$ and $H(s) = -\{\kappa_0 + [\kappa_1\eta_1/(s + \eta_1)]\}$. The linearized set is

$$\dot{\bar{x}}_0 = -h(\kappa_0\bar{x}_0 + \kappa_1\bar{y}_1)$$
$$\dot{\bar{y}}_1 = \eta_1(\bar{x}_0 - \bar{y}_1)$$

which is asymptotically stable for all values of $0 < h < \eta/|\kappa_0|$, hence $h_1 \to 0$ and $h_2 < \eta_1/|\kappa_0|$. We may point out here that $U_1(s)$ acquires a simple pole at $s = 0$ when $h_1 = 0$, and consequently, the integrals on the right-hand side of (17) become divergent separately. However, we shall see that this divergence cancels out when these two integrals are combined.

Let us assume for simplicity that only a flux perturbation is allowed, i.e., $x_0(0) \neq 0$ but $y_1(0) = 0$. Then, we find

$$
\begin{aligned}
K(s) &= x_0(0) \\
U_1(s) &= 1/s, \qquad V_1(s) = x_0(0)/s \\
U_2(s) &= H(s)/s, \qquad V_2(s) = x_0(0)H(s)/s \\
U_3(s) &= \delta\{[H(s)/s] - (1/h_2)\}, \qquad V_3(s) = \delta x_0(0)H(s)/s \\
W(s) &= -x_0(0)\{[H(s) + H(-s)]/s^2\} + \delta x_0(0)[H(s)/s] \\
\kappa(s) &= [H(s)/s^2] + \delta\{(1/h_2) - [H(s)/s]\}
\end{aligned}
$$

The condition (45a) reduces to

$$\operatorname{Re}\kappa(i\omega) = (1/\omega^2)\{\kappa_0 + [\kappa_1\eta_1^2/(\eta_1^2 + \omega^2)]\} + \delta\{(1/h_2) - [\kappa_1\eta_1/(\omega^2 + \eta_1^2)]\} > 0 \tag{48a}$$

For small ω^2, $\operatorname{Re}\kappa(i\omega)$ diverges as $(\kappa_0 + \kappa_1)/\omega^2$, indicating that $\kappa_0 + \kappa_1 > 0$ must hold. Clearly, (48a) cannot hold when $\kappa_0 < 0$ (as we have assumed) if $\delta = 0$. Hence, we must look for a $\delta > 0$. For this purpose, we rewrite (48a) as

$$
\begin{aligned}
&(h_2\omega^2/\delta)(\eta_1^2 + \omega^2)\operatorname{Re}\kappa(i\omega) \\
&\quad = \omega^4 + \omega^2[\eta_1^2 - \kappa_1\eta_1 h_2 + (h_2\kappa_0/\delta)] + (h_2/\delta)(\kappa_0 + \kappa_1)\eta_1^2 > 0
\end{aligned}
\tag{48b}
$$

and choose (this may not be the best choice, but is convenient)

$$\delta = h_2 |\kappa_0|^2/[2\eta_1^2(\kappa_0 + \kappa_1) - \kappa_0\eta_1(\eta_1 - \kappa_1 h_2)] \tag{49a}$$

Then, the above inequality reads

$$\{\omega^2 - [\eta_1^2(\kappa_0 + \kappa_1)/|\kappa_0|]\}^2 + [(\kappa_0 + \kappa_1)\kappa_1\eta_1^3/|\kappa_0|][(\eta_1/|\kappa_0|) - h_2] > 0 \tag{49b}$$

which is always positive, because $h_2 < \eta_1/|\kappa_0|$.

Since $\delta > 0$ and $\kappa_0 \neq 0$, condition (45b) is satisfied automatically, as we discussed previously.

The foregoing considerations are sufficient to secure the asymptotic stability of the null solution $x_0(t) \equiv y_1(t) \equiv 0$ of (47). To complete the analysis, we now must determine the region for the permissible initial disturbance $x_0(0)$.

We first consider the function $\alpha(x_0)$ defined by (41a), which becomes in the present case

$$\alpha(x_0) \equiv x_0 - \ln(1 + x_0) - (1/2h_2)(x_0)^2 \tag{50a}$$

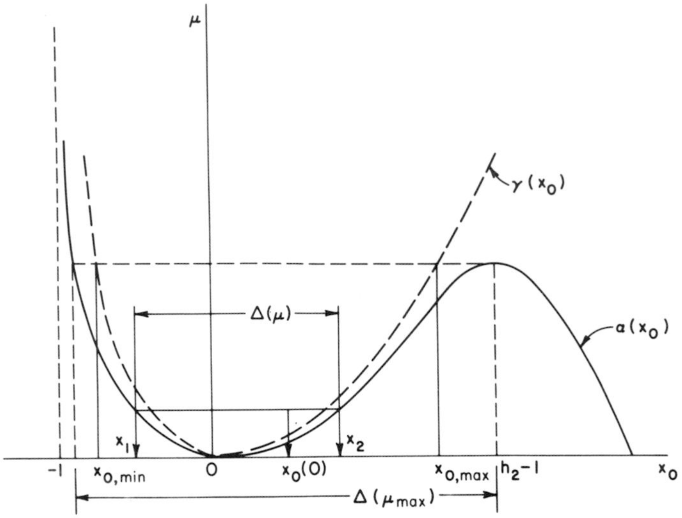

FIGURE 7.12.1. Determination of the permissible range of initial values in Popov's theorem.

because $P(x_0, y_0)$ is zero in the absence of delayed neutrons (cf Eq. 39). The variation of $\alpha(x_0)$ with x_0 is shown in Figure 7.12.1. This function has a positive maximum at $x_0 = h_2 - 1$ with a value

$$\mu_{\max} = [(h_2^2 - 1)/2h_2] - \ln h_2 \tag{50b}$$

We observe that the subdomain $\Delta(\mu)$ exists for all $0 \leqslant \mu \leqslant \mu_{\max}$ in which $\alpha(x) < \mu$, and the boundary points are given by $\alpha(x) = \mu$. There is no such subdomain if $\mu > \mu_{\max}$.

Secondly, we consider $\gamma(x_0, 0)$ which is defined by (41b), and reduces in the example under discussion to

$$\gamma(x_0) = x_0 - \ln(1 + x_0) + B(x_0, 0) \tag{51a}$$

where $B(x_0, 0)$ is defined in general by (36). Using the latter and the expressions of $U_j(s)$ and $V_j(s)$ given above, we find, after some complex algebra,

$$\beta(x_0, 0) \equiv \frac{x_0^2}{8\pi} \delta^2 \int_{-\infty}^{\infty} \frac{d\omega}{\omega^2} \frac{|H(i\omega)|^2 + (4/\delta h_2) \operatorname{Re} H(i\omega)}{\operatorname{Re} \kappa(i\omega)} \tag{51b}$$

The integrand in (51b) is finite at $\omega^2 = 0$, as can be verified by substituting $\operatorname{Re} \kappa(i\omega)$ from (48b). Furthermore, it vanishes as $1/\omega^2$ for

$\omega^2 \to \infty$, and thus the integral converges. Although it can be integrated in a closed form exactly by using complex integration techniques, its precise numerical value is not essential for our purpose. The only point to be made is that (51b) can be written as

$$\beta(x_0, 0) = x_0^2 A(\delta, h_2) \tag{51c}$$

where A depends on the choice of δ and h_2. If h_2 is sufficiently close to its maximum value $\eta_1/|\kappa_0|$, then A can be approximated by [note that the integrand in this case is a sharply peaked function at

$$\omega_0^2 = \eta_1^2(\kappa_0 + \kappa_1)/|\kappa_0|$$

and that Re $H(i\omega_0) = 0$]

$$A(\delta, h_2) \cong (\delta^2 |\kappa_0|/8)\{\kappa_1\eta_1[(\eta_1/|\kappa_0|) - h_2]\}^{-1/2} \tag{52}$$

Then, $\gamma(x_0)$ can be written as

$$\gamma(x_0) = x_0 - \ln(1 + x_0) + x_0^2 A$$

which is also plotted in Fig. 7.12.1. Suppose the initial value $x_0(0)$ is chosen such that

$$x_{\min} < x_0(0) < x_{\max}$$

where $x_{\min}$ and $x_{\max}$ are indicated in Figure 7.12.1. Then, there is a value μ such that $\gamma[x_0(0)] < \mu < \mu_{\max}$ holds. Furthermore, $x_0(0)$ also lies in $\Delta(\mu)$. Hence, any solutions starting at $x_0(0)$ will tend to the origin, and will satisfy $x_1 < x_0(t) < x_2$ for all $t > 0$. If $x_0(0) > x_{\max}$ or $x_0(0) < x_{\min}$, then $\mu > \mu_{\max}$, and the theorem does not yield any information.

It is noted that the theorem does not provide any bounds on $y_1(t)$ in the absence of delayed neutrons, because $P(x, y)$ in (40) is identically zero when delayed neutrons are not taken into account, and consequently $\alpha(x_0)$ does not depend on y_1. However, once the bounds for $x_0(t)$ are determined, one can always find the bound on $y_1(t)$ using the original equation, i.e., $\dot{y}_1 + \eta_1 y_1 = \eta_1 x_0(t)$. Since $y_1(0) = 0$ was assumed in this example, the solution of the latter is

$$y_1(t) = \int_0^t \eta_1 x_0(t - u)\, e^{-\eta_1 u}\, du$$

from which we get

$$|y_1(t)| \leqslant M(1 - e^{-\eta_1 t})$$

where M is the upper bound of $x_0(t)$ for all $t \geqslant 0$. This procedure was already used in Sec. 7.11 to establish bounds of $y(t)$ in terms of those of $x_0(t)$ in the case of $N > 1$.

In the presence of delayed neutrons, $\alpha(x_0, y_1)$ contains y_1, so that $\alpha(x_0, y_1) < \mu$ determines a bound of $y_1(t)$ as well as that of $x_0(t)$.

C. Concluding Remarks

The foregoing simple example should be sufficient to demonstrate the difficulties in numerically determining the domain of maximum allowable initial disturbance in more complicated cases. We shall therefore confine our discussion to two limiting cases: $h_1 \to 1$, $h_2 \to 1$; and $h_1 \to 0$, $h_2 \to \infty$.

In the former case, the domain in (44) narrows down the range of values of $x_0(0)$ to a point. Using $x_0(t) \equiv 0$ in (39), we get $P(x, y) = \sum_j (a_j/2\lambda_j)| x_j |^2$, and hence $\alpha(Z) = (l/\beta)F(x_0) + \sum_j (a_j/\lambda_j)[x_j - \ln(1 + x_j)]$. Since (44) also implies $-1 < x_j \leqslant 0^+$, $\mu_{\max} \to 0$.

Hence, it becomes impossible to establish, in this case, a finite domain of stability in the phase space, even though the condition (45a) is automatically satisfied because the term $\delta/(h_2 - h_1)$ with $\delta > 0$ approaches $+\infty$. The remaining conditions of the theorem tend toward the criterion obtained in the theory of linearization (cf Sec. 6.8).

The second case, $h_1 \to 0$ and $h_2 \to \infty$, implies linear stability at all power levels. The domain of stability becomes the entire phase space restricted to $x_j > -1, j = 0, 1, \ldots, I$. In the limits of $h_2 \to \infty$, $P(x, y) = 0$,

$$\alpha(Z) = (l/\beta)[x_0 - \ln(1 + x_0)] + \sum_{j=1}^{I} (a_j/\lambda_j)[x_j - \ln(1 + x_j)]$$

and the domain D in (44), as well as $\Delta(\mu_{\max})$, coincide with the entire phase space with the above restrictions. Furthermore, the condition (45a) reduces, when $h_2 \to \infty$ and $h_1 \to 0$, to

$$| Z(i\omega)|^2 \operatorname{Re}[H(i\omega)] + \delta \operatorname{Re}[Z(i\omega)H(i\omega)] < 0 \tag{53a}$$

from which we obtain

$$\operatorname{Re} H + \delta \operatorname{Re}(H/Z^*) < 0 \tag{53b}$$

This condition is sufficient for asymptotic stability in the large, provided the reactor is linearly stable at all power levels, i.e., $1 - hZ(s)H(s) = 0$ has no roots with positive real parts for any $h > 0$; all $\eta_j > 0$ and are distinct, and all $\kappa_j \neq 0$. The condition (c) of the theorem is automatically satisfied if (53b) holds for all ω because $Z(i\omega) \to \infty$ as $\omega \to 0$, and consequently (53b) implies $\operatorname{Re} H(0) = -(\kappa_0 + \kappa' I) < 0$.

The condition (53b) is a criterion for asymptotic stability in the large which takes into account explicitly the delayed neutron parameters. It is interesting to compare it to the frequency condition in the Kalman–Popov theory in various limiting cases in order to see whether it covers a larger region in (κ_0, κ_1) space for stability.

First, we consider $\delta \to 0$. In this case, Welton's criterion is obtained, i.e., $\operatorname{Re} H(i\omega) < 0$ without the equality sign, as opposed to (7.11, Eq. 60b) in the Kalman–Popov condition.

Secondly, suppose $\delta \to \infty$. Then, (53b) yields

$$(\operatorname{Re} H) + \left(\left\{(l/\beta) + \sum_{i=1}^{I} [a_i \lambda_i/(\omega^2 + \lambda_i^2)]\right\} \Big/ \sum_{i=1}^{I} [a_i/(\omega^2 + \lambda_i^2)]\right) [(\operatorname{Im} H)/\omega] \leqslant 0$$

$$\omega \neq 0 \qquad (54)$$

This result is precisely that obtained by Baron and Mayer [32]. It is important to note that we must now add back to (54) the condition (c), i.e., $\operatorname{Re} H(i0) = -(\kappa_0 + \kappa' I) < 0$, because (54) follows from (53b) only for $\omega \neq 0$. When $\omega = 0$, the first term in (53a) dominates no matter how large δ is.

We now consider the case in which δ is finite, and obtain from (53b) the following condition:

$$(\operatorname{Re} H) + \left(\delta\omega^2 \left\{(l/\beta) + \sum_{i=1}^{I} [a_i \lambda_i/(\omega^2 + \lambda_i^2)]\right\} \Big/ \left\{(l/\beta) + \delta\omega^2 \sum_{i=1}^{I} [a_i/(\omega^2 + \lambda_i^2)]\right\}\right)$$

$$\times [(\operatorname{Im} H)/\omega] < 0 \qquad (55)$$

We shall compare this to the Kalman–Popov frequency conditions in (7.11, Eq. 60) by finding conditions for asymptotic stability in the large of a reactor with $\kappa_0 = 0$.

A necessary condition for linear stability at all power levels can be obtained by expressing $1 - h_2 H(s) Z(s) = 0$ in a polynomial as

$$(l/\beta) S^{N+I+1} + \left\{(\operatorname{Tr} D)\left[1 + (l/\beta) \sum_{i=1}^{I} \lambda_i\right] + (l/2\beta)\left[(\operatorname{Tr} D)^2 - (\operatorname{Tr} D^2)\right.\right.$$

$$\left.\left. + \left(\sum_{i=1}^{I} \lambda_i\right)^2 - \sum_{i=1}^{I} \lambda_i^2 + h_2 \kappa' D I\right]\right\} S^{N+I} + \cdots = 0$$

or

$$(l/\beta)S^{N+I+1} + (A + h_2\kappa' DI)S^{N+I} + \cdots = 0$$

Since $A > 0$, there is always a finite value of $h_2 > 0$ that makes the coefficients of the second term <0 if $\kappa' DI < 0$. Hence, $\kappa' DI \geqslant 0$ if the reactor is to be linearly stable for all h_2.

Next, consider (55) as $\omega^2 \to 0$. Since $\kappa_0 = 0$, we get

$$\delta\kappa' DI/[1 + (\delta\beta/l)] < 0$$

This result can be reconciled with $\kappa' DI \geqslant 0$ only if $\delta = 0$. Hence, (55) can be satisfied only with Re $H(i\omega) < 0$. The latter now automatically guarantees linear stability at all power levels because $\text{Re}(1/Z) = h_2 \text{ Re } H(i\omega)$ cannot be satisfied for $\text{Re}(1/Z) \geqslant 0$. We observe that asymptotic stability in the large is guaranteed if Re $H(i\omega) < 0$ for all ω in the present case, whereas in the Kalman–Popov theory, it is guaranteed if Re $H(i\omega) \leqslant 0$, which includes equality.

It appears that, in the reactor model considered here, Popov's theorem yields a slightly more restrictive condition than the Kalman–Popov frequency condition in spite of the fact that it includes the delayed neutron parameters explicitly. Its real power, however, lies in the fact that bounds on perturbations can be discovered for which the system is asymptotically stable for finite h_2 whether $\kappa_0 \lesseqgtr 0$ as long as the other requirements of the theorem are satisfied.

7.13. Finite Escape Time and Boundedness

A. The Concept of Finite Escape Time in Reactor Kinetics

It was mentioned in Chapter 6 that a set of nonlinear differential equations of the form $\dot{\mathbf{x}} = \mathbf{f}(\mathbf{x})$ may have solutions which diverge at a finite time (cf 6.9B, Def. 4). We gave the simple example $\dot{x} = x^2$ as an illustration (cf 6.9, Eq. 8), and proved that if $\mathbf{f}(\mathbf{x})$ satisfies a global Lipschitz condition, the system described by $\dot{\mathbf{x}} = \mathbf{f}(\mathbf{x})$ cannot have solutions with finite escape time. In this section we discuss the concept of finite escape time in reactor kinetics and derive conditions for its nonexistence in terms of the reactor parameters with applications to simple reactor models.

The existence of solutions which diverge in a finite time was demonstrated by Chernick [33, 34] in 1951 using the single-temperature reactor model in the absence of delayed neutrons

$$\begin{aligned} l\dot{P} &= (k_0 + \alpha T)P \\ \mu\dot{T} &= P - hT \end{aligned} \tag{1}$$

where the constants have their usual meaning. He showed that for the shut-down equilibrium point this system of differential equations admits solutions which diverge at a finite time for any disturbance if $\alpha > 0$, i.e., if the temperature coefficient is positive. More recently, in 1963 Shotkin [35] carried out a thorough investigation of the two-temperature reactor model, and showed that solutions with finite escape time are possible when the reactor is operating at equilibrium power, provided the disturbance is large enough. In 1966 Akcasu and Noble [36] studied the question of existence, and the asymptotic behavior of solutions with finite escape time in reactor models that can be characterized by a linear feedback kernel. Quite recently, Smets [37] considered specific linear feedback models and verified the general conclusions obtained in [36] concerning the existence and the asymptotic behavior of such solutions, and provided alternative derivations.

The interest of pure mathematicians in the existence of solutions to ordinary differential equations that diverge at a finite time is quite old. Wintner [38] investigated this problem in 1945, and in fact gave the simple example $\dot{x} = x^2$ to illustrate the existence of such solutions (cf 6.9, Eq. 8). The term "finite escape time" seems to be of later origin [39]. The question of finite escape time and boundedness in certain class of integro-differential equations, which includes the reactor kinetic equations, was investigated recently by Londen [40, 41] in a highly abstract paper. Londen's work puts the main conclusions derived in [33–37] on a firmer mathematical basis. Here we follow reference [36].

In the absence of delayed neutrons (the effect of the delayed neutrons will be discussed later) and in the case of linear feedback, the reactor power satisfies

$$\dot{P}(t) = P(t)\left[\int_0^t G(t-u)\,P(u)\,du + I_1(t) - g(t)\right] \tag{2}$$

which is obtained from (7.1, Eq. 7) by neglecting the terms representing the effect of the delayed neutrons, choosing the unit of time such that $(l/\beta) = 1$, and measuring the power in units of P_0. The function $g(t)$ in (2) is defined as

$$g(t) \equiv \int_0^t G(u)\,du \tag{3}$$

The definition of $I_1(t)$ was given in (7.1, Eq. 8). Dividing (1) by $P(t)$, integrating the resulting equation in $(0, t)$ and using

$$\int_0^t dv \int_0^v du\, G(v-u)\, P(u) = \int_0^t du\, P(u) \int_u^t dv\, G(v-u)$$

$$= \int_0^t du\, P(u) \int_0^{t-u} dv\, G(v)$$

$$= \int_0^t du\, P(u)\, g(t-u) \tag{4}$$

we obtain the integral equation

$$\log[P(t)/P(0)] = \int_0^t du \{ g(u)\, P(t-u)\} + Q(t) \tag{5a}$$

where

$$Q(t) \equiv \int_0^t [I_1(u) - g(u)]\, du \tag{5b}$$

and where $P(0)$ is the initial power.

We look for a solution of the form

$$P(t) = [a_m/(t_e - t)^m] + Y(t) \tag{6a}$$

where a_m is a real number, m is a positive integer, and $Y(t)$ is an unknown function which is assumed to satisfy

$$\lim_{t \to t_e} (t_e - t)^m\, Y(t) = 0 \tag{6b}$$

We observe that $P(t)$ is a solution with a finite escape time t_e. Substitution of (6a) into (5a) yields

$$\log\{[a_m/P(0)(t_e - t)^m] + [Y(t)/P(0)]\} = Q(t)$$

$$+ \int_0^t du\, Y(t-u)\, g(u) + \int_0^t du \{a_m\, g(t-u)/(t_e - u)^m\} \tag{7}$$

The left-hand side of (7) diverges logarithmically as $-m \log(t_e - t)$ when $t \to t_e$. In order for (7) to hold, there must be a term on the right-hand side which also diverges logarithmically. Let us assume that $g(t)$ has a power series expansion as

$$g(t) = (1/n!)\, g^{(n)}(0)\, t^n + [1/(n+1)!]\, g^{(n+1)}(0)\, t^{n+1} + \cdots \tag{8}$$

where $g^{(n)}(0)$ is the first nonvanishing derivative of $g(t)$ at $t = 0$ (we shall comment on this point later). Substituting (8) into (7), we find that the first two terms on the right remain finite as $t \to t_e$. The last integral however gives rise to a term which diverges logarithmically

$$Z(t) \equiv [a_m g^{(n)}(0)/n!] \int_0^t du \, \{(t-u)^n/(t_e-u)^m\} \tag{9}$$

The limit of $Z(t)$ as $t \to t_e$ can be obtained from

$$\int_0^t du \, \{(t-u)^n/(t_e-u)^m\} = \int_0^{t_e-\epsilon} du \, \{(t_e-\epsilon-u)^n/(t_e-u)^m\}$$

$$= \int_0^{t_e-\epsilon} du \, (t_e-u)^{n-m} - \epsilon n \int_0^{t_e-\epsilon} du \, (t_e-u)^{n-m+1} + \cdots \tag{10}$$

where $\epsilon = t_e - t$. Thus, if we choose $m = n + 1$, then $Z(t)$ diverges as

$$-[a_m g^{(n)}(0)/n!] \log(t_e - t) \tag{11}$$

By equating (11) to $-m \log(t_e - t)$ representing the divergence of the left-hand side, we obtain a_m.

If follows that solutions of (5a) which diverge at a finite time t_e must be of the following form

$$P(t) = [(n+1)!/g^{(n)}(0)(t_e-t)^{n+1}] + Y(t) \tag{12}$$

The function $Y(t)$ and the finite escape time t_e are yet to be determined. Equation (12), which was obtained by Akcasu and Noble [36], displays the asymptotic behavior of solutions with finite escape time as t tends to t_e and yields a necessary condition for their existence. Since the first term in (12) becomes dominant as $t \to t_e$, it must be positive so that it may represent the reactor power. Hence we find that if

$$g^{(n)}(0) < 0 \tag{13}$$

there can be no solutions with finite escape time.

Using the definition of $g(t)$ in (3), we can express (13) directly in terms of the feedback kernel $G(t)$. When $G(0)$ is nonvanishing and finite, we find from (3) that $n = 1$, and $P(t)$ behaves as

$$P(t) \to [2/G(0)(t_e - t)^2] \tag{14}$$

as $t \to t_e$ provided $G(0) > 0$. Physically, the latter implies a prompt positive reactivity coefficient. It is obtained from the feedback transfer function $H(s)$ as

$$G(0) = \lim_{s\to\infty} sH(s) \tag{15}$$

In some reactor models, $G(t)$ is of the form

$$G(t) = \alpha_p \, \delta(t) + K(t) \tag{16}$$

where α_p is the power reactivity coefficient and $K(t)$ is a bounded function (cf 5.3, Eq. 28) representing other feedback effects, e.g., temperature feedback. In this case $g(0) = \alpha_p$ and hence $n = 0$. Thus, if $\alpha_p > 0$ we may have solutions with finite escape time, which behave as

$$P(t) \to [\alpha_p(t_e - t)]^{-1} \tag{17}$$

If $\alpha_p < 0$, there cannot be finite escape time regardless of the form of $K(t)$ in (16).

The foregoing necessary condition for the existence of a finite escape time and the asymptotic behavior as $t \to t_e$ are unchanged when delayed neutrons are taken into account. The reason is that the contribution of the delayed neutrons to the inverse period is small in the asymptotic region as can been seen from (7.1, Eq. 7) by dividing it by $P(t)$. However, the presence of the delayed neutrons will affect the minimum value of the initial perturbation that gives rise to finite escape time when $g^{(n)}(0) > 0$.

The above analysis proves that there can be no finite escape time if $g^{(n)}(0) < 0$. It does not imply that there are always solutions of finite escape time if $g^{(n)}(0) > 0$. The recent investigations [37, 40] indicate that the finite escape time solutions always exist if $G(0) > 0$ provided that the initial power disturbance $P(0)$ is sufficiently large.

The method described above also provides a crude estimate for the value of the finite escape time in terms of the initial perturbation. For this we substitute (12) into (7), and evaluate the resulting equation in the limit of $t \to t_e$. We present the result only for $g(0) \neq 0$ and $G(0) \neq 0$

$$\log[1/g(0)\, P(0)\, t_e] = Q(t_e) + (1/g(0)) \int_0^{t_e} du\, \{[g(t_e - u) - g(0)]/(t_e - u)\}$$

$$+ \lim_{t\to t_e} \int_0^t du\, g(u)\, Y(t-u), \qquad g(0) \neq 0 \tag{18}$$

and

$$\log[2/G(0)\,P(0)\,t_e^2] = Q(t_e) - [2g(t_e)/t_e G(0)] + [2/G(0)] \int_0^{t_e} du\, \{[G(t_e - u) - G(0)]/(t_e - u) + \lim_{t \to t_e} \int_0^t du\, g(u)\, Y(t - u), \qquad G(0) \neq 0 \tag{19}$$

It is observed from (18) and (19) that the relation between $P(0)$ and t_e involves all the values of $Y(t)$ in the interval $(0, t_e)$. Hence, it does not seem possible to determine t_e in terms of the initial perturbation $P(0)$ and the past history $Q(t_e)$ without actually solving the original equation. However we can still obtain an estimate of the dependence of t_e on $P(0)$ for small t_e. When t_e is small, we may ignore the right-hand side of (18) entirely and obtain

$$t_e \to [1/g(0)\,P(0)], \qquad g(0) > 0 \tag{20}$$

In the case of $G(0) > 0$, we must retain the term $[2g(t_e)/t_e G(0)]$ in (19), because it approaches 2 when t_e tends to zero. Then, (19) yields

$$t_e^2 \to [2e^2/G(0)\,P(0)], \qquad G(0) > 0 \tag{21}$$

where $e = 2.718$. Calculations similar to these yield in general

$$t_e^{(n+1)} \to \{(n+1)!\exp[-A(n+1)]/g^{(n)}(0)\,P(0)\} \tag{22a}$$

where

$$A \equiv \sum_{j=1}^{n} [(-1)^j\, n!/j(n-j)!\, j!] \tag{22b}$$

The applications indicate that the accuracy of the above estimates gets poorer for larger n.

Applications

In order to test the validity of the foregoing general results we now consider two reactor models for which the kinetic equations can be solved exactly.

The first model assumes a feedback kernel of the form [36]

$$G(t) = \alpha_f\, \delta(t) + \alpha_s\, \delta(t - \tau) \tag{23}$$

where α_f and α_s are the prompt and delayed power coefficients, and τ is the delay time. The condition for a finite equilibrium is $H(0) =$

$\alpha_f + \alpha_s < 0$. Integrating (23) we find $g(t) = \alpha_f + U(t - \tau)\,\alpha_s$ where $U(t)$ is a step function. Hence, $g(0) = \alpha_f$ which is assumed to be positive.

We assume for simplicity that the reactor is operated at a constant power for $t < 0$ so that $I_1(t)$ in (2) is zero (cf 7.1, Eq. 7). Substituting (23) into (2) we then obtain

$$[d \log P/dt] = \alpha_f[P(t) - 1] + \alpha_s U(t - \tau)[P(t - \tau) - 1] \tag{24}$$

We note that the second term in (24), representing the delayed feedback, does not contribute in the time interval $(0, \tau)$. The solution of (24) in this time interval is

$$P(t) = \{1 - \exp[\alpha_f(t - t_e)]\}^{-1} \tag{25}$$

We can easily show that $P(t) \to [1/\alpha_f(t_e - t)]$ as $t \to t_e$ which is in agreement with (17). The finite escape time is found in terms of the initial perturbation by evaluating (25) at $t = 0$

$$t_e = (1/\alpha_f) \log\{P(0)/[P(0) - 1]\} \tag{26}$$

For large values of $P(0)$, (26) reduces to $t_e = 1/\alpha_f P(0)$ which is identical to (20). It is recalled that these results are valid only if $t_e < \tau$. The minimum value of $P(0)$ for which $t_e < \tau$ can be found from (26) as

$$P_m(0) = [1 - \exp(-\tau\alpha_f)]^{-1} \tag{27}$$

We note that solutions of finite escape time may still exist even when $P(0) < P_m$ with an escape time greater than τ. These solutions can be investigated by considering the full kinetic equation in (24).

We consider next the adiabatic temperature feedback as a model corresponding to $G(0) \neq 0$ [37]. This model is characterized by

$$\dot{P}(t) = \alpha P(t) \int_0^t P(u)\, du \tag{28}$$

in which the temperature feedback is assumed to be proportional to the energy release in the past. Clearly, the feedback kernel associated with this model is $G(t) \equiv \alpha$. The solution of (28) for an initial perturbation $P(0)$ is readily obtained as

$$P(t) = P(0)\{\sin[(t_e - t)\,\pi/2t_e]\}^{-2} \tag{29}$$

where

$$t_e = (\pi/2)[2/\alpha P(0)]^{1/2} \tag{30}$$

The asymptotic behavior of $P(t)$ as $t \to t_e$ is obtained from (29) as $2/\alpha(t_e - t)^2$ which is identical to (14) with $G(0) = \alpha$. The finite escape time in (30) is inversely proportional to the initial power as predicted by (21). However, the proportionality constant $e = 2.718$ in (21) is replaced in the exact calculations by $\pi/2 = 1.57$.

Discussions

It is shown by the general analysis and illustrated by the examples above that the reactor kinetic equation contains solutions with finite escape time if the value of the feedback kernel at $t = 0$ is positive, i.e., $G(0) > 0$. This condition is replaced by $\alpha_p > 0$ when $G(t) = \alpha_p\, \delta(t) + K(t)$ where α_p is the prompt power coefficient. Depending on the magnitude of the initial perturbation $P(0)$, one may still have bounded solutions even when these conditions are fulfilled. There is no estimate of the critical value of $P(0)$ at present above which the reactor power diverges at a finite time. The value of the escape time t_e depends on the magnitude of the initial perturbation. However, the asymptotic behavior of the solutions with finite escape time while they are diverging is independent of the initial conditions. It is determinded only by the behavior of the feedback kernel near $t = 0$. These features of the solutions of finite escape time have been verified by numerical methods in more realistic reactor model in the references cited above.

It may seem that the existence of solutions of finite escape time is a shortcoming of the reactor kinetic equations because the total number of neutrons in a reactor is finite, and hence the reactor power can not become infinite at a finite time. However, we know that the kinetic equations have been successful in the interpretation of the kinetic behavior of a variety of reactor types. They cannot be regarded as unrealistic only because they admit solutions of finite escape time. The actual behavior of a reactor during rapid transients, such as reactor excursions, may well be represented over the time intervals of physical interest by solutions with finite escape time. In fact it is argued in reference [36] that the behavior of EBR-I following a ramp reactivity insertion [42] was better understood in terms of such solutions.

B. Boundedness

In some reactor models, the reactor power does not return to its equilibrium value following an initial perturbation, nor does it diverge from it at a finite or infinite time. Instead, it oscillates between two limits about the equilibrium power level as time unfolds. The solutions of the kinetic equations displaying this behavior are referred to as "bounded solutions" or "limit cycles." It is desirable to know under what conditions these solutions arise, and to determine, if possible, the upper bound of the power oscillation when they exist. Before discussing this problem mathematically, we consider as an illustration a reactor model in which bounded solutions have been observed. This model is characterized by xenon and prompt-power feedback, as discussed in Sections 6.3A and 7.4A. It was shown in Section 7.4A that the equilibrium state of this reactor is asymptotically stable for small perturbations but unstable for large perturbations in certain portions of the linear stability region of the parameter space. The unstable solutions do not diverge, but display oscillations with constant amplitude (cf 7.4, Fig. 2). We can easily show that all the solutions of the kinetic equations are bounded in this particular reactor model. When the delayed neutrons and the temperature feedback are treated adiabatically, a reactor which is controlled by xenon and temperature feedback is described by

$$(l^*/\beta)\dot{P} = [\delta_0 - (\sigma_{\mathrm{Xe}}\,\mathrm{Xe}/\beta c\sigma_{\mathrm{f}}) - \gamma P]P \tag{31}$$

where the symbols were defined in (6.3, Eq. 1a). It is observed from this equation that the reactor power can not exceed the value δ_0/γ because $\dot{P} < 0$ if $P > (\delta_0/\gamma)$. Thus, all the solutions of (31) are bounded regardless of the magnitude of the initial perturbation provided the prompt-power coefficient is stabilizing, i.e., $\gamma > 0$.

The demonstration of the boundedness is not as easy in general as in the foregoing reactor model when the feedback is described by an arbitrary linear kernel $G(t)$ or its Laplace transform $H(s)$. The remainder of this section is devoted to a brief discussion of the various criteria for boundedness in terms of $G(t)$.

The first criterion for boundedness was obtained by Smets [43] in 1960. Ignoring the delayed neutrons, he showed that "if

$$G(t) < 0 \tag{32}$$

holds for all $t \geqslant 0$, then all the solutions of the kinetic equation are bounded." Recently he obtained the same criterion for boundedness including the delayed neutrons [44]. It is interesting to note that (32) includes $G(0) < 0$ which is the condition for the nonexistence of solutions with finite escape time. It is clear that any criterion for boundedness must exclude the solutions with finite escape time which are not bounded.

A more general criterion for boundedness was proposed by Akcasu and Noble [36] in 1966, which is stated as follows: "If $H(s)$ does not vanish on the positive real axis, and if $H(0) < 0$, then all the Laplace transformable solutions of the kinetic equations are bounded." We shall present the derivation of this criterion after we discuss its implications using specific reactor models.

First we recall that the condition $H(0) < 0$ is required for the existence of a finite equilibrium power. If $H(s)$ does not vanish on the positive real axis, as the second condition of the above criterion, then $H(s) < 0$ must hold for all $\text{Re}\, s \geqslant 0$. When this conclusion is combined with $sH(s) \to G(0)$ as $s \to \infty$ (cf Eq. 15), it follows that the criterion implies $G(0) < 0$, and thereby excludes solutions with finite escape time.

As stated above, the criterion ensures the boundedness of the Laplace transformable solutions only. It does not apply to those solutions that are not transformable (e.g., solutions with finite escape time or those which behave as $\exp(\alpha t^2)$). Although it is possible that all such solutions are excluded automatically by the criterion, as in the case of solutions with finite escape time, one must provide other sufficient conditions to eliminate all the nontransformable solutions before applying the above criterion. This question was investigated by Akcasu and Noble [45] in the absence of the delayed neutrons as follows: we rewrite (5a) as

$$P(t) = P(0) \exp[I(t) + Q(t)] \tag{33}$$

where we have introduced

$$I(t) \equiv \int_0^t du\, g(u)\, P(t-u) \tag{34}$$

We can show that

$$|\, Q(t)| \equiv \left| \int_0^t dv\, [I_1(v) - g(v)] \right| \leqslant Q_0 t \tag{35}$$

holds for sufficiently long times. Here Q_0 is a positive number. Using the definition of $I_1(t)$ in (7.1, Eq. 8) we observe that

$$\left| \int_0^t I_1(v)\, dv \right| = \left| \int_0^\infty du \left\{ [P(-u) - 1] \int_0^t dv\, G(u+v) \right\} \right| \leqslant 2g_0 M \tag{36a}$$

where

$$M \equiv \int_0^\infty du\, |\, P(-u) - 1\, | < \infty \tag{36b}$$

$$|\, g(t)| \equiv \left| \int_0^t du\, G(u) \right| \leqslant g_0 < \infty \tag{36c}$$

We assume that the initial curves are restricted to the class of functions for which M in (36b) is finite. Similarly, we assume that the feedback kernel $G(t)$ is integrable in $(0, \infty)$ as implied by (36c). Combining (36) with (35) we find $|Q(t)| \leqslant 2g_0M + g_0t$ which proves the inequality in (35) for large t. Let us now suppose that $I(t)$ in (33) remains nonpositive for large times. Then, we would obtain from (33) that $P(t)/P(0) \leqslant \exp(Q_0t)$ holds for sufficiently large times, and that $P(t)$ is Laplace transformable. We can state these results as a criterion: "If

$$I(t) = \int_0^t du\, g(u)\, Y(t-u) \leqslant 0 \tag{37}$$

holds for all positive continuous test functions $Y(t)$ defined in $(0, \infty)$ and for sufficiently large times, and if $|g(t)| \leqslant g_0$ for all $t \geqslant 0$, then all the solutions of (2) corresponding to the initial curves satisfying (36b) are Laplace transformable."

It immediately follows that if $g(t) \leqslant 0$ for all t, (37) is satisfied. Furthermore, we can easily show that if $g(t) \leqslant 0$ holds for all t, $H(s) < 0$ for all Re $s \geqslant 0$. This follows from that the Laplace transform of $g(t)$ is $H(s)/s$, and is always negative for real and positive arguments if $g(t) \leqslant 0$. Thus, the latter condition implies the criterion for boundedness of the Laplace transformable solutions, and guarantees that all the solutions are transformable. Hence we obtain the following criterion for boundedness due to Akcasu and Noble [36, 45]: "If $|g(t)| \leqslant g_0 < \infty$ and

$$g(t) \equiv \int_0^t G(u)\, du \leqslant 0 \tag{38}$$

for all $t \geqslant 0$, then all the solutions of (2) satisfying (36b) are bounded."

Remarks

(a) Equation (38) implies $G(0) < 0$ and thus excludes solutions with finite escape time. (b) Since $G(t) \leqslant 0$ implies $g(t) \leqslant 0$, this criterion is more relaxed than that by Smets in (32). However, the latter is obtained with the delayed neutrons whereas (38) is applicable only in the absence of the delayed neutrons. (c) If $G(t) = \alpha_p\, \delta(t) + K(t)$ and $|K_1(t)| \equiv |\int_0^t K(u)\, du| \leqslant K_1 < \infty$, then (38) reduces to $\alpha_p + K_1(t) \leqslant 0$. In this case, a more relaxed conditions was obtained by Akcasu and Noble [45]: "If $\alpha_p + \bar{K}(s) < 0$ for all Re $s \geqslant 0$, and both $K(t)$ and $K_1(t)$ are bounded for all t, then all the solution of (2) satisfying (36b) are bounded." We note that $\alpha_p + K_1(t) \leqslant 0$ implies $\alpha_p + \bar{K}(s) < 0$ for all

Re $s \geqslant 0$. Hence the latter is less restrictive. This generalization is possible in this case because $\alpha_p < 0$ is sufficient for the solutions to be Laplace transformable as shown in [45]. The condition $\alpha_p + \bar{K}(s) < 0$ both guarantees boundedness and implies $\alpha_p < 0$ (note that $\bar{K}(s) \to 0$ as $s \to \infty$). To illustrate this point we point out that the kinetic Eq. (2) admits a Laplace transformable and yet unbounded solution if $\alpha_p < 0$ but $\alpha_p + \bar{K}(a) = 0$ is satisfied for some $a > 0$. The solution is $P(t) = P(0) \exp(at)(-\infty < t < +\infty)$ where a satisfies both $a + \alpha_p + \bar{K}(0) = 0$ and $\alpha_p + K(a) = 0$ provided they are compatible. (d) More relaxed conditions for the solutions to be transformable than $g(t) \leqslant 0$ may be obtained using (37) directly if the feedback kernel is specified.

Applications

We now investigate the question of boundedness in some specific reactor models to illustrate the application of the above criteria. Strictly speaking, the above conclusions are applicable in the absence of the delayed neutrons, although the presence of the delayed neutrons is not expected to modify the boundedness criteria.

(a) Two-Temperature Model. The linear stability of this model was discussed in Sect. 6.5. It is characterized by the following feedback kernel (cf 6.5, Eq. 8): $G(t) = -A_1 \exp(-\epsilon_1 t) - A_2 \exp(-\epsilon_2 t)$ where where we have used Shotkin's notations [35]. The condition (38) reduces to

$$g(t) = (A_1/\epsilon_1)(e^{-\epsilon_1 t} - 1) + (A_2/\epsilon_2)(e^{-\epsilon_2 t} - 1) \leqslant 0$$

from which we obtain the conditions for boundedness as $A_1 + A_2 > 0$ and $A_1\epsilon_2 + A_2\epsilon_1 > 0$. We can show that these conditions are necessary for $H(s) < 0$ to hold for Re $s \geqslant 0$. Hence, the condition (38) which was required for the solutions to be transformable is equivalent to that for boundedness. It is also interesting to note that if, in addition to the boundedness of solution, the linearized system is unstable as discussed in Sect. 6.5, there is at least one stable limit cycle.

(b) Two-Temperature Model with Time Delay. This model is characterized by

$$G(t) = -[A_1 + A_2 U(t - \tau)\, e^{\epsilon\tau}]\, e^{-\epsilon t}$$

It was used by Bethe [46] to analyze the oscillations of EBR-I [42]. The condition (38) for boundedness reduces in this case to

$$-\epsilon g(t) = A_1[1 - e^{-\epsilon t}] + A_2 U(t - \tau)[1 - e^{-\epsilon(t-\tau)}] \geqslant 0$$

which is satisfied if $A_1 > 0$ and $A_1 + A_2 > 0$ hold. The first condition is equivalent to $G(0) < 0$ and the second to $H(0) < 0$. In the analysis of the oscillations of EBR-I, it was assumed that $A_1 < 0$ and $A_2 > 0$, i.e., a prompt-positive and delayed-negative temperature coefficient, and $A_1 + A_2 > 0$. Since $A_1 < 0$, the condition for boundedness was not satisfied, and the solutions with finite escape was not excluded. The numerical computations of reference [36] for this model showed that the reactor power diverges at a finite time when the initial power disturbance $P(0)$ exceeds a critical value (see the comments at the end of 7.10B).

(c) Circulating Fuel Reactor. The feedback kernel of an idealized circulating fuel reactor was obtained (cf 5.2, Eq. 30) as

$$G(t) = \alpha\eta U(\theta - t)[1 - (t/\theta)]$$

We observe that if $\alpha < 0$, $G(t) \leqslant 0$ for all t. Hence $g(t) \leqslant 0$ is satisfied without any more conditions and all the solutions are bounded.

C. Derivation of the Boundedness Criterion

The criterion for the boundedness of the Laplace transformable solutions can be obtained with the delayed neutrons as discussed in [36]. Here we shall neglect the effect of the delayed neutrons for simplicity. Taking the Laplace transform of (5a), we obtain

$$\bar{X}(s) = [\bar{I}_1(s)/s] - [H(s)/s^2] + [H(s)\,\bar{P}(s)/s] \tag{39}$$

where $\bar{X}(s)$ and $\bar{I}_1(s)$ are the transforms of $X(t) \equiv \log[P(t)/P(0)]$ and $I_1(t)$, respectively. Solving for $\bar{P}(s)$, which is the transform of $P(t)$, we find

$$\bar{P}(s) = (1/s) + [s\bar{X}(s)/H(s)] - [\bar{I}_1(s)/H(s)] \tag{40}$$

We now investigate the analyticity of $\bar{P}(s)$ in the entire half complex plane $\operatorname{Re} s \geqslant 0$ by using (40), which expresses the Laplace transform of the reactor power in terms of the Laplace transform of its logarithm and the functions $H(s)$ and $\bar{I}_1(s)$. The latter functions are analytic for $\operatorname{Re} s \geqslant 0$ by virtue of their physical meaning.

The analysis is based on a theorem in the theory of Laplace transforms (Theorem 5b of Chapter II of Widder [47]), which states that the transform $\bar{P}(s)$ of a nonnegative function $P(t)$ must have a singularity at the real point of its axis of convergence. The latter is defined as the vertical line defined by $\operatorname{Re} s = \sigma$ such that the integral

$\int_0^\infty dt\, \{P(t) \exp(-st)\}$ diverges for any s with $\text{Re}\, s > \sigma$ and converges for $\text{Re}\, s < \sigma$.

Using this theorem we want to show that $\bar{P}(s)$ can have singularities in the half-plane $\text{Re}\, s > 0$ only at the zeroes of $H(s)$. Suppose that $H(s) \neq 0$ for $\text{Re}\, s > 0$, and that $\bar{P}(s)$ has its axis of convergence at $\text{Re}\, s = \sigma > 0$. The above theorem shows that $\bar{P}(s)$ has a singularity at $s = \sigma$. The order of $P(t)$ for large times is determined by this singularity. For definiteness, we assume that it is the only singularity of $\bar{P}(s)$ on the axis, and it is a simple pole (the extension to any type of singularity is possible with the same reasoning [36]). Then $P(t)$ behaves as $\exp(\sigma t)$ for large times.

We consider now $\bar{X}(s)$. Equation (39) indicates that $\bar{X}(s)$ and $\bar{P}(s)$ have the same singularity in the half-plane $\text{Re}\, s > 0$ if $H(s) \neq 0$ there. In particular, $\bar{X}(s)$ has a simple pole at $s = \sigma$, and hence $X(t)$ behaves as $C \exp(\sigma t)$ for large times, where C is the residue of $\bar{X}(s)$ at $s = \sigma$. Since $X(t) = \log[P(t)/P(0)]$, the foregoing behavior of $X(t)$ would imply $P(t) \to \exp[C \exp(\sigma t)]$. This conclusion is not compatible with the assumed exponential behavior of $P(t)$. Hence, $\bar{P}(s)$ can not have singularities with $\text{Re}\, s > 0$ if $H(s) \neq 0$.

The above contradiction does not arise if $\bar{P}(s)$ has a singularity at a zero of $H(s)$. Then, $H(s)\, \bar{P}(s)$ can be analytic for $\text{Re}\, s > 0$ even though $\bar{P}(s)$ has a pole with $\text{Re}\, s > 0$. Consequently, $\bar{X}(s)$ will also be analytic for $\text{Re}\, s > 0$, and can have a singularity at the origin to account for the singularity of $\bar{P}(s)$ at $s = \sigma > 0$.

It follows that the singularities of $\bar{P}(s)$ for $\text{Re}\, s > 0$ must coincide with the zeroes of $H(s)$ in the right half-plane. Since $\bar{P}(s)$ must have a singularity at the real point of its axis of convergence if it has singularities for $\text{Re}\, s > 0$ at all, and since this singularity must coincide with a zero of $H(s)$ on the real positive axis, we conclude that $\bar{P}(s)$ is necessarily analytic for $\text{Re}\, s > 0$ if $H(s)$ does not have positive real zeroes. This is the main condition of the criterion for boundedness discussed in 7.B.

We must now discuss the singularities of $\bar{P}(s)$ on the imaginary axis which is the axis of convergence with $\sigma = 0$. First we show that $\bar{P}(s)$ must have singularities on the imaginary axis if $\bar{P}(s)$ is analytic for $\text{Re}\, s > 0$, because $\bar{P}(s)$ cannot be analytic everywhere on the half-plane $\text{Re}\, s \geqslant 0$. If we assume the contrary, then $\bar{X}(s)$ is analytic for $\text{Re}\, s > 0$ and has a double pole at the origin since we assume $H(0) \neq 0$ (cf Eq. 39). This implies that $X(t) \to -H(0)t$, and thus $P(t) \to \exp[-H(0)t]$ for large t. If $H(0) < 0$ as assumed in the criterion for boundedness, $P(t)$ increases exponentially, contradicting the assumption that $\bar{P}(s)$ is analytic for $\text{Re}\, s \geqslant 0$. This argument shows that $\bar{P}(s)$ must have a singularity at the origin such that the contribution from $H(s)\,\bar{P}(s)/s$ will remove the contribution of the double pole in (39) arising from $H(s)/s^2$. A simple pole of $\bar{P}(s)$ at the origin with a residue of unity is an example of such a singularity. It is proven in [36] that if the singularities of $\bar{P}(s)$ on the imaginary axis are poles, then they must be simple poles located at $s = jnw_0$, $n = 0, \pm 1, \pm 2, \ldots$ (periodic solution). The discussion of the various types of singularities of $\bar{P}(s)$ on the imaginary axis, and the properties of the solutions $P(t)$ corresponding to these singularities (such as the question of existence of quasiperiodic solutions) is complicated. However, the arguments presented above show [36] that any singularity at the origin that gives rise to solutions increasing in time are ruled out, because the asymptotic behavior of $X(t)$ and $P(t)$ cannot be matched if such solutions existed.

PROBLEMS

1. Using (7.2, Eq. 7) find an upper bound for $|\ddot{V}(x)|$ where the conditions of Theorem 1 of Sect. 7.2 are fulfilled (Different bounds may be found depending on the accuracy of the inequalities used in the derivations).

2. If $G(t)$ is a causal function which does not contain an impulse at $t = 0$, and if its Fourier transform is $H(i\omega) = H_R(\omega) + iH_I(\omega)$, show that

$$G(t) = (2/\pi) \int_0^\infty d\omega\, H_R(\omega) \cos \omega t, \qquad t > 0$$

and

$$G(t) = (-2/\pi) \int_0^\infty d\omega\, H_I(\omega) \sin \omega t, \qquad t > 0$$

(Hint: First define

$$G_e(t) \equiv [G(t) + G(-t)]/2 \qquad \text{and} \qquad G_0(t) \equiv [G(t) - G(-t)]/2$$

and verify

$$G_e(t) = (1/2\pi) \int_{-\infty}^{+\infty} d\omega\, H_R(\omega) \exp(i\omega t)$$

$$G_o(t) = (i/2\pi) \int_{-\infty}^{+\infty} d\omega\, H_I(\omega) \exp(i\omega t)$$

then note that $G_e(t) = G(t)/2$, $G_0(t) = G(t)/2$ for $t > 0$ if $G(t) \equiv 0$ for $t < 0$.)

3. If $G(t)$ is a causal function without an impulse at $t = 0$, show that (cf Ref. [11a], p. 174) for $\text{Re}\, s > 0$

$$H(s) = (1/\pi) \int_{-\infty}^{+\infty} d\omega\, H_R(\omega)/(s - i\omega)$$

and

$$H(s) = (i/\pi) \int_{-\infty}^{+\infty} d\omega\, H_I(\omega)/(s - i\omega)$$

(Hint: First note that $U(t)\, G(t) = 2U(t)\, G_e(t)$ where $U(t)$ is a step function, and take the Fourier transform of $\exp(-\alpha t)\, U(t)\, G(t) = 2 \exp(-\alpha t)\, U(t)\, G_e(t)$. Use the convolution theorem to transform the product $G_e(t)$ and $U(t) \exp(-\alpha t)$ on the right hand side. Use $U(t)\, G(t) = 2U(t)\, G_o(t)$ to show the second formula.)

4. Show that the conditions (7.5, Eqs. 40d and 40f) are sufficient to guarantee linear stability. (Hint: Show that the characteristic equation $i\omega + [\gamma - \kappa(i\omega)]\, P_0 = 0$ cannot be satisfied with $P_0 > 0$) and $\omega > 0$ if the above conditions hold.)

5. Show that the conditions (7.5, Eqs. 40) imply $\gamma > 0$, or $K(0) < 0$ when $\gamma = 0$, and thus exclude solutions with finite escape time (cf 7.13A). (Hint: When $\gamma = 0$, first show that (7.5, Eq. 40b) can be written as $\text{Re}\, \mathbf{H}(i\omega) \leqslant 0$ where $\mathbf{H}(s)$ is the Laplace transform of $qK(t) - U(t) \int_t^\infty K(u)\, du$, then use (7.3, Eq. 19) to show that $\mathbf{H}(s) < 0$ on the positive real axis.)

6. Find the solution of (7.6, Eqs. 2 and 3) when $P_e(T) = P_0 = 0$ (adiabatic cooling) and $P_e(T) = 0$, $P_0 \neq 0$ (constant heat removal), and discuss the variation of $P(t)$ and $T(t)$ as a function of t.

7. Prove the following relation $\kappa' D^m I = \alpha' H^{m-1} b P_0$ where κ is defined by (7.7, Eqs. 13, 8c and 9a), α and b by (7.7, Eqs. 6), and D by (7.7, Eq. 5).

8. If

$$Q = \begin{pmatrix} G & P \\ S' & R \end{pmatrix}$$

where G is $N \times N$, R is $M \times M$, and P and S are $N \times M$ matrices, and if, in addition, G is nonsingular, then $|\, Q \,| = |\, G \,|\, |\, R - S'G^{-1}P \,|$

9. Using Theorem 1 of 7.10, show that if $\kappa_0 > 0$ and $N \leqslant 2$ in (7.10, Eqs. 1), then the entire region for linear stability at all power levels is also asymptotically stable in the large [8].

10. Show that $\dot{P}(t) = P(t) \int_0^\infty du\, G(u)[P(t - u) - 1]$ has a solution $P(t) = \exp(at)$ defined in $(-\infty, +\infty)$ where $a > 0$ satisfies both $a + H(0) = 0$ and $H(a) = 0$. It is assumed that $H(0) < 0$ (cf 7.13B).

REFERENCES

1. A. A. Lyapunov, "Problem Général de la Stabilité du Mouvement" (photo reproduction of the 1907 French transl.). Princeton Univ. Press, Princeton, New Jersey, 1949.
2. N. N. Krasovskii, "Stability of Motion." Stanford Univ. Press, Stanford, California, 1963.

2a. E. A. Barbasin and N. N. Krasovskii, On the existance of Lyapunov functions in the case of asymptotic stability in the large. *Prikl. Mat. Meh.* **18**, 345–350 (1954).

3. Z. Akcasu and A. Dalfes, A study of nonlinear reactor dynamics. *Nucl. Sci. Eng.* **8**, 2 (1960).

3a. A. Z. Akcasu and P. Akhtar, On the asymptotic stability of reactors with arbitrary feedback. *J. Math. Phys.* **11**, 155 (1970).

3b. F. DiPasquantonio and F. Kappel, Comments on theoretical and experimental criteria for reactor stability, Technical Note. *Nucl. Sci. Eng.* **39**, 133 (1970), see also *Energia Nucl.* **15**, 761 (1968).

4. H. S. Carslaw, "Introduction to the Theory of Fourier Series and Integrals." Dover, New York, 1930.
5. I. Barbâlat, *Rev. Math. Pure. Appl.* **4**, 267 (1959).
6. W. K. Ergen, Kinetics of the circulating fuel nuclear reactor. *J. Appl. Phys.* **25**, 6 (1954).
7. H. B. Smets, "Problems in Nuclear Power Reactor Stability." Presses Univ. de Bruxelles, Belgium, 1962.
8. V. M. Popov, Relaxing the sufficiency condition for absolute stability. *Avtomat. Telemekh.* **19**, 1 (1958).
9. V. M. Popov, Absolute stability of nonlinear systems of automatic control. *Avtomat. Telemekh.* **22**, 6 (1961).
10. T. A. Welton, A stability criterion for reactor systems. ORNL 1894. Oak Ridge Nat. Lab., Oak Ridge, Tennessee, 1955.
11. H. R. Pitt, "Integration, Measure and Probability." Oliver & Boyd, Edinburgh and London, 1963.

11a. A. Papoulis, "The Fourier Integral and its Applications," pp. 174 and 307, McGraw-Hill, New York, 1962.

12. V. M. Popov, A new criterion regarding the stability of systems containing nuclear reactors. *Stud. Cercet. Energet. Ser. A* (in Rumanian) **12**, (1962).
13. J. Chernick, G. S. Lellouche, and W. Wollman, The effect of temperature on xenon stability. *Nucl. Sci. Eng.* **10**, 120 (1966).
14. Z. Akcasu and P. Akhtar, *J. Nucl. Energy* **21**, 341 (1967).
15. G. S. Lellouche, *J. Nucl. Energy* **21**, 519 (1967).
16. C. Corduneanu, On an integral equation occurring in control theory. *C. R. Acad. Sci.* **256**, No. 17 (1963).
17. J. Lasalle and S. Lefschetz, "Stability by Lyapunov's Direct Method." Academic Press, New York, 1961.
18. W. Hahn, "Theory and Application of Lyapunov's Direct Method." Prentice-Hall, Englewood Cliffs, New Jersey, 1963.

18a. R. E. Kalman and J. E. Bertram, Control system analysis and design via the "second method" of Lyapunov. *J. Basic Engineering* **391**, 1960.

19. I. G. Malkin, Theory of stability of motion. AEC-tr-3352. 1958.
20. S. Lefschetz, "Stability of Nonlinear Control Systems." Academic Press, New York, 1965.

21. M. Marcus and H. Ming, "A Survey of Matrix Theory and Matrix Inequalities." Ally and Bacon, 1964.
22. A. I. Lurie, "Some Nonlinear Problems in the Theory of Automatic Controls" (in German). Akademie Verlag, Berlin, 1957.
23. A. M. Letov, "Stability in Nonlinear Control Systems." Princeton Univ. Press, Princeton, New Jersey, 1961.
24. H. B. Smets, Stability in the large and boundedness of some reactor models. OECD. 1963.
25. G. S. Lellouche, Reactor kinetics stability criteria. *Nucl. Sci. Eng.* **24**, 1 (1966).
26. S. Lefschetz, Some mathematical considerations of non-linear control systems. AFOSR 2501. 1962.
27. G. S. Lellouche, An application of Popov's asymptotic stability criterion. *J. Math. Phys.* (Cambridge, Mass.) **46**, 3 (1967).
28. R. E. Kalman, Lyapunov functions for the problem of Lurie in automatic control. *Proc. Nat. Acad. Sci. U. S.* **49**, 201 (1963).
29. R. E. Kalman, Canonical structure of linear dynamical systems. *Proc. Nat. Acad. Sci. U. S.* **49**, 201–205 (1963).
30. R. E. Kalman, Y. C. Ho, and K. S. Narendra, Controllability of linear dynamical systems. *Contrib. Differential Equations* **1**, 189–213 (1963).
31. M. Lax, *Phys. Rev.* **145**, 110 (1966).
32. N. Baron and K. Meyer, Effect of delayed neutrons on the stability of a nuclear power reactor. *Nucl. Sci. Eng.* **24**, 356 (1966).
33. J. Chernik, The dependence of reactor kinetics on temperature, *USAEC Report BNL-173*, Brookhaven National Laboratory, Upton New York, December 1951.
34. J. Chernick, Evolution of the finite-escape-time concept in reactor kinetics, *in* "Neutron Dynamics and Control." Proc. of the Symp. on Nucl. Eng., University of Arizona, 1965. *AEC Symposium Series* **7**, 1966.
35. L. M. Shotkin, Nonlinear kinetics of a two-temperature reactor. *BNL-815*, Brookhaven National Laboratory, Upton, New York, 1963.
36. A. Z. Akcasu and L. D. Noble, A nonlinear study of reactors with a linear feedback. *Nucl. Sci. Eng.* **25**, 47 (1966).
37. H. B. Smets, Unlimited power excursions in nuclear reactors. Paper presented at the Conference on Dynamics of Nuclear Systems, University of Arizona, April 1970.
38. A. Wintner, The non-local existence problem of ordinary differential equations. *Am. J. Math.* **67**, 284 (1945).
39. L. Markus, Escape times for ordinary differential equations. *Rend. Sem. Mat.* **11**, 271 (1951–2).
40. S.-O. London, On some nonlinear Volterra equations. *Ann. Acad. Sci. Fenn., Ser. A* **V1**, 317, Helsinki, 1969.
41. S.-O. Londen, On a Volterra integrodifferential equation. *Comm. Phys. Math.* **38**, Nr. 1 and 2, 1969.
42. R. O. Britton, Analysis of the EBR-I core meltdown. *A/Conf. 15/P/2156* (June, 1958).
43. H. B. Smets, *Nukleonik* **2**, 44 (1960).
44. H. B. Smets, *Nucl. Sci. Eng.* **39**, 289 (1970).
45. A. Z. Akcasu and L. D. Noble, *Nucl. Sci. Eng.* **25**, 427 (1966).
46. H. A. Bethe, Reactor safety and oscillator tests. *APDA-117*, Atomic Power Development Associates, Inc., Detroit, Mich. (1956).
47. D. V. Widder, "The Laplace Transform." Princeton Univ. Press, Princeton, New Jersey, 1946.

Index

D

E

F

M

N

O

P

R